(continued on inside back cover)

SECOND EDITION

FUNCTIONS AND CHANGE

A Modeling Approach to College Algebra

Bruce Crauder, Benny Evans, and **Alan Noell**

Oklahoma State University

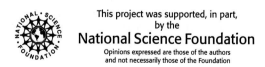

This project was supported, in part,
by the
National Science Foundation
Opinions expressed are those of the authors
and not necessarily those of the Foundation

HOUGHTON MIFFLIN COMPANY   Boston   New York

Gratitude and Thanks

*We are deeply grateful for our families
who supported us and believed in this project.*

For Douglas, Robbie, and, most of all, Anne—BCC
For Carla, Daniel, and Anna—BDE
For Evelyn, Philip, Laura, Stephen, and especially Liz—AVN

Editor-in-Chief, Mathematics: Jack Shira
Senior Sponsoring Editor: Lynn Cox
Development Manager: Maureen Ross
Editorial Assistant: Melissa Parkin
Senior Production/Design Coordinator: Carol Merrigan
Senior Manufacturing Coordinator: Marie Barnes
Senior Marketing Manager: Michael Busnach

Cover image: ©Tom and Pat Leeson

Printed in the U.S.A.

Library of Congress Control Number: 2001099010

ISBN: 0-618-21956-0

3456789-DOW-06 05 04 03

Contents

Preface

To Instructors

Mathematics is among the purest and most powerful of sciences, an art form of unsurpassed beauty, and a descriptive language that codifies ideas from many areas, including business, engineering, and the natural, physical, and social sciences, always showing that major concepts drawn from many different sources are in their essence the same. Mathematics pervades modern society and, to the knowledgeable eye, is everywhere evident in nature.

Practicing engineers, mathematicians, and scientists require a deep understanding of mathematics as well as a level of exactitude and facility with symbol manipulation that is sometimes difficult for entering students to master. For many, frustration with these aspects of mathematics obscures its power, beauty, and utility. But modern technology in the form of graphing calculators or computers can supplant much of the drudgery of mathematics, move the focus toward important concepts and ideas, and make mathematics more accessible.

> The goal of this text is to use technology and informal descriptions to empower entering students with mathematics as a descriptive problem-solving tool, and to reveal mathematics as an integral part of nature, science, and society.

Mathematics in Context: Style, Pedagogy, and Topics

This NSF-funded course arose through an effort to provide students with the mathematical tools they will need in courses that traditionally require college algebra and has been used successfully at many schools. The topics were selected after lengthy consultations with our colleagues from departments across the campus. The skills our colleagues wanted students to learn from an entry-level mathematics course included facility with graphs, tables of values, linear algebraic manipulations, and, most importantly, some level of confidence in relating sentences to formulas, tables, and graphs. Perhaps surprisingly, our colleagues also wanted a qualitative understanding of rates of change, leading us to make this one of the most important themes in the text. Rates of change is a unifying concept, following from the observation that knowledge of the initial value and how a function changes are sufficient to understand the function. This concept is fully developed and exploited for linear and exponential functions in Chapters 3 and 4 and is carried through the rest of the text.

Rates of change is a unifying concept.

Traditional college algebra courses focus on preparing students for an engineering-oriented calculus course that few of them will ever take.

Our consultations also indicated that these goals are met only marginally, if at all, by a traditional college algebra course—which, instead, focuses on preparing students for an engineering-oriented calculus course that few of them will ever take. The most common perception we got from our colleagues in other areas is that even when entering students have some facility with elementary mathematics, they may be afraid to apply it and indeed may see no relation at all between mathematics that they have been exposed to and applications of mathematics in other areas. As a result, compared with traditional college algebra texts, our treatment of topics is more geared toward applications in other disciplines. For example, we often use data tables to display functions, and students become proficient in recovering the formula for the function from data. Since the real-world applications frequently provide such rich opportunities for multi-step problems, we often continue the problems and examples into additional exercises. These exercises are marked with a ↻ icon.

Uses real-world problems and models almost exclusively.

This text uses real-world problems and models almost exclusively, including an incredible diversity of interesting applications and an early introduction to realistic multivariate problems. As well as being more interesting to students, the real-world context allows them to check their answers against intuition and common sense.

This text differs from traditional textbooks in many ways. A quick glance through the book shows that there are more words and fewer formulas than usual. Also evident are the extensive examples and problems. The choices of examples and exercises are part of an important theme in the text: Mathematics is easily learned from carefully chosen, realistic examples. In general, mathematical principles are developed through examples.

Develops mathematical principles through examples.

Only after the examples give good intuition and understanding are more general and abstract notions made explicit. In practice, this style has worked very well at Oklahoma State University, as well as at over 50 schools class testing the preliminary edition. Students are able to read the text and understand the examples. Students have opinions and bring their own independent understanding of the topics in the examples and exercises, so classroom discussions are more lively. Many exercises are conceptual, requiring thought and interpretation, not just quantitative answers.

Class tested at over 50 schools.

This text easily accommodates pedagogies other than traditional lecture. Because so much of the text focuses on examples, spending class time in discussions or working in groups is easy. The examples are realistic, which promotes bringing outside materials into the classroom and having students find examples from their other classes.

Provides students with the opportunity to succeed at sophisticated mathematics.

One of the most important goals of this text is to provide students with the opportunity to succeed at sophisticated mathematics. Our experience with the material in its seventh year of use is that the appropriately used calculator gives students the power needed to perform significant mathematical analyses. Many of our students report that success in mathematics is a new and finally pleasing experience. We also wanted students to see mathematics as part of their life experience, and we have anecdotal stories to indicate some success in that area as well.

This text was not designed as a prerequisite for calculus, but, as it develops, the text provides an excellent preparation for some of the newer reformed applied calculus texts. We are using it as such at Oklahoma State University and are pleased with the results.

Graphing Calculator Use and Reference

This text is designed to be used with a graphing calculator, and the calculator is an essential part of the presentation as well as the exercises. Throughout the text we employ TI-83 screens, but many graphing calculators and even some computer software can serve the purpose. Because the major graphing calculator producers have made their products remarkably powerful and easy to use, students should have no difficulty becoming proficient with the basic calculator operation. To ensure this, an accompanying *Technology Guide* provides TI-83/TI-83 Plus keystrokes for creating tables and graphs, entering expressions, and so on, in the Quick Reference section of the *Guide*. Keystrokes are cross-referenced using footnote boxes in the main text. For example, $\boxed{3.4}$ on page 211 of the text indicates that the keystrokes needed to create this graph are found in item number **3.4** of the Quick Reference pages of the *Guide*. We recommend removing the appropriate Quick Reference sheets from the *Technology Guide* for consultation while reading the text. The *Technology Guide* also includes generic instructions to help students become familiar with graphing calculators in general, along with related exercises.

Significant Changes Since the Preliminary Edition

A number of changes have been made from the Preliminary Edition in response to the helpful advice of students and faculty using this text. Many of these changes allow the text to accommodate different popular approaches to the material more easily. Some of the changes include:

- A great number of new exercises, including more practice setting up formulas, and a wider variety of nonscience applications have been added.

- A short set of skill-building exercises now precedes each exercise set.

- Another Look with Enrichment Exercises. Based on market feedback, this new feature has been added to the end of each section in the first six chapters. It provides a slight increase in the algebraic presentation and symbol manipulation, along with general enrichment or alternative treatments of the material. While written in the same spirit as the rest of the text, the Another Look feature contains more traditional coverage and is optional.

- New Lab Experiments, some of which utilize Calculator-Based Laboratory™ technology, are identified by a 🖳 icon. Some instructors may want to use these as group projects. The experiments are linked to a robust website, containing interactive spreadsheets and exercises. The experiments are found at

  ### http://math.college.hmco.com/students

  and then, under College Algebra, select "Functions and Change: A Modeling Approach to College Algebra, Crauder/Evans/Noell (©2003)" from the drop-down list, and click the Go button. Note that we recommend bookmarking this page so that you can easily find all subsequent experiments. From the book's web page, click on Lab Experiments and then select the appropriate experiment.

- Treatment of linear regression by hand calculations.

- More flexible coverage of exponential regression exercises, more easily allowing exponential regression to be performed either directly by calculator or via linear regression.

- A new section on logarithmic functions (Section 4.4).

- A new section on compositions of functions and piecewise-defined functions (Section 5.3).

- Expanded coverage of quadratic functions (Section 5.4).

- Expanded coverage of polynomial and rational functions, including treatment of asymptotes (Section 5.5).

- Greatly expanded coverage of periodic functions and trigonometry (Chapter 6).

Suggested Paths Through This Text

The eight chapters in this text provide more than enough material for a one-semester course with three credit hours; potentially there is enough material for a very satisfactory one-semester course with five credit hours or a full year course. Chapters 1 through 4 form the core material. Initially students are presented with many formulas in Section 1.1 so they may quickly see significant and important applications and so they won't be intimidated by formulas. Functions are viewed from many different approaches: by formulas, by words, by defining properties, by graphs, by tables of values, and, most importantly, by rates of change. As the course progresses we see the value of looking at functions simultaneously from several points of view. We study in detail the important examples of linear and exponential functions as part of the core material. After covering the core material, instructors have some options, as shown below.

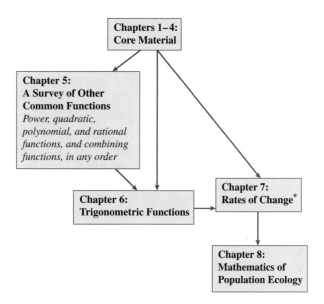

*Note: The discussion of graphical solutions for equations of change (i.e., differential equations) in Section 7.5 is the most challenging material in this text. (The authors believe it may also be the most rewarding.) It is necessary for parts of the presentation in Section 8.2 but otherwise may be omitted at the instructor's discretion. At Oklahoma State University we have found that students respond well to Chapter 7. Examining the texts used in courses from ecology to economics shows how useful the development of rates of change can be to students from a diverse collection of backgrounds.

Supplementary Materials

- *Technology Guide* includes:

  - TI-83 and TI-83 Plus instructions for creating tables and graphs, entering expressions, and so on, in the Quick Reference section of the *Guide*. Keystrokes are cross-referenced using footnote boxes in the main text. For example, ⟦ 3.4 ⟧ on page 211 of the text indicates that the keystrokes needed to create this graph are found in item number **3.4** of the Quick Reference pages of the *Guide*.

- generic instructions to help students become familiar with their graphing calculator or spreadsheet program, along with related exercises. Note that the Excel instructions are new to this edition.

- *Student Solutions Guide* includes complete solutions to all odd-numbered exercises.
- *Complete Solutions Guide* includes complete solutions to all exercises for instructors.
- *Instructor's Resource Guide with Tests* includes general teaching tips for adapting to this book's approach, specific section-by-section teaching tips, sample tests and quizzes, transparency masters, and sample syllabi. Note that the number of test questions has been significantly increased from the preliminary edition.
- *Computerized Test Bank*
- *Instructor's Class Prep CD*, which includes a computerized version of the *Instructor's Resource Guide with Tests*.
- New robust website with interactive spreadsheets, exercises, and group projects, plus instructions for spreadsheet applications, including a Quick Reference section.

To Students

Every effort has been made to show mathematics as you are likely to encounter it in other courses as well as in daily life. Learning mathematics requires effort. But learning is also fun, and success in mathematics can be rewarding in terms of personal accomplishment. It can also facilitate understanding in other courses as well as in everyday experience, and it may be a key to attaining your career goals. We intend that you reap these and other benefits from your experience with this text.

How to Learn with This Text

Effective use of this text requires that you actively participate in the presentation. You should read with your graphing calculator turned on and with the Quick Reference pages from the *Technology Guide* handy. (Please see the earlier description of this supplement to learn what it teaches and how it works.) It is not sufficient simply to read the examples in the text. Rather, you should work through each example yourself as it is presented, and when a calculator screen is shown, you should reproduce it on your own calculator. Key Idea boxes and Chapter Summaries are included as study tools.

As you begin, you will note that rarely is the final answer for an example presented simply as a number or a graph. Rather the answer is accompanied by sentences explaining how the answer was obtained and with appropriate conclusions. You should follow this pattern in solving the exercises. Your solution should include whatever calculations, graphs, or tables (copied from your calculator) you use as well as a clear statement of your conclusion accompanied by an explanation of your methods. A simple test of the clarity of your explanation is whether your peers can understand the solution by reading your work.

Our answers to the odd-numbered exercises are provided at the back of the text. They are of necessity brief and in general not acceptable as complete solutions. You should also be aware that for many of the exercises, there is no simple *right answer*. Instead, there is room for a number of conclusions, and any of them may be acceptable provided they are accompanied by a convincing argument. Sometimes you may have an answer that is different from that of one of your peers, and neither of your answers matches the one given in the back of the book, yet both of you may have correct solutions.

Mathematics is a tool that enhances your reasoning ability. It does not supplant that ability, and it is not a device that gives magical, unassailable answers. Whenever you are led to a conclusion that flies in the face of common sense, you should question the validity of your work and check carefully for a mistake.

There is a website for students using this text that contains group projects, lab exercises, interactive spreadsheets, and other helpful tools. Note that the lab experiments that you will find in the exercise sets are also linked to the web at

http://math.college.hmco.com/students

and then, under College Algebra, select "Functions and Change: A Modeling Approach to College Algebra, Crauder/Evans/Noell (©2003)" from the drop-down list, and click the Go button. Note that we recommend bookmarking this page so that you can easily find subsequent experiments. From the book's web page, click on Lab Experiments, and then select the appropriate experiment.

A Solicitation

This text is designed to be read by students, and while we are very much interested in input from instructors, the evolution of the text will be heavily influenced by what you have to say. We earnestly solicit any and all comments about the presentation. We would like your reactions to topics included or omitted as well as your estimation of the effectiveness of the presentation. We appreciate hearing about any errors, omissions, or inaccuracies. The best way to get information to us is through e-mail to **crauder**, **bevans**, or **noell**, each at **@math.okstate.edu**.

Thanks

Class Testers and Reviewers In addition to being used at our school, the preliminary edition and earlier drafts of *Functions and Change* were class tested at over 50 schools. We would like to thank the following schools, some of which have been using various drafts for several years, for being willing to try something different and for providing feedback based on their experiences. Thanks also to the students at these schools for their participation.

Adirondack Community College	John Brown University	Siena Heights University
Andrew College	Leeward Community College	St. Xavier University
Augusta State University	Linn Technical College	SUNY at Brockport
Ball State University	Longwood College	Thomas College
Big Bend Community College	Loras College	Tulsa Community College
Central GA Technical College	Loyola University	University of Central Arkansas
Columbia College	Martin Community College	University of Georgia
Columbus State University	Mercer University	University of Houston, Univ. Park
Edgewood College	Millikin University	University of Nevada at Reno
Floyd College	Monmouth University	University of Maine at Orono
Franklin Pierce College	Northampton City Area CC	University of Maryland
GA College & State University	Northeastern OK A & M College	University of MA, Dartmouth
Georgia Military College	Northern Michigan University	University of Northern Iowa
Glen Oaks Community College	Occidental College	University of Tulsa
Hannibal LaGrange College	Oklahoma State University	Wake Technical College
Hesston College	Prince William Sound CC	West Virginia Wesleyan College
Ithaca College	Randolph Macon Woman's College	
Jacksonville University	Ranken Technical College	

The following reviewers and class testers also deserve our thanks for their encouragement, criticisms, and suggestions.

Judith Ahrens	Pellissippi State Tech. College	Miles Hubbard	St. Cloud State University
Daniel Alexander	Drake University	Thomas P. Kline	University of Northern Iowa
Tony Bedenikovic	Bradley University	Kathy M.C. Ivey	Western Carolina University
Mark Clark	Palomar College	Larry S. Johnson	Metropolitan State College of Denver
Bill Coberly	University of Tulsa	Gina Kietzmann	Elmhurst College
Doug Colbert	University of Nevada at Reno	Jerry Kissick	Portland Community College
Diane Driscoll Schwartz	Ithaca College	John Lomax	Northeastern Oklahoma A & M
Richard Faulkenberry	University of MA, Dartmouth	Patty Monroe	Greenville Technical College
Dewey Furness	Ricks College	Mary Pratt-Cotter	Georgia College & State Univ.
John Gosselin	University of Georgia	Carol J. Rychly	Augusta State University
Joel K. Haack	University of Northern Iowa	Bernd Rossa	Xavier University
Joe Harkin	SUNY Brockport	Doran K. Samples	Georgia College & State Univ.
John Haverhals	Bradley University	Mary Jane Sterling	Bradley University
Judith Hector	Walters State CC	Andrius Tamulis	Cardinal Stritch University
Lisa M. Hodge	Wake Technical CC	Jeremy Underwood	Clayton College & State Univ.

In writing this book, we have also relied on the help of other mathematicians as well as specialists from agriculture, biology, business, chemistry, ecology, economics, engineering, physics, political science, and zoology. We offer our thanks to Bruce Ackerson, Brian Adam, Robert Darcy, Joel Haack, Stanley Fox, Adrienne Hyle, Smith Holt, Jerry Johnson, Lionel Raff, Scott Turner, and Gary Young. Errors and inaccuracies in applications are due to the authors' misrepresentation of correct information provided by our able consultants. We are grateful to the National Science Foundation for its foresight and support of initial development and to Oklahoma State University for its support. We very much appreciate Charles Hartford's continuing patience and good humor through some trying times.

The most important participants in the development of this work are the students at Oklahoma State University, particularly those in the fall of 1995 and spring of 1996, who suffered through very early versions of this text, and whose input has shaped the current version. This book is written for entering mathematics students, and further student reaction will direct the evolution of the text into a better product. Students and teachers at Oklahoma State University have had fun and learned with this material. We hope the same happens for others.

BRUCE CRAUDER
BENNY EVANS
ALAN NOELL

P

Prologue: Calculator Arithmetic

Graphing calculators are powerful tools for mathematical analysis, and this power has profound effects on how modern mathematics and its applications are done. Many mathematical applications that traditionally required sophisticated mathematical development can now be successfully analyzed at an elementary level. Indeed, modern calculating power enables entering students to attack problems that in the past would have been considered too complicated. The first step is to become proficient with arithmetic on the calculator. In this chapter we discuss key mathematical ideas associated with calculator arithmetic. Chapter 1 of the *Technology Guide* is intended to provide additional help for those who are new to the operation of the calculator and those who need a brief refresher on arithmetic operations.

Typing Mathematical Expressions

When we write expressions such as $\frac{71}{7} + 3^2 \times 5$ using pen and paper, the paper serves as a two-dimensional display, and we can express fractions by putting one number on top of another and exponents by using a superscript. When we enter such expressions on a computer, calculator, or typewriter, however, we must write them on a single line, using special symbols and (often) additional parentheses. The *caret* symbol $\wedge$ is commonly used to denote an exponent, so in *typewriter notation* $\frac{71}{7} + 3^2 \times 5$ comes out as

$$71 \div 7 + 3 \wedge 2 \times 5 \,.$$

In Figure P.1, we have entered $\boxed{0.1}$ this expression, and the resulting answer 55.14285714 is shown in Figure P.2. You should use your calculator to verify that we did it correctly. (The footnote symbol in a box $\boxed{0.1}$ indicates that the exact keystrokes for doing this are shown on the Quick Reference pages of the *Technology Guide*.)

Rounding

When a calculation yields a long answer such as the 55.14285714 shown in Figure P.2, we will commonly shorten it to a more manageable size by *rounding*. Rounding means that we keep a few of the digits after the decimal point, possibly changing the last one, and discard the rest. There is no set rule for how many digits after the decimal point you

FIGURE P.1 Entering
$\frac{71}{7} + 3^2 \times 5$

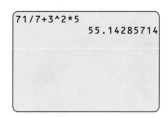

FIGURE P.2 The value of
$\frac{71}{7} + 3^2 \times 5$

should keep; in practice, it depends on how much accuracy you need in your answer, as well as on the accuracy of the data you input. As a general rule, in this text we will round to two decimal places. Thus for

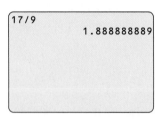

FIGURE P.3 An answer that will be reported as 1.71

$$\frac{71}{7} + 3^2 \times 5 = 55.14285714$$

we would report the answer as 55.14.

In order to make the abbreviated answer more accurate, it is standard practice to increase the last decimal entry by 1 if the following entry is 5 or greater. Verify with your calculator that

$$\frac{58.7}{6.3} = 9.317460317 \,.$$

In this answer the next digit after 1 is 7, which indicates that we should round up, so we would report the answer rounded to two decimal places as 9.32. Note that in reporting 55.14 as the rounded answer above, we followed this same rule. The next digit after 4 in 55.14285714 is 2, which does not indicate that we should round up.

To provide additional emphasis for this idea, Figure P.3 shows a calculation where rounding does not change the last reported digit, and Figure P.4 shows a calculation where rounding requires that the last reported digit be changed.

FIGURE P.4 An answer that will be reported as 1.89

KEY IDEA P.1

Rounding

When reporting complicated answers, we will adopt the convention of keeping two places beyond the decimal point. The last digit is increased by 1 if the next following digit is 5 or greater.

Although we generally round to two decimal places, there will be times when it is appropriate to use fewer or more decimal places. These circumstances will be explicitly noted in the text or will be clear from the context of the calculation.

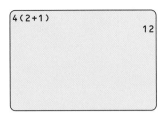

FIGURE P.5 A correct calculation of $4(2 + 1)$ when parentheses are properly entered

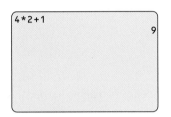

FIGURE P.6 An incorrect calculation of $4(2 + 1)$ caused by omitting parentheses

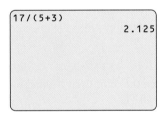

FIGURE P.7 Proper use of parentheses in the calculation of $\frac{17}{5+3}$

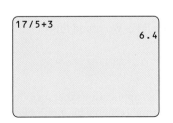

FIGURE P.8 An incorrect calculation of $\frac{17}{5+3}$ caused by omitting parentheses

Parentheses and Grouping

When parentheses appear in a calculation, the operations inside are to be done first. Thus $4(2 + 1)$ means that we should first add $2 + 1$ and then multiply the result by 4, getting an answer of 12. This is correctly entered 0.2 and calculated in Figure P.5. Where parentheses appear, their use is essential. If we had entered the expression as $4 \times 2 + 1$, leaving out the parentheses, the calculator would have interpreted it to mean first to multiply 4 times 2 and then to add 1 to the result, giving an incorrect answer of 9. This incorrect entry is shown in Figure P.6.

Sometimes parentheses do not appear, but we must supply them. For example, $\frac{17}{5+3}$ means $17 \div (5 + 3)$. The parentheses are there to show that the whole expression $5 + 3$ goes in the denominator. To do this on the calculator, we must supply 0.3 these parentheses. Figure P.7 shows the result. If the parentheses are not used, and $\frac{17}{5+3}$ is entered as $17 \div 5 + 3$, the calculator will interpret it to mean that only the 5 goes in the denominator of the fraction. This error is shown in Figure P.8. Similarly, $\frac{8+9}{7+2}$ means $(8 + 9) \div (7 + 2)$; the parentheses around $8 + 9$ indicate that the entire expression goes in the numerator, and the parentheses around $7 + 2$ indicate that the entire expression goes in the denominator. Enter 0.4 the expression on your calculator and check that the answer rounded to two places is 1.89. The same problem can occur with exponents. For example, $3^{2.7 \times 1.8}$ in typewriter notation is $3 \wedge (2.7 \times 1.8)$. Check 0.5 to see that the answer rounded to two places is 208.36.

In general, we advise that if you have trouble entering an expression into your calculator, or if you get an answer that you know is incorrect, go back and re-enter the expression after first writing it out in typewriter notation, and be careful to supply all needed parentheses.

Minus Signs

The minus sign used in arithmetic calculations actually has two different meanings. If you have \$9 in your wallet and spend \$3, then you will have $9 - 3 = 6$ dollars left. Here the minus sign means that we are to perform the operation of subtracting 3 from 9. Suppose in another setting that you receive news from the bank that your checking account is overdrawn by \$30, so your balance is -30 dollars. Here the minus sign is used to indicate that the number we are dealing with is negative; it does not signify an operation between two numbers. In everyday usage, the distinction is rarely emphasized and may go unnoticed. But most calculators actually have different keys 0.6 for the two operations, and they cannot be used interchangeably. Thus differentiating between the two becomes crucial to using the calculator correctly.

Once the problem is recognized, it is usually easy to spot when the minus sign denotes subtraction (when two numbers are involved) and when it indicates a change in sign (when only one number is involved). The following examples should help clarify the situation:

$$-8 - 4 \quad \text{means} \quad \textit{negative 8 subtract } 4 \boxed{0.7}$$

$$\frac{3 - 7}{-2 \times 3} \quad \text{means} \quad \frac{3 \textit{ subtract } 7}{\textit{negative } 2 \times 3} \boxed{0.8}$$

$$2^{-3} \quad \text{means} \quad 2^{\textit{negative } 3}.$$

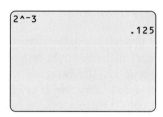

FIGURE P.9 Calculation of 2^{-3}
using the negative key

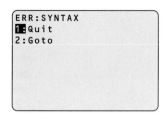

FIGURE P.10 Syntax error when
subtraction operation is used
in 2^{-3}

The calculation 0.9 of 2^{-3} is shown in Figure P.9. If we try to use the calculator's
subtraction key 0.10 , the calculator will not understand the input and will produce an
error message such as the one in Figure P.10.

EXAMPLE P.1   Some Simple Calculations

Make the following calculations and report the answer rounded to two digits beyond the
decimal point.

1. $\dfrac{\sqrt{11.4 - 3.5}}{26.5}$

2. $\dfrac{7 \times 3^{-2} + 1}{3 - 2^{-3}}$

```
√(11.4-3.5)/26.5

              .1060639194
```

FIGURE P.11 Solution to part 1

Solution to Part 1:   To make sure everything we want is included under the square root
symbol, we need to use parentheses. In typewriter notation, this looks like

$$\sqrt{}\,(11.4 - 3.5) \div 26.5\,.$$

We have calculated 0.11 this in Figure P.11. Since the third digit beyond the
decimal point, 6, is 5 or larger, we report the answer as 0.11.

```
(7*3^-2+1)/(3-2^
-3)
              .6183574879
```

FIGURE P.12 Solution to part 2

Solution to Part 2:   We need to take care to use parentheses to ensure that the numerator
and denominator are right, and we must use the correct keys for negative signs and
subtraction. In expanded typewriter notation,

$$\frac{7 \times 3^{-2} + 1}{3 - 2^{-3}} = (7 \times 3 \wedge \text{ negative } 2 + 1) \div (3 \text{ subtract } 2 \wedge \text{ negative } 3)\,.$$

The result 0.12 0.6183574879 is shown in Figure P.12. We round this to 0.62.

Special Numbers π and e

Two numbers, π and e, occur so often in mathematics and its applications that they deserve special mention. The number π is familiar from the formulas for the circumference
and area of a circle:

Area of a circle of radius $r = \pi r^2$

Circumference of a circle of radius $r = 2\pi r\,.$

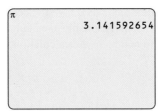

FIGURE P.13 A decimal approximation of π

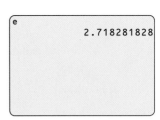

FIGURE P.14 A decimal approximation of e

The approximate value of π is 3.14159, but its exact value cannot be expressed by a simple decimal, and that is why it is normally written using a special symbol. Most calculators allow you to enter 0.13 the symbol π directly, as shown in Figure P.13. When we ask the calculator for a numerical answer 0.14 , we get the decimal approximation of π shown in Figure P.13.

The number e may not be as familiar as π, but it is just as important. Like π, it cannot be expressed exactly as a decimal, but its approximate value is 2.71828. In Figure P.14 we have entered 0.15 e, and the calculator responded with the decimal approximation shown. Often expressions that involve the number e include exponents, and most calculators have features to make entering such expressions easy. For example, when we enter 0.16 $e^{1.02}$ we obtain 2.77 after rounding.

Chain Calculations

Some calculations are most naturally done in stages. Many calculators have a special key 0.17 that accesses the result of the last calculation, allowing you to enter your work in pieces. To show how this works, let's look at

$$\left(\sqrt{13} - \sqrt{2}\right)^3 + \frac{17}{2 + \pi}.$$

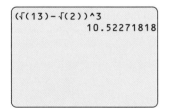

FIGURE P.15 The first step in a chain calculation

We will make the calculation in pieces. First we calculate $\left(\sqrt{13} - \sqrt{2}\right)^3$. Enter 0.18 this to get the answer in Figure P.15. To finish the calculation, we need to add this answer to $17/(2 + \pi)$:

$$\left(\sqrt{13} - \sqrt{2}\right)^3 + \frac{17}{2 + \pi} = \text{First answer} + \frac{17}{2 + \pi}.$$

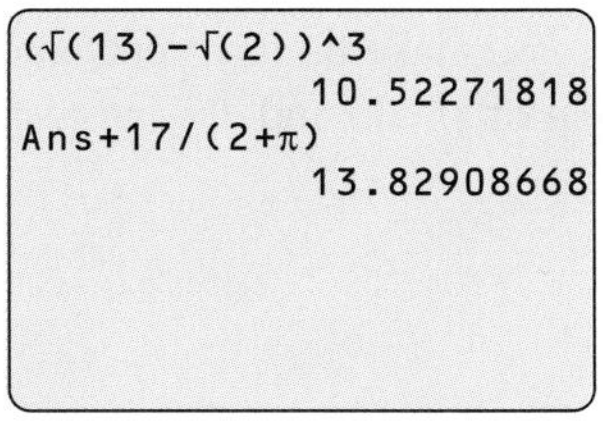

FIGURE P.16 Completing a chain calculation

In Figure P.16 we have used the answer 0.19 from Figure P.15 to complete the calculation.

Accessing the results of one calculation for use in another can be particularly helpful when the same thing appears several times in an expression. For example, let's calculate

$$\frac{3^{7/9} + 2^{7/9}}{5^{7/9} - 4^{7/9}}.$$

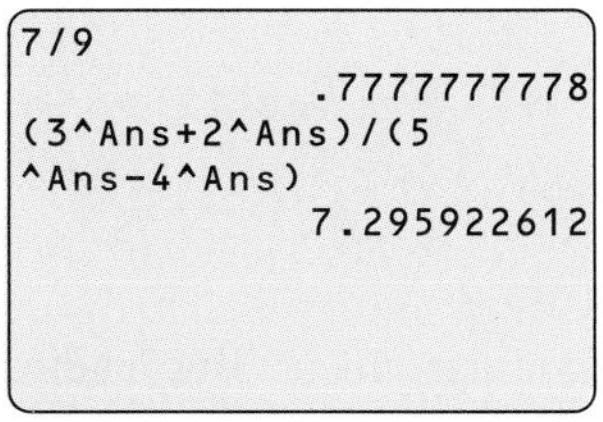

FIGURE P.17 Accessing previous results to get an accurate answer

Since 7/9 occurs several times, we have calculated it first in Figure P.17. Then we have used the results to complete 0.20 the calculation. We would report the final answer rounded to two decimal places as 7.30.

There is an additional advantage to accessing directly the answers of previous calculations. It might seem reasonable to calculate 7/9 first, round it to two decimal places, and then use that to complete the calculation. Thus we would be calculating

$$\frac{3^{0.78} + 2^{0.78}}{5^{0.78} - 4^{0.78}}.$$

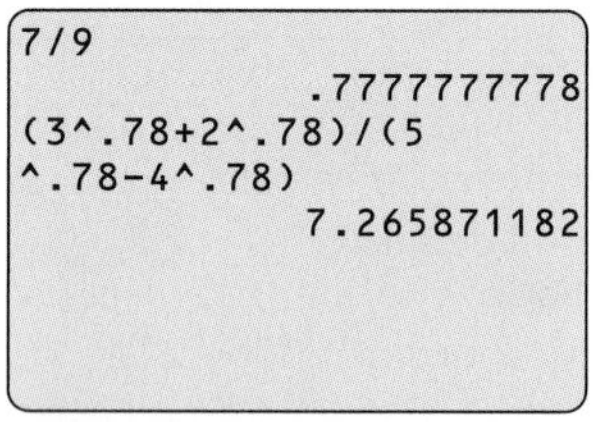

```
7/9
              .7777777778
(3^.78+2^.78)/(5
^.78-4^.78)
              7.265871182
```

FIGURE P.18 Inaccurate answer caused by early rounding

This is done in Figure P.18, which shows the danger in this practice. We got an answer, rounded to two decimal places, of 7.27, somewhat different from the more accurate answer, 7.30, that we got earlier. In many cases, errors caused by early rounding can be much more severe than is shown by this example. In general, if you are making a calculation in several steps, you should not round until you get the final answer. An important exception to this general rule occurs in applications where the result of an intermediate step must be rounded because of the context. For example, in a financial computation dollar amounts would be rounded to two decimal places.

EXAMPLE P.2   Compound Interest and APR

There are a number of ways in which lending institutions report and charge interest.

1. Paying *simple interest* on a loan means that you wait until the end of the loan before calculating or paying any interest. If you borrow $5000 from a bank that charges 7% simple interest, then after t years you will owe

$$5000 \times (1 + 0.07t) \text{ dollars}.$$

Under these conditions, how much money will you owe after 10 years?

2. Banks more commonly *compound the interest*. That is, at certain time periods the interest you have incurred is calculated and added to your debt. From that time on, you incur interest not only on your principal (the original debt) but on the added interest as well. Suppose the interest is compounded yearly, but you make no payments and there are no finance charges. Then, again with a principal of $5000 and 7% interest, after t years you will owe

$$5000 \times 1.07^t \text{ dollars}.$$

Under these conditions, how much will you owe after 10 years?

3. For many transactions such as automobile loans and home mortgages, interest is compounded monthly rather than yearly. In this case, the amount owed is calculated each month using the *monthly interest rate*. If r (as a decimal) is the monthly interest rate, then after m months, the amount owed is

$$5000 \times (1 + r)^m \text{ dollars},$$

assuming the principal is $5000.

The value of r is usually not apparent from the loan agreement. But lending institutions are required by the *Truth in Lending Act* to report the *annual percentage rate,* or APR, in a prominent place on all loan agreements. The same statute requires

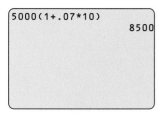

FIGURE P.19 Balance after 10 years using simple interest

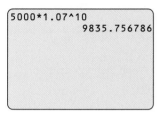

FIGURE P.20 Balance after 10 years using yearly compounding

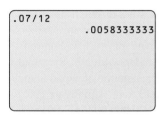

FIGURE P.21 Getting the monthly interest rate from the APR

that the value of r be calculated using the formula

$$r = \frac{APR}{12}.$$

If the annual percentage rate is 7%, what is the amount owed after 10 years?[1]

Solution to Part 1:   To find the amount owed after 10 years, we use $t = 10$ to get

$$5000 \times (1 + 0.07 \times 10).$$

Entering 0.21 this on the calculator as we have done in Figure P.19 reveals that the amount owed in 10 years will be $8500.

Solution to Part 2:   This time we use 0.22

$$5000 \times 1.07^{10}.$$

From Figure P.20 we see that, rounded to the nearest cent, the amount owed will be $9835.76. Comparison with part 1 shows the effect of compounding interest. We should note that at higher interest rates the effect is more dramatic.

Solution to Part 3:   The first step is to use the formula

$$r = \frac{APR}{12} = \frac{0.07}{12}$$

to get the value of r as we have done in Figure P.21. Ten years is 120 months, and this is the value we use for m. Using this value for m and incorporating the value of r that we just calculated, we entered 0.23 $5000 \times (1 + r)^{120}$ in Figure P.22, and we conclude that the amount owed will be $10,048.31. Comparing this with the answer from part 2, we see that the difference between yearly and monthly compounding is significant. It is important that you know how interest on your loan is calculated, and this may not be easy to find out from the paperwork you get from a lending institution. The APR will be reported, but the compounding periods may not be shown at all.   ▨ ▨ ▨

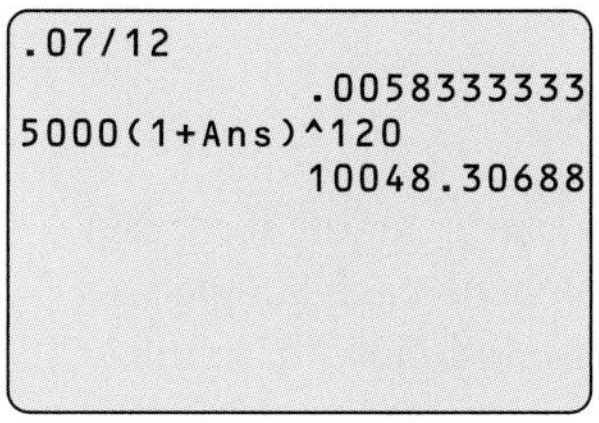

Scientific Notation

It is cumbersome to write down all the digits of some very large or very small numbers. A prime example of such a large number is *Avogadro's number*, which is the number of atoms in 12 grams of carbon 12. Its value is about

$$602,000,000,000,000,000,000,000.$$

FIGURE P.22 Balance after 10 years using monthly compounding

---

[1] Many consider the relationship between the monthly interest rate and the APR mandated by the Truth in Lending Act to be misleading. If, for example, you borrow $100 at an APR of 10%, then if no payments are made, you may expect to owe $110 at the end of 1 year. If interest is compounded monthly, then you will in fact owe somewhat more. For more information, see the discussion in Section 5.6 of *Fundamentals of Corporate Finance* by S. Ross, R. Westerfield, and B. Jordan (Chicago: Richard D. Irwin, 1995). See also Exercise 12.

An example of a small number that is awkward to write is the mass in grams of an electron:

$$0.000\ 000\ 000\ 000\ 000\ 000\ 000\ 000\ 000\ 000\ 911\ \text{gram}.$$

Scientists and mathematicians usually express such numbers in a more compact form using *scientific notation*. In this notation, numbers are written in a form with one nonzero digit to the left of the decimal point times a power of 10. Examples of numbers written in scientific notation are 2.7×10^4 and 2.7×10^{-4}. The power of 10 tells how the decimal point should be moved in order to write the number out in longhand. The 4 in 2.7×10^4 means that we should move the decimal point four places to the right. Thus

$$2.7 \times 10^4 = 27,000$$

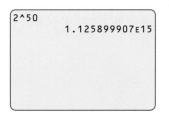

FIGURE P.23 Scientific notation for a large number

since we move the decimal point four places to the right. When the exponent on 10 is negative, the decimal point should be moved to the left. Thus

$$2.7 \times 10^{-4} = 0.00027$$

since we move the decimal point four places to the left. With this notation, Avogadro's number comes out as 6.02×10^{23}, and the mass of an electron as 9.11×10^{-31} gram.

Many times calculators display numbers like this but use a different notation for the power of 10. For example, Avogadro's number 6.02×10^{23} is displayed as 6.02E23, and the mass in grams of an electron 9.11×10^{-31} is shown as 9.11E-31. In Figure P.23 we have calculated 2^{50}. The answer reported by the calculator written in longhand is 1,125,899,907,000,000. In presenting the answer in scientific notation, it would in many settings be appropriate to round to two decimal places as 1.13×10^{15}. In Figure P.24 we have calculated $7/3^{20}$. The answer reported there equals 0.000 000 002 007 580 394. If we write it in scientific notation and round to two decimal places, we get 2.01×10^{-9}.

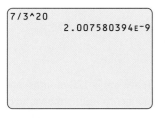

FIGURE P.24 Scientific notation for a small number

ANOTHER LOOK

Order of Operations

Once a mathematical expression has been written in typewriter notation, there is a collection of rules, the *order of operations*, that mathematicians have adopted so that expressions will be unambiguous. Calculators more or less follow these rules, but sometimes they exhibit inconsistencies. These inconsistencies can always be avoided by the proper use of parentheses.

Order of operations rule 1: Expressions inside parentheses are to be evaluated first. If parentheses are nested, innermost pairs are to be evaluated first.

Order of operations rule 2: If there are no parentheses to act as a guide, exponentials are calculated first, multiplications and divisions second, and sums and differences last.

Order of operations rule 3: If two or more from the list of multiplications, divisions, or subtractions occur in sequence with no parentheses to act as a guide, then the calculation should be completed working from left to right.

Order of operations rule 4: If two or more exponentials occur in sequence with no parentheses to act as a guide, then the calculation should be completed working from right to left.

When we follow these rules, there is only one way to evaluate a mathematical expression correctly. For example, if we are to evaluate $2 + 3 \times 4$, rule 2 requires that we calculate the product before evaluating the sum:

$$2 + 3 \times 4 = 2 + 12 = 14 .$$

You should check to see that if the sum is evaluated first, then we get an answer of 20. But according to the rules of operation, 14 is the correct answer, and 20 is incorrect.

The rules lead us correctly even through complicated calculations such as

$$3 + 4 \times 2 \wedge 3 - 12 \div 3 \times 4 .$$

There are no parentheses to act as a guide, so rule 2 applies and requires that we evaluate exponentials first:

$$3 + 4 \times 2 \wedge 3 - 12 \div 3 \times 4 = 3 + 4 \times 8 - 12 \div 3 \times 4 .$$

Next we evaluate multiplications and divisions, working left to right:

$$3 + 4 \times 8 - 12 \div 3 \times 4 = 3 + 32 - 12 \div 3 \times 4$$
$$3 + 32 - 12 \div 3 \times 4 = 3 + 32 - 4 \times 4$$
$$3 + 32 - 4 \times 4 = 3 + 32 - 16 .$$

We complete the calculation by evaluating additions and subtractions:

$$3 + 32 - 16 = 19 .$$

If there are parentheses present, we apply rules 2, 3, and 4 as required inside an innermost pair of parentheses. Let's look, for example, at

$$(2 + 3)((8 \div 2 + 5) \div 3) .$$

There are two innermost pairs of parentheses, $(2 + 3)$ and $(8 \div 2 + 5)$. We start inside these, applying rule 2:

$$(2 + 3)((8 \div 2 + 5) \div 3) = 5((4 + 5) \div 3)$$
$$5((4 + 5) \div 3) = 5(9 \div 3)$$
$$5(9 \div 3) = 5 \times 3 = 15 .$$

For an expression like $16 \div 4 \div 2$, rule 3 is needed. It says that we should work from left to right:

$$16 \div 4 \div 2 = 4 \div 2$$
$$4 \div 2 = 2 .$$

You should check to see that if we calculate from right to left, we obtain an incorrect answer of 8.

Finally, we look at $2 \wedge 3 \wedge 2$. Here we must apply rule 4 and work from right to left:

$$2 \wedge 3 \wedge 2 = 2 \wedge 9 = 512 \, .$$

It is worth pointing out that rules 3 and 4 are not often needed, and in ordinary usage they can be avoided by the judicious placement of parentheses. The two expressions $16 \div 4 \div 2$ and $(16 \div 4) \div 2$ mean the same thing, but the second leaves absolutely no room for miscommunication. It is always wise, when writing out mathematical expressions, to include parentheses that make the meaning absolutely clear.

Some of these rules are not consistently followed by calculators. In the case of rule 4, the TI-83 and TI-89 give different answers to

$$2 \wedge 3 \wedge 2 \, ,$$

for example. The TI-89 follows rule 4 and gives 512, whereas the TI-83 evaluates the expression from left to right and gives 64. Further, *implied* multiplication such as 2π is sometimes treated differently than explicit multiplication such as $2 \times \pi$. For example, the TI-82 and TI-83 give different answers to

$$1 \div 2\pi \, .$$

Since in these circumstances there may be doubt about how the calculator will process the information, you should use parentheses to make the meaning clear. All calculators will correctly handle the expressions $2 \wedge (3 \wedge 2)$ and $1 \div (2\pi)$. Both in writing expressions on paper and in entering them in a calculator, it is imperative to use parentheses appropriately for accurate communication.

Enrichment Exercises

E-1. **Order of operations when no parentheses are present:** Use the rules for order of operations to calculate the following by hand. You may wish to check your answers using your calculator.

a. $2 + 3 \times 4 + 5$ b. $6 \div 3 \times 3 \wedge 2$

c. $3 \times 16 \div 2 \wedge 2$ d. $5 - 4 \times 3 + 8 \div 2 \times 3$

e. $3 \wedge 2 \wedge 2$

E-2. **Order of operations when parentheses are present:** Use the rules for order of operations to calculate the following by hand. You may wish to check your answers using your calculator.

a. $(2 \wedge 3) \wedge 2$ b. $9 \div (2 + 1)$

c. $(8 - (3 \times 2))(4 + 2)$

P SKILL BUILDING EXERCISES

S-1. **Basic calculations:** $\dfrac{2.6 \times 5.9}{6.3}$

S-2. **Basic calculations:** $3^{3.2} - 2^{2.3}$

S-3. **Basic calculations:** $\dfrac{e}{\sqrt{\pi}}$

S-4. **Basic calculations:** $\dfrac{7.6^{1.7}}{9.2}$

S-5. **Parentheses and grouping:** $\dfrac{7.3 - 6.8}{2.5 + 1.8}$

S-6. **Parentheses and grouping:** $3^{2.4 \times 1.8 - 2}$

S-7. **Parentheses and grouping:** $\dfrac{\sqrt{6 + e} + 1}{3}$

S-8. **Parentheses and grouping:** $\dfrac{\pi - e}{\pi + e}$

S-9. **Subtraction versus sign:** $\dfrac{-3}{4 - 9}$

S-10. **Subtraction versus sign:** $-2 - 4^{-3}$

S-11. **Subtraction versus sign:** $-\sqrt{8.6 - 3.9}$

S-12. **Subtraction versus sign:** $\dfrac{-\sqrt{10} + 5^{-0.3}}{17 - 6.6}$

S-13. **Chain calculations:** All of the following can be done with a single entry, but they are intended to provide practice with chain calculations.

a. $\dfrac{3}{7.2 + 5.9} + \dfrac{7}{6.4 \times 2.8}$

b. $\left(1 + \dfrac{1}{36}\right)^{\left(1 - \frac{1}{36}\right)}$

S-14. **Scientific notation:** The following numbers are written in scientific notation. Express them as ordinary decimals.

a. 3.62×10^4

b. 7.19×10^6

c. 3.13×10^{-3}

d. 4.67×10^{-7}

P EXERCISES

1. **Arithmetic:** Calculate the following and report the answer rounded to two decimal places. For some of the calculations, you may wish to use the chain calculation facility of your calculator to help avoid errors.

a. $(4.3 + 8.6)(8.4 - 3.5)$

b. $\dfrac{2^{3.2} - 1}{\sqrt{3} + 4}$

c. $\sqrt{2^{-3} + e}$

d. $(2^{-3} + \sqrt{7} + \pi)\left(e^2 + \dfrac{7.6}{6.7}\right)$

e. $\dfrac{17 \times 3.6}{13 + \frac{12}{3.2}}$

2. **A good investment:** You have just received word that your original investment of $850 has increased in value by 13%. What is the value of your investment today?

3. **A bad investment:** You have just received word that your original investment of $720 has decreased in value by 7%. What is the value of your investment today?

4. **An uncertain investment:** Suppose you invested $1300 in the stock market 2 years ago. During the first year, the value of the stock increased by 12%. During the second year, the value of the stock decreased by 12%. How much money is your investment worth at the end of the two-year period? Did you earn money or lose money? (*Note:* The answer to the first question is *not* $1300.)

5. **Pay raise:** You receive a raise in your hourly pay from $7.25 per hour to $7.50 per hour. What percent increase in pay does this represent?

6. **Heart disease:** In a certain county, the number of deaths due to heart disease decreased from 235 in one year to 221 in the next year. What percent decrease in deaths due to heart disease does this represent?

7. **Trade discount:** Often retailers sell merchandise at a suggested retail price determined by the manufacturer. The *trade discount* is the percentage discount given to the retailer by the manufacturer. The resulting price is the retailer's net cost and so is called the *cost price*. For example, if the suggested retail price is $100.00 and the trade discount is 45%, then the cost price is $100.00 - 45\% \times 100.00 = 55.00$ dollars.

   a. If an item has a suggested retail price of $9.99 and the trade discount is 40%, what is the retailer's cost price?

   b. If an item has a cost price of $37.00 and a suggested retail price of $65.00, what trade discount was used?

8. **Series discount:** *This is a continuation of Exercise 7.* Sometimes manufacturers give more than one discount instead of a single trade discount—for example, in trading with large-volume retailers. Such a *series discount* is quoted as a sequence of discounts, taken one after another. Suppose a manufacturer normally gives a trade discount of 45%, but it has too much of the item in inventory and so wants to sell more. In this case, the manufacturer may give all retailers another discount of 15% and may perhaps extend yet *another* discount of 10% to a specific retailer it wants to land as a client. In this example, the series discount would be 45%, 15%, 10%, calculated one after another, like this: For an item with a suggested retail price of $100.00, applying the first discount gives $100.00 - 45\% \times 100.00 = 55.00$ dollars. The second discount of 15% is applied to the $55.00 as follows: $55.00 - 15\% \times 55.00 = 46.75$ dollars. Now the third discount gives the final cost price of $46.75 - 10\% \times 46.75 = 42.08$ dollars.

   a. Suppose an item has a suggested retail price of $80.00 and the manufacturer is giving a series discount of 25% and 10%. What is the resulting cost price?

   b. Suppose an item has a suggested retail price of $100.00 and the manufacturer is giving a series discount of 35%, 10%, 5%. What is the resulting cost price?

   c. What single trade discount would give the same cost price as a series discount of 35%, 10%, 5%? (*Note:* The answer is *not* 50%.)

   d. Explain why we could have calculated the same answer as in part b by multiplying

   $$100.00 \times 0.65 \times 0.90 \times 0.95.$$

   In this case, what do the 0.65, 0.90, and 0.95 represent?

9. **Present value:** *Present value* is the amount of money that must be invested now at a given rate of interest to produce a given future value. For a 1-year investment, the present value can be calculated using

   $$\text{Present value} = \frac{\text{Future value}}{1 + r},$$

   where r is the yearly interest rate expressed as a decimal. (Thus, if the yearly interest rate is 8%, then $1 + r = 1.08$.) If an investment yielding a yearly interest rate of 12% is available, what is the present value of an investment that will be worth $5000 at the end of 1 year? That is, how much must be invested today at 12% in order for the investment to have a value of $5000 at the end of a year?

10. **Future value:** Business and finance texts refer to the value of an investment at a future time as its *future value*. If an investment of P dollars is compounded yearly at an interest rate of r as a decimal, then the value of the investment after t years is given by

    $$\text{Future value} = P \times (1 + r)^t.$$

    In this formula, $(1 + r)^t$ is known as the *future value interest factor*, so the formula above can also be written

    $$\text{Future value} = P \times \text{Future value interest factor}.$$

    Financial officers normally calculate this (or look it up in a table) first.

    a. What future value interest factor will make an investment double? Triple?

    b. Say you have an investment which is compounded yearly at a rate of 9%. Find the future value interest factor for a 7-year investment.

    c. Use the results from part b to calculate the 7-year future value if your initial investment is $5000.

11. **The Rule of 72:** *This is a continuation of Exercise 10.* Financial advisors sometimes use a rule of thumb known as the *Rule of 72* to get a rough estimate of the time it takes for an investment to double in value. For an investment that is compounded yearly at an interest rate of $r\%$, this rule says it will take about $72/r$ years for the investment to double. In this calculation, r is the integer interest rate rather than a decimal. Thus if the interest rate is 8%, we would use $\frac{72}{8}$ rather than $\frac{72}{0.08}$.

    For the remainder of this exercise, we will consider an investment that is compounded yearly at an interest rate of 13%.

    a. According to the Rule of 72, how long will it take the investment to double in value?

    Parts b and c of this exercise will check to see how accurate this estimate is for this particular case.

    b. Using the answer you got from part a of this exercise, calculate the future value interest factor (as defined in Exercise 10). Is it exactly the same as your answer to the first question in part a of Exercise 10?

    c. If your initial investment was $5000, use your answer from part b to calculate the future value. Did your investment exactly double?

12. **The Truth in Lending Act:** Many lending agencies compound interest more often than yearly, and as we noted in Example P.2, they are required to report the annual percentage rate, or APR, in a prominent place on the loan agreement. Furthermore, they are required to calculate the APR in a specific way. If r is the monthly interest rate, then the APR is calculated using

    $$\text{APR} = 12 \times r.$$

    a. Suppose a credit card company charges a monthly interest rate of 1.9%. What APR must the company report?

    b. The phrase *annual percentage rate* leads some to believe that if you borrow $6000 from a credit card company which quotes an APR of 22.8%,

and if no payments are made, then at the end of 1 year interest would be calculated as 22.8% simple interest on $6000. How much would you owe at the end of a year if interest is calculated in this way?

c. If interest is compounded monthly (which is common), then the actual amount you would owe in the situation of part b is given by

$$6000 \times 1.019^{12}.$$

What is the actual amount you would owe at the end of a year?

13. **The size of the Earth:** The radius of the Earth is approximately 4000 miles.

    a. How far is it around the equator? (*Hint:* You are looking for the circumference of a circle.)

    b. What is the volume of the Earth? (*Note:* The volume of a sphere of radius r is given by $\frac{4}{3}\pi r^3$.)

    c. What is the surface area of the Earth? (*Note:* The surface area of a sphere of radius r is given by $4\pi r^2$.)

14. **When the radius increases:**

    a. A rope is wrapped tightly around a wheel with a radius of 2 feet. If the radius of the wheel is increased by 1 foot to a radius of 3 feet, by how much must the rope be lengthened to fit around the wheel?

    b. Consider a rope wrapped around the Earth's equator. We noted in Exercise 13 that the radius of the Earth is about 4000 miles. That is 21,120,000 feet. Suppose now that the rope is to be suspended exactly 1 foot above the equator. By how much must the rope be lengthened to accomplish this?

15. **The length of Earth's orbit:** The Earth is approximately 93 million miles from the sun. For this exercise we will assume that the Earth's orbit is a circle.[2]

    a. How far does the Earth travel in a year?

    b. What is the velocity in miles per year of the Earth in its orbit? (*Hint:* Recall that $\text{Velocity} = \frac{\text{Distance}}{\text{Time}}$.)

    $\longrightarrow$

---

[2]The orbit of the Earth is in fact an ellipse, but for many practical applications the assumption that it is a circle yields reasonably accurate results.

c. How many hours are there in a year? (*Note:* For this problem ignore leap years.)

d. What is the velocity in miles per hour of the Earth in its orbit?

16. **A population of bacteria:** Some populations, such as bacteria, can be expected under the right conditions to show *exponential growth*. If 2000 bacteria of a certain type are incubated under ideal conditions, then after t hours we expect to find 2000×1.07^t bacteria present. How many bacteria would we expect to find after 8 hours? How many after 2 days?

17. **Newton's second law of motion:** Newton's second law of motion states that the force F on an object is the product of its mass m with its acceleration a:

$$F = ma.$$

If mass is measured in kilograms and acceleration in meters per second per second, then the force is given in *newtons*. Another way to measure force is in *pounds*; in fact, 1 newton is 0.225 pound (which is about a quarter of a pound). In the case of an object near the surface of the Earth, the force due to gravity is its weight. Near the surface of the Earth, acceleration due to gravity is 9.8 meters per second per second. What is the weight in newtons of a man with a mass of 75 kilograms? What is his weight in pounds?

18. **Weight on the moon:** *This is a continuation of Exercise 17.* Acceleration due to gravity near the surface of the Earth's moon is only 1.67 meters per second per second. Thus an object has a different weight on the Earth than it would on the moon. What is the weight of the 75-kilogram man from Exercise 17 if he is standing on the moon? Give your answer first in newtons and then in pounds.

19. **Frequency of musical notes:** Counting sharps and flats, there are 12 notes in an octave on a standard piano. If one knows the frequency of a note, then one can find the frequency of the next higher note by multiplying by the 12th root of 2:

Frequency of next higher note

$$= \text{Frequency of given note} \times 2^{1/12}.$$

The frequency of middle C is 261.63 cycles per second. What is the frequency of the next higher note (which is C#) on a piano? What is the frequency of the D note just above middle C? (The D note is two notes higher than middle C.)

20. **Lean body weight in males:** A person's *lean body weight* L is the amount he or she would weigh if all body fat were magically to disappear. One text[3] gives the "equation that practitioners can use most feasibly in the field to predict lean body weight in young adult males." The equation is

$$L = 98.42 + 1.08W - 4.14A.$$

Here L is lean body weight in pounds, W is weight in pounds, and A is abdominal circumference in inches. Find the approximate lean body weight of a young adult male who weighs 188 pounds and has an abdominal circumference of 35 inches. What is the weight of his body fat? What is his body fat percent?

21. **Lean body weight in females:** *This is a continuation of Exercise 20.* The text cited in Exercise 20 gives a more complex method of calculating lean body weight for young adult females:

$$L = 19.81 + 0.73W + 21.2R - 0.88A$$
$$- 1.39H + 2.43F.$$

Here L is lean body weight in pounds, W is weight in pounds, R is wrist diameter in inches, A is abdominal circumference in inches, H is hip circumference in inches, and F is forearm circumference in inches. According to this formula, what is the approximate lean body weight of a young adult female who weighs 132 pounds and has wrist diameter of

---

[3]D. Kirkendall, J. Gruber, and R. Johnson, *Measurement and Evaluation for Physical Educators*, 2nd ed. (Champaign, IL: Human Kinetics Publishers, 1987).

2 inches, abdominal circumference of 27 inches, hip circumference of 37 inches, and forearm circumference of 7 inches? What is the weight of her body fat? What is her body fat percent?

22. **Manning's equation:** Hydrologists sometimes use *Manning's equation* to calculate the velocity v, in feet per second, of water flowing through a pipe. The velocity depends on the *hydraulic radius R* in feet, which is one-quarter of the diameter of the pipe when the pipe is flowing full; the *slope S* of the pipe, which gives the vertical drop in feet for each horizontal foot; and the *roughness coefficient n*, which depends on the material of which the pipe is made. The relationship is given by

$$v = \frac{1.486}{n} R^{2/3} S^{1/2} .$$

For a certain brass pipe, the roughness coefficient has been measured to be $n = 0.012$. The pipe has a diameter of 3 feet and a slope of 0.2 foot per foot. (That is, the pipe drops 0.2 foot for each horizontal foot.) If the pipe is flowing full, find the hydraulic radius of the pipe, and find the velocity of the water flowing through the pipe.

Summary

Modern graphing calculators are well designed for ease of use, but care must be taken when entering expressions. The most common errors occur when parentheses are omitted or misused. Also, rounding and scientific notation are significant concepts when you use a calculator. The special numbers e and π are important.

Typing Expressions and Parentheses

When entering an expression in the calculator, you must enter it not as one would write it on paper but, rather, in *typewriter notation*. If you have trouble getting an expression into the calculator properly, first write it out on paper in typewriter notation and then enter it into the calculator. Parentheses are essential when you need to tell the calculator that a certain operation is to be applied to a group of numbers.

Rounding

In order to do accurate calculations, the calculator uses decimals with many digits. Often only a few digits after the decimal point are needed for the final answer. We limit the number of decimal places by *rounding*. There is no set number of digits used in rounding; that depends on the accuracy of the data entered and on the accuracy needed for the answer. In general, however, answers are reported in this text rounded to two decimal places.

Rounding Convention for This Text: Unless otherwise specified, answers should be rounded to two decimal places. If the third digit beyond the decimal point is less than 5, discard all digits beyond the second. If the third digit is 5 or larger, increase the second digit by 1 before discarding additional digits.

Special Numbers

There are two special numbers, π and e, that occur so often in mathematics and its applications that their use cannot be avoided. Modern calculators allow for their direct entry. The number π is the familiar ratio of the circumference of a circle to its diameter. The number e is perhaps less familiar but is just as important, and it often arises in certain exponential contexts. Neither of these numbers can be expressed exactly as a finite decimal, but their approximate values are given below.

$$\pi \approx 3.14159$$
$$e \approx 2.71828$$

Scientific Notation

Some numbers use so many digits that it is more convenient to express them in *scientific notation*. This simply means to write the number using only a few digits and multiply by a power of 10 that tells how the decimal point should be adjusted. The adjustment required depends on the sign on the power of 10. Scientific notation can be entered in the calculator using 10 to a power, but when the calculator reports an answer in scientific notation, a special notation is common.

Entry	Calculator Display	Meaning
number $\times\ 10^{+k}$	*number* **E** $+$ **k**	Move decimal point k places right.
number $\times\ 10^{-k}$	*number* **E** $-$ **k**	Move decimal point k places left.

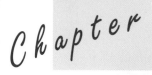

Chapter

1

Functions

A fundamental idea in mathematics and its applications is that of a *function*, which tells how one thing depends on others. One example of a function is the interest incurred on a loan after a certain number of years. In this case there is a formula[1] that allows you to calculate precisely how much you owe, and the formula makes explicit how the debt depends on time. Another example is the value of the Dow Jones Industrial Average at the close of each business day. In this case the value depends on the date, but there is no known formula.

This idea of a function is so basic that it is impossible to say where and when it originated, and it was almost certainly conceived independently by any number of people at different times and places. The idea remains today as the cornerstone to understanding and using mathematics.

In applications of mathematics, functions are often representations of real phenomena or events. Thus we say that they are *models*. Obtaining a function or functions to act as a model is commonly the key to understanding physical, natural, and social science phenomena. This applies to business and many other areas as well.

There are a number of common ways in which functions are presented and used. In this chapter we look at functions given by formulas, by tables, by graphs, and by words. Analyzing a function from each of these perspectives will be essential as we progress.

1.1 *Functions Given by Formulas*

We look first at functions given by formulas, since this provides a natural context for explaining how a function works.

Functions of One Variable

If your job pays $7.00 per hour, then the money M (in dollars) that you make depends on the number of hours h that you work, and the relationship is given by a simple formula:

$$\text{Money} = 7 \times \text{Hours worked}, \quad \text{or} \quad M = 7h \text{ dollars}.$$

---

[1] See Example P.2 of the Prologue.

The formula $M = 7h$ shows how the money M that you earn depends on the number of hours h that you work, and we say that M *is a function of* h. In this context we are thinking of h as a *variable* whose value we may not know until the end of the week. Once the value of h is known, the formula $M = 7h$ can be used to calculate the value of M. To emphasize that M is a function of h it is common to write $M = M(h)$ and to write the formula as $M(h) = 7h$.

Functions given in this way are very easy to use. For example, if you work 30 hours, then in *functional notation*, $M(30)$ is the money you earn.[2] To calculate that, you need only replace h in the formula by 30:

$$M(30) = 7 \times 30 = 210 \text{ dollars}.$$

It is important to remember that h is measured in hours and M is measured in dollars. You will not be very happy if your boss makes a mistake and pays you $7 \times 10 = 70$ cents for 10 hours worked. You may be happier if she pays you $7 \times 30 = 210$ dollars for 30 minutes of work, but both calculations are incorrect. The formula is not useful unless you state in words the units you are using. A proper presentation of the formula for this function would be $M = 7h$, where h is measured in hours and M is measured in dollars. The words that give the units are as important as the formula.

We should also note that you can use different letters for variables if you want. Whatever letters you use, it is critical that you explain in words what the letters mean. We could, for example, use the letter t instead of h to represent the number of hours worked. If we did that, we would emphasize the functional relationship with $M = M(t)$ and present the formula as $M = 7t$, where t is the number of hours worked, and M is the money earned in dollars.

Functions of Several Variables

Sometimes functions depend on more than one variable. Your grocery bill G may depend on the number a of apples you buy, the number s of sodas you buy, and the number p of frozen pizzas you put in your basket. If apples cost 60 cents each, sodas cost 50 cents each, and pizzas cost \$3.25 each, then we can express $G = G(a, s, p)$ as

Grocery bill = Total cost of apples + Total cost of sodas + Total cost of pizzas

$$G = 0.6a + 0.5s + 3.25p,$$

where G is measured in dollars. The notation $G = G(a, s, p)$ is simply a way of emphasizing that G is a function of the variables a, s, and p—that is, that the value of G depends on a, s, and p. We could also give a correct formula for the function as $G = 60a + 50s + 325p$, where G is measured this time in cents. Either expression is correct as long as we explicitly say what units we are using.

---

[2]The parentheses in functional notation indicate the dependence of the function on the variable. They do not represent multiplication. For example, $M(30)$ is not the same as $M \times 30$.

EXAMPLE 1.1   A Grocery Bill

Suppose your grocery bill is given by the function $G = G(a, s, p)$ above (with G measured in dollars).

1. Use functional notation to show the cost of buying 4 apples, 2 sodas, and 3 pizzas, and then calculate that cost.

2. Explain the meaning of $G(2, 6, 1)$.

3. Calculate the value of $G(2, 6, 1)$.

Solution to Part 1:   Since we are buying 4 apples, we use $a = 4$. Similarly, we are buying 2 sodas and 3 pizzas, so $s = 2$ and $p = 3$. Thus in functional notation our grocery bill is $G(4, 2, 3)$. To calculate this we use the formula $G = 0.6a + 0.5s + 3.25p$, replacing a by 4, s by 2, and p by 3:

$$G(4, 2, 3) = 0.6 \times 4 + 0.5 \times 2 + 3.25 \times 3$$
$$= 13.15 \text{ dollars}.$$

Thus the cost is $13.15.

Solution to Part 2:   The expression $G(2, 6, 1)$ is the value of G when $a = 2$, $s = 6$, and $p = 1$. It is your grocery bill when you buy 2 apples, 6 sodas, and 1 frozen pizza.

Solution to Part 3:   We calculate $G(2, 6, 1)$ just as we did in part 1, but this time we use $a = 2$, $s = 6$, and $p = 1$:

$$G(2, 6, 1) = 0.6 \times 2 + 0.5 \times 6 + 3.25 \times 1$$
$$= 7.45 \text{ dollars}.$$

Thus the cost is $7.45.

Even when the formula for a function is complicated, the idea of how you use it remains the same. Let's look, for example, at $f = f(x)$, where f is determined as a function of x by the formula

$$f = \frac{x^2 + 1}{\sqrt{x}}.$$

The value of f when x is 3 is expressed in functional notation as $f(3)$. To calculate $f(3)$, we simply replace x in the formula by 3:

$$f(3) = \frac{3^2 + 1}{\sqrt{3}}.$$

You should check $\boxed{1.1}$ to see that the calculator gives an answer of 5.773502692, which we round to 5.77. Do not allow formulas such as this one to intimidate you. With the aid of the calculator, it is easy to deal with them.

EXAMPLE 1.2   **Borrowing Money**

When you borrow money to buy a home or a car, you pay off the loan in monthly payments, but interest is always accruing on the outstanding balance. This makes the determination of your monthly payment on a loan more complicated than you might expect. If you borrow P dollars at a monthly interest rate[3] of r (as a decimal) and wish to pay off the note in t months, then your monthly payment $M = M(P, r, t)$ in dollars can be calculated using

$$M = \frac{Pr(1+r)^t}{(1+r)^t - 1}.$$

1. Explain the meaning of $M(7800, 0.0067, 48)$ and calculate its value.

2. Suppose you borrow $5000 to buy a car and wish to pay off the loan over 3 years. Take the prevailing monthly interest rate to be 0.58%. (That is an annual percentage rate, APR, of $12 \times 0.58 = 6.96\%$.) Use functional notation to show your monthly payment, and then calculate its value.

Solution to Part 1:   The expression $M(7800, 0.0067, 48)$ gives your monthly payment on a $7800 loan that you pay off in 48 months (4 years) at a monthly interest rate of 0.67%. (That is an APR of $12 \times 0.67 = 8.04\%$.) To get its value, we use the formula above, putting 7800 in place of P, 0.0067 in place of r, and 48 in place of t:

$$M(7800, 0.0067, 48) = \frac{7800 \times 0.0067 \times 1.0067^{48}}{1.0067^{48} - 1}.$$

```
1.0067^48-1
             .3778542919
```

FIGURE 1.1 The first step in calculating a loan payment

This can be entered all at once on the calculator, but to avoid typing errors, we do the calculation in pieces. The calculation [1.2] of the denominator $1.0067^{48} - 1$ is shown in Figure 1.1. To complete the calculation [1.3] we need to get

$$\frac{7800 \times 0.0067 \times 1.0067^{48}}{\text{Answer from first calculation}}.$$

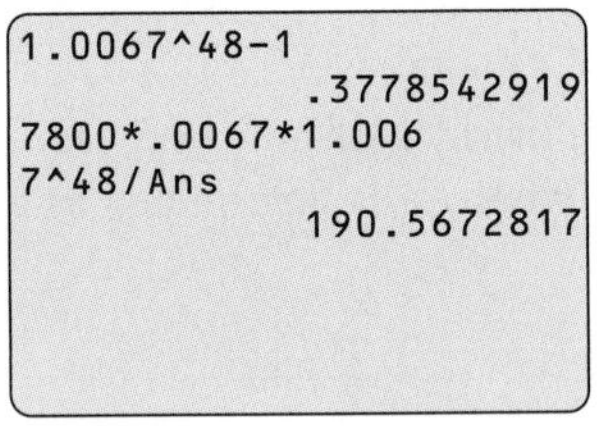

We round the answer shown in Figure 1.2 to get the monthly payment of $190.57.

```
1.0067^48-1
             .3778542919
7800*.0067*1.006
7^48/Ans
              190.5672817
```

FIGURE 1.2 Completing the calculation

Solution to Part 2:   We borrow $5000, so we use $P = 5000$. The monthly interest rate is 0.58%, so we use $r = 0.0058$, and we pay off the loan in 3 years, or 36 months, so $t = 36$. In functional notation, the monthly payment is $M(5000, 0.0058, 36)$. To calculate it we use

$$M(5000, 0.0058, 36) = \frac{5000 \times 0.0058 \times 1.0058^{36}}{1.0058^{36} - 1}.$$

---

[3]Here we are assuming monthly payment and interest compounding. If you use the annual percentage rate (APR) reported on your loan agreement, then you have $r = \frac{\text{APR}}{12}$. See also Exercise 12 at the end of the Prologue.

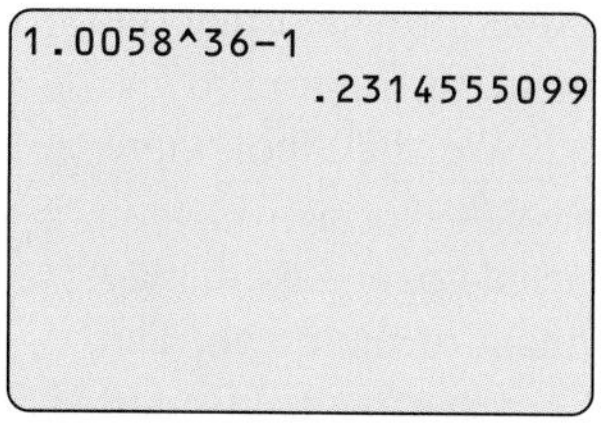

```
1.0058^36-1
           .2314555099
```

FIGURE 1.3 The first step in calculating the payment on a $5000 loan

Once again we make the calculation in two stages. First $\boxed{1.4}$ we get $1.0058^{36} - 1$ as shown in Figure 1.3. As before, we use this answer to complete the calculation $\boxed{1.5}$:

$$\frac{5000 \times 0.0058 \times 1.0058^{36}}{\text{Answer from the first calculation}}.$$

The result in Figure 1.4 shows that we will have to make a monthly payment of $154.29. ■ ■ ■

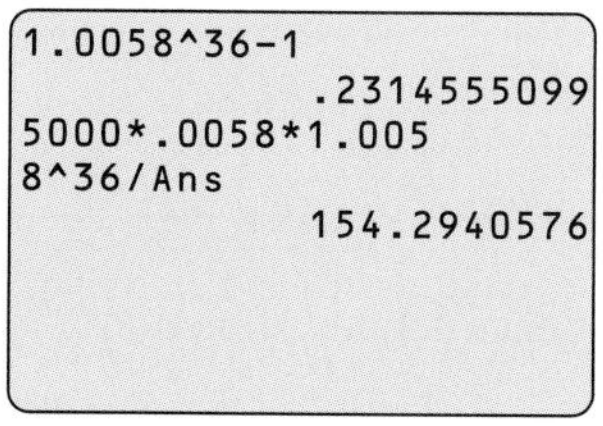

```
1.0058^36-1
           .2314555099
5000*.0058*1.005
8^36/Ans
           154.2940576
```

FIGURE 1.4 Completing the calculation

ANOTHER LOOK

Definition of a Function

At the most elementary level, it is sufficient to think of a function as a rule that tells how one thing depends on another, but in more advanced mathematics, it is important to give precise definitions of the terms we use. The precise definition of a function actually involves three parts: a set called the *domain*, a set called the *range*, and a correspondence that assigns to each element of the domain exactly one element of the range. For example, let the domain be the set $D = \{1, 2, 3, 4\}$, let the range be the set $R = \{2, 4, 6, 8\}$, and let the correspondence be as indicated below:

$$1 \longrightarrow 2$$
$$2 \longrightarrow 4$$
$$3 \longrightarrow 6$$
$$4 \longrightarrow 8.$$

This is a function since each element of the domain corresponds to exactly one element of the range. It is customary to give a function a name such as f and then to write $f : D \longrightarrow R$. It is also customary to denote by $f(x)$ the element assigned to x by the function f. Thus we could indicate the above correspondence denoted by f as follows:

$$f(1) = 2$$
$$f(2) = 4$$
$$f(3) = 6$$
$$f(4) = 8.$$

Another common way of presenting a function such as this one is with a *table of values*, as shown below.

x	1	2	3	4
$f(x)$	2	4	6	8

If we see a table such as this, we can interpret it as a function, provided that the elements of the top row are distinct. The top row is the domain, the bottom row is the range, and the columns give the function correspondence. We shall look more closely at functions given by tables in the next section.

It is not necessary that all the elements of the range be used up, and there may be several elements of the domain assigned to a single element of the range. The following correspondence also gives a function from D to R, even though 4 is used three times and 6 and 8 aren't used at all:

$$1 \longrightarrow 2$$
$$2 \longrightarrow 4$$
$$3 \longrightarrow 4$$
$$4 \longrightarrow 4.$$

In this case, it would also be correct to say that this correspondence defines a function with domain D and range $\{2, 4\}$.

On the other hand, a function is not allowed to leave out elements of the domain, nor may it assign the same element of the domain to more than one element of the range. Thus the following assignment is not a function from D to R. It fails to satisfy the definition on two counts, either of which would disqualify it. The number 4 is in the domain but is assigned to nothing, and 3 is assigned to both 6 and 8.

$$1 \longrightarrow 2$$
$$2 \longrightarrow 4$$
$$3 \longrightarrow 6$$
$$3 \longrightarrow 8.$$

When functions are given by formulas, the domain and range may not be specified. Consider the function f given by the formula $f(x) = 1/x$. If, as in this example, we do not specify the domain or the range, then the domain is assumed to be all real numbers for which the formula makes sense, and the range is assumed to be the all real numbers y such that $f(x) = y$ for some x. In order to find the domain for such a formula, it is usually best to work backwards. That is, we locate the numbers for which the formula does not make sense. In the case of $1/x$, the formula does not make sense when $x = 0$, because division by 0 is not defined. Thus the domain for this function is all real numbers except 0. In this case the range is also all real numbers except 0, but even for relatively simple formulas, the range may be difficult to determine.

As another example, consider $f(x) = \sqrt{x}$. This formula does not make sense as a real number if x is a negative number, so the domain is all real numbers greater than or equal to 0.

For many common formulas, we can find the domain by ruling out numbers that cause division by 0 or result in the square root of a negative number. As a final example, consider the formula

$$f(x) = \frac{x^2 + x + 1}{(x - 3)\sqrt{x - 1}}.$$

To find the domain, we look for trouble spots. The numerator causes no trouble since the formula $x^2 + x + 1$ makes sense for any number x. The denominator, however, needs a closer look. Using $x = 3$ or $x = 1$ will cause division by 0, and using any number less than 1 will result in the square root of a negative number. Thus the domain is the set of all real numbers greater than 1 excluding 3.

Enrichment Exercises

E-1. **Determining when a correspondence is a function:** Let $D = \{1, 2, 3, 4\}$, and let $R = \{5, 6, 7, 8\}$. Which of the following correspondences define a function f with domain D and range R? Be sure to explain your answers.

a. $1 \longrightarrow 8$
$2 \longrightarrow 7$
$3 \longrightarrow 5$
$4 \longrightarrow 6$

b. $1 \longrightarrow 8$
$2 \longrightarrow 8$
$3 \longrightarrow 5$
$4 \longrightarrow 6$

c. $1 \longrightarrow 8$
$1 \longrightarrow 5$
$2 \longrightarrow 7$
$3 \longrightarrow 5$
$4 \longrightarrow 6$

d. $1 \longrightarrow 8$
$2 \longrightarrow 7$
$4 \longrightarrow 6$

E-2. **Finding the domain:** Find the domain of the function given by each of the following formulas.

a. $f(x) = \dfrac{x^4 + x - 2}{7}$

b. $f(x) = \dfrac{x - 5}{(x - 4)(x + 6)}$

c. $f(x) = \dfrac{\sqrt{x - 8}}{x - 10}$

d. $f(x) = \sqrt{x - 3} + \sqrt{x - 4}$

E-3. **Functions on other sets:** Although the most common functions dealt with in mathematics involve numbers, some do not. The definition of a function remains the same. Each element of the domain must be assigned to a unique element of the range. Which of the following correspondences are functions?

a. Define D to be the set of all U.S. presidents and R the set of all last names. Let $f : D \to R$ be the assignment

$$f(\text{President}) = \text{Last name of president}.$$

b. Let D and R be as above, and let $f : R \to D$ be the assignment

$$f(\text{Last name}) = \text{President with that last name}.$$

c. Define D to be the collection of all new automobiles and R the collection of all colors. Let $f : D \to R$ be the assignment

$$f(\text{Automobile}) = \text{A color appearing on the car body}.$$

E-4. **Inverse functions:** We get the inverse of a function by simply reversing the direction of the arrows. The domain becomes the range, and the range becomes the domain. Sometimes the inverse relation is a function and sometimes it is not. For each of the following functions, determine whether the inverse relation is a function.

a. Define D to be the set of all positive integers and R the set of all positive even integers. Let $f : D \to R$ be given by $f(x) = 2x$.

b. Define D to be the set of all real numbers and R the set of all non-negative real numbers. Let $f : D \to R$ be given by $f(x) = x^2$.

c. Define D to be the set of all U.S. presidents and R the collection of all last names. Let $f : D \to R$ be given by

$$f(\text{President}) = \text{Last name of president}.$$

E-5. **Finding the range:** Find the minimal range of each of the following functions. The domain is all real numbers for which the formula makes sense.

a. $f(x) = x + 2$

b. $f(x) = x^2$

c. $f(x) = x^3$

d. $f(x) = x^8 + 7$

E-6. **One-to-one functions:** A function is *one-to-one,* or *injective,* if distinct points in the domain   are matched with distinct points in the range. For example, let both D and R be the set of all real numbers. Then $f(x) = x + 1$ is one-to-one since if $x \neq y$ then $f(x) \neq f(y)$. That is, distinct elements in the domain have distinct function values. On the other hand, $f(x) = x^2$ is not one-to-one since $2 \neq -2$, but $f(2) = f(-2)$. Two distinct elements of the domain have the same function value. Which of the following functions are one-to-one?

a. Let D and R be the set of all real numbers and $f(x) = 3x + 2$.

b. Let D and R be the set of all real numbers and $f(x) = x^4$.

c. Let $D = \{1, 2, 3, 4\}$ and $R = D$. Define $f : D \to R$ as follows:

$$1 \longrightarrow 4$$
$$2 \longrightarrow 3$$
$$3 \longrightarrow 2$$
$$4 \longrightarrow 1$$

d. Let D and R be as above, and define f as follows:

$$1 \longrightarrow 4$$
$$2 \longrightarrow 3$$
$$3 \longrightarrow 4$$
$$4 \longrightarrow 1$$

E-7. **Onto functions:** A function $f : D \to R$ is said to be *onto,* or *surjective,* if each element of R is a function value of some element of D. Thus, for example, if D and R are both the collection of all real numbers, then $f(x) = x + 1$ is onto since given a number y in the range, we have $f(y-1) = (y-1) + 1 = y$. Thus every element of the range is a function value. On the other hand, $f(x) = x^2$ is not onto since -1 is not the function value of any real number. Which of the following functions are onto?

a. Let D and R be the set of all real numbers and $f(x) = 2x$.

b. Let D and R be the set of all real numbers and $f(x) = x^8$.

c. Let D and R be the set of all real numbers and $f(x) = \dfrac{1}{x^2 + 1}$.

d. Let $D = R = \{1, 2, 3, 4\}$. Define $f : D \rightarrow R$ as follows:

$$1 \longrightarrow 4$$
$$2 \longrightarrow 3$$
$$3 \longrightarrow 2$$
$$4 \longrightarrow 1$$

e. Let D and R be as above, and define $f : D \rightarrow R$ as follows:

$$1 \longrightarrow 4$$
$$2 \longrightarrow 3$$
$$3 \longrightarrow 4$$
$$4 \longrightarrow 1$$

E-8. **Finding examples:** Give an example of a function that is one-to-one but not onto. Give an example of a function that is onto but not one-to-one. (See Exercises E-6 and E-7.)

E-9. **Bijections:** A function is a *bijection,* or *one-to-one correspondence,* if it is both one-to-one and onto. (See Exercises E-6 and E-7.) Which of the functions in Exercise E-7 are bijections?

E-10. **More on bijections:** *This is a continuation of Exercise E-9.* If D and R are sets, they are said to have the same *cardinality* if there is a bijection $f : D \rightarrow R$. When a child counts on his or her fingers, a bijection is established between a set of fingers and the set of objects being counted. Finite sets have the same cardinality precisely when they have the same number of elements. For infinite sets, the situation is a bit more complex. Let D be the set of all positive integers, and let R be the set of all positive even integers. Show that $f : D \rightarrow R$ defined by $f(k) = 2k$ is a bijection.

Note: This shows that the integers and the even integers have the same cardinality. Surprisingly, it can be shown that the integers have the same cardinality as the rational numbers (fractions), but not the same cardinality as the real numbers. This has the intriguing consequence that "most" real numbers are not rational numbers.

1.1 SKILL BUILDING EXERCISES

S-1. **Evaluating formulas:** Evaluate

$$f(x) = \frac{\sqrt{x+1}}{x^2+1}$$

at $x = 2$.

S-2. **Evaluating formulas:** Evaluate

$$f(x) = \left(3 + x^{1.2}\right)^{x+3.8}$$

at $x = 4.3$.

S-3. **Evaluating formulas:** Evaluate

$$g(x, y) = \frac{x^3 + y^3}{x^2 + y^2}$$

at $x = 4.1$, $y = 2.6$.

S-4. **Getting function values:** If $f(t) = 87.1 - e^{4t}$, calculate $f(1.3)$.

S-5. **Getting function values:** If $f(s) = \dfrac{s^2 + 1}{s^2 - 1}$, calculate $f(6.1)$.

S-6. **Evaluating functions of several variables:** If

$$f(r, s, t) = \sqrt{r + \sqrt{s + \sqrt{t}}},$$

calculate $f(2, 5, 7)$.

S-7. **Evaluating functions of several variables:** If $h(x, y, z) = x^y / z$, calculate $h(3, 2.2, 9.7)$.

S-8. **Using formulas:** If $p(t)$ is the profit (in dollars) that I expect my business to earn t years after opening, use functional notation to express expected profit after 2 years and 6 months.

S-9. **Using formulas:** If $c(p, s, h)$ is the cost of buying p bags of potato chips, s sodas, and h hot dogs, use functional notation to express the cost of buying 2 bags of potato chips, 3 sodas, and 5 hot dogs.

S-10. **What formulas mean:** If $s(L)$ is the top speed, in miles per hour, of a fish that is L inches long, what does $s(13)$ mean?

S-11. **Practicing calculations with formulas:** For each of the following functions $f = f(x)$, find the value of $f(3)$, reporting the answer rounded to two decimal places. Use your calculator where it is appropriate.

a. $f = 3x + \dfrac{1}{x}$

b. $f = 3^{-x} - \dfrac{x^2}{x + 1}$

c. $f = \sqrt{2x + 5}$

1.1 EXERCISES

Note: Some of the formulas below use the special number e, which was presented in the Prologue.

1. **Tax owed:** The income tax T owed in a certain state is a function of the taxable income I, both measured in dollars. The formula is

$$T = 0.11I - 500.$$

   a. Express using functional notation the tax owed on a taxable income of $13,000, and then calculate that value.

   b. If your taxable income increases from $13,000 to $14,000, by how much does your tax increase?

   c. If your taxable income increases from $14,000 to $15,000, by how much does your tax increase?

2. **Pole vault:** The height of the winning pole vault in the early years of the modern Olympic Games can be modeled as a function of time by the formula

$$H = 0.05t + 3.3.$$

   Here t is the number of years since 1900, and H is the winning height in meters. (One meter is 39.37 inches.)

   a. Calculate $H(4)$ and explain in practical terms what your answer means.

b. By how much did the height of the winning pole vault increase from 1900 to 1904? From 1904 to 1908?

3. **Flying ball:** A ball is tossed upward from a tall building, and its upward velocity V, in feet per second, is a function of the time t, in seconds, since the ball was thrown. The formula is

$$V = 40 - 32t$$

if we ignore air resistance. The function V is positive when the ball is rising and negative when the ball is falling.

a. Express using functional notation the velocity 1 second after the ball is thrown, and then calculate that value. Is the ball rising or falling then?

b. Find the velocity 2 seconds after the ball is thrown. Is the ball rising or falling then?

c. What is happening 1.25 seconds after the ball is thrown?

d. By how much does the velocity change from 1 to 2 seconds after the ball is thrown? From 2 to 3 seconds? From 3 to 4 seconds? Compare the answers to these three questions and explain in practical terms.

4. **Flushing chlorine:** City water, which is slightly chlorinated, is being used to flush a tank of heavily chlorinated water. The concentration $C = C(t)$ of chlorine in the tank t hours after flushing begins is given by

$$C = 0.1 + 2.78e^{-0.37t} \text{ milligrams per gallon}.$$

a. What is the initial concentration of chlorine in the tank?

b. Express the concentration of chlorine in the tank after 3 hours using functional notation, and then calculate its value.

5. **A population of deer:** When a breeding group of animals is introduced into a restricted area such as a wildlife reserve, the population can be expected to grow rapidly at first but to level out when the population grows to near the maximum that the environment can support. Such growth is known as *logistic population growth*, and ecologists sometimes use a formula to describe it. The number N of deer present at time t (measured in years since the herd was introduced) on a certain wildlife reserve has

been determined by ecologists to be given by the function

$$N = \frac{12.36}{0.03 + 0.55^t}.$$

a. How many deer were initially on the reserve?

b. Calculate $N(10)$ and explain the meaning of the number you have calculated.

c. Express the number of deer present after 15 years using functional notation, and then calculate it.

d. How much increase in the deer population do you expect from the 10th to the 15th year?

6. **A car that gets 32 miles per gallon:** The cost C of operating a certain car that gets 32 miles per gallon is a function of the price g, in dollars per gallon, of gasoline and the distance d, in miles, that you drive. The formula for $C = C(g, d)$ is $C = gd/32$ dollars.

a. Use functional notation to express the cost of operation if gasoline costs 98 cents per gallon and you drive 230 miles. Calculate the cost.

b. Calculate $C(1.03, 172)$ and explain the meaning of the number you have calculated.

7. **Radioactive substances** change form over time. For example, carbon 14, which is important for radiocarbon dating, changes through radiation into nitrogen. If we start with 5 grams of carbon 14, then the amount $C = C(t)$ of carbon 14 remaining after t years is given by

$$C = 5 \times 0.5^{t/5730}.$$

a. Express the amount of carbon 14 left after 800 years in functional notation, and then calculate its value.

b. How long will it take before half of the carbon 14 is gone? Explain how you got your answer. (*Hint:* You might use trial and error to solve this, or you might solve it by looking carefully at the exponent.)

8. **A roast** is taken from the refrigerator (where it had been for several days) and placed immediately in a preheated oven to cook. The temperature $R = R(t)$ of the roast t minutes after being placed in the oven is given by

$$R = 325 - 280e^{-0.005t} \text{ degrees Fahrenheit}.$$

$\longrightarrow$

a. What is the temperature of the refrigerator?

b. Express the temperature of the roast 30 minutes after being put in the oven in functional notation, and then calculate its value.

c. By how much did the temperature of the roast increase during the first 10 minutes of cooking?

d. By how much did the temperature of the roast increase from the first hour to 10 minutes after the first hour of cooking?

9. **What if interest is compounded more often than monthly?** Some lending institutions compound interest daily or even continuously. (The term *continuous compounding* is used when interest is being added as often as possible—that is, at each instant in time.) The point of this exercise is to show that, for most consumer loans, the answer you get with monthly compounding is very close to the right answer, even if the lending institution compounds more often. In part 1 of Example 1.2, we showed that if you borrow $7800 from an institution that compounds monthly at a monthly interest rate of 0.67% (for an APR of 8.04%), then in order to pay off the note in 48 months, you have to make a monthly payment of $190.57.

a. Would you expect your monthly payment to be higher or lower if interest were compounded daily rather than monthly? Explain why.

b. Which would you expect to result in a larger monthly payment, daily compounding or continuous compounding? Explain your reasoning.

c. When interest is compounded continuously, you can calculate your monthly payment $M = M(P, r, t)$, in dollars, for a loan of P dollars to be paid off over t months using

$$M = \frac{P(e^r - 1)}{1 - e^{-rt}},$$

where $r = \frac{APR}{12}$ if the APR is written in decimal form. Use this formula to calculate the monthly payment on a loan of $7800 to be paid off over 48 months with an APR of 8.04%. How does this answer compare with the result in Example 1.2?

10. **Present value:** The amount of money originally put into an investment is known as the *present value P* of the investment. For example, if you buy a $50 U.S. Savings Bond that matures in 10 years, the present

value of the investment is the amount of money you have to pay for the bond today. The value of the investment at some future time is known as the *future value F*. Thus, if you buy the savings bond mentioned above, its future value is $50.

If the investment pays an interest rate of r (as a decimal) compounded yearly, and if we know the future value F for t years in the future, then the present value $P = P(F, r, t)$, the amount we have to pay today, can be calculated using

$$P = F \times \frac{1}{(1 + r)^t}$$

if we measure F and P in dollars. The term $1/(1 + r)^t$ is known as the *present value factor*, or the *discount rate*, so the formula above can also be written as

$$P = F \times \text{ discount rate}.$$

a. Explain in your own words what information the function $P(F, r, t)$ gives you.

For the remainder of this problem, we will deal with an interest rate of 9% compounded yearly and a time t of 18 years in the future.

b. Calculate the discount rate.

c. Suppose you wish to put money into an account that will provide $40,000 to help your child attend college 18 years from now. How much money would you have to put into savings today in order to attain that goal?

11. **How much can I borrow?** The function in Example 1.2 can be rearranged to show the amount of money $P = P(M, r, t)$, in dollars, that you can afford to borrow at a monthly interest rate of r (as a decimal) if you are able to make t monthly payments of M dollars:

$$P = M \times \frac{1}{r} \times \left(1 - \frac{1}{(1 + r)^t}\right).$$

Suppose you can afford to pay $350 per month for 4 years.

a. How much money can you afford to borrow for the purchase of a car if the prevailing monthly interest rate is 0.75%? (That is 9% APR.) Express the answer in functional notation, and then calculate it.

b. Suppose your car dealer can arrange a special monthly interest rate of 0.25% (or 3% APR). How much can you afford to borrow now?

c. Even at 3% APR you find yourself looking at a car you can't afford, and you consider extending the period during which you are willing to make payments to 5 years. How much can you afford to borrow under these conditions?

12. **Financing a new car:** You are buying a new car, and you plan to finance your purchase with a loan you will repay over 48 months. The car dealer offers two options: either dealer financing with a low APR, or a $2000 rebate on the purchase price. If you use dealer financing, you will borrow $14,000 at an APR of 3.9%. If you take the rebate, you will reduce the amount you borrow to $12,000, but you will have to go to the local bank for a loan at an APR of 8.85%. Should you take the dealer financing or the rebate? How much will you save over the life of the loan by taking the option you chose? To answer the first question, you may need the formula

$$M = \frac{Pr(1+r)^{48}}{(1+r)^{48} - 1} .$$

Here M is your monthly payment (in dollars) if you borrow P dollars with a term of 48 months at a monthly interest rate of r (as a decimal), and $r = \frac{APR}{12}$.

13. **Brightness of stars:** The *apparent magnitude m* of a star is a measure of its apparent brightness as the star is viewed from Earth. Larger magnitudes correspond to dimmer stars, and magnitudes can be negative, indicating a very bright star. For example, the brightest star in the night sky is Sirius, which has an apparent magnitude of -1.45. Stars with apparent magnitude greater than about 6 are not visible to the naked eye. The magnitude scale is not *linear* in that a star that is double the magnitude of another does not appear to be twice as dim. Rather, the relation goes as follows: If one star has an apparent magnitude of m_1 and another has an apparent magnitude of m_2, then the first star is t times as bright as the

second, where t is given by

$$t = 2.512^{m_2 - m_1} .$$

The North Star, Polaris, has an apparent magnitude of 2.04. How much brighter than Polaris does Sirius appear?

14. **Stellar distances:** *This is a continuation of Exercise 13.* The *absolute magnitude M* of a star is a measure of its true brightness and is not dependent on its distance from Earth or on any other factors that might affect its apparent brightness. If both the absolute and the apparent magnitude of a star are known, the distance from Earth can be calculated as follows:

$$d(m, M) = 3.26 \times 10^{(m - M + 5)/5} .$$

Here m is the apparent magnitude, M is the absolute magnitude, and d is the distance measured in *light-years.*[4]

a. Explain in words the meaning of $d(2.2, 0.7)$.

b. In Exercise 13 we noted that Sirius has an apparent magnitude of -1.45. The absolute magnitude of Sirius is $+1.45$. How far away is Sirius from Earth?

15. **Parallax:** If we view a star now and then view it again 6 months later, our position will have changed by the diameter of the Earth's orbit around the sun. For *nearby* stars (within 100 light-years or so), the change in viewing location is sufficient to make the star appear to be in a slightly different location in the sky. Half of the angle from one location to the next is known as the *parallax angle* (see Figure 1.5). Parallax can be used to measure the distance to the star. An approximate relationship is given by

$$d = \frac{3.26}{p} ,$$

where d is the distance in light-years, and p is the parallax measured in seconds of arc.[5] Alpha Centauri is the star nearest to the sun, and it has a parallax angle of 0.751 second. How far is Alpha Centauri from the sun? →

---

[4]One light-year is the distance light travels in 1 year, or about 5,879,000,000,000 miles.

[5]One degree is divided into 60 minutes of arc, and 1 minute of arc is divided into 60 seconds of arc. One second of arc is very small indeed. If you hold a sheet of paper edgewise at arm's length, the thickness of the paper subtends an arc of about 30 seconds.

Note: Parallax is used not only to measure stellar distances. Our binocular vision actually provides the brain with a parallax angle that it uses to estimate distances to objects we see.

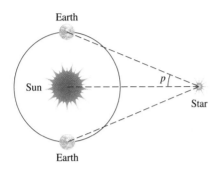

FIGURE 1.5

16. **Sound pressure and decibels:** Sound exerts a pressure P on the human ear.[6] This pressure increases as the loudness of the sound increases. If the loudness D is measured in *decibels* and the pressure P in dynes[7] per square centimeter, then the relationship is given by

$$P = 0.0002 \times 1.122^{D}.$$

a. Ordinary conversation has a loudness of about 65 decibels. What is the pressure exerted on the human ear by ordinary conversation?

b. A decibel level of 120 causes pain to the ear and can result in damage. What is the corresponding pressure level on the ear?

17. **Mitscherlich's equation:** An important agricultural problem is to determine how a quantity of nutrient, such as nitrogen, affects the growth of plants. We consider the situation wherein sufficient quantities of all but one nutrient are present. One *baule*[8] of a nutrient is the amount needed to produce 50% of maximum possible yield. In 1909 E. A. Mitscherlich proposed the following relation, which is known as Mitscherlich's equation:[9]

$$Y = 1 - 0.5^{b}.$$

Here b is the number of baules of nutrient applied, and Y is the percentage (as a decimal) of maximum yield produced.

a. Verify that the formula predicts that 50% of maximum yield will be produced if 1 baule of nutrient is applied.

b. Use functional notation to express the percentage of maximum yield produced by 3 baules of nutrient, and calculate that value.

c. The exact value of a baule depends on the nutrient in question. For nitrogen, 1 baule is 223 pounds per acre. What percentage of maximum yield will be produced if 500 pounds of nitrogen per acre is present?

18. **Yield response to several growth factors:** *This is a continuation of Exercise 17.* If more than one nutrient is considered, the formula for percentage of maximum yield is a bit more complex. For three nutrients, the formula is

$$Y(b, c, d) = \left(1 - 0.5^{b}\right)\left(1 - 0.5^{c}\right)\left(1 - 0.5^{d}\right).$$

Here Y is the percentage of maximum yield as a decimal, b is the number of baules of the first nutrient, c is the number of baules of the second nutrient, and d is the number of baules of the third nutrient.

a. Express using functional notation the percentage of maximum yield produced from 1 baule of the first nutrient, 2 baules of the second nutrient, and 3 baules of the third nutrient, and then calculate that value.

b. One baule of nitrogen is 223 pounds per acre, 1 baule of phosphorus is 45 pounds per acre, and 1 baule of potassium is 76 pounds per acre. What

---

[6]In fact, the human ear detects variation in pressure on the ear drum. The amplitude of this variation determines the intensity of the sound, and the perceived loudness of the sound is related to this intensity.

[7]The dyne is a very small unit of force. It takes 444,800 dynes to make a pound. A small insect egg has a weight of about 1 dyne.

[8]After the German mathematician who proposed the unit.

[9]A few years later the relation was noted independently by W. J. Spillman, and the relation is sometimes referred to as Spillman's equation. It should also be noted that the validity of Mitscherlich's equation is a source of controversy among modern agricultural scientists.

percentage of maximum yield will be obtained from 200 pounds of nitrogen per acre, 100 pounds of phosphorus per acre, and 150 pounds of potassium per acre?

19. **Thermal conductivity:** The heat flow Q due to conduction across a rectangular sheet of insulating material can be calculated using

$$Q = \frac{k(t_1 - t_2)}{d}.$$

Here Q is measured watts per square meter, d is the thickness of the insulating material in meters, t_1 is the temperature in degrees Celsius of the warm side of the insulating material, t_2 is the temperature of the cold side, and k is the *coefficient of thermal conductivity*, which is measured experimentally for different insulating materials. For glass, the coefficient of thermal conductivity is 0.85. Suppose the temperature inside a home is 24 degrees Celsius (about 75 degrees Fahrenheit), and that the temperature outside is 5 degrees Celsius (about 41 degrees Fahrenheit; see Figure 1.6).

a. What is the heat flow through a glass window that is 0.007 meter (about one-quarter of an inch) thick?

b. The total heat loss due to conduction is the product of the heat flow with the area of the insulating material. In the situation of part a, what is the total heat loss due to conduction if the window has an area of 2.5 square meters?

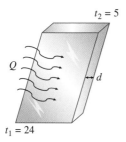

$t_2 = 5$

Q

d

$t_1 = 24$

FIGURE 1.6

20. **Reynolds number:** The *Reynolds number* is very important in such fields as fluid flow and aerodynamics. In the case of a fluid flowing through a pipe, the Reynolds number R is given by

$$R = \frac{vdD}{\mu}.$$

Here v is the velocity of the fluid in meters per second, d is the diameter of the pipe in meters, D is the density of the fluid in kilograms per cubic meter, and μ is the viscosity of the fluid measured in newton-seconds per square meter. Generally, when the Reynolds number is above 2000, the flow becomes turbulent, and rapid mixing occurs.[10] When the Reynolds number is less than 2000, the flow is streamline. Consider a fluid flowing through a pipe of diameter 0.05 meter at a velocity of 0.2 meter per second.

a. If the fluid in the pipe is toluene, its viscosity is 0.00059 newton-seconds per square meter, and its density is 867 kilograms per cubic meter. Is the flow turbulent or streamline?

b. If the toluene is replaced by glycerol, then the viscosity is 1.49 newton-seconds per square meter, and the density is 1216.3 kilograms per cubic meter. Is the glycerol flow turbulent or streamline?

21. **Fault rupture length:** Earthquakes can result in various forms of damage, but many result in fault ruptures, or cracks in the Earth's surface. In a 1958 study[11] of earthquakes in California and Nevada, D. Tocher found the following relationship between the fault rupture length L in kilometers and the magnitude M (on the Richter scale) of an earthquake:

$$L = 0.0000017 \times 10.47^M.$$

What fault rupture length would be expected from an earthquake measuring 6.5 on the Richter scale?

22. **Tubeworm:** A recent article in *Nature* reports on a study of the growth rate and life span of a marine *tubeworm*.[12] These tubeworms live near hydrocarbon seeps on the ocean floor and grow very slowly

---

[10]Reynolds numbers are dimensionless. That is, they have no units, such as grams or meters, associated with them.

[11]D. Tocher, "Earthquake energy and ground breakage," *Bull. Seism. Soc. Am.* **48** (1958), 48.

[12]D. Bergquist, F. Williams, and C. Fisher, "Longevity record for deep-sea invertebrate," *Nature* **403** (2000), 499–500.

indeed. Collecting data for creatures at a depth of 550 meters is extremely difficult. But for tubeworms living on the Louisiana continental slope, scientists developed a model for the time T (measured in years) required for a tubeworm to reach a length of L meters. From this model the scientists concluded that this tubeworm is the longest-lived noncolonial marine invertebrate known. The model is

$$T = 14e^{1.4L} - 20.$$

A tubeworm can grow to a length of 2 meters. How old is such a creature? (Round your answer to the nearest year.)

23. **Research project:** Look in a textbook for another class to find a function interesting to you that is given by a formula. Identify all the variables used in the formula, explaining the meaning of each variable. Explain how this formula is used.

1.2 *Functions Given by Tables*

Long before the idea of a function was formalized, it was used in the form of tables of values. Some of the earliest surviving samples of mathematics are from Babylon and Egypt and date from 2000 to 1000 B.C. They contain a variety of tabulated functions such as tables of squares of numbers. Functions given in this way nearly always leave gaps and are incomplete. In this respect they may appear less useful than functions given by formulas. On the other hand, tables often clearly show trends that are not easily discerned from formulas, and in many cases tables of values are much easier to obtain than a formula.

Reading Tables of Values

The population N of the United States depends on the date d. That is, $N = N(d)$ is a function of d. Table 1.1 was taken from the 1995 edition of the *Statistical Abstract of the United States*. It shows the population in millions each decade from 1950 through 1990. This is a common way to express functions when data are gathered by *sampling*, in this case by census takers.

In order to get the population in 1980, we look at the column corresponding to $d = 1980$ and read $N = 227.23$ million people. In functional notation this is $N(1980) = 227.23$ million people. Similarly, we read from the table that $N(1970) = 203.98$, which indicates that the U.S. population in 1970 was 203.98 million people.

Filling Gaps by Averaging

Functions given by tables of values have their limitations in that they nearly always leave gaps. Sometimes it is appropriate to fill these gaps by *averaging*. For example, Table 1.1 does not give the population in 1975. In the absence of further information, a reasonable guess for the value of $N(1975)$ would be the average of the populations in 1970 and 1980:

$$\frac{N(1970) + N(1980)}{2} = \frac{203.98 + 227.23}{2} = 215.61 \text{ million people}.$$

Thus the population in 1975 was approximately 215.61 million people. Population records show the actual population of the United States in 1975 as 215.47 million people, so it seems that the idea of averaging to estimate the value of $N(1975)$ worked pretty well in this case.

We emphasize here that, in the absence of further data, we have no way of determining the exact value of $N(1975)$, and so you should not assume that the answer we gave is the only acceptable one. If, for example, you had reason to believe that the population of the United States grew faster in the earlier part of the decade of the 1970s than it did

d = Year	1950	1960	1970	1980	1990
N = Population in millions	151.87	179.98	203.98	227.23	249.40

TABLE 1.1 Population of the United States

in the later part, then it would be reasonable for you to give a larger value for $N(1975)$. Such an answer, supported by an appropriate argument, would have as much validity as the one given here.

Average Rates of Change

A key tool in the analysis of functions is the idea of an *average rate of change*. To illustrate this idea, let's get the best estimate we can for $N(1972)$, the U.S. population in 1972. Since 1972 is not halfway between 1970 and 1980, it does not make sense here to average the two as we did above to estimate the population in 1975, but a simple extension of this idea will help. From 1970 to 1980 the population increased from 203.98 million to 227.23 million people. That is an increase of $227.23 - 203.98 = 23.25$ million people in 10 years. Thus, on average, during the decade of the 1970s the population was increasing by $\frac{23.25}{10} = 2.325$ million people per year. This is the *average yearly rate of change in N* during the 1970s, and there is a natural way to use it to estimate the population in 1972. In 1970 the population was 203.98 million people, and during this period the population was growing by about 2.325 million people per year. Thus in 1972 the population $N(1972)$ was approximately

$$\text{Population in 1972} = \text{Population in 1970} + \text{Two years of growth}$$

$$= 203.98 + 2 \times 2.325 = 208.63 \text{ million}.$$

Evaluating Functions Given by Tables

You can evaluate functions given by tables by locating the appropriate entry in the table. It is sometimes appropriate to fill in gaps in the table by using averages or average rates of change.

EXAMPLE 1.3 Women Employed Outside the Home

Table 1.2 shows the number W of women in the United States employed outside the home as a function of the date d. It was taken from the *1995 Statistical Abstract of the United States*.

d = Year	1960	1970	1980	1990
W = Number in millions	21.3	29.6	41.9	53.5

TABLE 1.2 Number of women employed outside the home in the United States

1. Explain the meaning of $W(1970)$ and give its value.

2. Explain the meaning of $W(1975)$ and estimate its value.

3. Express the number of women employed outside the home in 1972 in functional notation, and use the average yearly rate of change from 1970 to 1980 to estimate its value.

4. According to the *1950 Statistical Abstract of the United States*, in 1943 there were 18.7 million women employed outside the home, and in 1946 the number was 16.8 million. On the basis of this information, find the average yearly rate of change in W from 1943 to 1946 and the average decrease per year over this time interval. Use the results to estimate $W(1945)$.

Solution to Part 1:   The expression $W(1970)$ represents the number of women in the United States employed outside the home in 1970. Consulting the second column of the table, we find that $W(1970) = 29.6$ million women.

Solution to Part 2:   The expression $W(1975)$ represents the number of women in the United States employed outside the home in 1975. This value is not given in the table, so we must estimate it. Since 1975 is halfway between 1970 and 1980, it is reasonable to estimate $W(1975)$ as the average of $W(1970)$ and $W(1980)$:

$$\frac{W(1970) + W(1980)}{2} = \frac{29.6 + 41.9}{2} = 35.75 \text{ million}.$$

Thus we estimate that $W(1975)$ is about 35.8 million women.

Solution to Part 3:   In functional notation, the number of women employed outside the home in 1972 is $W(1972)$. To estimate this function value, we first calculate the average rate of change during the decade of the 1970s. From 1970 to 1980, the number of women employed outside the home increased from 29.6 million to 41.9 million, an increase of 12.3 million over the 10-year period. Thus in the decade of the 1970s the number increased on average by $\frac{12.3}{10} = 1.23$ million per year. In the 2-year period from 1970 to 1972, the increase was about $2 \times 1.23 = 2.46$ million. We estimate that

$$W(1972) = W(1970) + 2 \text{ years' growth}$$
$$= W(1970) + 2.46$$
$$= 29.6 + 2.46$$
$$= 32.06 \text{ million}.$$

Thus we estimate that the number of women employed outside the home in 1972 was about 32.1 million.

Solution to Part 4:   At the outset we observe that apparently the function is decreasing over this period, so we expect the rate of change to be negative. From 1943 to 1946 the change was $16.8 - 18.7 = -1.9$. Thus over this 3-year period the number changed on average by $\frac{-1.9}{3}$ million per year, or about -0.63 million per year. This is the average yearly rate of change from 1943 to 1946. The average decrease per year is obtained by dividing the amount of decrease, which is $18.7 - 16.8$, by the time interval, which is 3 years. The result is about 0.63 million per year. Note that the average decrease per year is just the magnitude of the average yearly rate of change.

To estimate $W(1945)$ we proceed as in part 3, using the average yearly rate of change. We expect that in the 2-year period from 1943 to 1945 the change was about $2 \times -0.63 = -1.26$ million. Thus we estimate that

$$W(1945) = W(1943) - 1.26$$

$$= 18.7 - 1.26$$

$$= 17.44 \text{ million}.$$

Our estimate for $W(1945)$ is 17.4 million women.   ■ ■ ■

There are features of this problem that are worth emphasis. First, the things we do with mathematics are neither, as some may have been led to believe, magic nor beyond the understanding of the average citizen. They are common-sense ideas that are used on a regular basis. Second, many times applications of mathematics do not involve a set way to solve the problem, and often there is not a simple "right answer." What is important is clear thinking leading to helpful information. Third, in applying mathematics we should be aware of the context. For example, in part 4 the time interval began in 1943, when the United States was heavily involved in World War II and thus many women were working outside the home. By 1946 (at the end of this interval) the war was over, and this is reflected in the decrease in the function W. Because the war continued well into 1945, we should be skeptical that our estimate of 17.4 million women for that year is reliable. In fact, the number was near its peak then, and the actual value of $W(1945)$ is 19.0 million (slightly higher than the value in 1943).

Spotting Trends

In some situations, tables of values show clear trends or *limiting values* for functions. A good example of this is provided by the famous[13] yeast experiment that Tor Carlson performed in the early 1900s. In this experiment, a population of yeast cells growing in a confined area was carefully monitored as it grew. Each day, the amount of yeast[14] was found and recorded. We let t be the number of hours since the experiment began and $N = N(t)$ the amount of yeast present. Carlson presented data through only 18 hours of growth. In Table 1.3 we have partially presented Carlson's data and added data that modern studies indicate he might have recorded if the experiment had continued. Thus, for example, $N(5) = 119$ is the amount of yeast present 5 hours after the experiment began.

We want to estimate the value of $N(35)$—that is, the amount of yeast present after 35 hours. Before we look at the data, let's think how our everyday experience tells us that a similar, but more familiar, situation would progress. The population of mold on a slice of bread[15] will show behavior much like that of yeast. On bread, the mold spreads rapidly

---

[13]This experiment is described by R. Pearl in "The growth of populations," *Quart. Rev. Biol* **2** (1927), 532–548. It served to illustrate the now widely used *logistic population growth model,* which we will examine in more detail in subsequent chapters.

[14]The actual unit of measure is unclear from the original reference. It is simply reported as *amount of yeast.*

[15]See also Exercise 21.

Time t	0	5	10	15	20	25	30
Amount of yeast N	10	119	513	651	662	664	665

TABLE 1.3 Amount of yeast

at first, but eventually the bread is covered and no room is left for further population growth. We would expect the amount of mold to stabilize at a *limiting value*. The same sort of growth seems likely for yeast growing in a confined area. In terms of the function N, that means we expect to see its values increase at first but eventually stabilize, and this is borne out by the data in Table 1.3, where evidently the yeast population is leveling out at about 665. Thus it is reasonable to expect that $N(35)$ is about 665.

   We note here that the significance of 665 is not just that it is the last entry in the table. Rather, this entry is indicative of the trend shown by the last few entries of the table: 662, 664, and 665. We also emphasize that we did not propose a value for $N(35)$ by looking at the data divorced from its meaning. Before we looked at the data, the physical situation led us to believe that N would level out at *some value*, and our everyday experience with moldy bread tells us that once mold begins to grow, this level will be reached in a relatively short time. The data simply enabled us to make a good guess at the limiting value we expected to see.

Limiting Values

Information about physical situations can sometimes show that limiting values are to be expected for functions that describe them. Under such conditions, the limiting value may be estimated from a trend established by the data.

   The next example shows how using rates of change can confirm the expectation that there is a limiting value.

EXAMPLE 1.4   A Skydiver

During the period of a skydiver's free fall, he is pulled downward by gravity, but his velocity is retarded by air resistance, which physicists believe increases as velocity does. Table 1.4 shows the downward velocity $v = v(t)$, in feet per second, of an average-size man t seconds after jumping from an airplane.

1. From the table, describe in words how the velocity of the skydiver changes with time.

2. Use functional notation to give the velocity of the skydiver 15 seconds into the fall. Estimate its value.

3. Explain how the physical situation leads you to believe that the function v will approach a limiting value.

Time t	0	10	20	30	40	50	60
Velocity v	0	147	171	175	175.8	176	176

TABLE 1.4 Velocity in feet per second t seconds into free fall

4. Make a table showing the average rate of change per second in velocity v over each of the 10-second intervals in Table 1.4. Explain how your table confirms the conclusion of part 3.

5. Estimate the *terminal velocity* of the skydiver—that is, the greatest velocity that the skydiver can attain.

Solution to Part 1: The velocity increases rapidly during the first part of the fall, but the rate of increase slows as the fall progresses.

Solution to Part 2: In functional notation, the velocity 15 seconds into the fall is $v(15)$. It is not given in Table 1.4, so we must estimate its value. One reasonable estimate is the average of the velocities at 10 and 20 seconds into the fall:

$$\frac{v(10) + v(20)}{2} = \frac{147 + 171}{2} = 159 \text{ feet per second}.$$

With the information we have, this is a reasonable estimate for $v(15)$, but a little closer look at the data might lead us to adjust this. Note that velocity increases very rapidly (by $147 - 0 = 147$ feet per second) during the first 10 seconds of the fall but increases by only $171 - 147 = 24$ feet per second during the next 10 seconds. It appears that the rate of increase in velocity is slowing dramatically. Thus it is reasonable to expect that the velocity increased more from 10 to 15 seconds than it did from 15 to 20 seconds. We might be led to make an upward revision in our estimate of $v(15)$. It turns out that the actual velocity at $t = 15$ is 164.44 feet per second, but with the information we have, we have no way of discovering that. What is important here is to obtain an estimate that is reasonable and is supported by an appropriate argument.

Solution to Part 3: The force of gravity causes the skydiver's velocity to increase, but this increase is retarded by air resistance, which increases along with velocity. When downward velocity reaches a certain level, we would expect the force of retardation to match the downward pull of gravity. From that point on, velocity will not change.

Solution to Part 4: Over the interval from $t = 0$ to $t = 10$, the change in v is $147 - 0 = 147$ feet per second, so the average rate of change per second in v over this 10-second interval is $\frac{147}{10} = 14.7$ feet per second per second. From $t = 10$ to $t = 20$, the change in v is $171 - 147 = 24$ feet per second, so the average rate of change per second in v over this interval is $\frac{24}{10} = 2.4$ feet per second per second. If we repeat this computation for each of the intervals, we get the values shown in Table 1.5. Here the average rate of change is measured in feet per second per second.

Note that Table 1.5 is consistent with our observations in parts 1 and 2. The table shows that the average rate of change in velocity decreases to zero as time goes on. This confirms our conclusion in part 3 that after a time, velocity will not change.

Time interval	0 to 10	10 to 20	20 to 30	30 to 40	40 to 50	50 to 60
Average rate of change in v	14.7	2.4	0.4	0.08	0.02	0

TABLE 1.5 Average rate of change in velocity v over each 10-second interval

Solution to Part 5: From parts 3 and 4, we have good reason to believe that the data should show a limiting value for velocity. From $t = 30$ to $t = 60$, the velocity seems to be inching up toward 176 feet per second, and it appears to level out there. This is confirmed by the fact that the average rate of change in velocity decreases to zero. The limiting value of 176 for v is the terminal velocity of the skydiver. You may be interested to know that this is about 120 miles per hour. ■ ■ ■

ANOTHER LOOK

More on Average Rates of Change

Average rates of change can be calculated not only for functions given by tables but also for functions given by formulas. In the latter case, we can calculate the rate of change over any interval in the domain that we choose. In general, an average rate of change in f from $x = a$ to $x = b$ is given by

$$\text{Average rate of change} = \frac{\text{Change in function value}}{\text{Change in } x \text{ value}} = \frac{f(b) - f(a)}{b - a}.$$

For example, if we take $f(x) = x^2$, we can calculate the average rate of change from 3 to 5 as follows:

$$\text{Average rate of change} = \frac{f(5) - f(3)}{5 - 3} = \frac{25 - 9}{2} = 8.$$

More important, using a little algebra, we can calculate average rates of change for indeterminate intervals. Let's calculate the average rate of change for $f(x) = x^2$ from x to $x + h$. (We can think of this as the average rate of change from x to a point h units to the right of x.) The calculation is as follows:

$$\text{Average rate of change} = \frac{f(x+h) - f(x)}{(x+h) - x} = \frac{(x+h)^2 - x^2}{h}.$$

To simplify this expression, we first apply the familiar squaring rule $(a + b)^2 = a^2 + 2ab + b^2$ to get

$$\frac{(x^2 + 2xh + h^2) - x^2}{h} = \frac{2xh + h^2}{h}.$$

Next we factor the common term h from the numerator and then cancel:

$$\frac{h(2x + h)}{h} = 2x + h.$$

In many settings, it is important to calculate the average rate of change over a very small interval—that is, when h is very near 0. For $f(x) = x^2$, when h is close to 0, the average rate of change $2x + h$ is close to $2x$. This value is known as the *instantaneous rate of change*, and it is a crucial idea in calculus. We shall not pursue it further here.

In calculating average rates of change algebraically as we have done here, there are two spots where students occasionally have trouble. We present them as cautions.

Caution 1: In calculating $f(x + h)$, simply replace each occurrence of x in the formula for f by $(x + h)$. Including the parentheses will help to avoid sign errors as well as other difficulties.

Caution 2: Do not cancel the h from the denominator until you have factored an h from each term of the numerator.

EXAMPLE 1.5   A Falling Rock

A rock dropped near the surface of the Earth travels $D(t) = 16t^2$ feet during the first t seconds of the fall. The average rate of change in $D(t)$ is the average velocity of the rock.

1. What is the average velocity of the rock from time $t = 2$ to time $t = 3$?

2. What is the average velocity of the rock over the time interval from t to $t + h$?

Solution to Part 1:   The average velocity we want is the average rate of change in D from $t = 2$ to $t = 3$:

$$\text{Average velocity} = \frac{D(3) - D(2)}{3 - 2} = \frac{16 \times 3^2 - 16 \times 2^2}{1} = 80 \text{ feet per second}.$$

Solution to Part 2:   The average velocity is the average rate of change in D from t to $t + h$:

$$\text{Average velocity} = \frac{D(t + h) - D(t)}{(t + h) - t}$$

$$= \frac{16(t + h)^2 - 16t^2}{h}$$

$$= \frac{16(t^2 + 2th + h^2) - 16t^2}{h}$$

$$= \frac{16t^2 + 32th + 16h^2 - 16t^2}{h}$$

$$= \frac{32th + 16h^2}{h}$$

$$= \frac{h(32t + 16h)}{h}$$

$$= 32t + 16h.$$

Enrichment Exercises

E-1.   **Average rates of change using numbers:** For each of the following functions, calculate the average rate of change for the indicated intervals.

   a. $f(x) = x^3$, $x = 2$ to $x = 3$ b. $f(x) = 3x + 4$, $x = 2$ to $x = 5$

   c. $f(x) = \dfrac{1}{x}$, $x = 2$ to $x = 4$ d. $f(x) = x^2 + x$, $x = 1$ to $x = 3$

E-2.   **Average rates of change using variables:** For each of the following functions, calculate the average rate of change from x to $x + h$.

   a. $f(x) = 3x + 1$

   b. $f(x) = x^3$ (Recall that $(a + b)^3 = a^3 + 3a^2b + 3ab^2 + b^3$.)

   c. $f(x) = x^2 + x$

E-3.   **Linear functions:** A function of the form $y = mx + b$ is known as a *linear function*. Show that for a linear function the average rate of change from $x = p$ to $x = q$ does not depend on either p or q.

E-4.   **Meaning of average rate of change:** We noted in Example 1.5 that if $s(t)$ denotes distance traveled in time t, then the average rate of change in s is the average velocity. If $v(t)$ denotes velocity at time t, what is the physical meaning of the average rate of change in v?

E-5.   **Sketching a graph:** For a certain function f, we have $f(0) = 1$. The average rate of change for f from $x = 0$ to $x = 4$ is 2. The average rate of change for f from $x = 4$ to $x = 6$ is -1. Make a graph of a function with these properties. *Note:* There are many correct graphs.

E-6.   **The effect of adding a constant:** How does the average rate of change from $x = a$ to $x = b$ for a function $f(x)$ compare with the average rate of change over the same interval for the function $g(x) = f(x) + c$? (Here c is a constant.)

E-7.   **The effect of adding a linear function:** Suppose the average rate of change for a function $f(x)$ from $x = a$ to $x = b$ is 7. Let $g(x) = f(x) + 3x + 5$. What is the average rate of change of $g(x)$ from $x = a$ to $x = b$?

E-8.   **Putting together rates of change:** Suppose the average rate of change for $f(x)$ from $x = a$ to $x = a + 2$ is 5 and the average rate of change for $f(x)$ from $x = a + 2$ to $x = a + 4$ is 3. What is the average rate of change for $f(x)$ from $x = a$ to $x = a + 4$?

x	f(x)
1	6
2	9
3	12
5	16

E-9.   **Central difference quotients:** Consider the accompanying table of values. If we wanted to know the rate of change of the function at $x = 3$, we could consider the average rate of change from $x = 2$ to $x = 3$ or the average rate of change from $x = 3$ to $x = 5$. The *central difference quotient* takes both of these into account and is the average of the average rates of change on either side. Calculate the central difference quotient at $x = 3$.

E-10.   **Calculating a derivative:** For $f(x) = x^3$, the average rate of change from x to $x + h$ is $3x^2 + 3xh + h^2$. The *derivative* is obtained by finding what the average rate of change is close to when h is close to 0. What is the derivative of x^3?

1.2   SKILL BUILDING EXERCISES

S-1. **A tabulated function:** The following table gives values for a function $N = N(t)$.

t	$N = N(t)$
10	17.6
20	23.8
30	44.6
40	51.3
50	53.2
60	53.7
70	53.9

Find the value of $N(20)$.

S-2. **Averaging:** Using the table in Exercise S-1, estimate the value of $N(25)$ by averaging.

S-3. **Average rate of change:** Using the table from Exercise S-1, find the average rate of change in N from $t = 20$ to $t = 30$.

S-4. **Using average rates of change:** Use your answer in Exercise S-3 to estimate the value of $N(23)$.

S-5. **Averaging:** Using the table in Exercise S-1, estimate the value of $N(35)$ by averaging.

S-6. **Average rate of change:** Using the table from Exercise S-1, find the average rate of change in N from $t = 30$ to $t = 40$.

S-7. **Using average rates of change:** Use your answer from Exercise S-6 to estimate the value of $N(37)$.

S-8. **Limiting values:** Assuming that the function N in Exercise S-1 describes a physical situation for which a limiting value is expected, estimate the limiting value of N.

S-9. **When limiting values occur:** Suppose $c(t)$ represents the number of cars your dealership sells in year t. If your car dealership sells almost exclusively to home town customers, explain why we expect c to have a limiting value.

S-10. **Another table:** The following is a partial table of values for $f = f(x)$.

x	$f = f(x)$
0	5.7
5	4.3
10	1.1
15	−3.6
20	−7.9

What is the value of $f(15)$?

S-11. **Average rate of change:** Using the table in Exercise S-10, find the average rate of change for f between $x = 15$ and $x = 20$.

S-12. **Using average rate of change:** Use the average rate of change calculated in Exercise S-11 to estimate the value of $f(17)$.

1.2   EXERCISES

1. **The American food dollar:** The following table shows the percentage $P = P(d)$ of the American food dollar that was spent on eating away from home (at restaurants, for example) as a function of the date d.

d = Year	P = Percent spent away from home
1970	24%
1980	29%
1988	33%

a. Find $P(1980)$ and explain what it means.

b. What does $P(1975)$ mean? Estimate its value.

c. What is the average rate of change per year in percentage of the food dollar spent away from home for the period from 1980 to 1988?

d. What does $P(1982)$ mean? Estimate its value. (*Hint:* Your calculation in part c should be useful.)

e. Estimate the value of $P(1990)$ and explain how you made your estimate.

2. **Gross domestic product:** The following table[16] shows the U.S. gross domestic product (GDP) G (in trillions of dollars) as a function of the year t.

t = Year	1992	1996	1998
G = GDP (trillions of dollars)	6.24	7.66	8.51

a. Explain in practical terms what $G(1992)$ means, and find its value.

b. Use functional notation to express the gross domestic product in 1994, and estimate that value.

c. What is the average yearly rate of change in G from 1996 to 1998? Give your answer to three decimal places.

d. Use your answer to part c to estimate the gross domestic product in the year 2003.

3. **Cable TV:** The following table gives the total dollars $C = C(t)$, in billions, spent by Americans on cable TV in year t.

t = Year	1984	1989	1994
C = Billions of dollars	6.98	11.71	16.55

a. Find $C(1989)$ and explain what it means.

b. Find the average rate of change per year during the period from 1984 to 1989.

c. Estimate the value of $C(1986)$. Explain how you got your answer.

4. **A cold front:** At 4 P.M. on a winter day, an arctic air mass moved from Kansas into Oklahoma, causing temperatures to plummet. The temperature $T = T(h)$ in degrees Fahrenheit h hours after 4 P.M. in Stillwater, Oklahoma, on that day is recorded in the following table.

h = Hours since 4 P.M.	T = Temperature
0	62
1	59
2	38
3	26
4	22

a. Use functional notation to express the temperature in Stillwater at 5:30 P.M., and then estimate its value.

b. What was the average rate of change per minute in temperature between 5 P.M. and 6 P.M.? What was the average decrease per minute over that time interval?

c. Estimate the temperature at 5:12 P.M.

d. At about what time did the temperature reach the freezing point? Explain your reasoning.

5. **A troublesome snowball:** One winter afternoon, unbeknownst to his mom, a child brings a snowball into the house, lays it on the floor, and then goes to watch TV. Let $W = W(t)$ be the volume of dirty water that has soaked into the carpet t minutes after the snowball was deposited on the floor. Explain in practical terms what the limiting value of W represents, and tell what has happened physically when this limiting value is reached.

6. **Falling with a parachute:** If an average-size man jumps from an airplane with a properly opening parachute, his downward velocity $v = v(t)$, in feet

---

[16]From the *1999 Statistical Abstract of the United States*.

per second, t seconds into the fall is given by the following table.

t = Seconds into the fall	v = Velocity
0	0
1	16
2	19.2
3	19.84
4	19.97

a. Explain why you expect v to have a limiting value and what this limiting value represents physically.

b. Estimate the terminal velocity of the parachutist.

7. **Carbon 14:** Carbon 14 is a radioactive substance that decays over time. One of its important uses is in dating relatively recent archaeological events. In the following table, time t is measured in thousands of years, and $C = C(t)$ is the amount, in grams, of carbon 14 remaining.

t = Thousands of years	C = Grams remaining
0	5
5	2.73
10	1.49
15	0.81
20	0.44

a. What is the average yearly rate of change of carbon 14 during the first 5000 years?

b. How many grams of carbon 14 would you expect to find remaining after 1236 years?

c. What would you expect to be the limiting value of C?

8. **Newton's law of cooling** says that a hot object cools rapidly when the difference between its temperature and that of the surrounding air is large, but it cools more slowly when the object nears room tem-

perature. Suppose a piece of aluminum is removed from an oven and left to cool. The following table gives the temperature $A = A(t)$, in degrees Fahrenheit, of the aluminum t minutes after it is removed from the oven.

t = Minutes	A = Temperature
0	302
30	152
60	100
90	81
120	75
150	73
180	72
210	72

a. Explain the meaning of $A(75)$ and estimate its value.

b. Find the average decrease per minute of temperature during the first half-hour of cooling.

c. Find the average decrease per minute of temperature during the first half of the second hour of cooling.

d. Explain how parts b and c support Newton's law of cooling.

e. Use functional notation to express the temperature of the aluminum after 1 hour and 13 minutes. Estimate the temperature at that time. (*Note:* Your work in part c should be helpful.)

f. What is the temperature of the oven? Express your answer using functional notation, and give its value.

g. Explain why you would expect the function A to have a limiting value.

h. What is room temperature? Explain your reasoning.

9. **Effective percentage rate for various compounding periods:** We have seen that in spite of its name, the annual percentage rate APR does not generally tell directly how much interest accrues on a loan in a year. That value, known as the *effective annual*

rate[17] or EAR, depends on how often the interest is compounded. Consider a loan with an annual percentage rate of 12%. The following table gives the EAR, $E = E(n)$, if interest is compounded n times each year. For example, there are 8760 hours in a year, so that column corresponds to compounding each hour.

n = Compounding periods	E = EAR
1	12%
2	12.36%
12	12.683%
365	12.747%
8760	12.750%
525,600	12.750%

a. State in everyday language the type of compounding that each row represents.

b. Explain in practical terms what $E(12)$ means and give its value.

c. Use the table to calculate the interest accrued in 1 year on an $8000 loan if the APR is 12% and interest is compounded daily.

d. Estimate the EAR if compounding is done continuously—that is, if interest is added at each moment in time. Explain your reasoning.

10. **New construction:** The following table[18] shows the value B, in billions of dollars, of new construction put in place in the United States during year t.

t = Year	B = Value (billions of dollars)
1989	469.8
1992	452.1
1995	537.4
1998	665.4

a. Make a table showing, for each of the 3-year periods, the average yearly rate of change in B.

b. Explain in practical terms what $B(1997)$ means, and estimate its value.

c. Over what period was the growth in value of new construction the greatest?

d. According to the table, in what year was the value of new construction the greatest?

11. **Growth in height:** The following table gives, for a certain man, his height $H = H(t)$ in inches at age t in years.

t = Age (years)	H = Height (inches)
0	21.5
5	42.5
10	55.0
15	67.0
20	73.5
25	74.0

a. Use functional notation to express the height of the man at age 13, and then estimate its value.

b. Now we study the man's growth rate.

  i. Make a table showing, for each of the 5-year periods, the average yearly growth rate—that is, the average yearly rate of change in H.

  ii. During which 5-year period did the man grow the most in height?

  iii. Describe the general trend in the man's growth rate.

c. What limiting value would you estimate for the height of this man? Explain your reasoning in physical terms.

12. **Growth in weight:** The table on the following page gives, for a certain man, his weight $W = W(t)$ in pounds at age t in years.   →

---

[17] The EAR is the same as the *annual percentage yield,* or APY. The APY is usually used in advertising investment returns. The APR, which is smaller, is used in advertising loan rates, although the EAR would be a more accurate representation.

[18] From the *1999 Statistical Abstract of the United States.*

t = Age (years)	W = Weight (pounds)
4	36
8	54
12	81
16	128
20	156
24	163

a. Make a table showing, for each of the 4-year periods, the average yearly rate of change in W.

b. Describe in general terms how the man's gain in weight varied over time. During which 4-year period did the man gain the most in weight?

c. Estimate how much the man weighed at age 30.

d. Use the average rate of change to estimate how much he weighed at birth. Is your answer reasonable?

13. **Tax owed:** The following table shows the income tax T owed in a certain state as a function of the taxable income I, both measured in dollars.

I = Taxable income	T = Tax owed
16,000	870
16,200	888
16,400	906
16,600	924

a. Make a table showing, for each of the intervals in the tax table above, the average rate of change in T.

b. Describe the general trend in the average rate of change. What does this mean in practical terms?

c. Would you expect T to have a limiting value? Be sure to explain your reasoning.

14. **Sales income:** The following table shows the net monthly income N for a real estate agency as a func-

tion of the monthly real estate sales s, both measured in dollars.

s = Sales	N = Net income
450,000	4000
500,000	5500
550,000	7000
600,000	8500

a. Make a table showing, for each of the intervals in the table above, the average rate of change in N. What pattern do you see?

b. Use the average rate of change to estimate the net monthly income for monthly real estate sales of $520,000. In light of your answer to part a, how confident are you that your estimate is an accurate representation of the actual income?

c. Would you expect N to have a limiting value? Be sure to explain your reasoning.

15. **Yellowfin tuna:** Data were collected comparing the weight W, in pounds, of a yellowfin tuna to its length L, in centimeters.[19] These data are presented in the table below.

L = Length	W = Weight
70	14.3
80	21.5
90	30.8
100	42.5
110	56.8
120	74.1
130	94.7
140	119
160	179
180	256

---

[19] See Michael R. Cullen, *Mathematics for the Biosciences* (Fairfax, VA: TechBooks, 1983).

a. What is the average rate of change, in weight per centimeter of length, in going from a length of 100 centimeters to a length of 110 centimeters?

b. What is the average rate of change, in weight per centimeter, in going from 160 to 180 centimeters?

c. Judging from the data in the table, does an extra centimeter of length make more difference in weight for a small tuna or for a large tuna?

d. Use the average rate of change to estimate the weight of a yellowfin tuna that is 167 centimeters long.

e. What is the average rate of change, in length per pound of weight, in going from a weight of 179 pounds to a weight of 256 pounds?

f. What would you expect to be the length of a yellowfin tuna weighing 225 pounds?

16. **Arterial blood flow:** Medical evidence shows that a small change in the radius of an artery can indicate a large change in blood flow. For example, if one artery has a radius only 5% larger than another, the blood flow rate is 1.22 times as large. Further information is given in the table below.

Increase in radius	Times greater blood flow rate
5%	1.22
10%	1.46
15%	1.75
20%	2.07

a. Use the average rate of change to estimate how many times greater the blood flow rate is in an artery that has a radius 12% larger than another.

b. Explain why if the radius is increased by 12% and then we increase the radius of the new artery by 12% again, the total increase in the radius is 25.44%.

c. Use parts a and b to answer the following question: How many times greater is the blood flow

rate in an artery that is 25.44% larger in radius than another?

d. Answer the question in part c using the average rate of change.

17. **Widget production:** The following table gives, for a certain manufacturing plant, the number W of widgets, in thousands, produced in a day as a function of n, the number of full-time workers.

n = Number of workers	W = Thousands of widgets produced
10	25.0
20	37.5
30	43.8
40	46.9
50	48.4

a. Make a table showing, for each of the 10-worker intervals, the average rate of change in W per worker.

b. Describe the general trend in the average rate of change. Explain in practical terms what this means.

c. Use the average rate of change to estimate how many widgets will be produced if there are 55 full-time workers.

d. Use your answer to part b to determine whether your estimate in part c is likely to be too high or too low.

18. **Sawtimber and land values:** The following table is adapted from an article by Marshall Thomas in the *F&W Forestry Letter*. The data refer to land and timber values in the South. Sawtimber prices are in dollars per thousand board feet.[20] Bare land prices are in dollars per acre and refer to land that has been cleared of timber.   →

---

[20]One board foot is 1 inch by 1 foot by 1 foot.

Year	Sawtimber price	Bare land price
1990	179	299
1994	310	321
1996	311	395
1998	391	446

a. What is the average rate of change per year in sawtimber prices from 1990 to 1994?

b. Use your answer in part a to estimate the sawtimber price in 1993. (The actual value was 240 dollars per thousand board feet.)

c. What is the average rate of change per year in bare land prices from 1996 to 1998?

d. Use your answer in part c to estimate the bare land price in 1999. (The actual value was 422 dollars per acre.)

e. Find the percentage increase in sawtimber and bare land prices from 1990 to 1998.

f. On the basis of your answer in part e, in the absence of other factors, would one be better advised to invest in bare land or in sawtimber?

19. **The Margaria-Kalamen test:** The Margaria-Kalamen test is used by physical educators as a measure of leg strength. An individual runs up a staircase, and the elapsed time from the third to the ninth step is recorded. The power score is calculated using a formula involving the individual's weight, the height of the stairs, and the running time. The table below[21] shows the power scores required of men for an excellent rating for selected ages.

Age	Power score for excellent rating
18	224
25	210
35	168
45	125
55	98

a. What is the average rate of change per year in excellence level from 25 years to 35 years old?

b. What power score would merit an excellent rating for a 27-year-old man?

c. During which 10-year period from age 25 to 55 would you expect to see the greatest decrease in leg power?

20. **ACRS and MACRS:** In 1981 Congress introduced the Accelerated Cost Recovery System (ACRS) tax depreciation. ACRS allowed for uniform recovery periods at fixed percentage rates each year. Part of the intent of the law was to encourage businesses to invest in new equipment by allowing faster depreciation. For example, a delivery truck is designated by ACRS as a 3-year property, and the depreciation percentages allowed each year are given in the table below.

Year	ACRS Depreciation
1	25%
2	38%
3	37%

Tax depreciation allows expenses in one year for durable goods, such as a truck, to be allocated and deducted from taxable income over several years. The depreciation percentage refers to percentage of the original expense. For example, if an item costs $10,000, then in the first year 25%, or $2500, is deducted; in the second year 38%, or $3800, is deducted; and in the third year 37%, or $3700, is deducted.

In the Tax Reform Act of 1986, the Modified Accelerated Cost Recovery System (MACRS) was mandated. This law designated light trucks, such as delivery trucks, as 5-year property and (curiously) required depreciation over 6 years, as shown in the following table.

---

[21]Adapted from D. K. Mathews and E. L. Fox, *The Physiological Basis of Physical Education and Athletics*, 2nd ed. (Philadelphia: Saunders, 1976).

Year	MACRS Depreciation
1	20%
2	32%
3	19.2%
4	11.52%
5	11.52%
6	5.76%

Suppose the delivery truck in question costs $14,500.

a. Under either ACRS or MACRS, what is the total depreciation allowed over the life of the truck?

b. Under ACRS, what dollar tax deduction for depreciation on the truck is allowed in each of the first 3 years?

c. Under MACRS, what dollar tax deduction for depreciation on the truck is allowed in each of the first 3 years?

d. Suppose your company has a 28% *marginal tax rate*. That is, a 1 dollar tax deduction results in a reduction of 28 cents in actual taxes paid. Under ACRS, what is the total tax savings due to depreciation of the truck during the second year?

21. **A home experiment:** In our discussion of Carlson's experiment with yeast, we indicated that there should be similarities between the growth of yeast and the growth of mold on a slice of bread. In this exercise, you will verify that. Begin with a slice of bread that has a few moldy spots on it. Put it in a plastic bag and leave it in a warm place such as your kitchen counter. Estimate the percentage of the bread surface that is covered by mold two or three times each day until the bread is covered with mold. (This may take several days to a week.) Record your data and provide a written report describing the growth of the mold.

22. **Research project:** Look in the latest edition of the *Statistical Abstract of the United States* for data of interest to you, and analyze the average rate of change of the data. Try to relate your findings to economic or social trends in the United States during the given period. This information is available in many libraries and also online at

**www.census.gov/prod/www/
statistical-abstract-us.html**

1.3   *Functions Given by Graphs*

The idea of picturing a function as a graph is usually credited to the 17th-century mathematicians Fermat and Descartes. But some give credit to Nicole Oresme, who preceded Fermat and Descartes by 300 years. In his study of velocity, he hit upon the idea of representing the velocity of an object using two dimensions. He drew a *base line* to represent time and then added *perpendiculars* from this line whose lengths represented the velocity at each instant. The velocity, he said, "cannot be known any better, more clearly, or more easily than by such mental images and relations to figures."[22] Oresme was expressing his belief, which modern mathematicians share, that the graph provides deeper insight into a function and makes mathematics easier. It is one more illustration of the old cliché "A picture is worth a thousand words." One of the best features of a graph is that it provides an overall view of a function and thus makes it easy to deduce important properties.

Reading Graphs

Figure 1.7 shows the percentage of unemployment $U = U(d)$ in Canada as a function of the date d. It is customary to describe this as a graph of U *versus* d or as a graph of U *against* d. Either of these expressions indicates that the horizontal axis corresponds to d and the vertical axis corresponds to U. To find the unemployment in 1970, we locate 1970 on the horizontal axis, move along the vertical line shown in Figure 1.8 up to the graph, and then move horizontally to locate the corresponding percentage on the vertical axis. In this case we see that it is about 5%, so in functional notation $U(1970) = 5$. As we see in this example, graphical presentations sometimes allow only rough approximation of function values, but interesting features of a function are often easy to spot when it is presented graphically. For example, in Figure 1.9 we have marked the *minimum*, or lowest point on the graph, as well as regions where the graph is *decreasing* and where it is *increasing*. This shows clearly that the unemployment rate declined from 1960 until it reached a minimum of about 4% in 1965 and that it generally increased during

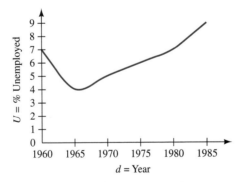

FIGURE 1.7 Unemployment in Canada

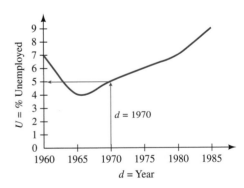

FIGURE 1.8 Getting *U*(1970) from the graph

---

[22]From Marshall Clagett, trans. and ed., *Nicole Oresme and the Medieval Geometry of Qualities and Motions* (Madison: University of Wisconsin Press, 1968).

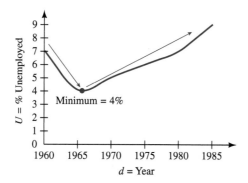

FIGURE 1.9 Region of decrease, region of increase, and the minimum value

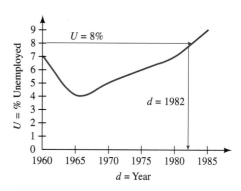

FIGURE 1.10 Finding when unemployment reached 8%

the ensuing 20 years. This is typical of the shape of a graph near a minimum value; the graph decreases down to the minimum value and increases after that. Similarly, the highest point, or *maximum*, typically occurs where the graph changes from increasing to decreasing. Sometimes, however, maximum or minimum values occur at the ends of a graph. For example, we see in Figure 1.7 that, during the period from 1960 to 1985, unemployment in Canada reached a maximum of about 9% in 1985, at the right-hand end of the graph.

A great deal more information is available from the graph. Suppose, for example, that we want to know when unemployment reached 8%. In functional notation, we want to solve the equation $U(d) = 8$ for d. To do so, we locate 8 on the vertical axis and move horizontally until we get to the graph. From that point we drop down to the horizontal axis as illustrated in Figure 1.10, and we see that unemployment reached 8% around 1982. In functional notation this is $U(1982) = 8$. We might also ask for the average yearly increase during the decade of the 1970s. From the graph we read $U(1970) = 5$ and $U(1980) = 7$. That is an increase of 2 percentage points over the 10-year period, so the average yearly increase is 0.2 percentage point per year.

Evaluating Functions Given by Graphs

To evaluate a function given by a graph, locate the point of interest on the horizontal axis, move vertically to the graph, and then move horizontally to the vertical axis. The function value is the location on the vertical axis.

EXAMPLE 1.6 **Currency Exchange**

Figure 1.11 shows the value $G = G(d)$, in American cents, of the German mark as a function of the date d.

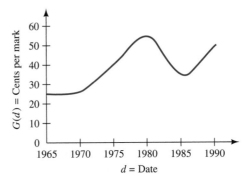

FIGURE 1.11 The value of the German mark in American cents

1. Explain the meaning of $G(1970)$ and estimate its value.

2. From 1965 through 1990, what was the largest value the German mark attained? When did that happen?

3. What was the average yearly increase in the value of the mark from 1975 to 1980?

4. During which year was the mark decreasing most rapidly in value?

5. During the period 1965 to 1990, what would have been the ideal buy date and the ideal sell date for an American investing in German marks? Explain your reasoning.

Solution to Part 1:   The expression $G(1970)$ is the value in cents of the German mark in 1970. From the graph we read that $G(1970)$ is about 26 cents.

Solution to Part 2:   The mark reaches its highest value where the graph reaches its highest point. That appears to be about 55 cents in 1980.

Solution to Part 3:   From 1975 to 1980 the mark increased in value from $G(1975) = 40$ cents to $G(1980) = 55$ cents. That is an increase of 15 cents over the 5-year period. Thus the value of the mark increased by about $\frac{15}{5} = 3$ cents per year.

Solution to Part 4:   The value of the mark is most rapidly decreasing where the graph is going downward the most steeply. That is difficult to discern exactly from the graph, but it surely occurred sometime between 1980 and 1985. The year 1982 appears to be fairly close.

Solution to Part 5:   There is more than one reasonable answer for this part. The ideal buy and sell dates for an American trading marks depend on the goals of the investment. If the goal was simply to get the maximum cumulative return on investment, the American should have bought the mark at its cheapest, about 25 cents in 1965, and sold it when it achieved its maximum value of 55 cents in 1980.

    Many times investors are concerned not only with cumulative profit but also with how long it takes the profit to accrue. With this in mind, one might prefer to have bought in 1970 for about 26 cents and sold in 1980 for 55 cents. There are a number of other answers that are acceptable if properly explained. ▪ ▪ ▪

Concavity and Rates of Change

Important features of a graph include places where it is increasing or decreasing, maxima and minima, and places where the graph may cross the horizontal axis. But a feature called *concavity* is often as important as any of these. A graph may be *concave up*, meaning that it has the shape of a wire whose ends are bent upward, or it may be *concave down*, meaning that it has the shape of a wire whose ends are bent downward. Figures 1.12 and 1.13 show graphs that are concave up, and Figures 1.14 and 1.15 show graphs that are concave down.

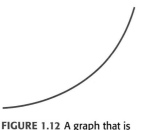

FIGURE 1.12 A graph that is concave up and increasing

Note that the concavity of a graph does not determine whether it is increasing or decreasing, but it does give additional information about the rate of increase or decrease. The graphs in Figures 1.12 and 1.14 are both increasing, but they increase in different ways. As we move to the right in Figure 1.12, the graph gets steeper. That is, the graph is increasing at a faster rate. As we move to the right in Figure 1.14, the graph becomes less steep, showing that the rate of increase is slowing. This is typical. Increasing graphs that are concave up increase at an increasing rate, whereas increasing graphs that are concave down increase at a decreasing rate. A similar analysis for Figures 1.13 and 1.15 shows that decreasing graphs that are concave up decrease at a decreasing rate, whereas decreasing graphs that are concave down decrease at an increasing rate. Concavity, if it persists, can have important long-term effects. For example, if prices were increasing, it would be a good deal more troubling in the long term if the graph of prices against time were concave up than if it were concave down.

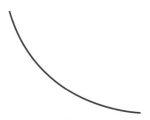

FIGURE 1.13 A graph that is concave up and decreasing

FIGURE 1.14 A graph that is concave down and increasing

FIGURE 1.15 A graph that is concave down and decreasing

EXAMPLE 1.7   Concavity and Currency Exchange

Consider once more the value of the German mark as given by the graph in Figure 1.11.

1. During what years is the graph concave up? Explain in practical terms what this means about the value of the mark during these periods.

2. During what years is the graph concave down? Explain in practical terms what this means about the value of the mark during this period.

Solution to Part 1:   The graph is clearly concave up from 1965 to about 1975. Since the graph is increasing and concave up, the value of the mark is increasing at an increasing rate during this period.

The graph is also concave up from about 1982 until 1990. We divide our analysis here into the places where the graph is decreasing and increasing. From 1982 to 1985, the graph is decreasing and concave up. The fact that it is concave up means that as we near 1985, the graph decreases at a slower rate. From 1985 to 1990, the graph is increasing and concave up. Thus as we approach 1990, the rate of increase is getting larger.

Solution to Part 2:   The graph is concave down only from about 1975 to 1982. As in part 1, we divide our analysis into the places where the graph is increasing and where it is decreasing. From 1975 to 1980, the graph is increasing and concave down. That means the value of the mark was increasing, but at a decreasing rate. From 1980 to 1982, the graph is decreasing and concave down. Thus during this period the value of the mark decreases more rapidly as we approach 1982.   ■ ■ ■

Inflection Points

We refer to points on a graph where concavity changes, up to down or down to up, as *inflection points*. In Figure 1.16 we have shown a graph where concavity changes. These changes in concavity determine the two inflection points marked in Figure 1.17. The first inflection point in Figure 1.17 is where the graph changes from concave down to concave up, and the second is where the graph changes from concave up to concave down. Note that the first inflection point in Figure 1.17 is also the point where the graph is decreasing most rapidly. Similarly, the second inflection point is where the graph is increasing most rapidly. Points of maximum increase or decrease are often found at inflection points, and in many applications this is where we should look. For example, in part 4 of Example 1.6, we found that the value of the mark was decreasing most rapidly in 1982. Also in Example 1.7 we identified this same point as the place where concavity changes from down to up—that is, as an inflection point. We can also observe that the graph in Figure 1.11 is increasing most rapidly at about 1975, and this corresponds to an inflection point where concavity changes from up to down.

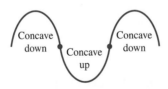

FIGURE 1.16 A graph with changing concavity

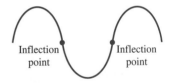

FIGURE 1.17 Inflection points

EXAMPLE *1.8*   **Driving Down a Straight Road**

You leave home at noon and drive your car down a straight road for a visit to a friend's house. Your distance D, in miles, from home as a function of time t, in hours, since noon is shown in the graph in Figure 1.18.

1. During what time period are you at your friend's house?

2. Over what time period is the graph decreasing? Explain in practical terms what this portion of the graph represents.

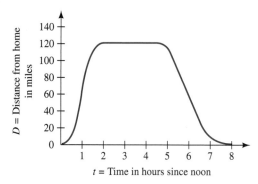

FIGURE 1.18 Distance from home versus time

3. At what value of *t* is the graph rising most steeply? Explain in terms of your speedometer what your answer means.

4. At what time do you get back home?

Solution to Part 1:   During the time you are at the friend's house, your distance from home is unchanging, or constant, so the graph of *D* versus *t* is level over that period. Since the graph is level between about $t = 2$ and $t = 4\frac{1}{2}$ hours after noon, you are at the friend's house between about 2:00 P.M. and 4:30 P.M. Note that the constant value of *D* over this period is about 120, so the friend lives about 120 miles from your home.

Solution to Part 2:   The graph is decreasing from around $t = 4\frac{1}{2}$ to around $t = 7\frac{3}{4}$, or roughly from 4:30 P.M. to 7:45 P.M. Since *D* is your distance from home, this portion of the graph represents your trip back home.

Solution to Part 3:   The steepest rise in the graph occurs at the first inflection point, where the graph changes from concave up to concave down. Examination of the graph shows that this point occurs at about $t = 1$. Before $t = 1$, the graph is concave up, so the distance *D* is increasing at an *increasing* rate; for a time after $t = 1$, the graph is concave down, and the distance is increasing at a *decreasing* rate. This means that the greatest rate of increase in distance occurs at $t = 1$. Now on the way to your friend's house, the speedometer reading tells you how fast you are going away from home; in other words, it gives the rate of increase in distance. Therefore, you interpret the time when the graph is rising most steeply as the time on your way to the friend's house when the speedometer reading is the greatest. In short, on the first part of your trip, you reach your maximum speed at about 1:00 P.M.

Solution to Part 4:   When you get back home, the distance *D* reaches 0, so the graph touches the horizontal axis. This occurs at around $t = 7\frac{3}{4}$ hours after noon, or 7:45 P.M.   ■ ■ ■

Drawing a Graph

A graph is a pictorial representation of a function, and it is generally not difficult to make such pictures. Here is an example. According to the *1996 Information Please Almanac*, in 1900 about 9.3 Americans per thousand were married each year. This number reached an all-time high of about 12.2 per thousand in 1945 (in the aftermath of World

War II). This number then decreased and reached a low of 8.5 per thousand in 1960. This information describes a function $M = M(d)$ that gives the number of marriages per thousand Americans in year d, and we want to draw its graph.

To make a graph of the function M means that we graph M against d. That is, the horizontal axis will correspond to the date, and the vertical axis will correspond to the number of marriages per thousand. These axes are so labeled in Figure 1.19. We note that the verbal description of the function gave us three data points, which we express in functional notation:

$$M(1900) = 9.3$$

$$M(1945) = 12.2$$

$$M(1960) = 8.5.$$

The next step is to mark these points on the graph as shown in Figure 1.19. To complete the graph, we join these points with a curve incorporating the additional information we have. In particular, the verbal description of M tells us that the graph should decrease after 1945. Our completed graph is shown in Figure 1.20.

It is important to note that there was a certain amount of guesswork done to produce the graph in Figure 1.20. The points we located in Figure 1.19 are not guesses, but the way we chose to join these points to complete the graph in Figure 1.20 does involve a good deal of guessing. For example, we were told that the marriage rate per thousand decreased from 1945 to 1960, but we were not told how it decreased. The true graph might, for example, be concave down in this region (rather than up, as we drew it), or it might be a jagged line (rather than the smooth curve we have drawn). When graphs are used to represent functions given by sampling data, some guessing is inevitably involved in connecting known data points. You should be aware that artistic flair can dramatically affect the appearance of such graphs. For example, if we had wanted to emphasize the high marriage rate at the end of World War II, we might have drawn the graph as in Figure 1.21. If we had wanted to de-emphasize it, we might have made the graph as in Figure 1.22. Note that both of these graphs agree with the verbal description we were given, but they convey quite different information to the eye. Further emphasis or de-emphasis of particular features of a function can be attained by adjusting the scale on the horizontal or vertical axis.

How do we know which of the pictures, if any, among Figures 1.20, 1.21, and 1.22 provide a true representation of the function M? The answer is that graphs give a more

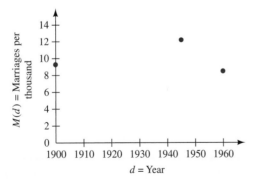

FIGURE 1.19 Locating given points for M versus d

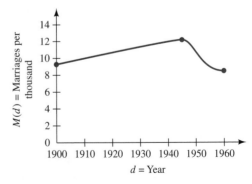

FIGURE 1.20 Marriages per thousand Americans

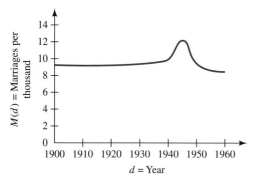

FIGURE 1.21 A representation of the marriage rate that emphasizes the peak in 1945

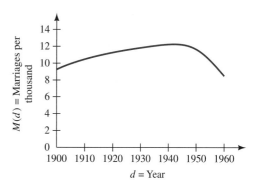

FIGURE 1.22 A representation of the marriage rate that de-emphasizes the peak in 1945

accurate representation of a function when more data points are used, and if your interest is in seeing the true nature of the American marriage rate in the first half of the 20th century, you should consult the almanac referred to earlier, where a good deal more information is available. See also Exercise 20 at the end of this section. As an informed citizen, you should be especially leery of free-hand graphs that do not provide references for data sources or for which the artist may be an advocate of a particular point of view.

EXAMPLE 1.9   **Education in the United States**

According to United States census figures, in 1940 about 62% of Americans in the 25- to 29-year age group had completed less than 12 years of school. This figure decreased to about 25% in 1970. From that point on, the decrease was much less pronounced, dropping to about 12% in 1990. Let $E = E(d)$ be the percentage of Americans in the 25- to 29-year age group who have completed less than 12 years of school in year d.

1. Three exact data points were given. Express them in functional notation.

2. Draw a graph of E against d. Be sure to label your axes and to make the graph as accurate as the data allow.

Solution to Part 1:   We are told that 62% of the target population had less than 12 years of school in 1940. Thus $E(1940) = 62$. Similarly, $E(1970) = 25$, and $E(1990) = 12$.

Solution to Part 2:   To graph E against d means that the date d goes on the horizontal axis and the percentage E goes on the vertical axis. In Figure 1.23 we labeled both years and percentage points in 10-year increments and noted what the numbers on the axes represent. Next we located the three given function values in Figure 1.23. To complete the graph as shown in Figure 1.24, we follow the verbal description and draw a decreasing curve through these points. Note that the fact that the decrease was less pronounced after 1970 is reflected by the fact that the graph does not go down so steeply from that point on. Remember, though, that there are a number of related graphs, all of which could be correct.   ■ ■ ■

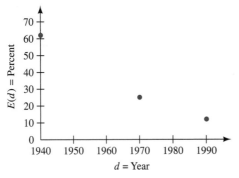

FIGURE 1.23 Locating three data points

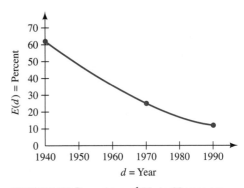

FIGURE 1.24 Percentage of 25- to 29-year age group with less than 12 years of school

ANOTHER LOOK

Secant Lines

When we use graphs, the average rate of change has an important geometric meaning that we explore here. The straight line joining two points on a graph is known as a *secant line* to the graph. In Figure 1.25 we have marked two points on a graph, and in Figure 1.26 we have added the corresponding secant line.

The secant line in Figure 1.26 passes through the points (2, 1) and (6, 3). Note that the slope[23] of the secant line is

$$\frac{\text{Vertical change}}{\text{Horizontal change}} = \frac{3 - 1}{6 - 2} = \frac{1}{2}.$$

This is also the average rate of change from $x = 2$ to $x = 6$. Using average rates of change to approximate function values as we did in Section 1.2 is the same as

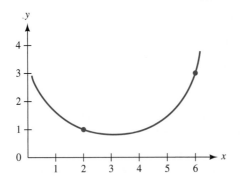

FIGURE 1.25 A graph with two points singled out

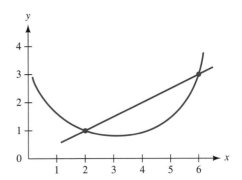

FIGURE 1.26 The secant line added

---

[23]This formula may be taken as the definition of slope. The concept of slope is extensively covered in Chapter 3.

approximating the graph by its secant line. The graph in Figure 1.26 is concave up, and it is easily seen that for such a graph, the secant line lies above the graph between the points (2, 1) and (6, 3) and below the graph outside those points. Thus, when a graph is concave up, approximating function values using average rates of change gives an answer that is too big between the points used and is too small outside those points. You will be asked in the enrichment exercises to analyze what happens for a graph that is concave down.

The graph in Figure 1.26 also shows why approximation using the average rate of change works best when we are dealing with points that are relatively close. The secant line between $x = 2$ and $x = 3$ lies much closer to the graph (at least between those two points) than does the secant line between $x = 2$ and $x = 6$.

If we put our finger on the mark at (6, 3) and move the marker along the graph back toward the point (2, 1), the secant line changes to follow our movement. If we push all the way back to (2, 1), the resulting line is known as the *tangent line*, and the slope of the tangent line is the instantaneous rate of change, or *derivative*, of the function at (2, 1). This is a critical idea in calculus.

Enrichment Exercises

E-1. **Secant lines for graphs that are concave down:** Sketch a graph that is concave down and add a secant line.

    a. Where is the secant line above the graph, and where is it below the graph?

    b. If you use an average rate of change to estimate function values for a graph that is concave down, where will the estimate be too large, and where will it be too small?

E-2. **Secant and tangent lines:** Define the function $f = f(x)$ by $f(x) = x^2 + 2x$.

    a. Calculate $f(0)$, $f(0.5)$, $f(1)$, $f(1.5)$, and $f(2)$. On the basis of these values, make a graph of f versus x.

    b. Find the average rate of change in f between $x = 0$ and $x = 2$. Sketch the secant line to the graph corresponding to this interval.

    c. Find the average rate of change in f between $x = 0$ and $x = 1$. Sketch the secant line to the graph corresponding to this interval.

    d. Find the average rate of change in f between $x = 0$ and $x = 0.1$, and between $x = 0$ and $x = 0.01$.

    e. What would you expect the slope of the tangent line to the graph at the point (0, 0) to be? Sketch this tangent line.

1.3   SKILL BUILDING EXERCISES

S-1. **A function given by a graph:** The following is the graph of a function $f = f(x)$.

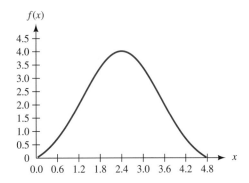

What is the value of $f(1.8)$?

S-2. **An x value for a given f value:** Using the graph from Exercise S-1, determine the smallest value of x for which $f(x) = 1.5$.

S-3. **The maximum:** Where does the graph in Exercise S-1 reach a maximum, and what is that maximum value?

S-4. **Increasing functions:** Where is the graph in Exercise S-1 increasing?

S-5. **Decreasing functions:** Where is the graph in Exercise S-1 decreasing?

S-6. **Concavity:** What is the concavity of the graph in Exercise S-1 between $x = 1.8$ and $x = 3$?

S-7. **Concavity again:** A certain graph is increasing, but at a decreasing rate. Discuss its concavity.

S-8. **Special points:** What is the mathematical term for points where concavity changes?

S-9. **Maximum and zero:** Sketch the graph of a function $f = f(x)$ if f has a maximum value of 3 when $x = 2$, and f has a zero at $x = 4$.

S-10. **Increasing and decreasing:** Sketch a graph that increases from $x = 0$ to $x = 3$ and decreases after that.

1.3   EXERCISES

1. **Sketching a graph with given concavity:**

   a. Sketch a graph that is always decreasing but starts out concave down and then changes to concave up. There should be a point of inflection in your picture. Mark and label it.

   b. Sketch a graph that is always decreasing but starts out concave up and then changes to concave down. There should be a point of inflection in your picture. Mark and label it.

2. **An investment:** In 1995 an investor put money into a fund. The graph in Figure 1.27 shows the value $v = v(d)$ of the investment, in dollars, as a function of the date d.

   a. Express the original investment using functional notation and give its value.

   b. Is the graph concave up or concave down? Explain what this means about the growth in value of the account.

   c. When will the value of the investment reach $55,000?

   d. What is the average yearly increase from 2035 to 2045?

   e. Which is larger, the average yearly increase from 2035 to 2045 or the average yearly increase from 1995 to 2005? Explain your reasoning.

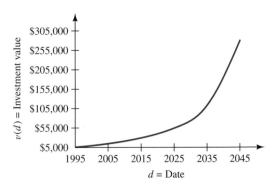

FIGURE 1.27 An investment

3. **A stock market investment:** A stock market investment of $10,000 was made in 1960. During the decade of the 1960s, the stock lost half its value. Beginning in 1970, the value increased until it reached $35,000 in 1980. After that its value has remained stable. Let $v = v(d)$ denote the value of the stock, in dollars, as a function of the date d.

   a. What are the values of $v(1960)$, $v(1970)$, $v(1980)$, and $v(1990)$?

   b. Make a graph of v against d. Label the axes appropriately.

   c. Estimate the time when your graph indicates that the value of the stock was most rapidly increasing.

4. **The value of the Canadian dollar:** The value $C = C(d)$, in American dollars, of the Canadian dollar is given by the graph in Figure 1.28.

   a. Describe how the value of the Canadian dollar fluctuated from 1974 to 1986. Give specific function values in your description where they are appropriate.

   b. When was the Canadian dollar worth 90 American cents?

   c. What was the average yearly decrease in the value of the Canadian dollar from 1976 to 1978?

   d. For an American investor trading in Canadian dollars, what would have been ideal buy and sell dates? Explain your answer.

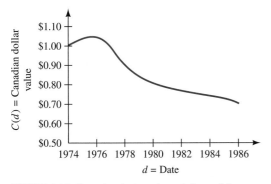

FIGURE 1.28 The value in American dollars of the Canadian dollar

5. **River flow:** The graph in Figure 1.29 shows the mean flow F for the Arkansas River, in cubic feet of water per second, as a function of the time t, in months, since the start of the year. The flow is measured near the river's headwaters in the Rocky Mountains.

   a. Use functional notation to express the flow at the end of July, and then estimate that value.

   b. When is the flow at its greatest?

   c. At what time is the flow increasing the fastest?

   d. Estimate the average rate of change per month in the flow during the first 2 months of the year.

   e. In light of the source of the Arkansas River, interpret your answers to parts b, c, and d.

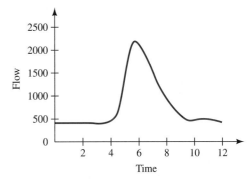

FIGURE 1.29 Flow for the Arkansas River

6. **Logistic population growth:** The graph in Figure 1.30 shows the population $N = N(d)$, in thousands of animals, of a particular species on a protected reserve as a function of the date d. Ecologists refer to growth of the type shown here as *logistic population growth*.

   a. Explain in general terms how the population N changes with time.

   b. When did the population reach 80 thousand?

   c. Your answer in part b is the solution of an equation involving $N(d)$. Which equation?

   d. During which period is the graph concave up? Explain what this means about population growth during this period.

   e. During which period is the graph concave down? Explain what this means about population growth during this period.

   f. When does a point of inflection occur on the graph? Explain how this point may be interpreted in terms of the growth rate. →

g. What is the *environmental carrying capacity* of this reserve for this species of animal? That is, what is the maximum number of individuals that the environment can support?

h. Make a new graph of $N(d)$ under the following new scenario. The population grew as in Figure 1.30 until 1980, when a natural disaster caused half of the population to die. After that the population resumed logistic growth.

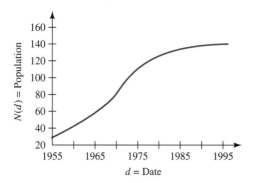

FIGURE 1.30 A population, in thousands, showing logistic population growth

7. **Cutting trees:** In forestry management it is important to know the *net stumpage value* of a stand (that is, a group) of trees. This is the commercial value of the trees minus the costs of felling, hauling, etc. The graph in Figure 1.31 shows the net stumpage value V, in dollars per acre, of a Douglas fir stand in the Pacific Northwest as a function of the age t, in years, of the stand.[24]

a. Estimate the net stumpage value of a Douglas fir stand that is 60 years old.

b. Estimate the age of a Douglas fir stand whose net stumpage value is $40,000 per acre.

c. At what age does the commercial value of the stand equal the costs of felling, hauling, etc.?

d. At what age is the net stumpage value increasing the fastest?

e. This graph shows V only up to age $t = 160$ years, but the Douglas fir lives for hundreds of years. Draw a graph to represent what you expect for V over the life span of the tree. Explain your reasoning.

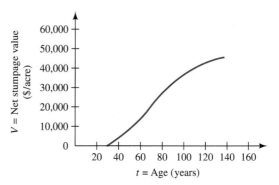

FIGURE 1.31 Net stumpage value of a Douglas fir stand

8. **Wind chill:** The graph in Figure 1.32 shows the temperature $T = T(v)$ adjusted for wind chill as a function of the velocity v of the wind when the thermometer reads 30 degrees Fahrenheit. The adjusted temperature T shows the temperature that has an equivalent cooling power when there is no wind.

a. At what wind speed is the temperature adjusted for wind chill equal to 0?

b. Your answer in part a is the solution of an equation involving $T(v)$. Which equation?

c. At what value of v would a small increase in v have the greatest effect on $T(v)$? In other words, at what wind speed could you expect a small increase in wind speed to cause the greatest change in wind chill? Explain your reasoning.

d. Suppose the wind speed is 45 miles per hour. Judging from the shape of the graph, how significant would you expect the effect on $T(v)$ to be if the wind speed increased?

---

[24]The normal-yield table used here (with site index 140) is from a study by R. E. McArdle et al., as presented by Thomas E. Avery and Harold E. Burkhart, *Forest Measurements*, 4th ed., (New York: McGraw-Hill, 1994).

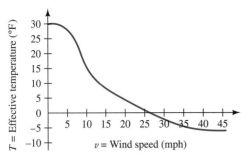

FIGURE 1.32 Temperature adjusted for wind chill when the thermometer reads 30 degrees Fahrenheit

9. **Tornadoes in Oklahoma:** The graph in Figure 1.33 shows the number $T = T(d)$ of tornadoes reported by the Oklahoma Climatological Survey.

   a. When were the most tornadoes reported? How many were reported in that year?

   b. When were the fewest tornadoes reported? How many were reported in that year?

   c. What was the average yearly rate of decrease in tornadic activity from 1986 to 1987?

   d. What was the average yearly rate of increase in tornadic activity from 1990 to 1991?

   e. What was the average yearly *rate of change* (that is, the average yearly rate of total increase or total decrease) in tornadic activity from 1987 to 1989?

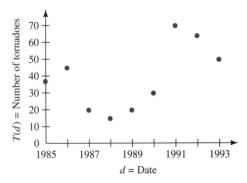

FIGURE 1.33 Tornadoes in Oklahoma

10. **Inflation:** During a period of high inflation, a political leader was up for re-election. Inflation had been increasing during his administration, but he announced that the *rate of increase* of inflation was decreasing. Draw a graph of inflation versus time that illustrates this situation. Would this announcement convince you that economic conditions were improving?

11. **Driving a car:** You are driving a car. The graph in Figure 1.34 shows your distance D, in miles, from home as a function of the time t, in minutes, since 1:00 P.M. Make up a driving story that matches the graph. Be sure to explain how your story incorporates the times when the graph is increasing, when it is decreasing, and when it is constant.

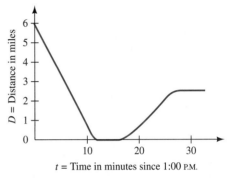

FIGURE 1.34 Distance from home versus time

12. **Walking to school:** You walk from home due east to school to get a book, and then you walk back west to visit a friend. Figure 1.35 on the following page shows your distance D, in yards, east of home as a function of the time t, in minutes, since you left home.

   a. How far away is school?

   b. At what time do you reach school?

   c. At what time(s) are you a distance of 900 yards from home?

   d. Compute the average rate of change in D over the intervals from $t = 0$ to $t = 3$, from $t = 3$ to $t = 6$, and from $t = 6$ to $t = 9$.

   e. What does your answer to part d tell you about how fast you are walking to school? How is this related to the shape of the graph over the interval from $t = 0$ to $t = 9$?

   f. At what time are you walking back west the fastest?   →

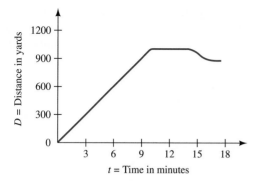

FIGURE 1.35 Distance east of home versus time

d. At which temperature (80 degrees, 60 degrees, or 40 degrees) is the net exchange of carbon dioxide least sensitive to light? Explain your reasoning.

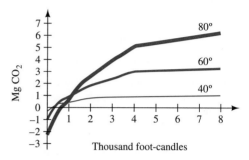

FIGURE 1.36 Net carbon dioxide exchange in young rice plants

13. **Photosynthesis:** During photosynthesis, plants both absorb and emit carbon dioxide. The net exchange (output or input) depends on many factors. Two of these factors are light and temperature. Figure 1.36 shows[25] the net carbon dioxide exchange rates for 5-week-old rice plants at 40 degrees, 60 degrees, and 80 degrees Fahrenheit. The vertical axis shows the net exchange, in milligrams, of carbon dioxide per hour per plant. Values above the horizontal axis indicate that the plant is taking in more carbon dioxide than it releases. Values below the horizontal axis indicate that the plant is releasing more carbon dioxide than it takes in. The horizontal axis shows amount of light, measured in thousands of foot-candles.[26] The following questions pertain to the 5-week-old rice plants at the indicated temperatures.

a. At a temperature of 80 degrees, about how many foot-candles of light will cause a plant to emit the same amount of carbon dioxide that it absorbs?

b. At about how many foot-candles will plants at 80 degrees and 40 degrees show the same net exchange of carbon dioxide?

c. At 0 foot-candles—that is, in the dark—which temperature (80 degrees, 60 degrees, or 40 degrees) will result in the largest emission of carbon dioxide?

14. **Protein content of wheat grain:** Protein content of wheat grain is affected by soil moisture and the amount of available nitrogen (among other things). Figure 1.37 shows[27] the percent of protein content of wheat grain versus pounds of nitrogen per acre applied in three separate situations. In each case soil moisture refers to moisture at the soil depth of 2 inches to 12 inches.

- **Situation 1:** Irrigation was used when soil moisture dropped to 49%.
- **Situation 2:** Irrigation was used when soil moisture dropped to 34%.
- **Situation 3**: Irrigation was used when soil moisture dropped to 1%.

a. If irrigation begins when soil moisture reaches 49%, what application of nitrogen will result in the lowest percentage of protein in wheat grain?

b. If irrigation begins when soil moisture reaches 34%, what application of nitrogen will result in the same protein content of wheat grain as beginning irrigation when soil moisture reaches 1%?

---

[25]Adapted from D. P. Ormrod, "Photosynthesis rates of young rice plants as affected by light intensity and temperature," *Agron. J.* **53** (1961), 93.

[26]The foot-candle is a measure of luminance. Modern physicists use the *candela*, which is defined in terms of black-body radiation.

[27]Adapted from R. Fernandez and R. J. Laird, "Yield and protein content of wheat in Central Mexico as affected by available soil moisture and nitrogen fertilization," *Agron. J.* **51** (1959), 33.

c. If you irrigate when soil moisture reaches 34%, how much nitrogen should you apply to achieve a 13% protein content in wheat grain?

d. Does Figure 1.37 indicate that, for nitrogen levels at 45 pounds per acre or higher, increased protein content in wheat grain is associated with higher or lower soil moisture?

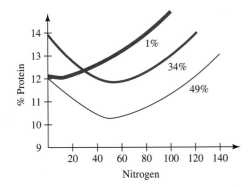

FIGURE 1.37 Protein content versus availability of nitrogen

15. **Carbon dioxide concentrations:** The amount of carbon dioxide in the air is influenced by several factors, including respiration of plants. Figure 1.38 shows[28] the concentration of carbon dioxide in parts per million (PPM) in a closed (that is, unventilated) greenhouse on a cold, clear day during a 24-hour period beginning at 6 A.M.

a. At what time is carbon dioxide in the greenhouse at its lowest concentration?

b. When is carbon dioxide in the greenhouse at its highest concentration?

c. Photosynthesis results in a carbon dioxide exchange between plants and the surrounding air. During what times do the plants in the greenhouse show a net absorption of carbon dioxide?

d. What is happening in terms of carbon dioxide exchange from 6 A.M. to 9 A.M.?

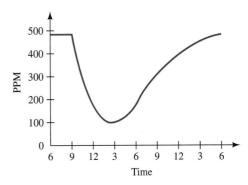

FIGURE 1.38 Carbon dioxide concentrations in a greenhouse

16. **Hydrographs:** When a rainfall brings more water than the soil can absorb, runoff occurs, and hydrologists refer to the event as a *rainfall excess*. The easiest way to envision runoff is to think of a watershed that drains into the mouth of a single stream. The runoff is the number of cubic feet per minute (cfpm) being dumped into the mouth of the stream. An important way of depicting runoff is the *hydrograph*, which is simply the graph of total discharge, in cubic feet per minute, versus time. A typical runoff hydrograph is shown in Figure 1.39 on the following page. The horizontal axis is hours since rainfall excess began. A hydrograph displays a number of important features.

a. *Time to peak* is the elapsed time from the start of rainfall excess to peak runoff. What is the time to peak shown by the hydrograph in Figure 1.39?

b. *Time of concentration* is the elapsed time from the end of rainfall excess to the inflection point after peak runoff. The end of rainfall excess is not readily apparent from a hydrograph, but it occurs before the peak. If the end of rainfall excess occurred 5 hours after the start of rainfall excess, estimate the time of concentration from Figure 1.39.

c. *Recession time* is the time from peak runoff to the end of runoff. Estimate the recession time for the hydrograph in Figure 1.39.

d. *Time base* is the time from beginning to end of surface runoff. What is the time base for the hydrograph in Figure 1.39? →

---

[28]Adapted from S. H. Wittwer and W. Robb, "Carbon dioxide enrichment of greenhouse atmospheres for vegetable crop production," *Econ. Botany* **18** (1964), 34.

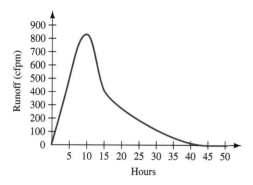

FIGURE 1.39 A runoff hydrograph

17. **Profit from fertilizer:** Fertilizer can increase crop production, but the cost of the fertilizer must be taken into account. The curved graph in Figure 1.40 shows[29] the added bushels per acre of corn that resulted from fertilizer application versus pounds per acre of nitrogen applied. The straight line shows the cost (in bushels of corn, not dollars) versus pounds per acre of nitrogen applied.

a. How many pounds per acre of nitrogen should be applied to produce maximum crop yield?

b. The profit is the crop yield minus the fertilizer cost. For each level of fertilizer application, how can the profit be read off the graph?

c. How many pounds per acre of nitrogen should be applied to achieve maximum profit?

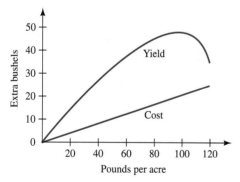

FIGURE 1.40 Effect of nitrogen on corn production

18. **H-R diagrams:** The *luminosity* of a star is the total amount of energy the star radiates (visible light

as well as x rays and all other wavelengths) in 1 second. In practice, astronomers compare the luminosity of a star with that of the sun and speak of *relative luminosity*. Thus a star of relative luminosity 5 is five times as luminous as the sun. One of the most important graphical representations in astronomy is the *Hertzsprung-Russell diagram*, or *H-R diagram*, which plots relative luminosity[30] versus surface temperature in thousands of *kelvins* (degrees on the Kelvin scale). Figure 1.41 shows an H-R diagram. Note that the temperature scale *decreases* as we read from left to right.

a. What is the surface temperature of a main sequence star that is 10,000 times as luminous as the sun?

b. What is the relative luminosity of the sun?

c. About 90% of all stars, including the sun, lie on or near the main sequence. Approximately what is the surface temperature of the sun?

d. The H-R diagram in Figure 1.41 shows that white dwarfs lie well below the main sequence. Are white dwarfs more or less luminous than main sequence stars of the same surface temperature?

e. If one star is three times as luminous as another, yet they have the same surface temperature, then the brighter star must have three times the surface area of the dimmer star. How would the surface area of a supergiant star with the same surface temperature as the sun compare with the surface area of the sun?

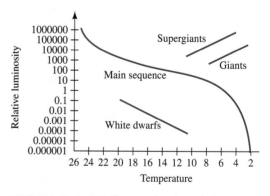

FIGURE 1.41 An H-R diagram showing relative luminosity versus surface temperature

---

[29]Adapted from an article by Allstetter in the *National Fertilizer Review* January-February-March, 1953.

[30]Strictly speaking, an H-R diagram plots absolute magnitude versus surface temperature.

19. **Laboratory experiment:** This lab uses a motion detector and a calculator-based laboratory (CBL^{TM}) unit. The motion detector measures the distance from the detector to an object in front of it, while the CBL records the data. After collecting the data, the CBL unit sends the data to the calculator, which displays a graph of the distance recorded with respect to time. In the lab, we will try to reproduce given graphs of distance versus time by walking in front of the detector while the CBL unit records our distance from the detector. For a detailed description, go to

http://math.college.hmco.com/students

and then, under College Algebra, select "Functions and Change: A Modeling Approach to College Algebra, Crauder/Evans/Noell (©2003)" from the drop-down list, and click the Go button. You should book-mark this page so that you can easily find the experiments in the rest of the book. From the book's web page, click on Lab Experiments and then select the Distance Graphs experiment.

20. **Research project:** In the discussion following Figures 1.21 and 1.22, we pointed out that the right way to get an accurate graph of marriage rates was to include more data points. On page 835 of the *1996 Information Please Almanac*, the marriage rate per thousand Americans is given in 5-year periods. This information will also be available in other reference sources in your library. Pursue an appropriate reference and, by producing your own graph, settle the question of which (if any) of the graphs we presented is an accurate representation of American marriage rates.

1.4   *Functions Given by Words*

Many times mathematical functions are described verbally. In order for us to work effectively with such functions, it is essential that the verbal description be clearly understood. Sometimes this is enough, but more often it is necessary to supplement the verbal description with a formula, graph, or table of values.

Comparing Formulas and Words

For example, there are initially 2000 bacteria in a petri dish. The bacteria reproduce by cell division, and each hour the number of bacteria doubles. This is a verbal description of a function $N = N(t)$, where N is the number of bacteria present at time t. It is common in situations like this to begin at time $t = 0$. Thus $N(0)$ is the number of bacteria we started with, 2000. One hour later the population doubles, so $N(1) = 4000$ bacteria. In one more hour the population doubles again, giving $N(2) = 8000$ bacteria. Continuing, we see that $N(3) = 16,000$ bacteria.

Note that although we can calculate $N(4)$ or $N(5)$, it is not so easy to figure out $N(4.5)$, the number of bacteria we have after 4 hours and 30 minutes. To make such a calculation, we need another description of the function, a formula. This situation is an example of *exponential growth*, which we will study in more detail in Chapter 4; there we will learn how to find formulas to match verbal descriptions of this kind. For now, we only want to make a comparison to indicate that the verbal description matches the formula $N = 2000 \times 2^t$ without showing where the formula came from. That is, we want to show that the formula gives the same answers as the verbal description. You can verify the following calculations by hand or using your calculator:

$$N(0) = 2000 \times 2^0 = 2000 \times 1 = 2000$$

$$N(1) = 2000 \times 2^1 = 2000 \times 2 = 4000$$

$$N(2) = 2000 \times 2^2 = 2000 \times 4 = 8000$$

$$N(3) = 2000 \times 2^3 = 2000 \times 8 = 16,000\,.$$

Note that these are the same values we got using the function's verbal description. You may wish to do further calculations to satisfy yourself that the formula agrees with the verbal description. This formula enables us to calculate the number of bacteria present after 4 hours and 30 minutes. Since that is 4.5 hours after the experiment began, we use $t = 4.5$:

$$N(4.5) = 2000 \times 2^{4.5} = 45,255 \text{ bacteria}\,.$$

We have rounded here to the nearest whole number since we don't expect to see fractional parts of bacteria.

We should point out that comparisons via the few calculations we made cannot establish for a certainty that the formula and the verbal description match. In fact, such a certain conclusion could not be drawn no matter how many individual calculations we made. But such calculations can establish patterns and provide partial evidence of agreement, and that is what we are after here.

EXAMPLE 1.10   **Purifying Water**

Water that is initially contaminated with a concentration of 9 milligrams of pollutant per liter of water is subjected to a cleaning process. The cleaning process is able to reduce the pollutant concentration by 25% each hour. Let $C = C(t)$ denote the concentration, in milligrams per liter, of pollutant in the water t hours after the purification process begins.

1. What is the concentration of pollutant in the water after 3 hours?

2. Compare the results predicted by the verbal description at the end of each of the first 3 hours with those given by the formula $C = 9 \times 0.75^t$.

3. The formula given in part 2 is in fact the correct one. Use it to find the concentration of pollutant after $4\frac{1}{4}$ hours of cleaning.

Solution to Part 1:   There is initially 9 milligrams per liter of pollutant in the water. One hour later, this has been reduced by 25%. That is a reduction of $0.25 \times 9 = 2.25$ milligrams per liter, leaving $9 - 2.25 = 6.75$ milligrams per liter in the water. During the second hour, the level is reduced by 25% again, leaving $6.75 - 0.25 \times 6.75 = 5.0625$ milligrams per liter. After one more hour there will be $5.0625 - 0.25 \times 5.0625 = 3.80$ milligrams per liter (rounded to two decimal places).

Solution to Part 2:   We calculated the appropriate concentrations in part 1. Use your calculator to verify the following and check part 1 for agreement:

$$9 \times 0.75^1 = 6.75$$

$$9 \times 0.75^2 = 5.0625$$

$$9 \times 0.75^3 = 3.80 \text{ (rounded to two decimal places)}.$$

Solution to Part 3:   To get the concentration after $4\frac{1}{4}$ hours, we put 4.25 into the formula 9×0.75^t:

$$9 \times 0.75^{4.25} = 2.65 \text{ milligrams per liter}. \qquad ■ ■ ■$$

EXAMPLE 1.11   **A Risky Stock Market Investment**

An entrepreneur invested $2000 in a risky stock. Unfortunately, over the next year the value of the stock decreased by 7% each month. Let $V = V(t)$ denote the value in dollars of the investment t months after the stock purchase was made.

1. Make a table of values showing the initial value of the stock and its value at the end of each of the first 5 months.

2. Make a graph of investment value versus time.

3. Verify that the formula $V = 2000 \times 0.93^t$ gives the same values you found in your table in part 1.

4. The formula given in part 3 is in fact a valid description of the function V. What is the value of the investment after $9\frac{1}{2}$ months?

5. If the trend for this stock continues—that is, if it continues to decrease in value by 7% each month—what will happen to the value in the long run?

Solution to Part 1:   We use the verbal description to find the needed function values. The original investment was 2000, so $V(0) = 2000$. During the first month, the investment lost 7% of its value. That is, its value decreased by $0.07 \times 2000 = 140$ dollars. Thus $V(1) = 2000 - 140 = 1860$ dollars. During the second month, the value again decreased by 7%—that is, by $0.07 \times 1860 = 130.20$ dollars. Thus $V(2) = 1860 - 130.20 = 1729.80$ dollars.

Continuing the calculation in this manner yields

$$V(3) = \$1608.71$$
$$V(4) = \$1496.10$$
$$V(5) = \$1391.38 .$$

Arranging these in a table of values, we get

t = Months	0	1	2	3	4	5
$V(t)$ = Investment value	$2000	$1860	$1729.80	$1608.71	$1496.10	$1391.38

Solution to Part 2:   Since we are graphing investment value versus time, we use time for the horizontal axis and investment value for the vertical axis. The first step is to locate the data points from the table as in Figure 1.42. Next we join the dots as in Figure 1.43.

Solution to Part 3:   We use $t = 0, 1, 2, 3, 4, 5$ to make the calculations

$$V(0) = 2000 \times 0.93^0 = 2000 \times 1 = 2000$$
$$V(1) = 2000 \times 0.93^1 = 2000 \times 0.93 = 1860$$
$$V(2) = 2000 \times 0.93^2 = 1729.80$$

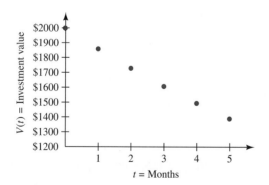

FIGURE 1.42 Plotting points

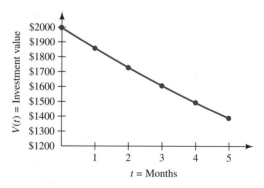

FIGURE 1.43 Completing the graph

$$V(3) = 2000 \times 0.93^3 = 1608.71$$

$$V(4) = 2000 \times 0.93^4 = 1496.10$$

$$V(5) = 2000 \times 0.93^5 = 1391.38 \,.$$

We note that these calculations agree with the table of values we calculated in part 1 using the verbal description of the function. This is evidence that $V = 2000 \times 0.93^t$ gives a correct formula for this function.

Solution to Part 4:   We use the formula from part 3 with $t = 9.5$:

$$V(9.5) = 2000 \times 0.93^{9.5} = 1003.73 \text{ dollars} \,.$$

Solution to Part 5:   The value of the investment decreases by 7% each month. By calculating $V(t)$ for some large values of t using the formula from part 3, we can see whether a trend emerges. Here we use $t = 10$, $t = 50$, and $t = 100$ months:

$$V(10) = 2000 \times 0.93^{10} \ = 967.96 \text{ dollars}$$

$$V(50) = 2000 \times 0.93^{50} \ = 53.11 \text{ dollars}$$

$$V(100) = 2000 \times 0.93^{100} = \ 1.41 \text{ dollars} \,.$$

In the long run, the investment will have virtually no value at all.   ■ ■ ■

Getting Formulas from Words

Many times verbal descriptions can be directly translated into formulas for functions. Suppose, for example, that an engineering firm invests $78,000 in the design and development of a more efficient computer hard disk drive. For each disk drive sold, the engineering firm makes a profit of $98. We want to get a formula that shows the company's net profit as a function of the number of disk drives sold, taking into account the initial investment. This is not a difficult problem, and you may find that you can do it without further help, but we want to show a general method that may be useful for more complicated descriptions.

Step 1:   *Identify the function and the things on which it depends, and write the relationships you know in a formula using words.* The function we are interested in is net profit. That is the profit on disk drives sold minus money spent on initial development. The profit on drives sold depends in turn on the number of sales:

$$\text{Net profit} = \text{Profit from sales} - \text{Initial investment} \tag{1.1}$$

$$= \text{Profit per item} \times \text{Number sold} - \text{Initial investment} \,. \tag{1.2}$$

Step 2:   *Select and record letter names for the function and for each of the variables involved, and state their units.* We can use any letters we want as long as we identify them and state clearly what they mean. To say that we want to express net profit as a function of the number of disk drives sold means that net profit is the function and the number of disk drives is the variable. We will use N to denote our function, the net profit in dollars, and d to represent the variable, the number of disk drives sold:

$$N = \text{Net profit in dollars}$$

$$d = \text{Number of drives sold}.$$

Step 3:   *Replace the words in step 1 by the letters identified in step 2 and appropriate information from the verbal description.* Here we simply replace the words *Net profit* in Equation (1.2) by N, *Profit per item* by 98, *Number sold* by d, and *Initial investment* by 78,000:

$$\underset{N}{\text{Net profit}} = \underset{98}{\text{Profit per item}} \times \underset{d}{\text{Number sold}} - \underset{78,000}{\text{Initial investment}}$$

Thus we get the formula as $N = 98d - 78{,}000$, where d is the number of disk drives sold and N is the net profit in dollars. We can use this formula to provide information about the engineering firm's disk drive project. For example, in functional notation, $N(1200)$ represents the net profit in dollars if 1200 disk drives are sold. We can calculate its value by replacing d by 1200: $N(1200) = 98 \times 1200 - 78{,}000 = 39{,}600$ dollars.

EXAMPLE 1.12   Cutting a Diamond

The total investment a jeweler has in a gem-quality diamond is the price paid for the rough stone plus the amount paid to work the stone. Suppose the gem cutter earns $40 per hour.

1. Choose variable and function names, and give a formula for a function that shows the total investment in terms of the cost of the stone and the number of hours required to work the stone.

2. Use functional notation to express the jeweler's investment in a stone that costs $320 and requires 5 hours and 15 minutes of labor by the gem cutter.

3. Use the formula you made in part 1 to calculate the value from part 2.

Solution to Part 1:   To say that we want the total investment as a function of the cost of the stone and the hours of labor means that *total investment* is our function and that *cost of stone* and *hours of labor* are our variables. The total investment is the cost of the stone plus the cost of labor, which in turn depends on the number of hours needed to work the stone. We write out the relationship using a formula with words. We will accomplish this in two steps:

$$\text{Investment} = \text{Cost of stone} + \text{Cost of labor} \qquad (1.3)$$

$$= \text{Cost of stone} + \text{Hourly wage} \times \text{Hours of labor}. \qquad (1.4)$$

The next step is to choose letters to represent the function and the variables. We let c be the cost in dollars of the rough stone and h the number of hours of labor required. (It is fine to use any letters you like as long as you say in words what they represent.) We let $I = I(c, h)$ be the total investment in dollars in the diamond. To get the formula we want, we use Equation (1.4), replacing *Investment* by *I*, *Cost of stone* by *c*, *Hourly wage* by 40, and *Hours of labor* by *h*:

$$\underbrace{\text{Investment}}_{I} = \underbrace{\text{Cost of stone}}_{c} + \underbrace{\text{Hourly wage}}_{40} \times \underbrace{\text{Hours of labor}}_{h}$$

The result is $I = c + 40h$, where I is the total investment in dollars, c is the cost in dollars of the rough stone, and h is the number of hours required to work the stone.

Note that, in our final presentation of the formula, we were careful to identify the meaning of each letter, noting the appropriate units. Such descriptions are as important as the formula. Also, in this case you may find it easy enough to bypass much of what we presented here and go directly to the final answer. You are very much encouraged, however, to put in the intermediate step using words, just as it appears here. You will encounter many *word problems*, and this practice will make things much easier for you as the course progresses.

Solution to Part 2: The cost of the rough stone is $320, so we use $c = 320$. To get the value of h, we need to remember that h is measured in hours. Because 5 hours and 15 minutes is $5\frac{1}{4}$ hours, $h = 5.25$ hours, and that is what we will put in for h. Thus in functional notation, the total investment is $I(320, 5.25)$.

Solution to Part 3: In the formula we put 320 in place of c and 5.25 in place of h:

$$I(320, 5.25) = 320 + 40 \times 5.25 = 530 \text{ dollars} .$$

■ ■ ■

Proportion

It is common in many applications of mathematics to use the term *proportional* to indicate that one thing is a multiple of another. For example, the total salary of an hourly wage earner can be calculated using

$$\text{Salary} = \text{Hourly wage} \times \text{Hours worked} .$$

Another way of expressing this is to say that salary is *proportional to*[31] the number of hours worked. If we use S to denote the salary, w for the hourly wage, and h for the hours worked, then the formula is

$$S = wh .$$

Many texts use the special symbol $\propto$ to denote a proportionality relation. Thus $S \propto h$ is shorthand notation for the phrase "S is proportional to h." In this context, the hourly wage w would be termed the *constant of proportionality*. It is what we need to multiply by in order to change the proportionality relation into an equation. The statement that S is proportional to h usually carries with it the implicit understanding that h is the variable but that the constant of proportionality w does not change. Thus we would be discussing a worker with a fixed hourly wage but whose working hours might differ from pay period to pay period.

---

[31] Sometimes the phrase *directly proportional to* is used.

Since the salary S is a multiple of the hourly wage w, it is also correct to say that S is proportional to w or, in symbols, that $S \propto w$. Looking at it this way, we are thinking of w as the variable and of h (the number of hours worked) as the constant of proportionality. Implicit in this description is that h does not change but w does. We might, for example, be discussing the long-term salary of someone who always works 40 hours each week but is anticipating raises.

When a function is given verbally, it can sometimes be a challenge to find a formula that represents it. But when such verbal descriptions are in terms of a proportion, it is always an easy task to get the formula. This can be done even without understanding the meaning of the letters involved. For example, if you are told that a quantity denoted by the letter T is proportional to another quantity denoted by z, then you know that $T \propto z$ and that there is a constant of proportionality k that makes the equation $T = kz$ true. Determining the value of k and making sense of the equation do, of course, depend on an understanding of what the letters mean.

EXAMPLE 1.13   Total Cost

You are paying for a group of people to visit the zoo. Your total cost is proportional to the number of people in the group. Let c denote your total cost in dollars and n the number of people in the group. Assume that the price of admission is $7.50 per person.

1. What is your total cost if there are 12 people in the group?

2. Write an equation that shows the proportionality relation. What is the constant of proportionality? Explain in practical terms what it represents.

3. Use the equation you found in part 2 to calculate your total cost if there are 15 people in the group.

Solution to Part 1:   Since the price of admission is $7.50 per person and there are 12 people in the group, the total cost is $7.50 \times 12 = 90$ dollars. The purpose of this part of the example is to illustrate that to get the total cost we multiply the admission price of $7.50 by the number of people in the group.

Solution to Part 2:   The total cost c is proportional to the number n of people in the group. That is, $c \propto n$. We noted in part 1 that to get the total cost we multiply $7.50 by the number of people in the group. Written as an equation, this is $c = 7.50n$. Thus the constant of proportionality is 7.50 dollars per person. In practical terms it is the admission price per person. Note that the units fit together here: The total cost in dollars is the product of the admission price in dollars per person with the number of people. It is a good idea to check the units in proportionality relations like this.

Solution to Part 3:   We use the formula we got in part 2, putting in 15 for n:

$$c = 7.50n = 7.50 \times 15 = 112.50 \text{ dollars}.$$

The total cost is $112.50.

Geometric Constructions

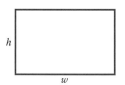

FIGURE 1.44 A rectangle with width w and height h

Many classic problems in mathematics involve geometric figures, and solving these problems often requires expressing one geometric property in terms of another. Some basic formulas are needed, and the most common are listed below, where the variables are as shown in Figures 1.44 and 1.45.

Perimeter of a rectangle: A rectangle with width w and height h has perimeter $P = 2w + 2h$.

Area of a rectangle: A rectangle with width w and height h has area $A = hw$.

Area of a triangle: A triangle with base b and height h has area $A = \frac{1}{2}b \times h$.

Circumference of a circle: A circle with radius r has circumference $C = 2\pi r$.

FIGURE 1.45 A triangle with base b and height h

Area of a circle: A circle with radius r has area $A = \pi r^2$.

Thus if a rectangle has width 4 and height 2, then its perimeter is $P = 2 \times 4 + 2 \times 2 = 12$, and its area is $A = 2 \times 4 = 8$. A rectangle with width 5 and height 1 has perimeter $P = 2 \times 5 + 2 \times 1 = 12$ and area $A = 1 \times 5 = 5$. Note that these two rectangles have the same perimeter but their areas are different.

EXAMPLE 1.14 **Building a Pen**

One hundred feet of wire are used to construct a rectangular pen. Suppose the width of the rectangle is w feet.

1. Express the height of the rectangle in terms of w.

2. Express the area of the rectangle in terms of w.

Solution to Part 1: We use the formula $P = 2w + 2h$. We know in this case that $P = 100$, so we obtain $100 = 2w + 2h$. To get the height, we solve this equation for h:

$$100 = 2w + 2h$$

$$100 - 2w = 2h$$

$$\frac{100 - 2w}{2} = h$$

$$50 - w = h.$$

We will review the algebra skills needed to solve such linear equations in Section 2.3.

Solution to Part 2: To get the area in terms of w, we replace h in the formula $A = hw$ by the expression we obtained in part 1:

$$A = hw = (50 - w)w.$$

■ ■ ■

EXAMPLE *1.15*   **Comparing Triangles**

Two triangles have the same base, but one has twice the height of the second. How do their areas compare?

Solution: One triangle has base b and height h. The second triangle has base b and height $2h$. We calculate the area of each triangle:

$$\text{First area} = \frac{1}{2}b \times h$$

$$\text{Second area} = \frac{1}{2}b \times 2h = bh \,.$$

Thus doubling the height leads to a doubling of the area. This makes sense: The area is proportional to the height.   ■ ■ ■

EXAMPLE *1.16*   **New Formulas Involving the Area of a Circle**

1. Show that the area A of a circle with circumference C can be calculated using

$$A = \frac{C^2}{4\pi} \,.$$

2. Show that the radius r of a circle with area A can be calculated using

$$r = \sqrt{\frac{A}{\pi}} \,.$$

Solution to Part 1:   We first solve for the radius r in the formula giving the circumference:

$$C = 2\pi r$$

$$\frac{C}{2\pi} = r \,.$$

Next we put this expression into the formula for area:

$$A = \pi r^2 = \pi \left(\frac{C}{2\pi} \right)^2 = \pi \frac{C^2}{4\pi^2} = \frac{C^2}{4\pi} \,.$$

This is the desired formula.

Solution to Part 2:   We solve for the radius r in the formula giving the area:

$$A = \pi r^2 \,, \qquad \frac{A}{\pi} = r^2 \,, \qquad \sqrt{\frac{A}{\pi}} = r \,.$$

This is the desired formula.   ■ ■ ■

Enrichment Exercises

E-1. **A fence next to a river:** Two hundred yards of fence are to be used to enclose a rectangular area next to a straight river (see Figure 1.46). The river bank acts as one side of the rectangle, and the fence is used to make the other three sides of the rectangle. Suppose the width w in yards of the rectangle is along the river bank.

a. Express the height of the rectangle in terms of w.

b. Express the area of the rectangle in terms of w.

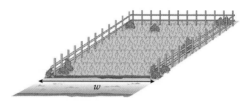

FIGURE 1.46

E-2. **A rectangle with given area:** A rectangle has an area of 20 square inches. Let w denote the width in inches of the rectangle.

a. Express the height of the rectangle in terms of w.

b. Express the perimeter of the rectangle in terms of w.

E-3. **Circumference in terms of area:** A circle has area A. Use the first formula from Example 1.16 to express the circumference C of the circle in terms of A.

E-4. **Unusual areas:**

a. Find the area of the region shown in Figure 1.47. (The segment of length 3 is perpendicular to the segment of length 4.)

b. Find the area of the region shown in Figure 1.48. (The region removed from the top of the figure is half of a circle.)

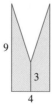

FIGURE 1.47 A rectangle with a triangle removed

FIGURE 1.48 A rectangle with a semi-circle removed

E-5. **Making a box:** Squares of side x are cut from each corner of a 5×10 rectangle as shown in Figure 1.49 on the following page. The resulting tabs on each side are folded up to make a box with no top. (The first fold is shown in Figure 1.50.)

a. Express the width of the box in terms of x.   →

b. Express the length of the box in terms of x.

c. Express the height of the box in terms of x.

d. The volume of a box is width $\times$ length $\times$ height. Express the volume of the box in terms of x.

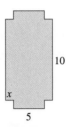

10

x

5

FIGURE 1.49 A rectangle with squares cut from the corners

FIGURE 1.50 One tab folded

FIGURE 1.51

E-6. **A soda can with given surface area:** A soda can is made from 40 square inches of aluminum. Let x denote the radius of the top of the can, and let h denote the height (see Figure 1.51), both in inches.

a. The area of the cylindrical part of the can is the circumference of the can times the height. Express the total surface area of the can using x and h. *Note:* The total surface area is the area of the top plus the area of the bottom plus the area of the cylinder.

b. Using the fact that the total area is 40 square inches, express h in terms of x.

c. The volume of the can is the area of the top times the height. Express the volume of the can in terms of x.

E-7. **A soda can with given volume:** A soda can has a volume of 25 cubic inches. Let x denote its radius and h its height, both in inches.

a. Using the fact that the volume of the can is 25 cubic inches, express h in terms of x. (See part c of Exercise E-6 if you need help expressing the volume of the can in terms of x and h.)

b. Express the surface area of the can in terms of x. (See part a of Exercise E-6 if you need help expressing the surface area of the can in terms of x and h.)

1.4   SKILL BUILDING EXERCISES

S-1. **A description:** You have $5000 in a cookie jar. Each month you spend half of the balance. How much do you have after 4 months?

S-2. **Light:** It is 93,000,000 miles from the Earth to the sun. Light travels 186,000 miles per second. How long does it take light to travel from the sun to the Earth?

S-3. **A description:** The initial value of a function $f = f(x)$ is 5. That is, $f(0) = 5$. Each time x is increased by 1, the value of f triples. What is the value of $f(4)$? Verify that the formula $f(x) = 5 \times 3^x$ gives the same answer.

S-4. **Getting a formula:** You sell lemonade for 25 cents per glass. You invested $2.00 in the ingredients. Write a formula that gives the profit $P = P(n)$ as a function of the number n of glasses you sell.

S-5. **Getting a formula:** Each time a certain balding gentleman showers, he loses 67 strands of hair down the shower drain. Write a formula that gives the total number $N = N(s)$ of hairs lost by this man after s showers.

S-6. **Getting a formula:** You pay $500 to rent a special area in a restaurant. In addition, you pay $10 for each guest. Write a formula that gives your total cost C, in dollars, as a function of the number n of dinner guests.

S-7. **Getting a formula:** You currently have $500 in a piggy bank. You add $37 to the bank each month. Find a formula that gives the balance B, in dollars, in the piggy bank after t months.

S-8. **Getting a formula:** An object is removed from a hot oven and left to cool. After t minutes, the difference between the temperature of the object and room temperature, 75 degrees, is 325×0.07^t degrees. Find a formula for the temperature T in degrees of the object t minutes after it is removed from the oven.

S-9. **Proportionality:** For a certain function $f = f(x)$, we know that f is proportional to x and that the constant of proportionality is 8. Find a formula for f.

S-10. **Constant of proportionality:** If $g(t) = 16t$, then g is proportional to t. What is the constant of proportionality?

1.4   EXERCISES

1. **United States population growth:** In 1960 the population of the United States was about 180 million. Since that time the population has increased by approximately 1.2% each year. This is a verbal description of the function $N = N(t)$, where N is the population, in millions, and t is the number of years since 1960.

   a. Express in functional notation the population of the United States in 1963. Calculate its value.

   b. Use the verbal description of N to make a table of values that shows U.S. population in millions from 1960 through 1965.

   c. Make a graph of U.S. population versus time. Be sure to label your graph appropriately.

   d. Verify that the formula 180×1.012^t million people, where t is the number of years since 1960,

gives the same values as those you found in the table in part b. (*Note:* Because t is the number of years since 1960, you would use $t = 2$ to get the population in 1962.)

   e. Assuming that the population has been growing at the same percentage rate since 1960, what value does the formula above give for the population in 1995? (*Note:* The actual population in 1995 was about 263 million.)

2. **Education and income:** According to the *1995 Digest of Education Statistics*, the median (middle of the range) annual income of a high school graduate in 1993 was about $27,000 per year. Assuming that it takes 4 years to earn a bachelor's degree and 2 additional years to earn a master's degree, the median annual income increases by 12.1% for each year that an individual attends college. This describes a

function $I = I(y)$, where I is the median income in 1993 for a person with y years of college education.

a. Express in functional notation the median income in 1993 of an individual with 2 years of college. Calculate the value.

b. Make a table of values that shows the median income in 1993 for individuals completing 0 through 4 years of college.

c. Make a graph of median income versus years of college completed. Be sure to provide appropriate labels.

d. Verify that the formula $I = 27 \times 1.121^y$ thousand dollars, where y is the number of years of college completed, gives the same values as those you found in the table you made.

e. Using the formula given in part d, find the median income in 1993 of an individual who has a master's degree.

f. Assuming it takes 3 years beyond the master's degree to complete a Ph.D., and assuming the formula in part d applies, what was the median income of a Ph.D. in 1993?

g. In fact, the same rate of increase does *not* apply for the years spent on a Ph.D. The actual median income for a Ph.D. in 1993 was $63,149. Does that mean that the increase in median income for years spent earning a Ph.D. is higher or lower than 12.1% per year?

3. **Altitude:** A helicopter takes off from the roof of a building that is 200 feet above the ground. The altitude of the helicopter increases by 150 feet each minute.

a. Use a formula to express the altitude of a helicopter as a function of time. Be sure to explain the meaning of the letters you choose and the units.

b. Express using functional notation the altitude of the helicopter 90 seconds after takeoff, and then calculate that value.

c. Make a graph of altitude versus time covering the first 3 minutes of the flight. Explain how the description of the function is reflected in the shape of the graph.

4. **Swimming records:** The world record time for a certain swimming event was 63.2 seconds in 1950.

Each year thereafter, the world record time decreased by 0.4 second.

a. Use a formula to express the world record time as a function of the time since 1950. Be sure to explain the meaning of the letters you choose and the units.

b. Express using functional notation the world record time in the year 1955, and then calculate that value.

c. Would you expect the formula to be valid indefinitely? Be sure to explain your answer.

5. **A rental:** A rental car agency charges $29.00 per day and 6 cents per mile.

a. Calculate the rental charge if you rent a car for 2 days and drive 100 miles.

b. Use a formula to express the cost of renting a car as a function of the number of days you keep it and the number of miles you drive. Identify the function and each variable you use, and state the units.

c. It is about 250 miles from Dallas to Austin. Use functional notation to express the cost to rent a car in Dallas, drive it to Austin, and return it in Dallas 1 week later. Use the formula from part b to calculate the cost.

6. **Preparing a letter:** You pay your secretary $6.25 per hour. A stamped envelope costs 38 cents, and paper costs 3 cents per page.

a. How much does it cost to prepare and mail a 3-page letter if your secretary spends 2 hours on typing and corrections?

b. Use a formula to express the cost of preparing and mailing a letter as a function of the number of pages in the letter and the time it takes your secretary to type it. Identify the function and each of the variables you use, and state the units.

c. Use the function you made in part b to find the cost of preparing and mailing a 2-page letter that it takes your secretary 25 minutes to type. (*Note:* 25 minutes is $\frac{25}{60}$ hour.)

7. **Preparing a letter, continued:** *This is a continuation of Exercise 6.* You pay your secretary $6.25 per hour. A stamped envelope costs 38 cents, and regu-

lar stationery costs 3 cents per page, but fancy letterhead stationery costs 16 cents per page. Assume that a letter requires fancy letterhead stationery for the first page but that regular paper will suffice for the rest of the letter.

a. How much does the stationery alone cost for a 3-page letter?

b. How much does it cost to prepare and mail a 3-page letter if your secretary spends 2 hours on typing and corrections?

c. Use a formula to express the cost of the stationery alone for a letter as a function of the number of pages in the letter. Identify the function and each of the variables you use, and state the units.

d. Use a formula to express the cost of preparing and mailing a letter as a function of the number of pages in the letter and the time it takes your secretary to type it. Identify the function and each of the variables you use, and state the units.

e. Use the function you made in part b to find the cost of preparing and mailing a 2-page letter that it takes your secretary 25 minutes to type.

8. **A car that gets m miles per gallon:** The cost of operating a car depends on the gas mileage m that your car gets, the cost g per gallon of gasoline, and the distance d that you drive.

a. How much does it cost to drive 100 miles if your car gets 25 miles per gallon and gasoline costs 99 cents per gallon?

b. Find a formula that gives the cost C as a function of m, g, and d. Be sure to state the units of each variable.

c. Use functional notation to show the cost of driving a car that gets 28 miles per gallon a distance of 138 miles if gasoline costs $1.17 per gallon. Use the formula from part b to calculate the cost.

9. **Stock turnover rate:** In a retail store the stock turnover rate of an item is the number of times that the average inventory of the item needs to be replaced as a result of sales in a given time period. It is an important measure of sales demand and merchandising efficiency. Suppose a retail clothing store maintains an average inventory of 50 shirts of a particular brand.

a. Suppose that the clothing store sells 350 shirts of that brand each year. How many orders of 50 shirts will be needed to replace the items sold?

b. What is the annual stock turnover rate for that brand of shirt if the store sells 350 shirts each year?

c. What would be the annual stock turnover rate if 500 shirts were sold?

d. Write a formula expressing the annual stock turnover rate as a function of the number of shirts sold. Identify the function and the variable, and state the units.

10. **Stock turnover rate, continued:** *This is a continuation of Exercise 9.* As we saw earlier, the stock turnover rate of an item is the number of times that the average inventory of the item needs to be replaced as a result of sales in a given time period. Suppose that a hardware store sells 80 shovels each year.

a. Suppose that the hardware store maintains an average inventory of 5 shovels. What is the annual stock turnover rate for the shovels? How is this related to the yearly number of orders to the wholesaler needed to restock inventory?

b. What would be the annual stock turnover rate if the store maintained an average inventory of 20 shovels?

c. Write a formula expressing the annual stock turnover rate as a function of the average inventory of shovels. Identify the function and the variable, and state the units.

11. **Total cost:** The *total cost C* for a manufacturer during a given time period is a function of the number N of items produced during that period. To determine a formula for the total cost, we need to know the manufacturer's *fixed costs* (covering things such as plant maintenance and insurance), as well as the cost for each unit produced, which is called the *variable cost*. To find the total cost, we multiply the variable cost by the number of items produced during that period and then add the fixed costs.

   Suppose that a manufacturer of widgets has fixed costs of $9000 per month and that the variable cost is $15 per widget (so it costs $15 to produce 1 widget). →

a. Use a formula to express the total cost C of this manufacturer in a month as a function of the number of widgets produced in a month. Be sure to state the units you use.

b. Express using functional notation the total cost if there are 250 widgets produced in a month, and then calculate that value.

12. **Total revenue and profit:** *This is a continuation of Exercise 11.* The *total revenue* R for a manufacturer during a given time period is a function of the number N of items produced during that period. To determine a formula for the total revenue, we need to know the selling price per unit of the item. To find the total revenue, we multiply this selling price by the number of items produced.

The *profit* P for a manufacturer is the total revenue minus the total cost. If this number is positive, then the manufacturer *turns a profit*, whereas if this number is negative, then the manufacturer *has a loss*. If the profit is zero, then the manufacturer is at a *break-even point*.

Suppose the manufacturer of widgets in Exercise 11 sells the widgets for $25 each.

a. Use a formula to express this manufacturer's total revenue R in a month as a function of the number of widgets produced in a month. Be sure to state the units you use.

b. Use a formula to express the profit P of this manufacturer as a function of the number of widgets produced in a month. Be sure to state the units you use.

c. Express using functional notation the profit of this manufacturer if there are 250 widgets produced in a month, and then calculate that value.

d. At the production level of 250 widgets per month, does the manufacturer turn a profit or have a loss? What about the production level of 1000 widgets per month?

13. **More on revenue:** *This is a continuation of Exercises 11 and 12.* In general, the highest price p per unit of an item at which a manufacturer can sell N items is not constant but is rather a function of N. The total revenue R is still the product of p and N, but the formula for R is more complicated when p depends on N.

Suppose the manufacturer of widgets in Exercises 11 and 12 no longer sells widgets for $25 each. Rather, the manufacturer has developed the following table showing the highest price p, in dollars, of a widget at which N widgets can be sold.

N = Number of widgets sold	p = Price
100	49
200	48
300	47
400	46
500	45

a. Verify that the formula $p = 50 - 0.01N$, where p is the price in dollars, gives the same values as those in the table.

b. Use the formula from part a and the fact that R is the product of p and N to find a formula expressing the total revenue R as a function of N for this widget manufacturer.

c. Express using functional notation the total revenue of this manufacturer if there are 450 widgets produced in a month, and then calculate that value.

14. **More on profit:** *This is a continuation of Exercises 11, 12, and 13.* In this exercise we use the formula for the total cost of the widget manufacturer found in Exercise 11 and the formula for the total revenue found in Exercise 13.

a. Use a formula to express the profit P of this manufacturer as a function of N.

b. Consider the three production levels: $N = 200$, $N = 700$, and $N = 1200$. For each of these, determine whether the manufacturer has a loss, turns a profit, or is at a break-even point.

15. **Renting motel rooms:** You own a motel with 30 rooms and have a pricing structure that encourages rentals of rooms in groups. One room rents for $85.00, two for $83.00 each, and in general the

group rate per room is found by taking $2 off the base of $85 for each extra room rented.

a. How much money do you charge per room if a group rents 3 rooms? What is the total amount of money you take in?

b. Use a formula to give the rate you charge for each room if you rent n rooms to an organization.

c. Find a formula for a function $R = R(n)$ that gives the total revenue from renting n rooms to a convention host.

d. Use functional notation to show the total revenue from renting a block of 9 rooms to a group. Calculate the value.

16. **A cattle pen:** A rancher wants to use a fence as an enclosure for a rectangular cattle pen with area 400 square feet.

a. Suppose he decides to make one side of the pen 40 feet long. Draw a picture and label the length of each side of the pen. What is the length of each side? What is the total amount of fence needed?

b. What would be the total amount of fence needed if the pen were a square?

c. Can two rectangles with the same area have different perimeters?

d. Find a formula for a function $F = F(l)$ that gives total amount of fence, in feet, required in terms of the length l, in feet, of one of its sides. (*Hint:* First draw a picture of the pen and label one side l. Next figure out the lengths of the other sides in terms of l.)

17. **Catering a dinner:** You are having a dinner catered. You pay a rental fee of $150 for the dining hall, and you pay the caterer $10 for each person who attends the dinner.

a. Suppose you just want to break even.

  i. How much should you charge *per ticket* if you expect 50 people to attend?

  ii. Use a formula to express the amount you should charge per ticket as a function of the number of people attending. Be sure to explain the meaning of the letters you choose and the units.

  iii. You expect 65 people to attend the dinner. Use your answer to part ii to express in functional notation the amount you should charge per ticket, and then calculate that amount.

b. Suppose now that you want to make a profit of $100 from the dinner. Use a formula to express the amount you should charge per ticket as a function of the number of people attending. Again, be sure to explain the meaning of the letters you choose and the units.

18. **A car:** The distance d, in miles, that a car travels on a 3-hour trip is proportional to its speed s (which we assume remains the same throughout the trip), in miles per hour.

a. What is the constant of proportionality in this case?

b. Write a formula that expresses d as a function of s.

19. **Production rate:** The total number t of items that a manufacturing company can produce is directly proportional to the number n of employees.

a. Choose a letter to denote the constant of proportionality, and write an equation that shows the proportionality relation.

b. What in practical terms does the constant of proportionality represent in this case?

20. **Density:** The total weight of a rock depends on its size and is proportional to the *density*. In this context, density is the weight per cubic inch. Let w denote the weight of the rock in pounds, s the size of the rock in cubic inches, and d the density of the rock in pounds per cubic inch.

a. What is the total weight of a 3-cubic-inch rock that weighs 2 pounds per cubic inch?

b. Write an equation that shows the proportionality relation. What is the constant of proportionality?

c. Use the equation you found in part b to find the total weight of a 14-cubic-inch rock with density 0.3 pound per cubic inch.

21. **Head and pressure:** Determining the water pressure at a given location employs the concept of the *head*, which is the vertical distance, in feet, from the surface of a source body of water to the location.

→

The pressure exerted by water is proportional to the head. If we measure head in feet and pressure in pounds per square inch, then the constant of proportionality is the weight of a column of water that is 1 foot high and 1 inch square at the base. That much water weighs 0.434 pound. (See Figure 1.52.)

a. Write an equation that expresses the proportionality relationship between pressure p and head h.

b. For a pumper truck pumping water to a fire, the *back pressure* is the additional pressure on the pump caused by the height of the nozzle. Consider a pumper at street level pumping water through a hose to firefighters on the top of the eighth floor of a building. If each floor is 12 feet high, what is the head of water at the mouth of the nozzle? What is the back pressure on the pumper? (Another way of thinking of back pressure is as the minimum pressure the pumper must produce in order to make water flow out the end of the nozzle.)

c. Head (and therefore back pressure) depends only on the height of the nozzle above the pumper. It is affected neither by the volume of the water nor by horizontal distance. A pumper in a remote location is pumping water to firefighters on the far slope of a hill. At its peak, the hill is 185 feet higher than the pumper. The hose goes over the hill and then down the hill to a point 40 feet below the peak. Find the head and the back pressure on the pumper.

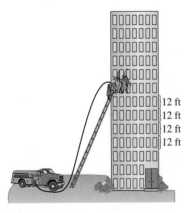

FIGURE 1.52

22. **Head and aquifers:** *This is a continuation of Exercise 21.* In underground water supplies such as aquifers, the water normally permeates some other medium such as sand or gravel. The head for such water is determined by first drilling a well down to the water source. When the well reaches the aquifer, pressure causes the water to rise in the well. The head is the height to which the water rises. In this setting, we get the pressure using

$$\text{Pressure} = \text{Density} \times 9.8 \times \text{Head}.$$

Here density is in kilograms per cubic meter, head is in meters, and pressure is in newtons per square meter. (One newton is about a quarter of a pound.) A sandy layer of soil has been contaminated with a dangerous fluid at a density of 1050 kilograms per cubic meter. Below the sand there is a rock layer that contains water at a density of 990 kilograms per cubic meter. This aquifer feeds a city water supply. Test wells show that the head in the sand is 4.3 meters, whereas the head in the rock is 4.4 meters. A liquid will flow from higher pressure to lower pressure. Is there a danger that the city water supply will be polluted by the material in the sand layer?

23. **Darcy's law:** The French hydrologist Henri Darcy discovered that the velocity V of underground water is proportional to the magnitude of the slope S of the water table (see Figure 1.53). The constant of proportionality in this case is the *permeability* of the medium through which the water is flowing. This proportionality relationship is known as *Darcy's law*, and it is important in modern hydrology.

a. Using K as the constant of proportionality, express Darcy's law as an equation.

b. Sandstone has a permeability of about 0.041 meter per day. If an underground aquifer is seeping through sandstone, and if the water table drops 0.03 vertical meter for each horizontal meter, what is the velocity of the water flow? Be sure to use appropriate units for velocity and keep all digits.

c. Sand has a permeability of about 41 meters per day. If the aquifer from part a were flowing through sand, what would be its velocity?

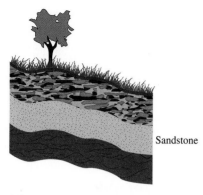

FIGURE 1.53

24. **Hubble's constant:** Astronomers believe that the universe is expanding and that stellar objects are moving away from us at a radial velocity V proportional to the distance D from Earth to the object.

a. Write V as a function of D using H as the constant of proportionality.

b. The equation in part a was first discovered by Edwin Hubble in 1929 and is known as *Hubble's law*. The constant of proportionality H is known as *Hubble's constant*. The currently accepted value of Hubble's constant is 70 kilometers per second per megaparsec. (One megaparsec is about 3.085×10^{19} kilometers.) With these units for H, the distance D is measured in megaparsecs, and the velocity V is measured in kilometers per second. The galaxy **G2237 + 305** is about 122.7 megaparsecs from Earth. How fast is **G2237 + 305** receding from Earth?

c. One important feature of Hubble's constant is that scientists use it to estimate the age of the universe. The approximate relation is

$$y = \frac{10^{12}}{H},$$

where y is time in years. Hubble's constant is extremely difficult to measure, and Edwin Hubble's best estimate in 1929 was about 530 kilometers per second per megaparsec. What is the approximate age of the universe when this value of H is used?

d. The calculation in part c would give scientists some concern since Earth is thought to be about

4.6 billion years old. What estimate of the age of the universe does the more modern value of 70 kilometers per second per megaparsec give?

25. **The $3x + 1$ problem:** Here is a mathematical function $f(n)$ that applies only to whole numbers n. If a number is even, divide it by 2. If it is odd, triple it and add 1. For example, 16 is even, so we divide by 2: $f(16) = \frac{16}{2} = 8$. On the other hand, 15 is odd, so we triple it and add 1: $f(15) = 3 \times 15 + 1 = 46$.

a. Apply the function f repeatedly beginning with $n = 1$. That is, calculate $f(1)$, f(the answer from the first part), f(the answer from the second part), and so on. What pattern do you see?

b. Apply the function f repeatedly beginning with $n = 5$. How many steps does it take to get to 1?

c. Apply the function f repeatedly beginning with $n = 7$. How many steps does it take to get to 1?

d. Try several other numbers of your own choosing. Does the process always take you back to 1? (*Note:* We can't be sure what your answer will be here. Every number that anyone has tried so far leads eventually back to 1, and it is conjectured that this happens no matter what number you start with. This is known to mathematicians as the $3x + 1$ *conjecture*, and it is, as of the writing of this book, an unsolved problem. If you can find a starting number that does not lead back to 1, or if you can somehow show that the path *always* leads back to 1, you will have solved a problem that has eluded mathematicians for a number of years. Good hunting!)

26. **Research project:** For this project you are to find and describe a function that is commonly used. Find a patient person whose job is interesting to you. Ask that person what types of calculations he or she makes. These calculations could range from how many bricks to order for building a wall to lifetime wages lost for a wrongful-injury settlement to how much insulin to inject. Be creative and persistent—don't settle for "I look it up in a table." Write a description, in words, of the function and how it is calculated. Then write a formula for the function, carefully identifying variables and units.

Summary

The idea of a *function* is as old as mathematics itself, and it is central to mathematics and applications. It is certainly a key topic in this text, where it occurs in one form or another throughout. A function is nothing more than a clear description of how one thing depends on another (or on several other things), and functions are presented in various ways. The most common ways of presenting functions are the section topics of this chapter.

Functions Given by Formulas

This is perhaps the way in which most people think of a function, and it is an important one. An example of a function given by a formula is

$$M = 7h,$$

where h is the number of hours an employee may work, and M is the money earned, in dollars. The formula simply says that one can calculate the money earned by multiplying the number of hours worked by 7. In other words, the formula describes the pay of an employee who earns $7.00 an hour. Sometimes formulas for functions are quite complicated. The calculator makes functions given by formulas easy to deal with in spite of their apparent complexity.

In applications of mathematics, functions are often representations of real phenomena or events. Thus we say that they are *models*. Obtaining a function or functions to act as a model is commonly the key to understanding physical, natural, and social science phenomena. This applies to business and many other areas as well.

Functions Given by Tables

One of the most common ways in which functions are encountered in daily life is in terms of *tables of values*. Such tables can be found everywhere, presenting census data, payment schedules, college enrollment statistics, lists of species occupying a certain region, and a myriad of other familiar types of data. An example is the U.S. census data that appears in the following table.

d = Year	1950	1960	1970	1980	1990
N = Population in millions	151.87	179.98	203.98	227.23	249.40

This table gives U.S. population $N = N(d)$ as a function of the date. Tables are almost always incomplete; that is, some information is left out of the table. Here, for example, the population in 1954 is not reported. A common way of estimating function values that are not given in a table is by using the *average rate of change*. For example, from 1950 to 1960, the U.S. population grew from 151.87 million to 179.98 million. That is an increase of 28.11 million over a 10-year period. Thus from 1950 to 1960, the U.S. population grew at a rate of approximately

$$\frac{28.11}{10} = 2.811 \text{ million per year}.$$

This is the average yearly rate of change during the 1950s. It is reasonable to estimate the population in 1954 by

$$N(1954) = 151.87 + 4 \times 2.811 = 163.114 \text{ million},$$

or about 163.11 million.

Sometimes, a function has a *limiting value* that may be estimated from the tabular form of a function. For example, the following table shows the amount of yeast present in an enclosed area t hours after observations began.

Time t	0	5	10	15	20	25	30
Amount of yeast N	10	119	513	651	662	664	665

Since an enclosed area is being observed, we expect that there is a limit to the amount of yeast that will ever be present. Looking at the table, it is reasonable to expect that this limiting value is about 665.

Functions Given by Graphs

Another way in which functions are commonly presented is with graphs. Generally, it is more difficult to get exact function values from a graph than from a formula or table, but a graph has the advantage of clearly showing certain overall features of a function. It clearly shows, for example, when a function is increasing or decreasing, when it reaches maxima or minima, any concavity of the graph, and (often) limiting values of the function. These function properties are displayed by the graph according to the following table.

Function Property	Graph Display
Increasing	Rising graph
Decreasing	Falling graph
Maximum	Graph reaches a peak
Minimum	Graph reaches a valley
Concave up	Graph is bent upward (holds water)
Concave down	Graph is bent downward (spills water)
Has a limiting value	Graph levels out on the right-hand side

The concavity of a graph gives important information about the rate of change of the function. Increasing graphs that are concave up represent functions that increase at an increasing rate, whereas increasing graphs that are concave down represent functions that increase at a decreasing rate. Similarly, decreasing graphs that are concave up represent functions that decrease at a decreasing rate, whereas decreasing graphs that are concave down represent functions that decrease at an increasing rate.

Functions Given by Words

Very often a function is presented with a verbal description, and the key to understanding it may well be to translate this verbal description into a formula, table, or graph. For example, suppose that a company invests $78,000 in the design and development of a more efficient computer hard drive and that, for each drive sold, the firm makes a profit of $98. This can be thought of as a verbal description of the net profit $N = N(d)$ as a function of the number d of drives sold. Furthermore, it is not difficult to translate this verbal description into a formula:

$$\text{Net profit} = \text{Profit from sales} - \text{Initial investment}$$

$$N = 98d - 78{,}000 \text{ dollars}.$$

One way in which verbal descriptions of functions are commonly given is in terms of *proportion*. This simply indicates that one thing is a multiple of another. For example, money earned by a wage employee is proportional to the number of hours worked. In terms of a formula, this proportionality statement means

$$\text{Money earned} = \text{Hourly wage} \times \text{Hours worked}.$$

In this context, the hourly wage is known as the *proportionality constant*.

2

Graphical and Tabular Analysis

Each type of function presentation that we studied in Chapter 1 has advantages and disadvantages, and one of the most useful methods of analysis is to look at functions in more than one way. Tables and graphs often show information that is difficult to obtain directly from formulas. The graphing calculator makes it easy to go from a formula to a table or graph and hence becomes a tool for solving significant problems.

2.1   *Tables and Trends*

The advantage of functions given by formulas is that they allow for the calculation of any function value. Tables of values always leave gaps, but they may be more helpful than formulas for seeing trends, predicting future values, or discerning other interesting information about the function.

Getting Tables from Formulas

Let's look at an example to illustrate this. Proper management of wildlife depends on the ability of ecologists to monitor and predict population growth. In many situations, it is reasonable to expect animal populations to exhibit *logistic growth*. A special formula that is studied extensively by ecologists describes this type of growth.

The circumstances surrounding the *George Deer Reserve* in Michigan have made it particularly easy for ecologists to monitor accurately the growth of the deer population on the reserve and to develop a logistic growth formula for the number $N = N(t)$ of deer expected to be present after t years:[1]

$$N = \frac{6.21}{0.035 + 0.45^t} \text{ deer}.$$

When a breeding group of animals is introduced into a limited area, one expects that it will over time grow to the largest size that the environment can support. Wildlife managers refer to this as the *environmental carrying capacity*. Let's find the carrying

---

[1] Dale R. McCullough, *The George Reserve Deer Herd* (Ann Arbor: University of Michigan Press, 1979).

91

capacity for deer of the George Reserve. That is, we want to know the deer population after a long period of time. This is not an easy question to answer by looking at the formula, but if we use the formula to make a table of values, then the trend will become apparent. In the accompanying table, we have calculated $N(0)$, $N(5)$, $N(10)$, ..., $N(30)$. (These values represent the initial population, the population after 5 years, the population after 10 years, etc.)

Scanning down the right-hand column of the table, we see the growth of the deer population with time. From the first row of the table, we see that there were initially 6 deer on the reserve. During the first 10 years, the population increases rapidly, but the rate of increase slows down dramatically after that. It appears that after about 20 years, the population levels out at approximately 177 deer. (We have rounded to the nearest whole number since we don't expect to see parts of deer on the reserve.) This is the carrying capacity of the reserve.

The key idea here is that, many times, needed information is difficult to obtain directly from a formula. Supplementing the formula with a table of values can provide deeper insight into what is happening.

Tables of values can be made by calculating each wanted function value one at a time, but many calculators have a built-in feature that generates such tables automatically. Making tables of values using a calculator is a skill that will be needed often in what follows, so you are strongly encouraged to consult the *Technology Guide* for instructions on how to do this. To become familiar with the procedure, you should work through the practice problems that are presented there.

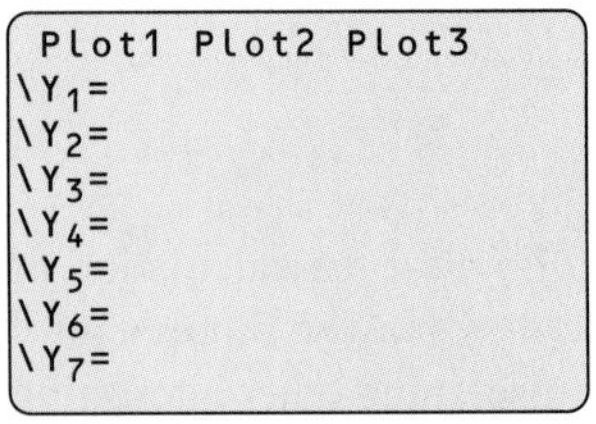

When you make a table of values with a calculator, there are three key bits of information that you must input. You must tell your calculator which function you want to use; you must decide on a place to begin the table, the *table starting value;* and you must decide on the periods, the *table increment value*, when you want to see additional data. In making the table above for deer population, we used the function $6.21/(0.035 + 0.45^t)$ with a table starting value of $t = 0$, and we viewed the data in 5-year periods. That is, we used a table increment value of 5. In what follows, we will refer to these latter two items as the *table setup*. Thus, if we wanted to instruct you to make the table above in exactly the same way we did, we would say "Enter [2.1] the function $6.21/(0.035 + 0.45^t)$, and for table setup [2.2] use a starting value of $t = 0$ and a table increment value of 5." The function entry screen on a graphing calculator will typically appear as in Figure 2.1 and the completed function entry as in Figure 2.2, but displays will vary from calculator to calculator. The table setup will typically appear as in Figure 2.3, and the completed table of values will typically appear as in Figure 2.4.

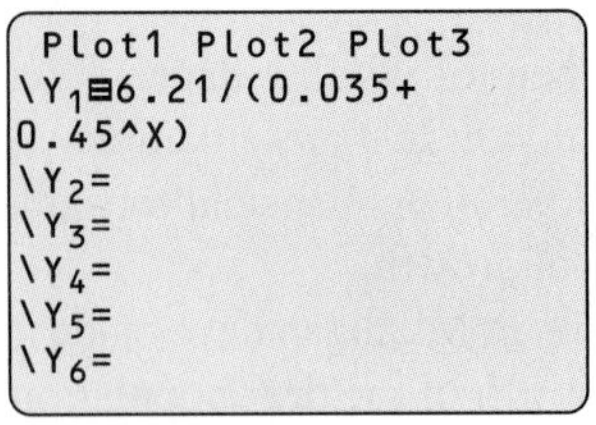

We are working with a function whose name is N and with a variable t, but as we see in Figures 2.2 and 2.4, the calculator has chosen its own name, Y_1, for the function N and X for the variable t. Your calculator may use other letters to represent the function

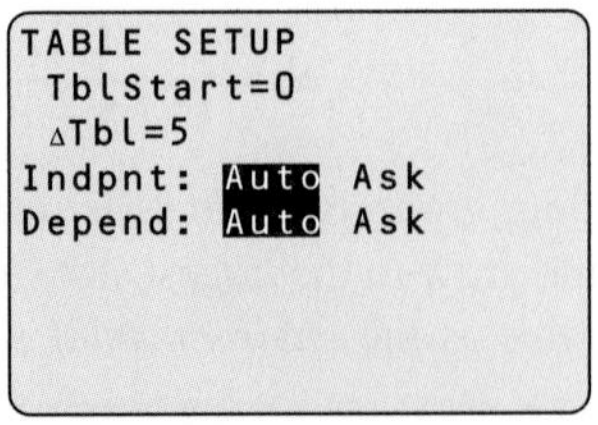

Year t	$N(t)$ Population in Year t
0	6
5	116.18
10	175.72
15	177.4
20	177.43
25	177.43
30	177.43

FIGURE 2.1 A typical function entry screen

FIGURE 2.2 The properly entered function

FIGURE 2.3 A typical TABLE SETUP menu

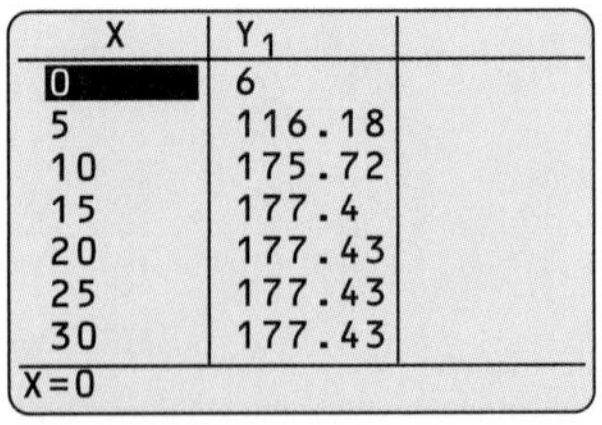

FIGURE 2.4 A table of values for the deer population

and the variable, but whatever letters your calculator uses, it is important to keep track of the proper associations. It is good practice to write down the correspondence, and we will always do that in the examples we present:

$$Y_1 = N, \text{ population} \tag{2.1}$$

$$X = t, \text{ time in years}. \tag{2.2}$$

Now if we want to use the table in Figure 2.4 to find the value of N when t is 20, we note from Equation (2.2) that we should look in the X column to find 20 and from Equation (2.1) that the corresponding Y_1 value 177.43 is the function value for N. That is, $N(20) = 177.43$.

EXAMPLE 2.1   A Skydiver

A falling object is pulled downward by gravity, but its fall is retarded by air resistance, which under appropriate conditions is directly proportional to velocity. When a skydiver jumps from an airplane, her downward velocity $v = v(t)$ before she opens her parachute is given by

$$v = 176(1 - 0.834^t) \text{ feet per second},$$

where t is the number of seconds that have elapsed since she jumped from the airplane.

1. Express the velocity of the skydiver 2 seconds into the fall using functional notation, and calculate its value.

2. Describe how the velocity of the skydiver changes with time. Include in your description the average rate of increase in velocity during the first 5 seconds and the average rate of increase in velocity during the next 5 seconds.

3. What is the *terminal velocity*? That is, what is the maximum speed the skydiver can attain?

4. How long does it take the skydiver to reach 99% of terminal velocity?

Solution to Part 1:   In functional notation, the downward velocity 2 seconds into the fall is $v(2)$. To make the calculation, we put 2 in for t:

$$v(2) = 176(1 - 0.834^2) = 53.58 \text{ feet per second}.$$

Solution to Part 2:   We want to see how the velocity increases with time. This is difficult to see from the formula, but a table of values showing velocity in 5-second intervals will give us the information we need. Thus we want to enter ⟨2.3⟩ the function, and we want a table setup ⟨2.4⟩ with a starting value of 0 and an increment of 5. The correctly entered function is shown in Figure 2.5, and the correctly configured table setup menu is shown in Figure 2.6.

Once again, the calculator is using its own choices of letters, so we record the appropriate correspondences:

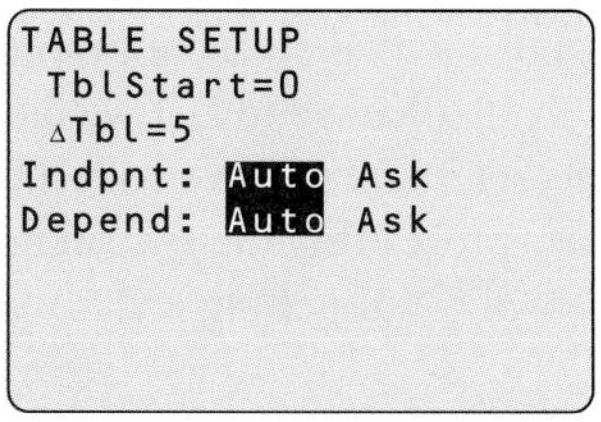

$$Y_1 = v, \text{ velocity}$$

$$X = t, \text{ time in seconds}.$$

```
Plot1 Plot2 Plot3
\Y₁▪176(1-0.834^
X)
\Y₂=
\Y₃=
\Y₄=
\Y₅=
\Y₆=
```

FIGURE 2.5 Entering the function for velocity

```
TABLE SETUP
 TblStart=0
 ΔTbl=5
Indpnt: Auto Ask
Depend: Auto Ask
```

FIGURE 2.6 Setting up the table

X	Y_1
0	0
5	104.99
10	147.35
15	164.44
20	171.34
25	174.12
30	175.24

X=0

FIGURE 2.7 A table of values for velocity

When we view 2.5 the table shown in Figure 2.7, we can read down the right-hand column to see that the velocity increases rapidly to begin with but seems to be leveling off near 30 seconds into the fall. It appears that the downward pull of gravity makes the skydiver accelerate rapidly at first, but air resistance seems to have a greater effect at high velocities. This is a consequence of the fact that air resistance is in this case directly proportional to velocity.

The table shows that the velocity increased from 0 to 104.99 feet per second during the first 5 seconds of the fall. Thus, during this period, velocity increased at an average rate of $\frac{104.99}{5} = 21$ feet per second per second. During the next 5 seconds of the fall, the velocity increased by $147.35 - 104.99 = 42.36$ feet per second. That gives an average increase in velocity of $\frac{42.36}{5} = 8.47$ feet per second per second. As we should have expected, the rate of increase in velocity is much less during the second 5-second period than in the first 5 seconds. What would this say about the concavity of the graph of velocity against time?

Solution to Part 3: To get the terminal velocity, we want to know what the velocity of the skydiver would be if she continued in free fall for a long time without opening her parachute. That is, we want to look at the table for large values of t. Your graphing calculator may have a feature that lets you extend 2.6 the table without going back to the table setup menu. We have shown the table for $t = 35$ seconds to $t = 65$ seconds in Figure 2.8. This table shows clearly that velocity levels out at 176 feet per second. (Look further down the table for more evidence of this.) This is the terminal velocity, where the downward pull of gravity matches air resistance.

X	Y_1
35	175.69
40	175.88
45	175.95
50	175.98
55	175.99
60	176
65	176

X=65

FIGURE 2.8 Extending the table

Solution to Part 4: Now 99% of terminal velocity is $0.99 \times 176 = 174.24$ feet per second. Consulting the table in Figure 2.7, we see that this velocity is reached about 25 seconds into the fall. You may wish to improve the accuracy of this answer by changing the table setup so that the increment is 0.5.

We should also note that questions about terminal velocity in particular—and about limiting values of functions in general—can lead to unexpected difficulties that require more advanced mathematical analysis than is appropriate here. See Exercise 24 at the end of this section.

It is worth emphasizing the significance of what we have done here. The table of values enabled us to make an in-depth analysis of what happens in the free-fall period of a skydiver's fall. None of the conclusions we drew in parts 2 and 3 are apparent from the formula, but they are easily discernible from the table of values. The technology we are using gives us the power to attack and resolve real problems.

Optimizing with Tables of Values

We can also use tables of values to find maximum and minimum values for functions. To illustrate this, let's suppose we roll several dice hoping to get exactly 3 sixes, not more or less. The probability that this will occur depends on how many dice we roll. If we use only 4 dice, then it seems unlikely that we will get as many as 3 sixes. If we roll 100 dice, then we will expect to get more than 3 sixes. Elementary probability theory can be used to show that if we roll N dice, then the probability $p = p(N)$ of getting exactly 3 sixes is given by the formula

$$p = \frac{N(N-1)(N-2)}{750} \times \left(\frac{5}{6}\right)^N.$$

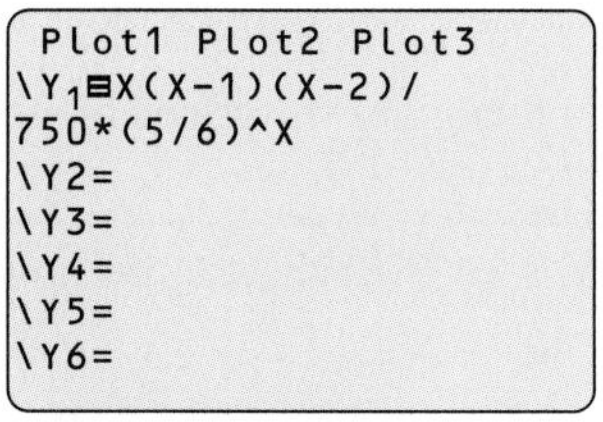

```
Plot1 Plot2 Plot3
\Y₁⬛X(X-1)(X-2)/
750*(5/6)^X
\Y2=
\Y3=
\Y4=
\Y5=
\Y6=
```

FIGURE 2.9 Entering a probability function

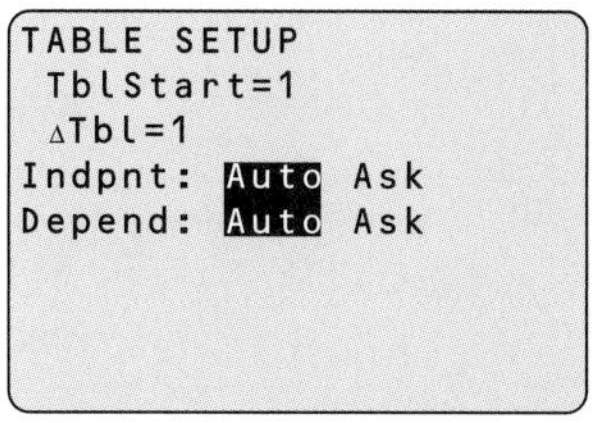

```
TABLE SETUP
 TblStart=1
 ⊿Tbl=1
Indpnt: Auto Ask
Depend: Auto Ask
```

FIGURE 2.10 Setting up the table

Thus if we want to know the probability of getting exactly 3 sixes when we roll 10 dice, we put $N = 10$ into the formula for p:

$$p(10) = \frac{10 \times 9 \times 8}{750} \times \left(\frac{5}{6}\right)^{10} = 0.15505.$$

If we round this to two decimal places, we get $p = 0.16$. This means that if you roll 10 dice, you can expect to get exactly 3 sixes about 16 times in 100 rolls.

How many dice should we roll so that we have the best possible chance of getting exactly 3 sixes? To answer this, we want to make a table of values showing the probability of getting exactly 3 sixes for various values of N. First we enter 2.7 the function as shown in Figure 2.9. We record the letter correspondences:

$$Y_1 = p, \text{ probability of exactly 3 sixes}$$

$$X = N, \text{ number of dice}.$$

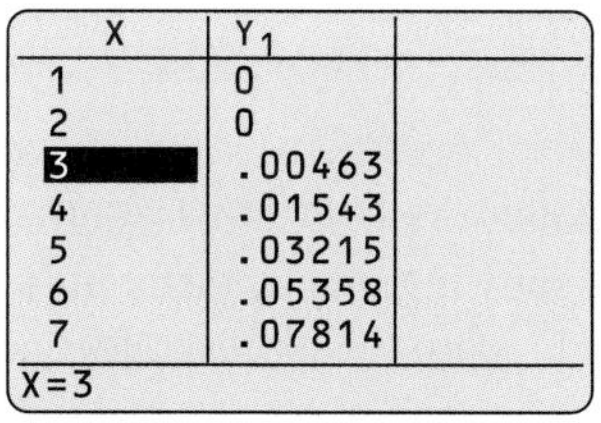

X	Y₁	
1	0	
2	0	
3	.00463	
4	.01543	
5	.03215	
6	.05358	
7	.07814	

X=3

FIGURE 2.11 Probability of 3 sixes when 1 through 7 dice are used

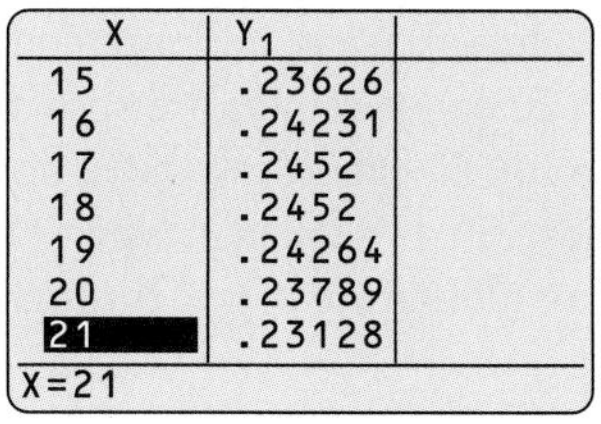

X	Y₁	
15	.23626	
16	.24231	
17	.2452	
18	.2452	
19	.24264	
20	.23789	
21	.23128	

X=21

FIGURE 2.12 Probability of 3 sixes when 15 through 21 dice are used

Next we set up the table using a starting value 2.8 of 1 and an increment of 1. The correctly configured table setup menu is shown in Figure 2.10. When we view the table of values, we get the display in Figure 2.11. We see that if we use 1 or 2 dice, the probability is 0, indicating (as we expected) that it is impossible to get 3 sixes by rolling fewer than 3 dice. If we roll 3 dice, we see that the probability of getting exactly 3 sixes is 0.00463, or 0.0046 rounded to four decimal places. This means that we would expect to get 3 sixes only 46 times out of 10,000 rolls. As the number of dice increases from 3 up through 7, the probability of getting exactly 3 sixes gets larger. In Figure 2.12 we have extended 2.9 the table so that we can see what happens when we roll between 15 and 21 dice. This table shows us that the probability of 3 sixes increases as the number of dice increases up to 17 but decreases for more than 18 dice. The decrease occurs because if we roll too many dice, we would expect to get more than 3 sixes. Thus the probability is at its largest, 0.2452, if we use either 17 or 18 dice. You should look both forward and backward in the table to ensure that there are not larger values outside the ranges shown in Figures 2.11 and 2.12.

EXAMPLE 2.2 Renting Canoes

A small business has 20 canoes that it rents for float trips down the Illinois River. The pricing structure offers a discount for group rentals. One canoe rents for $35, two rent for $34 each, three rent for $33 each, and in general the group rate per canoe is found by taking $1 off the base of $35 for each extra canoe rented.

1. How much money is taken in if 4 canoes are rented to a group?

2. Write a formula that gives the price charged for each canoe if n canoes are rented to a group.

3. Find a formula $R = R(n)$ that shows how much money is taken in from renting n canoes to a group.

4. How large a rental to a single group will bring the most income?

Solution to Part 1: One canoe rents for \$35, two rent for \$34 each, and three rent for \$33 each. We expect that four rent for \$32 each. If four are rented, then $4 \times 32 = 128$ dollars will be collected.

Solution to Part 2: We want a formula expressing the amount charged per canoe in terms of the number of canoes rented to a group. If the business rents 1 canoe, the charge is $35 = 36 - 1$ dollars per canoe. If the group rental is 2 canoes, the charge is $34 = 36 - 2$ dollars per canoe; for 3, the charge is $33 = 36 - 3$ dollars per canoe. If n canoes are rented to a group, the charge is $36 - n$ dollars per canoe.

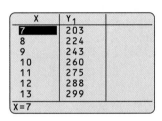

X	Y$_1$	
7	203	
8	224	
9	243	
10	260	
11	275	
12	288	
13	299	

X=7

FIGURE 2.13 Renting to groups that number 7 to 13

Solution to Part 3: We know that n canoes are rented at the rate we found in part 2. Thus if the business rents n canoes, then it charges $36 - n$ dollars for each canoe. That means $n(36 - n)$ dollars are taken in:

$$R(n) = n(36 - n).$$

Solution to Part 4: We want a table that shows how much money is taken in from renting canoes in groups. Thus we need a table of values for R showing money taken in for $n = 1$ through $n = 20$. We enter $\boxed{2.10}$ the function and record the correspondence for function and variable names:

$$Y_1 = R, \text{ revenue}$$

$$X = n, \text{ canoes rented}.$$

X	Y$_1$	
14	308	
15	315	
16	320	
17	323	
18	324	
19	323	
20	320	

X=20

FIGURE 2.14 Renting to groups that number 14 to 20

Next we set up the table $\boxed{2.11}$ using a table starting value of 1 and an increment value of 1. The tables for 7 through 13 and 14 through 20 canoe rentals are shown in Figure 2.13 and Figure 2.14. Since the business has only 20 canoes, we need not look further down the table. It shows that the most money, \$324, is taken in when 18 canoes are rented to a group. You should view the entire table from $n = 1$ through $n = 20$ to make sure that nothing has been overlooked. ▨ ▨ ▨

ANOTHER LOOK

Limits

We have used tables to spot trends or limiting values. More formally, we are looking at a mathematical concept known as *limits*. In Example 2.1, we looked at the velocity of a skydiver t seconds into the fall. The velocity is given by

$$v = 176(1 - 0.834^t).$$

In part 3 of that example, we considered the notion of *terminal velocity*. That is, we asked what would be the velocity of the skydiver after a long period of falling. Formally, this limiting value is denoted by $\lim_{t\to\infty} v$, and it represents the value that v gets close to when t increases toward infinity. In that example, we estimated the limiting velocity to be 176 feet per second. We will see that, in this case, this is more than an estimate and is in fact the exact limiting value.

In many cases the exact limiting value can be calculated directly from a formula. There are two cases that occur often and enable us to calculate many limiting values directly. You will be asked in the Enrichment Exercises to supply numerical evidence for the following two basic limits.

Basic exponential limit: If $a < 1$ is a positive number, then

$$\lim_{t\to\infty} a^t = 0.$$

Basic power limit: If n is a positive number, then

$$\lim_{t\to\infty} \frac{1}{t^n} = 0.$$

Let's see how to use the basic exponential limit to calculate $\lim_{t\to\infty} 176(1 - 0.834^t)$. Since $0.834 < 1$, the basic exponential limit tells us that 0.834^t is near 0 when t is very large. Thus

$$\lim_{t\to\infty} 176(1 - 0.834^t) = 176(1 - 0) = 176.$$

Similar reasoning works for the number of deer on the George Deer Reserve after t years. The number N is given by

$$N = \frac{6.21}{0.035 + 0.45^t}.$$

If we want to know the number of deer that will eventually be on the reserve, we need to calculate $\lim_{t\to\infty} N$. Once again, since 0.45 is less than 1, the basic exponential limit tells us that $\lim_{t\to\infty} 0.45^t = 0$. Thus

$$\lim_{t\to\infty} \frac{6.21}{0.035 + 0.45^t} = \frac{6.21}{0.035 + 0} = 177.43.$$

The basic power limit applies to expressions such as $5 + 4/t^3$:

$$\lim_{t\to\infty} \left(5 + \frac{4}{t^3}\right) = \lim_{t\to\infty} \left(5 + 4\frac{1}{t^3}\right) = 5 + 4 \times 0 = 5.$$

The basic power limit, together with a standard trick, enables us to calculate limits that may at first appear difficult. Let's look, for example, at $\lim_{t\to\infty}(t^2 + t + 1)/(3t^2 + 1)$. For limits of fractions such as this one, it often helps to divide both the top and the bottom of the fraction by the highest power of t that appears. In this case, that is t^2. Note that, in what follows, we multiply the fraction by

$$\frac{\frac{1}{t^2}}{\frac{1}{t^2}} = 1.$$

Thus we are changing only the appearance of the fraction, not its value. We have

$$\lim_{t\to\infty}\frac{t^2+t+1}{3t^2+1}=\lim_{t\to\infty}\frac{t^2+t+1}{3t^2+1}\times\frac{\dfrac{1}{t^2}}{\dfrac{1}{t^2}}$$

$$=\lim_{t\to\infty}\frac{1+\dfrac{1}{t}+\dfrac{1}{t^2}}{3+\dfrac{1}{t^2}}.$$

Now we apply the basic power limit to the terms $1/t$ and $1/t^2$:

$$\lim_{t\to\infty}\frac{1+\dfrac{1}{t}+\dfrac{1}{t^2}}{3+\dfrac{1}{t^2}}=\frac{1+0+0}{3+0}=\frac{1}{3}.$$

Enrichment Exercises

E-1. Verifying the basic exponential limit: Choose several values of a that are less than 1, and for each choice make a table of values to give evidence that $\lim_{t\to\infty} a^t = 0$.

E-2. Verifying the basic power limit: Choose several values of n larger than 0, and for each choice make a table of values to give evidence that $\lim_{t\to\infty} 1/t^n = 0$.

E-3. Calculating with the basic exponential limit: Use the basic exponential limit to calculate the following:

a. $\lim_{t\to\infty} 15(1-3\times 0.5^t)$

b. $\lim_{t\to\infty}\dfrac{6+0.4^t}{3-0.7^t}$

c. $\lim_{t\to\infty}\sqrt{7+0.5^t}$

d. $\lim_{t\to\infty}\dfrac{2^t}{3^t}$

E-4. Calculating with the basic power limit: Use the basic power limit to calculate the following:

a. $\lim_{t\to\infty} 5\left(6+\dfrac{4}{t^2}\right)$

b. $\lim_{t\to\infty}\dfrac{2+\dfrac{1}{t}}{3+\dfrac{4}{t^2}}$

c. $\lim_{t\to\infty} 2^{1/t}$

d. $\lim_{t\to\infty}(5+t^{-1/3})$

E-5. Calculating using the division trick: Calculate the following limits by first dividing top and bottom by the largest power of t that occurs and then applying the basic power limit.

a. $\lim_{t\to\infty}\dfrac{4t+5}{t+1}$

b. $\lim_{t\to\infty}\dfrac{6t^3+5}{2t^3+3t+1}$

c. $\lim_{t\to\infty}\dfrac{t^2+1}{t^3+1}$

E-6. **Carrying capacity for the general logistic function:** In general, a logistic function can be written in the form

$$\frac{k}{1 + pa^t},$$

where $0 < a < 1$. If this function describes population growth for a certain species in a certain area, what is the carrying capacity?

E-7. **Calculating using both exponential and power limits:** Calculate the following limits. Assume that $b \neq 0$ in parts a and c.

a. $\displaystyle\lim_{t \to \infty} \frac{a + \dfrac{1}{t}}{b + 0.2^t}$

b. $\displaystyle\lim_{t \to \infty} \left(b \times 0.8^t - \frac{a}{t^2} \right)$

c. $\displaystyle\lim_{t \to \infty} \frac{at^2 + 0.3^t}{bt^2 + 1}$

E-8. **Water in a tank:** The volume V of water in a tank at time t is given by $V = a - bc^t$ cubic feet, where $c < 1$. How much water will be in the tank after a long period of time?

E-9. **Concentration of salt:** The concentration of salt in a certain chemical solution varies with time. The concentration c is given by $c = a + b/t$ pounds per gallon of solution. What is the eventual concentration of salt in the solution?

2.1 SKILL BUILDING EXERCISES

S-1. **Making a table:** Make a table for $f(x) = x^2 - 1$ showing function values for $x = 4, 6, 8, \ldots$.

S-2. **Comparing functions:** Make a table that shows a comparison of the values of f from Exercise S-1 with those of $g = 2^x$. (Use the same table setup values as in Exercise S-1.)

S-3. **Making a table:** Make a table for $f(x) = 16 - x^3$ showing function values for $x = 3, 7, 11, \ldots$.

S-4. **Comparing functions:** Make a table that shows a comparison of the values of f from Exercise S-3 with those of $g = 23 - 2^x$. (Use the same table setup values as in Exercise S-3.)

S-5. **Finding a limiting value:** It is a fact that the function $(4x^2 - 1)/(7x^2 + 1)$ has a limiting value. Use a table of values to estimate the limiting value. (*Suggestion:* We suggest starting the table at 0 and using a table increment of 20.)

S-6. **Finding a limiting value:** It is a fact that the function $(2 + 3^{-x})/(5 - 3^{-x})$ has a limiting value. Use a table of values to estimate the limiting value. (*Suggestion:* We suggest starting the table at 0 and using a table increment of 2.)

S-7. **Finding a minimum:** Suppose the function $f = x^2 - 8x + 21$ describes a physical situation that makes sense only for whole numbers between 0 and 20 (such as family expense as a function of number of children). For what value of x does

f reach a minimum, and what is that minimum value? (*Suggestion:* We suggest a table starting at 0 with a table increment of 1.)

S-8. **Finding a minimum:** Suppose the function $f = x^4/30 - 9x + 50$ describes a physical situation that makes sense only for whole numbers between 0 and 10 (such as family expense as a function of number of children). For what value of x does f reach a minimum, and what is the minimum value? (*Suggestion:* We suggest beginning with a table starting at 0 with a table increment of 1 and then panning farther down the table.)

S-9. **Finding a maximum:** Suppose the function $f = 9x^2 - 2^x + 1$ describes a physical situation that makes sense only for whole numbers between 0 and 10 (such as family expense as a function of number of children). For what value of x does f reach a maximum, and what is the maximum value? (*Suggestion:* We suggest beginning with a table starting at 0 with a table increment of 1 and then panning farther down the table.)

S-10. **Finding a maximum:** Suppose the function $f = 3 \times 2^x - 2.15^x$ describes a physical situation that makes sense only for whole numbers between 0 and 15 (such as family expense as a function of number of children). For what value of x does f reach a maximum, and what is that maximum value? (*Suggestion:* We suggest a table starting at 0 with a table increment of 1.)

2.1 EXERCISES

1. **Harvard Step Test:** The Harvard Step Test was developed[2] in 1943 as a physical fitness test, and modifications of it remain in use today. The candidate steps up and down a bench 20 inches high 30 times per minute for 5 minutes. The pulse is counted three times for 30 seconds: at 1 minute, 2 minutes, and 3 minutes after the exercise is completed. If P is

the sum of the three pulse counts, then the *physical efficiency index E* is calculated using

$$E = \frac{15{,}000}{P}.$$

The following table shows how to interpret the results of the test.

---

[2]Adapted from L. Brouha, "The Step Test," *Research Quarterly* **14(1)** (1943), 31–35.

Efficiency index	Interpretation
Below 55	Poor condition
55 to 64	Low average
65 to 79	High average
80 to 89	Good
90 & above	Excellent

a. Does the physical efficiency index increase or decrease with increasing values of P? Explain in practical terms what this means.

b. Express using functional notation the physical efficiency index of someone whose total pulse count is 200, and then calculate that value.

c. What is the physical condition of someone whose total pulse count is 200?

d. What pulse counts will result in an excellent rating?

2. **Public high school enrollment:** One model for the number of students enrolled in U.S. public high schools as a function of time since 1986 is

$$N = 0.05t^2 - 0.42t + 12.33.$$

Here N is the enrollment in millions of students, t is the time in years since 1986, and the model is relevant from 1986 to 1996.

a. Use functional notation to express the number of students enrolled in U.S. public high schools in the year 1989, and then calculate that value.

b. Explain in practical terms what $N(8)$ means and calculate that value.

c. In what year was the enrollment the smallest?

3. **Earlier public high school enrollment:** Here is a model for the number of students enrolled in U.S. public high schools as a function of time since 1965:

$$N = -0.02t^2 + 0.44t + 11.65.$$

In this formula N is the enrollment in millions of students, t is the time in years since 1965, and the model is applicable from 1965 to 1985.

a. Calculate $N(7)$ and explain in practical terms what it means.

b. In what year was the enrollment the largest? What was the largest enrollment?

c. Find the average yearly rate of change in enrollment from 1965 to 1985. Is the result misleading, considering your answer to part b?

4. **Species-area relation:** The number of species of a given taxonomic group within a given habitat (often an island) is a function of the area of the habitat. For islands in the West Indies, the formula

$$S(A) = 3A^{0.3}$$

approximates the number S of species of amphibians and reptiles on an island in terms of the island area A in square miles. This is an example of a *species-area relation*.

a. Make a table giving the value of S for islands ranging in area from 4000 to 40,000 square miles.

b. Explain in practical terms what $S(4000)$ means and calculate that value.

c. Use functional notation to express the number of species on an island whose area is 8000 square miles, and then calculate that value.

d. Would you expect a graph of S to be concave up or concave down?

5. **Competition:** Two friends enjoy competing with each other to see who has the best time in running a mile. Initially (before they ever raced each other), the first friend runs a mile in 7 minutes, and for each race that they run, his time decreases by 13 seconds. Initially, the second friend runs a mile in 7 minutes and 20 seconds, and for each race that they run, his time decreases by 16 seconds. Which will be the first race in which the second friend beats the first?

6. **Profit:** The profit P, in thousands of dollars, that a manufacturer makes is a function of the number N of items produced in a year, and the formula is

$$P = -0.2N^2 + 3.6N - 9.$$

a. Express using functional notation the profit at a production level of 5 items per year, and then calculate that value.

b. Determine the two break-even points for this manufacturer—that is, the two production levels at which the profit is zero.

c. Determine the maximum profit if the manufacturer can produce at most 20 items in a year.

7. **Counting when order matters:** The *factorial function* occurs often in probability and statistics. For a non-negative integer n, the factorial is denoted $n!$ (which is read "n factorial") and is defined as follows: First, $0!$ is defined to be 1. Next, if n is 1 or larger, then $n!$ means $n(n-1)(n-2)\cdots 3 \times 2 \times 1$. Thus $3! = 3 \times 2 \times 1 = 6$. Consult the *Technology Guide* to see how to enter the factorial operation on the calculator.

In some counting situations, order makes a difference. For example, if we arrange people into a line (first to last), then each different ordering is considered a different arrangement. The number of ways in which you can arrange n individuals in a line is $n!$.

a. In how many ways can you arrange 5 people in a line?

b. How many people will result in more than 1000 possible arrangements for a line?

c. Suppose you remember that your four-digit bank card PIN number uses 7, 5, 3, and 1, but you can't remember in which order they come. How many guesses would you need to ensure that you got the right PIN number?

d. There are 52 cards in an ordinary deck of playing cards. How many possible shufflings are there of a deck of cards?

8. **Counting when order does not matter:** *This is a continuation of Exercise 7.* In many situations, the number of possibilities is not affected by order. For example, if a group of 4 people is selected from a group of 20 to go on a trip, then the order of selection does not matter. In general, the number C of ways to select a group of k things from a group of n things is given by

$$C = \frac{n!}{k!(n-k)!}$$

if k is not greater than n.

a. How many different groups of 4 people could be selected from a group of 20 to go on a trip?

b. How many groups of 16 could be selected from a group of 20?

c. Your answers in parts a and b should have been the same. Explain why this is true.

d. What group size chosen from among 20 people will result in the largest number of possibilities? How many possibilities are there for this group size?

9. **APR and EAR:** Recall that the APR (the *annual percentage rate*) is the percentage rate on a loan that the Truth in Lending Act requires lending institutions to report on loan agreements. It does not tell directly what the interest rate really is. If you borrow money for 1 year and make no payments, then in order to calculate how much you owe at the end of the year, you must use another interest rate, the EAR (the *effective annual rate*), which is not normally reported on loan agreements. The calculation is made by adding the interest indicated by the EAR to the amount borrowed.

The relationship between the APR and the EAR depends on how often interest is compounded. If you borrow money at an annual percentage rate APR (as a decimal), and if interest is compounded n times per year, then the effective annual rate EAR (as a decimal) is given by

$$\text{EAR} = \left(1 + \frac{\text{APR}}{n}\right)^n - 1.$$

For the remainder of this problem, we will assume an APR of 10%. Thus in the formula above, we would use 0.1 in place of APR.

a. Would you expect a larger or a smaller EAR if interest is compounded more often? Explain your reasoning.

b. Make a table that shows how the EAR depends on the number of compounding periods. Use your table to report the EAR if interest is compounded once each year, monthly, and daily. (*Note:* The formula will give the EAR as a decimal. You should report your answer as a percent with three decimal places.)

c. If you borrow $5000 and make no payments for 1 year, how much will you owe at the end of a year if interest is compounded monthly? If interest is compounded daily?

d. If interest is compounded as often as possible—that is, continuously—then the relationship between APR and EAR is given by

$$\text{EAR} = e^{\text{APR}} - 1.$$

Again using an APR of 10%, compare the EAR when the interest is compounded monthly with the EAR when the interest is compounded continuously.

10. **An amortization table:** Suppose you borrow P dollars at a monthly interest rate of r (as a decimal) and wish to pay off the loan in t months. Then your monthly payment can be calculated using

$$M = \frac{Pr(1+r)^t}{(1+r)^t - 1} \text{ dollars}.$$

Remember that for monthly compounding, you get the monthly rate by dividing the APR by 12. Suppose you borrow $3500 at 9% APR (meaning that you use $r = 0.09/12$ in the formula above) and pay it back in 2 years.

a. What is your monthly payment?

b. Let's look ahead to the time when the loan is paid off.

   i. What is the total amount you paid to the bank?

   ii. How much of that was interest?

c. The amount B that you still owe the bank after making k monthly payments can be calculated using the variables r, P, and t. The relationship is given by

$$B = P \times \left(\frac{(1+r)^t - (1+r)^k}{(1+r)^t - 1} \right) \text{ dollars}.$$

   i. How much do you still owe the bank after 1 year of payments?

   ii. An *amortization table* is a table that shows how much you still owe the bank after each payment. Make an amortization table for this loan.

11. **An amortization table for continuous compounding:** *This is a continuation of Exercise 10.* Suppose you have borrowed P dollars from a lending institution that compounds interest as often as possible—that is, continuously. If the loan is to be paid off in

t months, then you would calculate your monthly payment in dollars using

$$M = \frac{P(e^r - 1)}{1 - e^{-rt}},$$

where $r = \frac{\text{APR}}{12}$ if the APR is written in decimal form. Under these circumstances, the balance B in dollars that you owe the bank after k monthly payments is given by

$$B = \frac{P(e^{rt} - e^{rk})}{e^{rt} - 1}.$$

Suppose you borrow $3500 at an APR of 9% and pay off the note in 2 years.[3]

a. Calculate your monthly payment, and compare your answer with the answer you obtained in Exercise 10.

b. Make an amortization table and compare it with the answer you got in Exercise 10.

12. **Renting motel rooms:** You own a motel with 30 rooms and have a pricing structure that encourages rentals of rooms in groups. One room rents for $85, two rent for $83 each, and in general the group rate per room is found by taking $2 off the base of $85 for each extra room rented.

a. How much money do you take in if a family rents two rooms?

b. Use a formula to give the rate you charge for each room if you rent n rooms to an organization.

c. Find a formula for a function $R = R(n)$ that gives the revenue from renting n rooms to a convention host.

d. What is the most money you can make from rental to a single group? How many rooms do you rent?

13. **Inventory:** For retailers who buy from a distributor or manufacturer and sell to the public, a major concern is the cost of maintaining unsold inventory. You must have appropriate stock to do business, but

---

[3]It is worth pointing out that the formulas we see in this exercise are simpler than the corresponding ones in Exercise 10. Continuous compounding may appear at first sight to be more complicated than monthly compounding, but it is in fact easier to handle. And as you will see when you complete this exercise, in many applications it does not give significantly different answers.

if you order too much at a time, your profits may be eaten up by storage costs. One of the simplest tools for analysis of inventory costs is the *basic order quantity model*. It gives the yearly inventory expense $E = E(c, N, Q, f)$ when the following inventory and restocking cost factors are taken into account:

- The *carrying cost c*, which is the cost in dollars per year of keeping a single unsold item in your warehouse.

- The number N of this item that you expect to sell in 1 year.

- The number Q of items you order at a time.

- The fixed costs f in dollars of processing a restocking order to the manufacturer. (*Note:* This is not the cost of the order; the price of an item does not play a role here. Rather, f is the cost you would incur with any order of any size. It might include the cost of processing the paperwork, fixed costs you pay the manufacturer for each order, shipping charges that do not depend on the size of the order, the cost of counting your inventory, or the cost of cleaning and rearranging your warehouse in preparation for delivery.)

The relationship is given by

$$E = \left(\frac{Q}{2}\right)c + \left(\frac{N}{Q}\right)f \text{ dollars per year}.$$

A new-car dealer expects to sell 36 of a particular model car in the next year. It costs $850 per year to keep an unsold car on the lot. Fixed costs associated with preparing, processing, and receiving a single order from Detroit total $230 per order.

a. Using the information provided, express the yearly inventory expense $E = E(Q)$ as a function of Q, the number of automobiles included in a single order.

b. What is the yearly inventory expense if 3 cars at a time are ordered?

c. How many cars at a time should be ordered to make yearly inventory expenses a minimum?

d. Using the value of Q you found in part c, determine how many orders to Detroit will be placed this year.

e. What is the average rate of increase in yearly inventory expense from the number you found in part c to an order of 2 cars more?

14. **A population of foxes:** A breeding group of foxes is introduced into a protected area and exhibits logistic population growth. After t years the number of foxes is given by

$$N(t) = \frac{37.5}{0.25 + 0.76^t} \text{ foxes}.$$

a. How many foxes were introduced into the protected area?

b. Calculate $N(5)$ and explain the meaning of the number you have calculated.

c. Explain how the population varies with time. Include in your explanation the average rate of increase over the first 10-year period and the average rate of increase over the second 10-year period.

d. Find the carrying capacity for foxes in the protected area.

e. As we saw in the discussion of terminal velocity for a skydiver, the question of when the carrying capacity is reached may lead to an involved discussion. We ask the question differently. When is 99% of carrying capacity reached?

15. **Falling with a parachute:** If an average-size man jumps from an airplane with an open parachute, his downward velocity t seconds into the fall is $v(t) = 20(1 - 0.2^t)$ feet per second.

a. Use functional notation to express the velocity 2 seconds into the fall, and then calculate it.

b. Explain how the velocity increases with time. Include in your explanation the average rate of change from the beginning of the fall to the end of the first second and the average rate of change from the fifth second to the sixth second of the fall.

c. Find the terminal velocity.

d. Compare the time it takes to reach 99% of terminal velocity here with the time it took to reach 99% of terminal velocity in Example 2.1. On the basis of the information we have, which would you expect to reach 99% of terminal velocity first, a feather or a cannonball?

16. **Rolling 4 sixes:** If you roll N dice, then the probability $p = p(N)$ that you will get exactly 4 sixes is given by

$$p = \frac{N(N-1)(N-2)(N-3)}{24} \times \left(\frac{1}{6}\right)^4 \left(\frac{5}{6}\right)^{N-4}.$$

a. What is the probability, rounded to three decimal places, of getting exactly 4 sixes if 10 dice are rolled? How many times out of 1000 rolls would you expect this to happen?

b. How many dice should be rolled so that the probability of getting exactly 4 sixes is the greatest?

17. **Profit:** *The background for this exercise can be found in Exercises 11 and 12 in Section 1.4.* A manufacturer of widgets has fixed costs of $150 per month, and the variable cost is $50 per widget (so it costs $50 to produce 1 widget). Let N be the number of widgets produced in a month.

a. Find a formula for the manufacturer's total cost C as a function of N.

b. The manufacturer sells the widgets for $65 each. Find a formula for the total revenue R as a function of N.

c. Use your answers to parts a and b to find a formula for the profit P of this manufacturer as a function of N.

d. Use your formula from part c to determine the break-even point for this manufacturer.

18. **Profit with varying price:** *The background for this exercise can be found in Exercises 11, 12, 13, and 14 in Section 1.4.* A manufacturer of widgets has fixed costs of $1200 per month, and the variable cost is $40 per widget (so it costs $40 to produce 1 widget). Let N be the number of widgets produced in a month.

a. Find a formula for the manufacturer's total cost C as a function of N.

b. The highest price p, in dollars, of a widget at which N widgets can be sold is given by the formula $p = 53 - 0.01N$. Using this, find a formula for the total revenue R as a function of N.

c. Use your answers to parts a and b to find a formula for the profit P of this manufacturer as a function of N.

d. Use your formula from part c to determine the two break-even points for this manufacturer. Assume here that the manufacturer produces the widgets in blocks of 50, so a table setup showing N in multiples of 50 is appropriate.

e. Use your formula from part c to determine the production level at which profit is maximized if

the manufacturer can produce at most 1500 widgets in a month. As in part d, assume that the manufacturer produces the widgets in blocks of 50.

19. **A precocious child and her blocks:** A child has 64 blocks that are 1-inch cubes. She wants to arrange the blocks into a solid rectangle h blocks long and w blocks wide. There is a relationship between h and w that is determined by the restriction that all 64 blocks must go into the rectangle. A rectangle h blocks long and w blocks wide uses a total of $h \times w$ blocks. Thus $hw = 64$. Applying some elementary algebra, we get the relationship we need:

$$w = \frac{64}{h}. \tag{2.3}$$

a. Use a formula to express the perimeter P in terms of h and w.

b. Using Equation (2.3), find a formula that expresses the perimeter P in terms of the height only.

c. How should the child arrange the blocks if she wants the perimeter to be the smallest possible?

d. Do parts b and c again, this time assuming that the child has 60 blocks rather than 64 blocks. In this situation the relationship between h and w is $w = 60/h$. (*Note:* Be careful when you do part c. The child will not cut the blocks into pieces!)

20. **Renting paddleboats:** An enterprise rents out paddleboats for all-day use on a lake. The owner knows that he can rent out all 27 of his paddleboats if he charges $1 for each rental. He also knows that he can rent out only 26 if he charges $2 for each rental and that, in general, there will be 1 less paddleboat rental for each extra dollar he charges per rental.

a. What would the owner's total revenue be if he charged $3 for each paddleboat rental?

b. Use a formula to express the number of rentals as a function of the amount charged for each rental.

c. Use a formula to express the total revenue as a function of the amount charged for each rental.

d. How much should the owner charge to get the largest total revenue?

21. **Growth in length of haddock:** In 1933, Riatt found that the length L of haddock in centimeters as a

function of the age t in years is given approximately by the formula

$$L = 53 - 42.82 \times 0.82^t.$$

a. Calculate $L(4)$ and explain what it means.

b. Compare the average yearly rate of growth in length from age 5 to 10 years with the average yearly rate of growth from age 15 to 20 years. Explain in practical terms what this tells you about the way haddock grow.

c. What is the longest haddock you would expect to find anywhere?

22. **Discharge from a fire hose:** The discharge of a fire hose depends on the diameter of the nozzle. Nozzle diameters are normally in multiples of $\frac{1}{8}$ inch. Sometimes it is important to replace several hoses with a single hose of equivalent discharge capacity. Hoses with nozzle diameters $d_1, d_2, \ldots, d_n$ have the same discharge capacity as a single hose with nozzle diameter D, where

$$D = \sqrt{d_1^2 + d_2^2 + \ldots + d_n^2}.$$

a. A nozzle of what diameter has the same discharge capacity as three combined nozzles of diameters $1\frac{1}{8}$ inches, $1\frac{5}{8}$ inches, and $1\frac{7}{8}$ inches? You should report your answer as an available nozzle size, that is, in multiples of $\frac{1}{8}$.

b. We have two 1-inch nozzles and wish to use a third so that the combined discharge capacity of the three nozzles is the same as the discharge capacity of a $2\frac{1}{4}$-inch nozzle. What should be the diameter of the third nozzle?

c. If we wish to use n hoses each with nozzle size d in order to have the combined discharge capacity of a single hose with nozzle size D, then we must use

$$n = \left(\frac{D}{d}\right)^2 \text{ nozzles}.$$

How many half-inch nozzles are needed to attain the discharge capacity of a 2-inch nozzle?

d. We want to replace a nozzle of diameter $2\frac{1}{4}$ inches with 4 hoses each of the same nozzle diameter. What nozzle diameter for the 4 hoses will produce the same discharge capacity as the single hose?

23. **California earthquakes:** We most often hear of the power of earthquakes given in terms of the *Richter scale*, but this tells only the power of the earthquake at its epicenter. Of more immediate importance is how an earthquake affects the location where we are. Seismologists measure this in terms of ground movement, and for technical reasons they find the acceleration of ground movement most useful. For the purpose of this problem, a *major earthquake* is one that produces a ground acceleration of at least 5% of g, where g is the acceleration[4] due to gravity near the surface of the Earth. In California, the probability $p(n)$ of one's home being affected by exactly n major earthquakes over a 10-year period is given approximately[5] by

$$p(n) = 0.379 \times \frac{0.9695^n}{n!}.$$

See Exercise 7 for an explanation of $n!$.

a. What is the probability of a California home being affected by exactly 3 major earthquakes over a 10-year period?

b. What is the limiting value of $p(n)$? Explain in practical terms what this means.

c. What is the probability of a California home being affected by no major earthquakes over a 10-year period?

d. What is the probability of a California home being affected by *at least* one major earthquake over a 10-year period?
Hint: It is a certainty that an event either will or will not occur, and the probability assigned to a certainty is 1. Expressed in a formula, this is

Probability of an event occurring
+ Probability of an event not occurring $= 1$.

---

[4]The value of g is 9.8 meters per second per second, or 32 feet per second per second.

[5]The probability here is calculated using earthquake frequency data and intensity distributions given on pages 80 and 81 of *Earthquake Engineering*, edited by Robert L. Wiegel (Englewood Cliffs, NJ: Prentice-Hall, 1970). The calculation also assumes that earthquakes are equally likely at any location in California. It is thus too low for highly seismic regions and too high for some less seismic regions.

This, in conjunction with part c, may be helpful for part d.

24. **Terminal velocity revisited:** In one of the early "Functions and Change" pilot courses at Oklahoma State University, the instructor asked the class to determine when in Example 2.1 terminal velocity would be reached. Three students gave the following three answers:

    Student 1: 58 seconds into the fall.

    Student 2: 147 seconds into the fall.

    Student 3: Never.

    Each student's answer was accompanied by what the instructor judged to be an appropriate supporting argument, and each student received full credit for the problem. What supporting arguments might the students have used to convince the instructor that these three different answers could all be deserving of full credit? (*Hint:* Consider the formula given in Example 2.1. For student 1, look at a table of values where the entries are rounded to two decimal places. For student 2, look at a table of values made by using all the digits beyond the decimal point that the calculator can handle. In this case that was nine. For student 3, consider what value 0.834^t must have to make $176(1 - 0.834^t)$ equal to 176.)

25. **Research project:** Find a function given by a formula in one of your textbooks for another class or some other handy source. If the formula involves more than one variable, assign reasonable values for all the variables except one so that your formula involves only one variable. Now make a table of values, using an appropriate starting value and increment value so that the table shows some interesting aspect of the function, such as a trend or significant values. Carefully describe the function, formula, variables (including units), and table, and explain how the table is useful.

2.2   *Graphs*

A graph is a picture of a function, and just as tables do, graphs can show features of a function that are difficult to see by looking at formulas. For many applications, the graph is more useful than a table of values. The graphing calculator can generate the graph of a function as easily as it makes tables.

Hand-Drawn Graphs from Formulas

If we have a function given by a formula, there is a standard procedure for generating the graph. We will make a hand-drawn picture of the graph of $f = f(x)$, where $f = x^2 - 1$. The first step is to make a table of values.

x	$f = x^2 - 1$
−2	3
−1	0
0	−1
1	0
2	3

This table tells us that the points $(-2, 3)$, $(-1, 0)$, $(0, -1)$, $(1, 0)$, $(2, 3)$ lie on the graph. We plot the individual points as shown in Figure 2.15. We complete the graph by joining the dots with a smooth curve as shown in Figure 2.16.

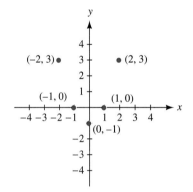

FIGURE 2.15 Plotting points

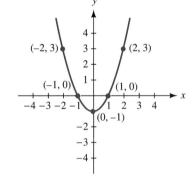

FIGURE 2.16 Completing the graph: $x^2 - 1$

Graphing with the Calculator

For more complicated formulas this process is tedious and subject to inaccuracy. But the graphing calculator can make graphs easily. It does exactly what we did above, but it doesn't mind doing all that arithmetic and can accurately plot many more points than we used. For calculator-specific instructions on how to make graphs from formulas, see Chapter 2 of the *Technology Guide*.

There are two steps involved in using a calculator to get a graph from a formula. We will look at them in the context of making the graph of $f = x^2 - 1$.

Step 1, Entering the function: First we have to tell the calculator which function we want to use. We use the function entry screen to do this exactly as we did in Section 2.1 when we made tables of values. In Figure 2.17 we have cleared old formulas from the function entry screen and entered <u>2.12</u> the new one. As expected, the calculator

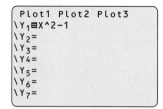

FIGURE 2.17 Entering
$f = x^2 - 1$

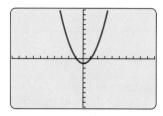

FIGURE 2.18 The standard view
of the graph

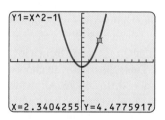

FIGURE 2.19 Tracing the graph

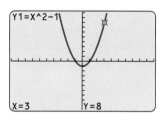

FIGURE 2.20 Getting $f(3)$

has chosen its own names for functions and variables, and it is important to record the associations:

$$Y_1 = f, \text{ corresponding to the vertical axis}$$

$$X = x, \text{ corresponding to the horizontal axis}.$$

Step 2, Selecting the viewing window: Like any picture, the graph looks different from different points of view. If you are sure that your viewing window is set up to show the graph as you wish to see it, you can skip this step and go directly to the graph $\boxed{2.13}$. If this produces an unsatisfactory view, then it will be necessary to make adjustments in the viewing window. Figure 2.18 illustrates what we will refer to as the *standard view* $\boxed{2.14}$, which shows the graph in a window extending from -10 to 10 in the horizontal direction and from -10 to 10 in the vertical direction. This standard display is satisfactory for some graphs, and we will use it occasionally in what follows. Below we will discuss what to do when the standard view is unsatisfactory.

Tracing the Graph

Once we have a graph on the screen, there are several ways to adjust the view or to get information from it. We will continue working with $f = x^2 - 1$, and, to follow the discussion, it is important for you to have a graph on your screen that matches the one in Figure 2.18. If you are having difficulty getting that picture, see Chapter 2 of the *Technology Guide*.

   Most calculators allow you to *trace* $\boxed{2.15}$ a graph that appears on the screen. This means to put on the screen a movable cursor that follows the graph and is controlled by left and right arrow keys. In Figure 2.19 we have used this feature to add a cursor to the screen and move it along the graph. As the cursor moves, its location is recorded on the screen. In Figure 2.19 the X=2.3404255 prompt at the bottom of the screen shows where we are relative to the horizontal axis, and the Y=4.4775917 prompt shows where we are relative to the vertical axis. This tells us that the cursor is located at the point (2.3404255, 4.4775917). Since this point lies on the graph, it also tells us that $f(2.3404255) = 4.4775917$, or, rounded to two decimal places, $f(2.34) = 4.48$.

   The trace feature cannot directly show all function values. For example, on our calculator we were unable, using the arrow keys, to make the cursor land exactly on X=3. Most calculators allow alternative ways $\boxed{2.16}$ to locate the cursor at X=3. We have used this feature in Figure 2.20, and we read from the prompt at the bottom of the screen that $f(3) = 8$ as expected. It is important that you be able to move the graphing cursor to any point on the graph, and you are encouraged to consult Chapter 2 of the *Technology Guide* for information on how to do this.

Choosing a Viewing Window in Practical Settings

The key to making a usable graph on the calculator is to make a proper choice of a viewing window. Finding a viewing window that shows a good picture can sometimes be a bit frustrating, but often appropriate settings can be determined from practical considerations.

   Consider, for example, a leather craftsman who has produced 25 belts that he intends to sell at an upcoming art fair for $22.75 each. He has invested a total of $300 in leather, buckles, and other accessories for the belts. We want to look at his net profit $p = p(n)$ in dollars as a function of the number n of belts that he sells:

$$\begin{aligned} \text{Net profit} &= \text{Profit from sales} - \text{Investment} \\ &= \text{Price per item} \times \text{Number sold} - \text{Investment} \\ p &= 22.75n - 300\,. \end{aligned}$$

```
 Plot1 Plot2 Plot3
\Y1■22.75X-300
\Y2=
\Y3=
\Y4=
\Y5=
\Y6=
\Y7=
```

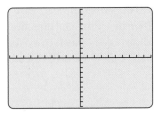

FIGURE 2.21 Entering a function for net profit on the sale of belts

Let's make a graph of p versus n—that is, a graph that shows net profit as a function of the number of belts sold. First we enter 2.17 the function as shown in Figure 2.21. The proper variable associations are

$$\begin{aligned} \mathsf{Y}_1 &= p, \text{ net profit on the vertical axis} \\ \mathsf{X} &= n, \text{ number sold on the horizontal axis}\,. \end{aligned}$$

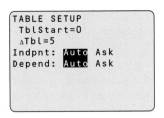

FIGURE 2.22 The standard view of the graph is unsatisfactory

If we look at the standard 2.18 view of the function as shown in Figure 2.22, we see no graph at all! We need to choose a different viewing window that will show the graph. To do that, we note first of all that there are 25 belts available for sale, so we are interested in the function only for values of n between 0 and 25. As a consequence, the horizontal axis should extend from 0 to 25 rather than from the standard setting of -10 to 10. But it is not immediately apparent how to choose a vertical span. A good way to handle this is to look at a table of values, which we already know how to make. As shown in Figure 2.23, we choose a starting value of 0 and a table increment of 5. When we look at the table in Figure 2.24, we see that the possible values for net profit range from a low of $-\$300$ for no sales to \$268.75 for the sale of all 25 belts. This tells us how to choose the vertical span of the viewing window; we want it to go from -300 to 268.75. Allowing a little extra margin, we show in Figure 2.25 a window setup 2.19 where the horizontal span goes from 0 to 25 and the vertical span goes from -325 to 300. Now when we graph 2.20 we get a very good picture of the function, as shown in Figure 2.26. If we now trace the graph, we can get usable information from it. For example, in Figure 2.27 we have put the cursor 2.21 at X=10, and we see that if 10 belts are sold, the net profit is -72.50 dollars. That is, the craftsman will lose \$72.50. In Figure 2.28 we have put the cursor at X=20, and we see that if 20 belts are sold, there will be a net profit of \$155. What is the practical significance of the place where the graph crosses the horizontal axis?

```
TABLE SETUP
 TblStart=0
 ΔTbl=5
Indpnt: Auto Ask
Depend: Auto Ask
```

FIGURE 2.23 Setting up the table

X	Y_1
0	-300
5	-186.3
10	-72.5
15	41.25
20	155
25	268.75
30	382.5

X=0

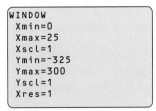

FIGURE 2.24 Table for belt sales

The example of the leather craftsman illustrates that, in many situations, the horizontal span can be determined from practical considerations. We noted in the example that the reasonable values of the variable n were between 0 and 25. Traditionally this has been called the *domain* of the function, so the horizontal span of the graphing window is

```
WINDOW
 Xmin=0
 Xmax=25
 Xscl=1
 Ymin=-325
 Ymax=300
 Yscl=1
 Xres=1
```

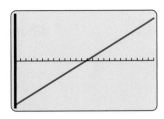

FIGURE 2.25 Setting up the viewing window

FIGURE 2.26 Graph of profit versus sales

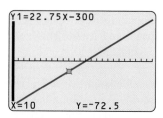

FIGURE 2.27 $72.50 lost if only 10 belts are sold

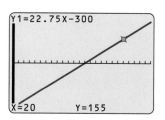

FIGURE 2.28 A net profit of $155 if 20 belts are sold

a way of visualizing the domain. The corresponding values taken by the function form what has traditionally been called the *range* of the function. We found this in the example by using a table for the function p to get the vertical span of -325 to 300. Again, the vertical span of the graphing window gives a helpful way of visualizing the range of a function.

EXAMPLE 2.3 Labor Productivity

For a manufacturing company with w workers working h hours per day and producing a total of I items, *labor productivity* $P = P(w, h, I)$ is measured by the function

$$P = \frac{I}{wh} \text{ items per worker hour}.$$

1. Suppose a company produces 750 items per day and each worker works 8 hours per day. The company budget allows for no more than 11 workers.

   a. Find a formula for labor productivity P as a function of the number of workers w, and make its graph.

   b. Use the graph to find labor productivity if there are 7 workers.

   c. What happens to labor productivity if the number of workers is increased? What happens if the number is decreased?

2. Suppose your company has 5 workers each of whom works 8 hours each day. Company resources and product demand dictate that between 500 and 1000 items per day must be produced. Use a graph to show how labor productivity changes when I increases.

```
Plot1 Plot2 Plot3
\Y₁冒750/(8X)
\Y₂=
\Y₃=
\Y₄=
\Y₅=
\Y₆=
\Y₇=
```

FIGURE 2.29 Entering the labor productivity function

X	Y₁
1	93.75
3	31.25
5	18.75
7	13.393
9	10.417
11	8.5227
13	7.2115
X=1	

FIGURE 2.30 A table of values for labor productivity

Solution to Part 1a: Under the given conditions we have $I = 750$ and $h = 8$. Thus labor productivity is given by the formula $P = 750/8w$. To make the graph, we first enter [2.22] the function as shown in Figure 2.29 and record the variable correspondences:

$$\mathbf{Y}_1 = P, \text{ productivity on the vertical axis}$$
$$\mathbf{X} = w, \text{ number of workers on the horizontal axis}.$$

Next we need to choose a window for graphing. Since the company budget allows for at most 11 workers, we will set the horizontal span of the window from 1 to 11. To find the vertical span, we made the table of values shown in Figure 2.30 with a starting value of 1 and an increment [2.23] value of 2. Consulting the table and adding a little margin

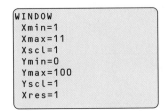

FIGURE 2.31 Setting up the window

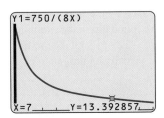

FIGURE 2.32 Graph of labor productivity versus number of workers

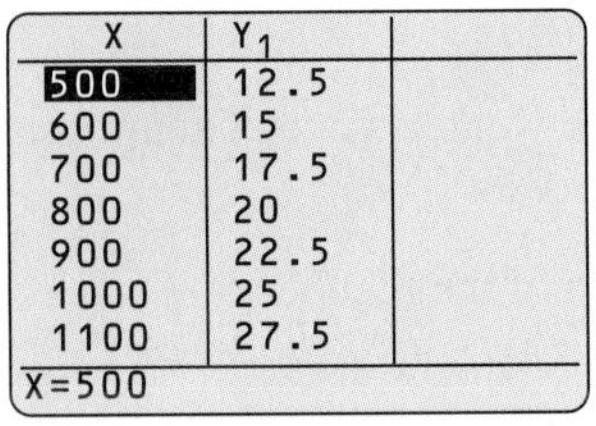

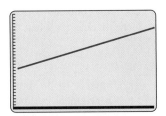

FIGURE 2.33 A table of values for productivity as a function of items produced

at the top and bottom, we use a window setup $\boxed{2.24}$ with a horizontal span from 1 to 11 and a vertical span from 0 to 100. This is shown in Figure 2.31. These settings give the graph $\boxed{2.25}$ in Figure 2.32.

Solution to Part 1b: We want to get $P(7)$ from the graph. To do this, we locate the cursor $\boxed{2.26}$ at X=7. We read from the Y= prompt at the bottom of the screen in Figure 2.32 that using 7 workers gives a labor productivity of about 13.39 items per worker hour.

Solution to Part 1c: The graph in Figure 2.32 is decreasing. This shows that as the number of workers increases, labor productivity decreases. This is not a surprise, since we are increasing the number of worker hours but holding the number of items produced constant at 750 items. Thus each worker is producing less. We see from Figure 2.32 that if we decrease the number of workers, labor productivity increases. Once again, this is not surprising, since we are decreasing worker hours but holding the number of items produced constant. Thus each worker is producing more.

Solution to Part 2:   In this scenario, we are holding the number of workers constant at 5 and the number of hours constant at 8. We are looking at labor productivity as a function of the number of items produced:

$$P = \frac{I}{5 \times 8} = \frac{I}{40} \text{ items per worker hour}.$$

We enter $\boxed{2.27}$ this function and record the variable correspondences:

$$Y_1 = P, \text{ productivity on the vertical axis}$$
$$X = I, \text{ number of items produced on the horizontal axis}.$$

Since the company must produce between 500 and 1000 items per day, we will set the horizontal span of the graphing window from 500 to 1000. To get the vertical span for the window, we made the table of values in Figure 2.33 with a starting value of 500 and a table increment of 100. From this table, we choose a window setup $\boxed{2.28}$ with a horizontal span of 500 to 1000 and a vertical span of 0 to 30. This gives the graph in Figure 2.34. It is interesting to note that if productivity is viewed as a function of the number of items produced, then its graph is a straight line, but if productivity is viewed as a function of the number of worker hours, then the graph is curved. ■ ■ ■

Getting Limiting Values from Graphs

We have seen that tables of values can be helpful in determining limiting values. Graphs can do that as well, and in addition they provide an informative picture of how limiting values may be approached. The height $h = h(t)$ of some plants as a function of time t closely follows a *logistic formula*. Interestingly enough, this is the same type of formula that is often used to study population growth. For a certain variety of sunflower[6] growing

FIGURE 2.34 A graph of productivity versus items produced

---

[6]This example is adapted from the presentation on pages 42–43 of *Differential Equations* by D. Lomen and D. Lovelock (New York: John Wiley, 1999).

under ideal conditions, and starting at a time when the plant is already a few centimeters tall, its height may be given by the function

$$h = \frac{13}{0.93^t + 0.05},$$

where h is measured in centimeters, and t is measured in days. We want to make a graph of h versus t and see what information we can gain from it. In particular, we would like to be able to describe how the sunflower grows and figure out its maximum height.

First we enter $\boxed{2.29}$ the function as shown in Figure 2.35 and record the variable correspondences:

$$Y_1 = h, \text{ height on the vertical axis}$$

$$X = t, \text{ time on the horizontal axis}.$$

Next we need to determine a viewing window. In contrast with what happened in Example 2.3, we are not provided with explicit information on the values of t to help us set up the window. But our everyday experience with annual plants can give us the information we need. To get an idea of what is happening over a possible 4-month growing period, we have in Figure 2.36 made a table of values showing the height of the sunflower from 60 to 120 days at 10-day intervals. We see that in 120 days the sunflower will be just over 259 centimeters tall, and if you pan farther down the table, you will see that almost no further growth occurs. Adding a bit of extra room as usual, we have in Figure 2.37 plotted the graph in a window where the horizontal span is from 0 to 120 and the vertical span is from 0 to 300. Also in this figure, we have traced and moved the cursor toward the right-hand end of the graph, where the height is just over 259 centimeters. From the graph and from the table of values, it appears that the maximum height of the sunflower is about 260 centimeters.

Limiting values can be shown to good advantage if the corresponding horizontal line is added to the graph. We added $\boxed{2.30}$ $Y_2 = 260$ to the function list, and in Figure 2.38

```
Plot1 Plot2 Plot3
\Y₁≡13/(0.93^X+0
.05)
\Y₂=
\Y₃=
\Y₄=
\Y₅=
\Y₆=
```

FIGURE 2.35 Entering a logistic growth function

X	Y₁
60	206.83
70	231.23
80	245.23
90	252.64
100	256.38
110	258.24
120	259.14

X=60

FIGURE 2.36 A table of values for four months of growth

Y1=13/(0.93^X+0.05)

X=118.7234 Y=259.0611

FIGURE 2.37 Sunflower growth

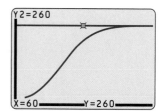

Y2=260

X=60 Y=260

FIGURE 2.38 Limiting height added

both graphs are shown. Note the inflection point in the graph of h, where the graph changes from concave up to concave down. This represents the point of most rapid growth. This picture shows a growth model for sunflowers that agrees with our everyday experience. Growth is slow when the plant is very young, but once a healthy plant is established, it grows rapidly. When maturity is reached, growth slows and little if any additional height is attained.

EXAMPLE 2.4   Blood Cholesterol

X	Y_1	
0	130	
1	157.21	
2	177.37	
3	192.31	
4	203.37	
5	211.57	
6	217.64	
X=0		

FIGURE 2.39 A table for blood cholesterol in 1-month intervals

X	Y_1	
6	217.64	
12	232.13	
18	234.53	
24	234.92	
30	234.99	
36	235	
42	235	
X=6		

FIGURE 2.40 A table for blood cholesterol in 6-month intervals

```
WINDOW
 Xmin=0
 Xmax=36
 Xscl=1
 Ymin=100
 Ymax=250
 Yscl=1
 Xres=1
```

FIGURE 2.41 Setting up the graphing window

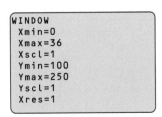

FIGURE 2.42 Blood cholesterol versus time

The amount $C = C(t)$ of cholesterol, in milligrams per deciliter, in the blood of a certain man on an unhealthful diet is given by

$$C = 235 - 105e^{-0.3t},$$

where t is time measured in months.

1. Make a graph that shows the blood cholesterol level as a function of time if the unhealthful diet is continued.

2. The doctor has issued a warning that this man may experience severe health problems if cholesterol levels in excess of 200 milligrams per deciliter of blood are reached. Is there a danger of exceeding this level?

3. If the unhealthful diet is continued indefinitely, what eventual cholesterol level will be reached?

4. Is the graph of C concave up or concave down? Explain in practical terms what your answer means.

Solution to Part 1:   We first enter 2.31 the function and record the variable correspondences:

$$Y_1 = C, \text{ blood cholesterol on vertical axis}$$

$$X = t, \text{ time on horizontal axis}.$$

We have very little information beyond the formula itself to help us set up the graphing window. Thus we experiment a little with tables of values. First we make a table of values starting at $t = 0$ with a table increment of 1. This is shown in Figure 2.39, where we see that, after 6 months, cholesterol levels are still increasing. If you pan farther down the table, you will see the levels continue to increase. If you are patient and pan down far enough, you can see what eventually happens. A quicker way is to return to the table setup and increase the table increment value. In Figure 2.40 we have changed the starting value to 6 and viewed the table in 6-month intervals. We see that blood cholesterol has leveled out at 235 milligrams per deciliter by 36 months. As usual, we allow some margin and set up the graphing window 2.32 in Figure 2.41 with a horizontal span from 0 to 36 and a vertical span from 100 to 250. The graph with these settings is shown in Figure 2.42.

Solution to Part 2:   We have already seen from our table of values that a blood cholesterol level of 200 will certainly be exceeded. But we can use the graph to show this in a striking way. In Figure 2.43 we have added 2.33 the horizontal line corresponding

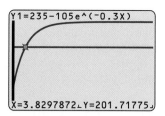

FIGURE 2.43 When blood cholesterol reaches 200 milligrams per deciliter

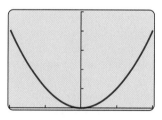

FIGURE 2.44 Limiting value for blood cholesterol

to a blood level of 200 milligrams of cholesterol per deciliter. Note that the point where these graphs cross gives the time when the danger level will be exceeded. In Figure 2.43 we have used the trace option and moved the graphing cursor as close as we could to the crossing point, and we see from the prompt at the bottom of the screen that this individual may incur health risks in about $3\frac{1}{2}$ months.

We will return to this problem in Section 2.4, where we will show how to locate accurately crossing points such as this one.

Solution to Part 3: The table of values shows that the limiting value for blood cholesterol in this case is about 235 milligrams per deciliter. In Figure 2.44 we have added 2.34 the horizontal line corresponding to this limiting value.

Solution to Part 4: The graph is concave down, so the function is increasing at a decreasing rate. Cholesterol levels increase rapidly at first, but the rate of increase slows near the limiting value. ■ ■ ■

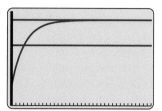

FIGURE 2.45 The graph of x^2

ANOTHER LOOK

Shifting and Stretching

It is important to understand how certain parameters affect graphs. Perhaps the most important of these are *shifts* and *stretches*.

We look first at shifts. In Figure 2.45 we have used the calculator to get the familiar parabolic graph of $f(x) = x^2$. (We have used a horizontal span from -2 to 2 and a vertical span from 0 to 5.) We want to know how the graph is affected if we add 1 to the function. In Figure 2.46 we have included the graph of $g(x) = x^2 + 1$. We see that the new graph is obtained from the old one by moving it up 1 unit. This is not a surprise since we get the graph of $g(x)$ by adding 1 unit to each point on the graph of $f(x)$. That is, we move the graph of $f(x)$ up 1 unit to get the graph of $g(x)$. The same idea works in general if we add a constant to a function.

Vertical shifts: Let a be a positive constant. The graph of $g(x) = f(x) + a$ is obtained by shifting the graph of $f(x)$ up a units. The graph of $h(x) = f(x) - a$ is obtained by shifting the graph of $f(x)$ down a units.

We see another kind of shift if we look at the graph of $(x - 1)^2$. In Figure 2.47 we have included this graph with the graph of x^2. We see that the new graph is

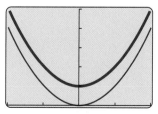

FIGURE 2.46 Shifting up 1 unit to get the graph of $x^2 + 1$

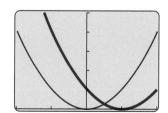

FIGURE 2.47 Shifting 1 unit to the right to get the graph of $(x - 1)^2$

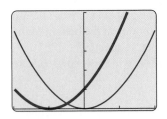

FIGURE 2.48 Shifting 1 unit to the left to get the graph of $(x + 1)^2$

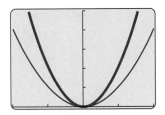

FIGURE 2.49 Stretching vertically to obtain the graph of $2x^2$

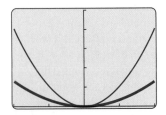

FIGURE 2.50 Stretching horizontally to obtain the graph of $(x/2)^2$

obtained from the old by shifting to the right 1 unit. This is as expected since the value of x^2 is the same as the value of $(x - 1)^2$ at the point 1 unit to the right. Similarly, we see in Figure 2.48 that we get the graph of $(x + 1)^2$ by moving the graph of x^2 to the left 1 unit.

As with vertical shifts, this works with complete generality.

Horizontal shifts: Let a be a positive constant. The graph of $g(x) = f(x - a)$ is obtained by shifting the graph of $f(x)$ to the right a units. The graph of $h(x) = f(x + a)$ is obtained by shifting the graph of $f(x)$ to the left a units.

If instead of adding a constant we multiply by a constant, then we get a vertical rescaling or stretching of the graph. Let's look, for example, at the graph of $2x^2$, which is shown in Figure 2.49. We see that we get the new graph by stretching vertically by a factor of 2. That is because the value of $2x^2$ is twice that of x^2.

We get a different kind of stretch if we look at $(2x)^2$. Note that the value of $(2x)^2$ is the same as the value of x^2 twice as far down the horizontal axis. Thus we get the graph of $(2x)^2$ by compressing the graph of x^2 horizontally by a factor of 2. Similarly, we get the graph of $(x/2)^2$ by stretching the graph horizontally by a factor of 2. This stretch is shown in Figure 2.50.

Vertical stretches: If $a > 0$, then we get the graph of $g(x) = af(x)$ by stretching the graph of $f(x)$ vertically by a factor of a.

Horizontal stretches: Let $a > 1$ be a constant. We get the graph of $g(x) = f(ax)$ by compressing horizontally by a factor of a. We get the graph of $h(x) = f(x/a)$ by stretching horizontally by a factor of a.

Many times, we need to use a combination of shifts and stretches to get a graph. Suppose, for example, that the graph of $f(x)$ is as shown in Figure 2.51. We want to sketch the graph of $2f(x + 1) - 1$. We do this in three steps. First we get the graph of $f(x + 1)$ by shifting the graph of $f(x)$ to the left 1 unit as shown in Figure 2.52.

Next we stretch the graph of $f(x + 1)$ vertically by a factor of 2 to get the graph of $2f(x + 1)$ shown in Figure 2.53. Finally, we move the graph of $2f(x + 1) - 1$ down 1 unit to get the graph of $2f(x + 1) - 1$ shown in Figure 2.54.

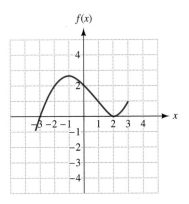

FIGURE 2.51 The graph of $f(x)$

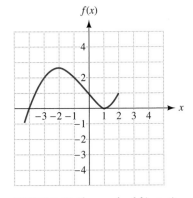

FIGURE 2.52 The graph of $f(x + 1)$

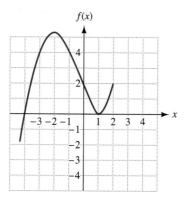

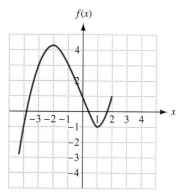

FIGURE 2.53 The graph of $2f(x + 1)$

FIGURE 2.54 The graph of $2f(x + 1) - 1$

Enrichment Exercises

E-1. **Shifting:** The graph of $f(x)$ is shown in Figure 2.55. Sketch the graphs of the following functions.

a. $f(x + 2)$

b. $f(x) + 2$

c. $f(x - 2)$

d. $f(x) - 2$

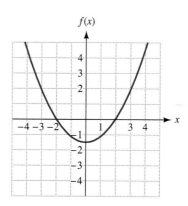

FIGURE 2.55 The graph of $f(x)$

E-2. **Stretching:** The graph of $f(x)$ is shown in Figure 2.55. Sketch the graphs of the following functions.

a. $2f(x)$

b. $\dfrac{1}{2}f(x)$

c. $f(2x)$

d. $f\left(\dfrac{x}{2}\right)$

E-3. **Combinations of shifting and stretching:** The graph of $f(x)$ is shown in Figure 2.55. Sketch the graphs of the following functions.

a. $f\left(\dfrac{x}{2}\right)$

b. $2f\left(\dfrac{x}{2}\right)$

c. $2f\left(\dfrac{x}{2}\right) + 1$

E-4. **Periodic functions:** Some functions are *periodic* in that they repeat the same pattern over and over. The *period* of such a function is the time required for it to repeat. The graph over one period of a periodic function f is shown in Figure 2.56. Figure 2.57 shows two periods of the function. This function is periodic with period 2. Formally, f is *periodic* with *period p* if p is the smallest non-negative number such that $f(x + p) = f(x)$ for all x. The following questions refer to the function graphed in Figure 2.56.

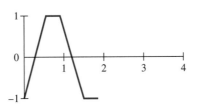

FIGURE 2.56 One period of f **FIGURE 2.57** Two periods of f

    a. Sketch the graph of $f(x) + 1$. What is the period of $f(x) + 1$?

    b. Sketch the graph of $f(x + 1)$. What is the period of $f(x + 1)$?

    c. Sketch the graph of $3f(x)$. What is the period of $3f(x)$?

    d. Sketch the graph of $f(3x)$. What is the period of $f(3x)$?

E-5. **Adding functions:** Use your calculator to plot on the same screen the graphs of x, x^2, and $x + x^2$. In general, explain how you get the graph of $f(x) + g(x)$ from the graphs of $f(x)$ and $g(x)$.

E-6. **Reflections:** The graph of $f(x)$ is shown in Figure 2.55.

    a. Sketch the graph of $-f(x)$. If you do this properly, you will produce the reflection of the graph through the horizontal axis.

    b. Sketch the graph of $f(-x)$. If you do this properly, you will see the reflection of the graph through the vertical axis.

    c. Sketch the graph of $-f(-x)$. Describe the result in terms of reflections of the graph of f.

2.2 SKILL BUILDING EXERCISES

S-1. **Graphs and function values:** Get the standard view of the graph of $f = 2 - x^2$. Use the graph to get the value of $f(3)$.

S-2. **Graphs and function values:** Get the standard view of the graph of $f = x^3/30 + 1$. Use the graph to get the value of $f(3)$.

S-3. **Graphs and function values:** Get the standard view of the graph of $f = (x^2 + 2^x)/(x + 10)$. Use the graph to get the value of $f(3)$.

S-4. **Zooming in:** Get the standard view of $x - x^4/75$. Then zoom in once near the peak of the graph.

S-5. **Finding a window:** Find an appropriate window setup that will show a good graph of $x^3/500$ with a horizontal span of -3 to 3.

S-6. **Finding a window:** Find an appropriate window setup that will show a good graph of $2^x - x^2$ with a horizontal span of 0 to 5.

S-7. **Finding a window:** Find an appropriate window setup that will show a good graph of $(x^4 + 1)/(x^2 + 1)$ with a horizontal span of 0 to 300.

S-8. **Finding a window:** Find an appropriate window setup that will show a good graph of $\sqrt{x^2 + 10} - \sqrt{x^2 + 5}$ with a horizontal span of 0 to 10.

S-9. **Finding a window:** Find an appropriate window setup that will show a good graph of $1/(x^2 + 1)$ with a horizontal span of -2 to 2.

S-10. **Two graphs:** Show the graphs of $f = x + 1$ and $g = 3 - x$ together on the same screen. (Use the standard viewing window.)

2.2 EXERCISES

In each of the following exercises, you are asked to produce a graph that you should turn in as part of the solution. Ideally you would transfer the graph via a computer link to a printer. If such technology is not available to you, you should provide hand-drawn copies of calculator-generated graphs. Be sure to label your graphs, to include identifying names for the horizontal and vertical axes, and to indicate the graphing window you use. Note also that the graphing windows you choose may not be the same as those we used to make the odd-answer key. Thus you should not expect your graphs to match ours exactly.

1. **Weekly cost:** The weekly cost of running a small firm is a function of the number of employees. Every week there is a fixed cost of $2500, and each employee costs the firm $350. For example, if there are 10 employees, then the weekly cost is $2500 + 350 \times 10 = 6000$ dollars.

   a. What is the weekly cost if there are 3 employees?

   b. Find a formula for the weekly cost as a function of the number of employees. (You need to choose variable and function names. Be sure to state the units.)

   c. Make a graph of the weekly cost as a function of the number of employees. Include values of the variable up to 10 employees.

   d. For what number of employees will the weekly cost be $4250?

2. **Average speed:** A commuter regularly drives 70 miles from home to work, and the amount of time required for the trip varies widely as a result of road and traffic conditions. The average speed for such a trip is a function of the time required. For example, if the trip takes 2 hours, then the average speed is $\frac{70}{2} = 35$ miles per hour.

   a. What is the average speed if the trip takes an hour and a half?

   b. Find a formula for the average speed as a function of the time required for the trip. (You need to choose variable and function names. Be sure to state the units.)

   c. Make a graph of the average speed as a function of the time required. Include trips from 1 hour to 3 hours in length.

   d. Is the graph concave up or concave down? Explain in practical terms what this means.

3. **Resale value:** The resale value V, in dollars, of a certain car is a function of the number of years t since the year 2000. In the year 2000 the resale value is $18,000, and each year thereafter the resale value decreases by $1700.

   a. What is the resale value in the year 2001?

   b. Find a formula for V as a function of t.

   c. Make a graph of V versus t covering the first 4 years since the year 2000.

   d. Use functional notation to express the resale value in the year 2003, and then calculate that value.

4. **Profit:** The yearly profit P for a widget producer is a function of the number n of widgets sold. The formula is

   $$P = -180 + 100n - 4n^2 .$$

   Here P is measured in thousands of dollars, n is measured in thousands of widgets, and the formula is valid up to a level of 20 thousand widgets sold.

   a. Make a graph of P versus n.

   b. Calculate $P(0)$ and explain in practical terms what your answer means.

   c. What profit will the producer make if 15 thousand widgets are sold?

   d. The break-even point is the sales level at which the profit is 0. Approximate the break-even point for this widget producer.

   e. What is the largest profit possible?

5. **Baking a potato:** A potato is placed in a preheated oven to bake. Its temperature $P = P(t)$ is given by

   $$P = 400 - 325e^{-t/50} ,$$

   where P is measured in degrees Fahrenheit and t is the time in minutes since the potato was placed in the oven.

   a. Make a graph of P versus t. (*Suggestion:* In choosing your graphing window, it is reasonable to look at the potato over no more than a 2-hour period. After that, it will surely be burned to a crisp. You may wish to look at a table of values to select a vertical span.)

   b. What was the initial temperature of the potato?

   c. Did the potato's temperature rise more during the first 30 minutes or the second 30 minutes of bak-

ing? What was the average rate of change per minute during the first 30 minutes? What was the average rate of change per minute during the second 30 minutes?

   d. Is this graph concave up or concave down? Explain what that tells you about how the potato heats up, and relate this to part c.

   e. The potato will be done when it reaches a temperature of 270 degrees. Approximate the time when the potato will be done.

   f. What is the temperature of the oven? Explain how you got your answer. (*Hint:* If the potato were left in the oven for a long time, its temperature would match that of the oven.)

6. **Functional response:** The amount C of food consumed in a day by a sheep is a function of the amount V of vegetation available, and a model is

   $$C = \frac{3V}{50 + V} .$$

   Here C is measured in pounds and V in pounds per acre. This relationship is called the *functional response*.

   a. Make a graph of C versus V. Include vegetation levels up to 1000 pounds per acre.

   b. Calculate $C(300)$ and explain in practical terms what your answer means.

   c. Is the graph concave up or concave down? Explain in practical terms what this means.

   d. From the graph it should be apparent that there is a limit to the amount of food consumed as more and more vegetation is available. Find this limiting value of C.

7. **Population growth:** The growth G of a population over a week is a function of the population size n at the beginning of the week. If both n and G are measured in thousands of animals, the formula is

   $$G = -0.25n^2 + 5n .$$

   a. Make a graph of G versus n. Include values of n up to 25 thousand animals.

   b. Use functional notation to express the growth over a week if the population at the beginning is 4 thousand animals, and then calculate that value.

c. Calculate $G(22)$ and explain in practical terms what your answer means.

d. For what values of n is the function G increasing? Determine whether the graph is concave up or concave down for these values, and explain in practical terms what this means.

8. **Ohm's law** says that when electric current is flowing across a resistor, the current i, measured in amperes, can be calculated from the voltage v, measured in volts, and the resistance R, measured in ohms. The relationship is given by

$$i = \frac{v}{R} \text{ amperes}.$$

a. A resistor in a radio circuit is rated at 4000 ohms.

   i. Find a formula for the current as a function of the voltage.

   ii. Plot the graph of i versus v. Include values of the voltage up to 12 volts.

   iii. What happens to the current when voltage increases?

b. The lights on your car operate on a 12-volt battery.

   i. Find a formula for the current in your car lights as a function of the resistance.

   ii. Plot the graph of i versus R. We suggest a horizontal span here of 1 to 25.

   iii. What happens to the current when resistance increases?

9. **The economic order quantity model** tells a company how many items at a time to order so that inventory costs will be minimized. The number $Q = Q(N, c, h)$ of items that should be included in a single order depends on the demand N per year for the product, the fixed cost c in dollars associated with placing a single order (not the price of the item), and the carrying cost h in dollars. (This is the cost of keeping an unsold item in stock.) The relationship is given by

$$Q = \sqrt{\frac{2Nc}{h}}.$$

a. Assume that the demand for a certain item is 400 units per year and that the carrying cost is $24 per unit per year. That is, $N = 400$ and $h = 24$.

   i. Find a formula for Q as a function of the fixed ordering cost c, and plot its graph. For this

particular item, we do not expect the fixed ordering costs ever to exceed $25.

   ii. Use the graph to find the number of items to order at a time if the fixed ordering cost is $6 per order.

   iii. How should increasing fixed ordering cost affect the number of items you order at a time?

b. Assume that the demand for a certain item is 400 units per year and that the fixed ordering cost is $14 per order.

   i. Find a formula for Q as a function of the carrying cost h and make its graph. We do not expect the carrying cost for this particular item ever to exceed $25.

   ii. Use the graph to find the optimal order size if the carrying cost is $15 per unit per year.

   iii. How should an increase in carrying cost affect the optimal order size?

   iv. What is the average rate of change per dollar in optimal order size if the carrying cost increases from $15 to $18?

   v. Is this graph concave up or concave down? Explain what that tells you about how optimal order size depends on carrying costs.

10. **Monthly payment for a home:** If you borrow $120,000 at an APR of 6% in order to buy a home, and if the lending institution compounds interest continuously, then your monthly payment $M = M(Y)$, in dollars, depends on the number of years Y you take to pay off the loan. The relationship is given by

$$M = \frac{120000\left(e^{0.005} - 1\right)}{1 - e^{-0.06Y}}.$$

a. Make a graph of M versus Y. In choosing a graphing window, you should note that a home mortgage rarely extends beyond 30 years.

b. Express in functional notation your monthly payment if you pay off the loan in 20 years, and then use the graph to find that value.

c. Use the graph to find your monthly payment if you pay off the loan in 30 years.

d. From part b to part c of this problem, you increased the debt period by 50%. Did this decrease your monthly payment by 50%?   →

e. Is the graph concave up or concave down? Explain your answer in practical terms.

f. Calculate the average decrease per year in your monthly payment from a loan period of 25 to a loan period of 30 years.

11. **An annuity:** Suppose you are able to find an investment that pays a monthly interest rate of r as a decimal. You want to invest P dollars that will help support your child. If you want your child to be able to withdraw M dollars per month for t months, then the amount you must invest is given by

$$P = M \times \frac{1}{r} \times \left(1 - \frac{1}{(1+r)^t}\right) \text{ dollars}.$$

A fund such as this is known as an *annuity*. For the remainder of this problem, we suppose that you have found an investment with a monthly interest rate of 0.01 and that you want your child to be able to withdraw $200 from the account each month.

a. Find a formula for your initial investment P as a function of t, the number of monthly withdrawals you want to provide, and make a graph of P versus t. Be sure your graph shows up through 40 years (480 months).

b. Use the graph to find out how much you need to invest so that your child can withdraw $200 per month for 4 years.

c. How much would you have to invest if you wanted your child to be able to withdraw $200 per month for 10 years?

d. A *perpetuity* is an annuity that allows for withdrawals for an indefinite period. How much money would you need to invest so that your descendants could withdraw $200 per month from the account forever? Be sure to explain how you got your answer.

12. **Alexander's formula:** One interesting problem in the study of dinosaurs is to determine from their tracks how fast they ran. The scientist R. McNeill Alexander developed a formula giving the velocity of any running animal in terms of its stride length and the height of its hip above the ground.[7]

The stride length of a dinosaur can be measured from successive prints of the same foot, and the hip height (roughly the leg length) can be estimated on the basis of the size of a footprint, so Alexander's formula gives a way of estimating from dinosaur tracks how fast the dinosaur was running. See Figure 2.58.

If the velocity v is measured in meters per second, and the stride length s and hip height h are measured in meters, then Alexander's formula is

$$v = 0.78s^{1.67}h^{-1.17}.$$

(For comparison, a length of 1 meter is 39.37 inches, and a velocity of 1 meter per second is about 2.2 miles per hour.)

a. First we study animals with varying stride lengths but all with a hip height of 2 meters (so $h = 2$).

i. Find a formula for the velocity v as a function of the stride length s.

ii. Make a graph of v versus s. Include stride lengths from 2 to 10 meters.

iii. What happens to the velocity as the stride length increases? Explain your answer in practical terms.

iv. Some dinosaur tracks show a stride length of 3 meters, and a scientist estimates that the hip height of the dinosaur was 2 meters. How fast was the dinosaur running?

b. Now we study animals with varying hip heights but all with a stride length of 3 meters (so $s = 3$).

i. Find a formula for the velocity v as a function of the hip height h.

ii. Make a graph of v versus h. Include hip heights from 0.5 to 3 meters.

iii. What happens to the velocity as the hip height increases? Explain your answer in practical terms.

[7]See his article "Estimates of speeds of dinosaurs," *Nature* **261** (1976), 129–130. See also his book *Animal Mechanics*, 2nd ed. (Oxford: Blackwell, 1983).

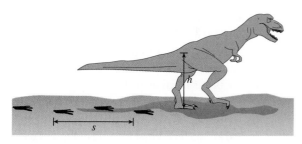

FIGURE 2.58

13. **Artificial gravity:** To compensate for weightlessness in a space station, artificial gravity can be produced by rotating the station.[8] The required number N of rotations per minute is a function of two variables: the distance r to the center of rotation, and a, the desired acceleration (or magnitude of artificial gravity). See Figure 2.59. The formula is

$$N = \frac{30}{\pi} \times \sqrt{\frac{a}{r}}.$$

We measure r in meters and a in meters per second per second.

a. First we assume that we want to simulate the gravity of Earth, so $a = 9.8$ meters per second per second.

   i. Find a formula for the required number N of rotations per minute as a function of the distance r to the center of rotation.

   ii. Make a graph of N versus r. Include distances from 10 to 200 meters.

   iii. What happens to the required number of rotations per minute as the distance increases? Explain your answer in practical terms.

   iv. What number of rotations per minute is necessary to produce Earth gravity if the distance to the center is 150 meters?

b. Now we assume that the distance to the center is 150 meters (so $r = 150$).

   i. Find a formula for the required number N of rotations per minute as a function of the desired acceleration a.

   ii. Make a graph of N versus a. Include values of a from 2.45 (one-quarter of Earth gravity) to 9.8 meters per second per second.

   iii. What happens to the required number of rotations per minute as the desired acceleration increases?

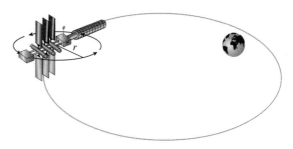

FIGURE 2.59

14. **Plant growth:** The amount of growth of plants in an ungrazed pasture is a function of the amount of plant biomass already present and the amount of rainfall.[9] For a pasture in the arid zone of Australia, the formula

$$Y = -55.12 - 0.01535N - 0.00056N^2 + 3.946R$$

gives an approximation of the growth. Here R is the amount of rainfall (in millimeters) over a 3-month period, N is the plant biomass (in kilograms per hectare) at the beginning of that period, and Y is the growth (in kilograms per hectare) of the biomass over that period. (For comparison, 100 millimeters is about 3.9 inches, and 100 kilograms per hectare is about 89 pounds per acre.)

   For this exercise, assume that the amount of plant biomass initially present is 400 kilograms per hectare, so $N = 400$.

a. Find a formula for the growth Y as a function of the amount R of rainfall.

b. Make a graph of Y versus R. Include values of R from 40 to 160 millimeters.

c. What happens to Y as R increases? Explain your answer in practical terms. →

---

[8]This exercise is based on B. Kastner's *Space Mathematics*, published by NASA in 1985.

[9]This exercise and the next are based on the work of G. Robertson, "Plant dynamics," in G. Caughley, N. Shepherd, and J. Short, eds., *Kangaroos* (Cambridge, England: Cambridge University Press, 1987).

d. How much growth will there be over a 3-month period if initially there are 400 kilograms per hectare of plant biomass and the amount of rainfall is 100 millimeters?

15. **More on plant growth:** *This is a continuation of Exercise 14.* Now we consider the amount of growth Y as a function of the amount of plant biomass N already present.

   a. First we assume that the rainfall is 100 millimeters, so $R = 100$.

      i. Find a formula for the growth Y as a function of the amount N of plant biomass already present.

      ii. Make a graph of Y versus N. Include biomass levels N from 0 to 800 kilograms per hectare.

      iii. What happens to the amount of growth Y as the amount N of plant biomass already present increases? Explain your answer in practical terms.

   b. Next we assume that the rainfall is 80 millimeters, so $R = 80$.

      i. With this lower rainfall level, find a formula for the growth Y as a function of the amount N of plant biomass already present.

      ii. Add the graph of Y versus N for the lower rainfall to the graph you found in part a.

      iii. According to the graph you just found, what is the effect of lowered rainfall on plant growth? Is your answer consistent with that in part c of Exercise 14?

16. **Viewing Earth:** Astronauts looking at Earth from a spacecraft can see only a portion of the surface.[10] See Figure 2.60. The fraction F of the surface of Earth that is visible at a height h, in kilometers, above the surface is given by the formula

$$F = \frac{0.5h}{R + h}.$$

Here R is the radius of Earth, about 6380 kilometers. (For comparison, 1 kilometer is about 0.62 mile, and the moon is about 380,000 kilometers from Earth.)

a. Make a graph of F versus h covering heights up to 100,000 kilometers.

b. A value of F equal to 0.25 means that 25%, or one-quarter, of Earth's surface is visible. At what height is this fraction visible?

c. During one flight of a space shuttle, astronauts performed an extravehicular activity at a height of 280 kilometers. What fraction of the surface of Earth is visible at that height?

d. Is the graph of F concave up or concave down? Explain your answer in practical terms.

e. Determine the limiting value for F as the height h gets larger. Explain your answer in practical terms.

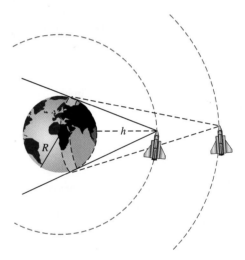

FIGURE 2.60

17. **Magazine circulation:** The circulation C of a certain magazine as a function of time t is given by the formula

$$C = \frac{5.2}{0.1 + 0.3^t}.$$

Here C is measured in thousands, and t is measured in years since the beginning of 1992, when the magazine was started.

a. Make a graph of C versus t covering the first 6 years of the magazine's existence.

---

[10]This exercise is based on B. Kastner's *Space Mathematics*, published by NASA in 1985.

b. Express using functional notation the circulation of the magazine 18 months after it was started, and then find that value.

c. Over what time interval is the graph of C concave up? Explain your answer in practical terms.

d. At what time was the circulation increasing the fastest?

e. Determine the limiting value for C. Explain your answer in practical terms.

18. **Growth:** The length L, in inches, of a certain flatfish is given by the formula

$$L = 15 - 19 \times 0.6^t,$$

and its weight W, in pounds, is given by the formula

$$W = (1 - 1.3 \times 0.6^t)^3.$$

Here t is the age of the fish, in years, and both formulas are valid from the age of 1 year.

a. Make a graph of the length of the fish against its age, covering ages 1 to 8.

b. To what limiting length does the fish grow? At what age does it reach 90% of this length?

c. Make a graph of the weight of the fish against its age, covering ages 1 to 8.

d. To what limiting weight does the fish grow? At what age does it reach 90% of this weight?

e. One of the graphs you made in parts a and c should have an inflection point, whereas the other is always concave down. Identify which is which, and explain in practical terms what this means. Include in your explanation the approximate location of the inflection point.

19. **Buffalo:** Waterton Lakes National Park of Canada, where the Great Plains dramatically meet the Rocky Mountains in Alberta, has a migratory buffalo (bison) herd that spends falls and winters in the Park. The herd is currently managed and so kept small; however, if it were unmanaged and allowed to grow, then the number N of buffalo in the herd could be estimated by the logistic formula

$$N = \frac{315}{1 + 14e^{-0.23t}}.$$

Here t is the number of years since the beginning of 2002, the first year the herd is unmanaged.

a. Make a graph of N versus t covering the next 30 years of the herd's existence (corresponding to dates up to 2032).

b. How many buffalo are in the herd at the beginning of 2002?

c. When will the number of buffalo first exceed 300?

d. How many buffalo will there eventually be in the herd?

e. When is the graph of N, as a function of t, concave up? When is it concave down? What does this mean in terms of the growth of the buffalo herd?

20. **Doppler effect:** Motion toward or away from us distorts the pitch of sound, and it also distorts the wavelength of light. This phenomenon is known as the *Doppler effect*. In the case of light, the distortion is measurable only for objects moving at extremely high velocities. Motion of objects toward us produces a *blue shift* in the spectrum, whereas motion of objects away from us produces a *red shift*. Quantitatively, the red shift S is the change in wavelength divided by the unshifted wavelength, and thus red shift is a pure number that has no units associated with it. Cosmologists believe that the universe is expanding and that many stellar objects are moving away from Earth at *radial velocities* sufficient to produce a measurable red shift. Particularly notable among these are quasars, which have a number of important properties (some of which remain poorly understood).[11] Quasars are moving rapidly away from us and thus produce a large red shift. The radial velocity V can be calculated from the red shift S using

$$V = c \times \left(\frac{(S+1)^2 - 1}{(S+1)^2 + 1} \right),$$

where c is the speed of light. →

---

[11] The universe is thought to be 15 to 20 billion years old. Quasars have been located as far away as 15 billion light years, so their light reaching Earth shows the universe when it was relatively young.

a. Most known quasars have a red shift greater than 1. What would be the radial velocity of a quasar showing a red shift of 2? Report your answer as a multiple of the speed of light.

b. Make a graph of the radial velocity (as a multiple of the speed of light) versus the red shift. Include values of the red shift from 0 to 5.

c. The quasar **3C 48** shows a red shift of 0.37. How fast is **3C 48** moving away from us?

d. Find approximately the red shift that would indicate a radial velocity of half the speed of light.

e. What is the maximum theoretical radial velocity that a quasar could achieve?

21. **Research project:** Find a function given by a formula in one of your textbooks for another class or some other handy source. If the formula involves more than one variable, assign reasonable values for all the variables except one so that your formula involves only one variable. Now make a graph, using an appropriate horizontal and vertical span so that the graph shows some interesting aspect of the function, such as a trend, significant values, or concavity. Carefully describe the function, formula, variables (including units), and graph, and explain how the graph is useful.

2.3 *Solving Linear Equations*

Many times we will need to find when two functions are equal or when a function value is zero. To do this we will need to solve an equation. For our purposes, equations come in two types, *linear equations* and *nonlinear equations*. Linear equations are the simplest kind of equation; they are those that do not involve powers, square roots, or other complications of the variable for which we want to solve.

The Basic Operations

As you may recall from your elementary algebra course, linear equations can always be solved using two basic operations: subtraction (or addition) and division (or multiplication). The following is presented as a reminder.

> **Adding to both sides of an equation:** You may add (or subtract) the same thing to (or from) both sides of an equation. This can be thought of as moving a term from one side of the equation to the other, provided that you change its sign, positive to negative or negative to positive.

> **Divide (or multiply) both sides of an equation by any nonzero number:** The same thing may be accomplished if you divide (or multiply) each term of an equation by any nonzero number; however, the results may sometimes appear to be different.

We will show how to use these operations to solve $5x + 7 = 18 - 2x$. The steps in solving a linear equation are always the same.

Step 1: Move all terms that include the variable to one side of the equation and all terms that do not involve the variable to the other side of the equation, changing the sign of each moved term. This is accomplished by adding (or subtracting) the same thing to (or from) both sides of an equation.

In our case we need to move $-2x$ to the left side of the equation and 7 to the right side. We can do this by adding $2x$ to both sides of the equation and subtracting 7 from both sides. We get

$$\underbrace{5x + 7}_{\text{Add } 2x \text{ and subtract } 7} = \underbrace{18 - 2x}_{\text{Add } 2x \text{ and subtract } 7}$$

$$5x + 2x = 18 - 7 .$$

Step 2: Combine terms. Combining $5x + 2x$ into a single term is an easy task. For example, 5 cars + 2 cars is 7 cars, and it works the same with x's: $5x + 2x = 7x$. We get

$$\underbrace{5x + 2x}_{\text{Combine to } 7x} = \underbrace{18 - 7}_{\text{Combine to } 11}$$

$$7x = 11 .$$

Step 3: Divide by the coefficient of x. In our case, we want to divide both sides of the equation by 7:

$$\underbrace{7x}_{\text{Divide by 7}} = \underbrace{11}_{\text{Divide by 7}}$$

$$x = \frac{11}{7} = 1.57 \,,$$

rounding to two decimal places.

A word of caution is in order concerning division in equations. Sometimes we may have more than two terms in our equation when we need to divide, and the results will *appear* to be different, depending on whether we divide both sides of the equation or divide each term of the equation. For example, suppose that in step 3 above we had $7x = 11 + 8y$, and our goal was to solve for x. We could divide both sides of the equation by 7 to get $x = (11 + 8y)/7$. Alternatively, we could divide each term by 7, but in that case we must remember to divide both 11 and $8y$. The result is $x = 11/7 + 8y/7$, or (rounding to two decimal places) $x = 1.57 + 1.14y$. Even though the two answers, $x = (11 + 8y)/7$ and $x = 1.57 + 1.14y$, may appear to be different, they are the same (allowing for round-off error).

EXAMPLE 2.5 Rental Cars

A car rental company charges $38.00 in insurance and other fees plus a flat rate of $12.00 per day.

1. Use a formula to express the cost of renting a car as a function of the number of days that you keep it.

2. Your expense account allows you $200 to spend on a rental car. How many days can you keep the car without exceeding your expense account?

Solution to Part 1: We need to choose variable and function names. We let d be the number of days that we keep the car and C the cost (in dollars) of renting the car.

The cost is the initial fee of $38 plus $12 per day:

$$\text{Cost} = \text{Initial fee} + 12 \times \text{Number of days}\,, \quad \text{or} \quad C = 38 + 12d \,.$$

Solution to Part 2: We want to know how many rental days will make the cost be $200. That is, we want to solve the following equation for d:

$$C = 200\,, \quad \text{or} \quad 38 + 12d = 200\,.$$

We solve it using the steps described above.

$$12d = 200 - 38 \qquad \text{Subtract 38 from both sides.}$$

$$12d = 162$$

$$d = \frac{162}{12} = 13.5\,. \quad \text{Divide by 12.}$$

Thus you can rent the car for $13\frac{1}{2}$ days without exceeding your expense account. Since the rental company will probably charge you a full day's price for a half day's rental, you may only be able to rent the car for 13 days.   ▪ ▪ ▪

Reversing the Roles of Variables

Sometimes when we are given formulas expressing one thing in terms of another, we can gain additional information by changing the roles of the variables. In formal mathematical terms, we are finding the *inverse* of a function.

For example, the relationship between the temperature in degrees Fahrenheit F and the temperature in degrees Celsius C is given by the formula

$$F = 1.8C + 32 . \tag{2.4}$$

This formula tells us that a room where the temperature is 30 degrees Celsius has a temperature of $F = 1.8 \times 30 + 32 = 86$ degrees Fahrenheit.

But Equation (2.4) does not immediately tell us the temperature in degrees Celsius of a room that is 75 degrees Fahrenheit. To find this, we need to rearrange the formula so that it shows C in terms of F. That is, we need to solve Equation (2.4) for C. Since the equation is linear, we know how to do that.

$$F = \ 1.8C + 32$$

$$F - 32 = 1.8C \quad \text{Subtract 32 from both sides.}$$

$$\frac{F - 32}{1.8} = C . \quad \text{Divide by 1.8.}$$

We can use this formula to find out the temperature in degrees Celsius of a room where the temperature is 75 degrees Fahrenheit:

$$C = \frac{75 - 32}{1.8} = 23.89 \text{ degrees Celsius} .$$

In the derivation above, we divided both sides of the equation by 1.8. Alternatively, we could have divided each term by 1.8, giving

$$C = \frac{F}{1.8} - \frac{32}{1.8} .$$

The two expressions for C in terms of F look different, but they are in fact the same, and either is acceptable.

EXAMPLE 2.6   **A Moving Car**

If a car moves at a constant velocity, then the distance traveled d can be expressed as a function of velocity v and the time traveled t. If we measure d in miles, v in miles per hour, and t in hours, then the relationship is

$$\text{Distance} = \text{Velocity} \times \text{Time}$$

$$d = vt .$$

1. What distance have you traveled if you drive 55 miles per hour for $2\frac{1}{2}$ hours?

2. Use a formula to express v as a function of d and t. Use your function to find your velocity if it takes you 3 hours to drive 172 miles.

3. Use a formula to express time as a function of velocity and distance traveled. Use your function to find the time it takes to travel 230 miles at a velocity of 40 miles per hour.

Solution to Part 1: Simply plug in the given numbers for v and t:

$$d = 55 \times 2.5 = 137.5 \text{ miles}.$$

Solution to Part 2: You need to solve the equation $d = vt$ for v. Divide both sides by t to obtain

$$v = \frac{d}{t} \text{ miles per hour}.$$

Note that this is the familiar formula that says velocity is distance divided by time:

$$\text{Velocity} = \frac{\text{Distance}}{\text{Time}}.$$

If you take 3 hours to drive 172 miles, then your velocity is $v = \frac{172}{3} = 57.33$ miles per hour.

Solution to Part 3: This time you want to solve the equation $d = vt$ for t. Divide both sides by v and obtain

$$t = \frac{d}{v}.$$

This says

$$\text{Time} = \frac{\text{Distance}}{\text{Velocity}}.$$

Driving 230 miles at 40 miles per hour, it takes $\frac{230}{40} = 5.75$ hours to complete the trip. ■ ■ ■

EXAMPLE 2.7 Car Rentals Again

Company Alpha charges an initial fee of $28.00, a daily rate of $4.00, and a rate of 29 cents per mile. Company Beta charges an initial fee of $32.00, a daily rate of $6.00, and a rate of 14 cents per mile. You need a rental car for 3 days.

1. Use a formula to express the cost of renting a car from Company Alpha as a function of the number of miles you drive.

2. Use a formula to express the cost of renting a car from Company Beta as a function of the number of miles you drive.

3. For what number of miles driven are the cost of renting a car from Company Alpha and the cost of renting a car from Company Beta the same?
4. On the same screen, plot the graphs of the cost of renting from Company Alpha and the cost of renting from Company Beta.
5. How do you decide from which company to rent?

Solution to Part 1: First choose variable and function names: Let m be the number of miles driven and A the cost, in dollars, of renting a car from Company Alpha. The initial fee for Company Alpha is \$28, and you pay \$12 for the 3-day rental. In addition, you pay \$0.29 for each mile you drive.

$$\begin{aligned}\text{Cost for Alpha} &= \text{Initial fee} + \text{3-day price} + 0.29 \times \text{miles driven}\\ A &= 28 + 12 + 0.29m\\ &= 40 + 0.29m\end{aligned}$$

Solution to Part 2: Let B be the cost, in dollars, of renting a car from Company Beta. For Company Beta you pay the \$32 initial fee, plus \$18 for the 3-day rental, plus \$0.14 for each mile you drive.

$$\begin{aligned}\text{Cost for Beta} &= \text{Initial fee} + \text{3-day price} + 0.14 \times \text{miles driven}\\ B &= 32 + 18 + 0.14m\\ &= 50 + 0.14m\end{aligned}$$

Solution to Part 3: We want to find when the cost for the two companies is the same.

$$\begin{aligned}\text{Cost for Alpha} &= \text{Cost for Beta}\\ A &= B\\ 40 + 0.29m &= 50 + 0.14m\\ 0.29m - 0.14m &= 50 - 40 && \text{Subtract 40 and } 0.14m \text{ from both sides.}\\ 0.15m &= 10\\ m &= \frac{10}{0.15} = 66.67 && \text{Divide by 0.15.}\end{aligned}$$

Thus the costs for the two companies will be the same if you drive 66.67 miles.

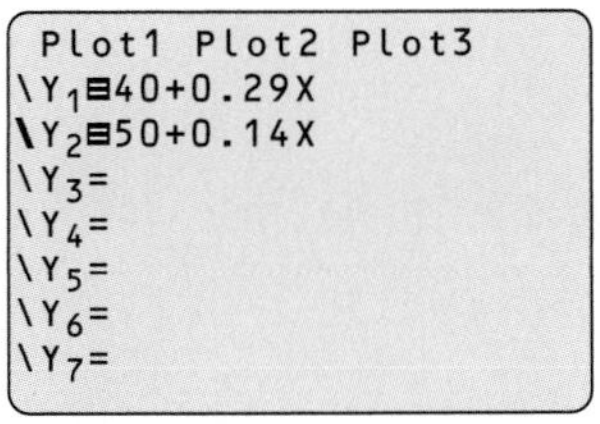
Plot1 Plot2 Plot3
\Y1=40+0.29X
\Y2=50+0.14X
\Y3=
\Y4=
\Y5=
\Y6=
\Y7=

FIGURE 2.61 Entering both functions

Solution to Part 4: We need to enter [2.35] both functions in the calculator. This is shown in Figure 2.61, and we list the appropriate correspondences:

$$\begin{aligned}\mathsf{Y_1} &= A, \text{ cost for Alpha on vertical axis}\\ \mathsf{Y_2} &= B, \text{ cost for Beta on vertical axis}\\ \mathsf{X} &= m, \text{ miles driven on horizontal axis}.\end{aligned}$$

Before we make the graphs, let's think about how to set up the window. We surely want the picture to show where the cost is the same, and so we want to include 66.67 in the horizontal span. This will show nicely if the horizontal span is from

X	Y$_1$	Y$_2$
0	40	50
20	45.8	52.8
40	51.6	55.6
60	57.4	58.4
80	63.2	61.2
100	69	64
120	74.8	66.8
X=0		

FIGURE 2.62 A table of values for car rentals

```
WINDOW
 Xmin=0
 Xmax=100
 Xscl=1
 Ymin=20
 Ymax=90
 Yscl=1
 Xres=1
```

FIGURE 2.63 Configuring the window

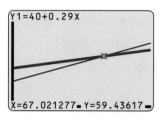

Y1=40+0.29X

X=67.021277∎ Y=59.43617∎

FIGURE 2.64 Thin graph for Company Alpha, thick graph for Company Beta

0 to 100 miles. To choose the vertical span, we look at the table of values shown in Figure 2.62. We used a starting value of 0 and a table increment of 20. Note that in this table, the Y$_1$ column corresponds to Company Alpha and the Y$_2$ column corresponds to Company Beta. Allowing a little extra room, we choose a vertical span of 20 to 90.

The properly configured window 2.36 is shown in Figure 2.63. When we graph, both plots appear as in Figure 2.64. In this figure, the thinner line represents Company Alpha, and the thicker line 2.37 represents Company Beta.

Solution to Part 5: We want to spend only as much money as is necessary. Figure 2.64 shows that the graph of the cost of Alpha is initially on the bottom, indicating that it costs less. But after the graphs cross at about $m = 67$ miles, the graph of the cost of Beta is on the bottom, indicating that it costs less. We found the exact value of this crossing point in part 3. Thus, if we are going to drive for less than 66.67 miles, we should rent from Company Alpha. If we need to drive farther, we should rent from Company Beta. ∎ ∎ ∎

ANOTHER LOOK

Equations That Are Linear in One Variable

Sometimes, equations that contain more than one variable are linear in one variable but not linear in other variables. For example, the equation $xy^2 = 4$ is linear in x but is not linear in y since it contains an y^2. Nonetheless, we can solve the equation for x in the same way we would solve $xA = 4$ for x. The coefficient of x is y^2, so we divide each side by y^2:

$$xy^2 = 4$$

$$x = \frac{4}{y^2}.$$

Many times, in order to solve such equations, it is necessary to do some elementary factoring. The key factoring rule here is

$$xA + xB = x(A + B).$$

We can use this rule, for example, to solve

$$xy^3 = xy^2 + z$$

for x. Note that the equation is linear in x and z but is not linear in y. We can solve the equation for either x or z. We solve it for x just as we would solve $xA = xB + z$. You are encouraged to solve that equation and compare your work with the steps below:

$$xy^3 = xy^2 + z$$
$$xy^3 - xy^2 = z$$
$$x\left(y^3 - y^2\right) = z$$
$$x = \frac{z}{y^3 - y^2} \,.$$

If instead we want to solve for z, then the procedure is much simpler:

$$xy^3 = xy^2 + z$$
$$xy^3 - xy^2 = z \,,$$

so

$$z = xy^3 - xy^2 \,.$$

Solving an equation that contains nonlinear terms may involve some messy work, but the basic ideas are the same as though we were solving an equation as simple as $ax = b$. If confusion arises, it may be helpful to replace the nonlinear terms with other letters so that the steps can be clearly seen. For example, the equation $x\sqrt{y+1} = 5x + (y-2)^3$ is linear in x (but not in y), and we can solve it for x. The equation may look a little less intimidating if we replace $\sqrt{y+1}$ by A and $(y-2)^3$ by B:

$$xA = 5x + B$$
$$xA - 5x = B$$
$$x(A - 5) = B$$
$$x = \frac{B}{A - 5} \,.$$

Thus the solution is

$$x = \frac{(y-2)^3}{\sqrt{y+1} - 5} \,.$$

Note in what follows that the work is exactly the same if we do not do the replacements:

$$x\sqrt{y+1} = 5x + (y-2)^3$$
$$x\sqrt{y+1} - 5x = (y-2)^3$$
$$x\left(\sqrt{y+1} - 5\right) = (y-2)^3$$
$$x = \frac{(y-2)^3}{\sqrt{y+1} - 5} \,.$$

The key in solving equations that are linear in one variable but nonlinear in other variables is to think of terms such as $\sqrt{y+1}$ or $(y-2)^3$ as single items such as A or B. If we have difficulties in the process of solving, it may be helpful actually to do the replacements as we did in the example above.

One useful application of solving linear equations is finding the formula for the inverses of certain functions. The *inverse* of a function $f(x)$ is an expression y such that $f(y) = x$. For example, if $f(x) = x - 1$ and $y = x + 1$, then y is the inverse of $f(x)$ because

$$f(y) = y - 1 = (x + 1) - 1 = x.$$

Some functions do not have inverses, and for some that do, the formula may be difficult or impossible to find. But for certain functions the procedure is relatively simple. Let's look, for example, at $f(x) = 2x + 7$. We are looking for an expression y such that $f(y) = x$; that is, we want to solve the equation $2y + 7 = x$ for y. Subtracting 7 from each side and then dividing by 2, we see that $y = (x - 7)/2$. Thus the inverse function of f is $y = (x - 7)/2$.

A more complex problem is to find the inverse function of $f(x) = (2x - 1)/(3x + 1)$. We are looking for an expression y such that $f(y) = x$. That is, we want to solve the equation $(2y - 1)/(3y + 1) = x$ for y. If we first multiply by the denominator of the left-hand side, we get an equation that is linear in y:

$$\frac{2y - 1}{3y + 1} = x$$

$$2y - 1 = x(3y + 1).$$

Then we solve for y:

$$2y - 1 = 3xy + x$$

$$2y - 3xy = x + 1$$

$$y(2 - 3x) = x + 1$$

$$y = \frac{x + 1}{2 - 3x}.$$

Thus the inverse function of f is $y = (x + 1)/(2 - 3x)$.

Enrichment Exercises

E-1. **Solving equations linear in one variable:** Each of the following equations is linear in x. Solve the equations for x.

    a. $xy^3 = xy + y$ b. $3x\sqrt{y} = 2x + \sqrt{y}$

    c. $yx + z^2x = z - y^2x$ d. $\sqrt{9 + y} + x\sqrt{7 + y} = yz + xz$

E-2. **Detecting and solving equations linear in one variable:** Each of the following equations is linear in one or more variables. Solve each equation for each variable in which it is linear.

    a. $xy + 4 = y + 7x$ b. $xy^2z = yx + xz + 1$

    c. $x^2 + y = x^3 + xyz$ d. $x + y + z = x^2 + y^2 + 1$

E-3. **Finding inverse functions:** Find a formula for the inverse of each of the following functions.

a. $f(x) = 7x + 5$

b. $f(x) = \dfrac{3x + 1}{2x - 5}$

c. $f(x) = \dfrac{2x - 4}{x}$

d. $f(x) = \dfrac{x}{4x - 3}$

E-4. **Finding inverse functions:** Find a formula for the inverse of each of the following functions.

a. $f(x) = ax + b$ with $a \neq 0$

b. $f(x) = \dfrac{a^3 x}{a^2 x + 1}$ with $a \neq 0$

c. $f(x) = \dfrac{ax + b}{cx + d}$ with $ad - bc \neq 0$

E-5. **Making equations linear:** Some equations are not linear but become linear after certain algebraic operations. For example, the equation $2/x = 3$ is not linear, but if we multiply both sides of the equation by x, we get the linear equation $2 = 3x$. Transform each of the following equations into a linear equation by applying the indicated operation. Then solve the equation for x.

a. $\dfrac{7}{x + 1} = 3$. Multiply each side by $x + 1$.

b. $\dfrac{9}{2x} = \dfrac{7}{3}$. Multiply each side by $6x$.

c. $3\sqrt{x} = a$. Square both sides of the equation.

d. $3\sqrt{x} = 2\sqrt{x} + \dfrac{4}{\sqrt{x}}$. Multiply both sides by $\sqrt{x}$.

E-6. **Substitutions:** Some equations are not linear but become linear when a proper substitution is made. For example, $2x^2 = 8$ is not linear, but if we put $y = x^2$, then we get the linear equation $2y = 8$. The solution is $y = 4$. Hence $x^2 = 4$, so $x = \pm 2$. Make the suggested substitution in each of the following equations and then solve.

a. $2x^3 = 4x^3 - 1$. Substitute $y = x^3$.

b. $\dfrac{4}{x} + \dfrac{7}{x} = 5$. Substitute $y = \dfrac{1}{x}$.

c. $(x^2 + 1)^2 = a(x^2 + 1)^2 + b$. Substitute $y = (x^2 + 1)^2$.

2.3   *SKILL BUILDING EXERCISES*

S-1. **Linear equations:** Solve $3x + 7 = x + 21$ for x.

S-2. **Linear equations:** Solve $2x + 4 = 5x - 12$ for x.

S-3. **Linear equations:** Solve $3 - 5x = 23 + 4x$ for x.

S-4. **Linear equations:** Solve $13x + 4 = 5x + 33$ for x.

S-5. **Linear equations:** Solve $12x + 4 = 55x + 42$ for x.

S-6. **Linear equations:** Solve $6 - x = 16 + x$ for x.

S-7. **Linear equations:** Solve for x: $cx + d = 12$.

S-8. **Linear equations:** Solve for p: $cp = r$.

S-9. **Linear equations:** Solve for k: $2k + m = 5k + n$.

S-10. **Linear equations:** Solve for t: $tx^2 = t + 1$.

2.3   *EXERCISES*

1. **Gross domestic product:** The *gross domestic product* is used by economists as a measure of the nation's economic position. The gross domestic product P is calculated as the sum of final[12] sales C to consumers, final sales and inventory changes B to businesses, final sales G to government, and net sales E to foreigners (exports minus imports). All of these are measured in billions of dollars.

   a. Find a formula that gives the gross domestic product P in terms of C, B, G, and E.

   b. In 1992, final sales to consumers was 4095.8 billion dollars, final sales and inventory changes for businesses was 770.4 billion dollars, final sales to government was 1114.9 billion dollars, and net sales to foreigners was -30.4 billion dollars.

      i. Which was larger in 1992, U.S. imports or exports?

      ii. Calculate the gross domestic product for 1992.

   c. Solve the equation in part a for E.

   d. Suppose that in another year the values for C, B, and G remain the same as the 1992 figures, but the gross domestic product is 5997.3 billion dollars. What is the net sales to foreigners for this year?

2. **Juice sales:** The number J, in thousands, of cans of frozen orange juice sold weekly is a function of the price P, in dollars, of a can. In a certain grocery store, the formula is

$$J = 11 - 2.5P.$$

   a. Express using functional notation the number of cans sold weekly if the price of a can is $1.40, and then calculate that value.

   b. At what price will there be 7.25 thousand cans sold weekly?

   c. Solve for P in the formula above to obtain a formula expressing P as a function of J.

   d. At what price will there be 7.75 thousand cans sold weekly?

3. **Resale value:** The resale value V, in thousands of dollars, of a boat is a function of the number of years t since the start of 2001, and the formula is

$$V = 12.5 - 1.1t.$$

   a. Calculate $V(3)$ and explain in practical terms what your answer means.

   b. In what year will the resale value be 7 thousand dollars?

   c. Solve for t in the formula above to obtain a formula expressing t as a function of V.

---

[12]When a home is constructed, the builder makes a number of purchases, such as lumber and heating and cooling units. The final sale price is the cost to the consumer of the completed home, and intermediate sales from businesses to the builder are not counted.

d. In what year will the resale value be 4.8 thousand dollars?

4. **Aerobic power:** Aerobic power can be thought of as the maximum oxygen consumption attainable per kilogram of body mass. There are a number of ways in which physical educators estimate this. One method uses the *Queens College Step Test*.[13] In this test, males step up and down a 16-inch bleacher step 24 times per minute for 3 minutes, and females do 22 steps per minute for 3 minutes. Five seconds after the exercise is complete, a 15-second pulse count P is taken. Maximum oxygen consumption M, in milliliters per kilogram, for males is approximated by

$$M = 111.3 - 1.68P.$$

For females the recommended formula is

$$F = 65.81 - 0.74P.$$

a. Calculate the maximum oxygen consumption for both a male and a female who show a 15-second pulse count of 40.

b. What 15-second pulse count for a male will indicate a maximum oxygen consumption of 35.7 milliliters per kilogram?

c. What 15-second pulse count will indicate the same maximum oxygen consumption for a male as for a female?

d. What maximum oxygen consumption is associated with your answer in part c?

5. **Gas mileage:** The distance d, in miles, that you can travel without stopping for gas depends on the number of gallons g of gasoline in your tank and the gas mileage m, in miles per gallon, that your car gets. The relationship is

$$d = gm. \qquad (2.5)$$

a. How far can you drive if you have 12 gallons of gas in your tank and your car gets 24 miles per gallon?

b. Solve Equation (2.5) for m.

i. Explain in everyday terms what this new equation means.

ii. Use this equation to determine the gas mileage of your car if you can drive 335 miles on a full 13-gallon tank of gas.

c. A Detroit engineer wants to be sure that the car she is designing can go 425 miles on a full tank of gas, and she must design a gas tank to ensure that. She does not yet know what gas mileage this new-model car will get, and so she decides to make a graph of the size of the gas tank as a function of gas mileage.

i. Solve Equation (2.5) for g using 425 for the distance.

ii. Make the graph that the engineer made. Is it a straight line?

6. **Isocost equation:** We are to buy quantities of two items: n_1 units of item 1 and n_2 units of item 2.

a. If item 1 costs \$3.50 per unit and item 2 costs \$2.80 per unit, find a formula that gives the total cost C, in dollars, of the purchase.

b. An *isocost equation* shows the relationship between the number of units of each of two items to be purchased when the total purchase price is predetermined. If the total purchase price is predetermined to be $C = 162.40$ dollars, find the isocost equation for items 1 and 2 from part a.

c. Solve for n_1 the isocost equation you found in part b.

d. Using the equation from part c, determine how many units of item 1 are to be purchased if 18 units of item 2 are purchased.

7. **Supply and demand:** The quantity S of barley, in billions of bushels, that barley suppliers in a certain country are willing to produce in a year and offer for sale at a price P, in dollars per bushel, is determined by the relation

$$P = 1.9S - 0.7. \qquad \longrightarrow$$

---

[13]Adapted from F. I. Katch and W. D. McArdle, *Nutrition, Weight Control, and Exercise* (Philadelphia: Lea & Febiger, 1983).

The quantity D of barley, in billions of bushels, that barley consumers are willing to purchase in a year at price P is determined by the relation

$$P = 2.8 - 0.6D.$$

The *equilibrium price* is the price at which the quantity supplied is the same as the quantity demanded. Find the equilibrium price for barley.

8. **Stock turnover at retail:** In retail sales an important marker of retail activity is the *stock turnover at retail*. This figure is calculated for a specific period of time as the total net sales divided by the retail value of the average stock during that time, where both are measured in dollars.[14] As a formula this is written

$$\text{Stock turnover} = \frac{\text{Net sales}}{\text{Average stock at retail}}.$$

This formula expresses stock turnover as a function of net sales and average stock at retail.

a. Suppose that your store had net sales of $682,000 in men's shoes over the past six months and that the retail value of the average stock of men's shoes was $163,000. What was the stock turnover at retail for that time period?

b. Suppose that in a certain month your store's net sales of women's dresses were $83,000 and that the usual stock turnover at retail is 0.8 per month. What do you estimate to be your store's average stock at retail?

c. Solve the equation for average stock at retail— that is, write a formula giving average stock at retail as a function of stock turnover and net sales.

d. Suppose that in a certain time period your store had an average stock of socks with a retail value of $45,000 and a stock turnover at retail of 1.6. What were the store's net sales of socks during that time period?

e. Solve the equation for net sales—that is, write a formula giving net sales as a function of stock turnover and average stock at retail.

9. **Fire engine pressure:** For firefighting an important concept is that of *engine pressure*, which is the pressure that the pumper requires to overcome friction loss from the hose and to maintain the proper pressure at the nozzle of the hose. One formula[15] for engine pressure is

$$EP = NP(1.1 + K \times L),$$

where EP is the engine pressure in pounds per square inch (psi), NP is the nozzle pressure in psi, K is the "K" factor (which depends on which tip is used on the nozzle), and L is the length of the hose in feet divided by 50. This formula expresses EP as a function of NP, K, and L.

a. Find the engine pressure if the nozzle pressure is 80 psi, the "K" factor is 0.51, and the length of the hose is 80 feet.

b. Find the nozzle pressure if the "K" factor is 0.73, the length of the hose is 45 feet, and the engine pressure is 150 psi.

c. Solve the equation for NP—that is, write a formula expressing NP as a function of EP, K, and L.

d. Find the "K" factor for a nozzle tip if the length of the hose is 190 feet, the engine pressure is 160 psi, and the nozzle pressure is 90 psi.

e. Solve the equation for K—that is, write a formula expressing K as a function of NP, EP, and L.

10. **Sales strategy:** A small business is considering hiring a new sales representative (sales rep) to market its product in a nearby city. Two pay scales are under consideration.

   Pay scale 1: Pay the sales rep a base yearly salary of $10,000 plus a commission of 8% of total yearly sales.

   Pay scale 2: Pay the sales rep a base yearly salary of $13,000 plus a commission of 6% of total yearly sales.

a. For each of the pay scales above, use a formula to express the total yearly earnings for the sales rep as a function of total yearly sales. Be sure to identify clearly what the letters that you use mean.

---

[14] See A. P. Kneider, *Mathematics of Merchandising*, 4th ed. (Englewood Cliffs, NJ: Regents/Prentice-Hall, 1994).

[15] See Eugene Mahoney, *Fire Department Hydraulics* (Boston: Allyn and Bacon, 1980).

b. What amount of total yearly sales would result in the same total yearly earnings for the sales rep no matter which of the two pay scales is used?

c. On the same screen, plot the graphs of the functions you made in part a. Copy the picture onto the paper you turn in. Be sure to label the horizontal and vertical axes, and be sure your picture includes the number you found in part b.

d. If you were a sales rep negotiating for the new position, under what conditions would you prefer pay scale 1? Under what conditions would you prefer pay scale 2?

11. **Net profit:** Suppose you pay R dollars per month to rent space for the production of dolls. You pay c dollars in material and labor to make each doll, which you then sell for d dollars.

a. If you produce n dolls per month, use a formula to express your net profit p per month as a function of R, c, d, and n. (*Suggestion:* First make a formula using the words *rent*, *cost of a doll*, *selling price*, and *number of dolls*. Then replace the words by appropriate letters.)

b. What is your net profit per month if the rent is $1280 per month, it costs $2 to make each doll, which you sell for $6.85, and you produce 826 dolls per month?

c. Solve the equation you got in part a for d.

d. Your accountant tells you that you need to make a net profit of $4000 per month. Your rent is $1200 per month, it costs $2 to make each doll, and your production line can make only 700 of them in a month. Under these conditions, what price do you need to get for each doll?

12. **Ohm's law** says that when electric current is flowing across a resistor, then the voltage v, measured in volts, is the product of the current i, measured in amperes, and the resistance R, measured in ohms. That is, $v = iR$.

a. What is the voltage if the current is 20 amperes and the resistance is 15 ohms?

b. Find a formula expressing resistance as a function of current and voltage. Use your function to find the resistance if the current is 15 amperes and the voltage is 12 volts.

c. Find a formula expressing current as a function of voltage and resistance. Use your function to find the current if the voltage is 6 volts and the resistance is 8 ohms.

13. **Temperature conversion:** In everyday experience, the measures of temperature most often used are Fahrenheit F and Celsius C. Recall that the relationship between them is given by

$$F = 1.8C + 32.$$

Physicists and chemists often use the Kelvin temperature scale.[16] You can get *kelvins* K from degrees Celsius by using

$$K = C + 273.15.$$

a. Explain in practical terms what $K(30)$ means, and then calculate that value.

b. Find a formula expressing the temperature C in degrees Celsius as a function of the temperature K in kelvins.

c. Find a formula expressing the temperature F in degrees Fahrenheit as a function of the temperature K in kelvins.

d. What is the temperature in degrees Fahrenheit of an object that is 310 kelvins?

14. **The ideal gas law:** A *mole* of a chemical compound is a fixed number,[17] like a dozen, of molecules (or atoms in the case of an element) of that compound. A mole of water, for example, is about 18 grams, or just over a half an ounce in your kitchen. Chemists often use the mole as the measure of the amount of a chemical compound.

A mole of carbon dioxide has a fixed mass, but the volume V that it occupies depends on pressure p and temperature T; greater pressure tends to

[16]With this temperature scale, physicists traditionally do not use the term *degrees*. Rather, if the temperature is 100 on the Kelvin scale, they say that it is "100 kelvins." The temperature scale is named after Lord Kelvin, and its name is capitalized, but the unit of temperature, the kelvin, is not.

[17]You may be interested to know that this number is 6.02217×10^{23}. It may appear to be a strange number, but this is a natural unit of measure for chemists to use.

compress the gas into a smaller volume, whereas increasing temperature tends to make the gas expand into a larger volume. If we measure the pressure in atmospheres (1 atm is the pressure exerted by the atmosphere at sea level), the temperature in kelvins, and the volume in liters, then the relationship is given by the *ideal gas law*:

$$pV = 0.082T.$$

a. Solve the ideal gas law for the volume V.

b. What is the volume of 1 mole of carbon dioxide under 3 atm of pressure at a temperature of 300 kelvins?

c. Solve the ideal gas law for pressure.

d. What is the pressure on 1 mole of carbon dioxide if it occupies a volume of 0.4 liter at a temperature of 350 kelvins?

e. Solve the ideal gas law for temperature.

f. At what temperature will 1 mole of carbon dioxide occupy a volume of 2 liters under a pressure of 0.3 atm?

15. **Running ants:** A scientist observed that the speed S at which certain ants ran was a function of T, the ambient temperature.[18] He discovered the formula

$$S = 0.2T - 2.7,$$

where S is measured in centimeters per second and T is in degrees Celsius.

a. Using functional notation, express the speed of the ants when the ambient temperature is 30 degrees Celsius, and calculate that speed using the formula above.

b. Solve for T in the formula above to obtain a formula expressing the ambient temperature T as a function of the speed S at which the ants run.

c. If the ants are running at a speed of 3 centimeters per second, what is the ambient temperature?

16. **Tax owed:** According to an Oklahoma Income Tax table for 2001, the income tax owed by an Oklahoma resident who is married and filing jointly with a taxable income above $50,000 is $3850 plus 10% of the taxable income over $50,000.

a. Use a formula to express the tax owed as a function of the taxable income, assuming that the taxable income is above $50,000. Be sure to explain the meaning of the letters you choose and the units.

b. What is the tax on a taxable income of $58,650?

c. For what taxable income would the tax be $4865?

17. **Profit:** *The background for this exercise can be found in Exercises 11 and 12 in Section 1.4.* A manufacturer of widgets has fixed costs of $200 per month, and the variable cost is $55 per thousand widgets (so it costs $55 to produce a thousand widgets). Let N be the number, in thousands, of widgets produced in a month.

a. Find a formula for the manufacturer's total cost C as a function of N.

b. The manufacturer sells the widgets for $58 per thousand widgets. Find a formula for the total revenue R as a function of N.

c. Use your answers to parts a and b to find a formula for the profit P of this manufacturer as a function of N.

d. Use your formula from part c to determine the break-even point for this manufacturer.

18. **Growth in weight and height:** Between the ages of 7 and 11 years, the weight w, in pounds, of a certain girl is given by the formula

$$w = 8t.$$

Here t represents her age in years.

a. Use a formula to express the age t of the girl as a function of her weight w.

b. At what age does she attain a weight of 68 pounds?

c. The height h, in inches, of this girl during the same period is given by the formula

$$h = 1.8t + 40.$$

i. Use your answer to part b to determine how tall she is when she weighs 68 pounds.

ii. Use a formula to express the height h of the girl as a function of her weight w.

---

[18] See the study by H. Shapley, "Note on the thermokinetics of Dolichoderine ants," *Proc. Nat. Acad. Sci.* **10** (1924), 436–439.

iii. Answer the question in part i again, this time using your answer to part ii.

19. **Plant growth:** The amount of growth of plants in an ungrazed pasture is a function of the amount of plant biomass already present and the amount of rainfall.[19] For a pasture in the arid zone of Australia, the formula

$$Y = -55.12 - 0.01535N$$
$$-0.00056N^2 + 3.946R \qquad (2.6)$$

gives an approximation of the growth. Here R is the amount of rainfall (in millimeters) over a 3-month period, N is the plant biomass (in kilograms per hectare) at the beginning of that period, and Y is the growth (in kilograms per hectare) of the biomass over that period. (For comparison, 100 millimeters is about 3.9 inches, and 100 kilograms per hectare is about 89 pounds per acre.)

a. Solve Equation (2.6) for R.

b. Ecologists are interested in the relationship between the amount of rainfall and the initial plant biomass if there is to be no plant growth over the period. Put $Y = 0$ in the equation you found in part a to get a formula for R in terms of N that describes this relationship.

c. Use the formula you found in part b to make a graph of R versus N (again with $Y = 0$). Include values of N from 0 to 800 kilograms per hectare. This graph is called the *isocline for zero growth*. It shows the amount of rainfall needed over the 3-month period just to maintain a given initial plant biomass.

d. With regard to the isocline for zero growth that you found in part c, what happens to R as N increases? Explain your answer in practical terms.

e. How much rainfall is needed just to maintain the initial plant biomass if that biomass is 400 kilograms per hectare?

f. A point below the zero isocline graph corresponds to having less rainfall than is needed to sustain the given initial plant biomass, and in this situation the plants will die back. A point above the zero isocline graph corresponds to having more rainfall than is needed to sustain the given initial plant biomass, and in this situation the plants will grow. If the initial plant biomass is 500 kilograms per hectare and there are 40 millimeters of rain, what will happen to the plant biomass over the period?

20. **Collision:** In reconstructing an automobile accident, investigators study the *total momentum*, both before and after the accident, of the vehicles involved. The total momentum of two vehicles moving in the same direction is found by multiplying the weight of each vehicle by its speed and then adding the results.[20] For example, if one vehicle weighs 3000 pounds and is traveling at 35 miles per hour, and another weighs 2500 pounds and is traveling at 45 miles per hour in the same direction, then the total momentum is $3000 \times 35 + 2500 \times 45 = 217,500$.

In this exercise we study a collision in which a vehicle weighing 3000 pounds ran into the rear of a vehicle weighing 2000 pounds.

a. After the collision, the larger vehicle was traveling at 30 miles per hour, and the smaller vehicle was traveling at 45 miles per hour. Find the total momentum of the vehicles after the collision.

b. The smaller vehicle was traveling at 30 miles per hour before the collision, but the speed V, in miles per hour, of the larger vehicle before the collision is unknown. Find a formula expressing the total momentum of the vehicles before the collision as a function of V.

c. The *principle of conservation of momentum* states that the total momentum before the collision equals the total momentum after the collision. Using this principle with parts a and b, determine at what speed the larger vehicle was traveling before the collision.

21. **Competition between populations:** In this exercise we consider the question of competition between two populations that vie for resources but do not prey on each other. Let m be the size of the first

---

[19]This exercise is based on the work of G. Robertson, "Plant dynamics," in G. Caughley, N. Shepherd, and J. Short, eds. *Kangaroos* (Cambridge, England: Cambridge University Press, 1987).

[20]See J. C. Collins, *Accident Reconstruction* (Springfield, IL: Charles C Thomas Publisher, 1979). In physics the mass is used instead of the weight, but in this exercise we ignore the distinction.

population and n the size of the second (both measured in thousands of animals), and assume that the populations coexist eventually. Here is an example of one common model for the interaction:

Per capita growth rate for $m = 5(1 - m - n)$,

Per capita growth rate for $n\ = 6(1 - 0.7m - 1.2n)$.

a. An *isocline* is formed by the points at which the per capita growth rate for m is zero. These are the solutions of the equation $5(1 - m - n) = 0$. Find a formula for n in terms of m that describes this isocline.

b. The points at which the per capita growth rate for n is zero form another isocline. Find a formula for n in terms of m that describes this isocline.

c. At an *equilibrium point* the per capita growth rates for m and for n are both zero. If the populations

reach such a point, they will remain there indefinitely. Use your answers to parts a and b to find the equilibrium point.

22. **Research project:** Find a function from a textbook in auditing, marketing, or merchandising given by a formula that involves several variables. For each of the variables, try to reverse the role of variable and function by appropriately rewriting the formula. For example, if the function f is given by a formula involving three variables x, y, and z, then you are to try to write three new formulas: one for x involving f, y, and z; one for y involving f, x, and z; and one for z involving f, x, and y. For some formulas this task is too difficult; in that case find another formula to use. Carefully describe the original formula and its variables (including units), and explain how these new formulas could be used.

2.4 **Solving Nonlinear Equations**

Nonlinear equations—those that involve powers, square roots, or other complicating factors—are in general much more difficult to solve than linear equations. In fact, most nonlinear equations cannot be solved by simple hand calculation. There are some notable exceptions. For instance, you may recall from elementary algebra that the *quadratic formula* can be used to solve *second-degree* equations. We will nonetheless treat all nonlinear equations the same and solve them using the graphing calculator.

Let's look, for example, at an aluminum bar that must be heated before it can be properly worked. The bar is placed in a preheated oven, and its temperature $A = A(t)$, t minutes later, is given by

$$A = 800 - 730e^{-0.06t} \text{ degrees Fahrenheit}.$$

The aluminum bar will be ready for bending and shaping when it reaches a temperature of 600 degrees. How long should the bar remain in the oven?

We want to know when the temperature A is 600 degrees. That is, we want to solve the equation

$$800 - 730e^{-0.06t} = 600$$

for t. This equation is not a linear equation like the ones we solved in Section 2.3. Rather, it is *nonlinear*, and it cannot be solved via the methods of Section 2.3. We will show two methods for solving equations of this type using the graphing calculator.

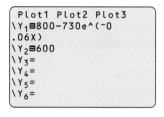

FIGURE 2.65 Entering left-hand and right-hand sides of the equation for the crossing-graphs method

The Crossing-Graphs Method

We illustrate the crossing-graphs method using the equation

$$800 - 730e^{-0.06t} = 600.$$

For this method we will make two graphs, the graph of the left-hand side of the equation, $800 - 730e^{-0.06t}$, and the graph of the right-hand side of the equation, the function with a constant value of 600. Then we will find out where they are the same—that is, where the graphs cross.

The first step is to enter $\boxed{2.38}$ both functions as shown in Figure 2.65. We record the appropriate correspondences:

$$Y_1 = A, \text{ temperature on vertical axis}$$

$$Y_2 = 600, \text{ target temperature}$$

$$X = t, \text{ minutes on horizontal axis}.$$

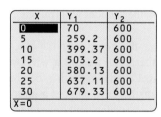

FIGURE 2.66 A table of values for temperature

To get a rough idea of when the temperature will reach 600 degrees, we look at the table of values for $800 - 730e^{-0.06t}$ in Figure 2.66, where we used a starting value of 0 and an increment of 5 minutes. We see that the temperature will reach 600 degrees between 20 and 25 minutes after the bar is placed in the oven.

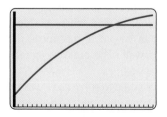

FIGURE 2.67 Graphs of temperature and the target temperature

On the basis of the table in Figure 2.66, we use a window setup $\boxed{2.39}$ with a horizontal span from 0 to 30 minutes and a vertical span from 0 to 700 degrees. The graphs are shown in Figure 2.67. We want the point where the graphs cross, or intersect. We can get reasonably close to the answer if we trace one graph and move the cursor as close as we

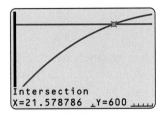

FIGURE 2.68 When the bar reaches 600 degrees

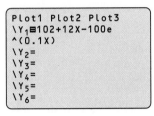

FIGURE 2.69 Entering the function for the single-graph method

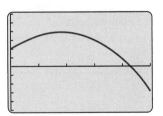

FIGURE 2.70 Getting an estimate with a table of values

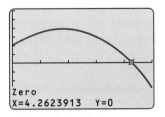

FIGURE 2.71 Picture for the single-graph method

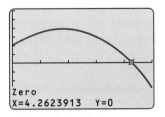

FIGURE 2.72 Finding the root

can to the crossing point, but a calculator can locate the crossing point quite accurately, and only a few keystrokes 2.40 are required. The result is shown in Figure 2.68, and we see from the prompt at the bottom of the screen that the temperature will be 600 degrees at the time 21.58 minutes after the bar is placed in the oven. Consult the *Technology Guide* to find out exactly what keystrokes are required to execute the crossing-graphs method on your calculator.

The Single-Graph Method

A slightly different method for solving nonlinear equations, the *single-graph method*, finds where a function equals 0. To illustrate it, we consider the account for a health care plan a firm offers its employees. When the firm instituted the plan, it pledged that for 5 years there would be no increase in the insurance premium, but since then health expenses have risen dramatically. The plan manager estimates that the plan's account balance B, in millions of dollars, as a function of the number of years t since the plan was instituted, is given by the formula

$$B = 102 + 12t - 100e^{0.1t}.$$

When will the plan's account run out of money? Will this happen before the 5-year period is over? To answer these questions, we need to know when the balance B is 0, so we must solve the equation

$$102 + 12t - 100e^{0.1t} = 0$$

for t. This is not a linear equation, and we will use the calculator to solve it.

We want to see where $102 + 12t - 100e^{0.1t}$ is zero, and this is where the graph crosses the horizontal axis. Such points are known as *zeros* or *roots*. We enter 2.41 the function as shown in Figure 2.69. As in the crossing-graphs method, we first look at a table of values. We are interested in the period between 0 and 5 years after the plan was instituted, and from the table in Figure 2.70, we see that the account will run out of money before the end of the 5-year period.

Now we set up the graphing window. We want to see the graph near the place where it crosses the horizontal axis, so, on the basis of the table, we use a window setup 2.42 with a horizontal span from 0 to 5 and a vertical span from −5 to 6. The graph is shown in Figure 2.71. As in the crossing-graphs method, we can get a pretty good estimate of where the graph crosses the horizontal axis by tracing. But once again, your calculator can find 2.43 the point quickly and accurately, as shown in Figure 2.72. Consult the *Technology Guide* for specific instructions regarding your calculator. We see from the prompt at the bottom of the screen in Figure 2.72 that the account will run out of money 4.26 years, or about 4 years and 3 months, after the plan is instituted.

The single-graph method applies directly when we want to find where a single function is equal to zero. Since more general equations can be put in this form, the single-graph method can be applied in those situations as well. The crossing-graphs method is widely applicable and is often easier to use.

We should note that whenever the calculator executes a routine to solve an equation, it is seldom able to find exactly the value it is looking for. Rather, it provides an approximation that is normally so accurate that we need not be concerned. But the presentation of approximate rather than exact answers is not a failing of the calculator. It is, rather, a mathematical necessity reflected in how the calculator functions.

EXAMPLE 2.8 **The George Deer Reserve Again**

```
Plot1 Plot2 Plot3
\Y₁◼6.21/(0.035+
0.45^X)
\Y₂=
\Y₃=
\Y₄=
\Y₅=
\Y₆=
```

FIGURE 2.73 Entering a logistic function for population growth

X	Y₁
0	6
5	116.18
10	175.72
15	177.4
20	177.43
25	177.43
30	177.43
X=0	

FIGURE 2.74 Searching for a good window size

```
WINDOW
 Xmin=0
 Xmax=15
 Xscl=1
 Ymin=0
 Ymax=200
 Yscl=1
 Xres=1
```

FIGURE 2.75 Selecting the window size

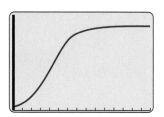

FIGURE 2.76 The graph of deer population versus time

For this example we look again at the logistic population growth formula introduced in Section 2.1 for deer on the George Reserve. The number N of deer expected to be present on the reserve after t years has been determined by ecologists to be

$$N = \frac{6.21}{0.035 + 0.45^t} \text{ deer}.$$

1. Plot the graph of N versus t and explain how the deer population increases with time.

2. For planning purposes, the wildlife manager for the reserve needs to know when to expect there to be 85 deer on the reserve. The answer to that is the solution of an equation. Which equation?

3. Solve the equation you found in part 2.

Solution to Part 1:   We enter 2.44 the population function as shown in Figure 2.73 and record the proper variable associations:

$$Y_1 = N, \text{ population on vertical axis}$$

$$X = t, \text{ years on horizontal axis}.$$

To determine a good window size, we look at the table of values in Figure 2.74. This table leads us to choose the window setup in Figure 2.75. Now when we graph, we see in Figure 2.76 the classic S-shaped curve that is characteristic of logistic population growth. This graph shows that the deer population increases rapidly for the first few years, but as population size nears the *carrying capacity* of the reserve, the rate of increase slows, and the population levels off at around 177 deer.

Solution to Part 2:   We want to know when the deer population will reach 85. We first write the equation in words and then replace the words by the appropriate letters:

$$\text{Deer population} = 85$$

$$N = 85$$

$$\frac{6.21}{0.035 + 0.45^t} = 85.$$

The time when the deer population will reach 85 is the value of t that makes this equation true. That is, we need to solve this equation for t.

Solution to Part 3:   To solve this equation we will use the crossing-graphs method. Since we have already entered the left-hand side of the equation, we need only enter 2.45 the right-hand side of the equation, 85. Now we graph to see the picture in Figure 2.77. We use the calculator to find 2.46 the crossing point shown in Figure 2.78. We see that 85 deer can be expected to be present in about 4.09 years, or about 4 years and 1 month.

   We should note that this equation is difficult to solve by hand calculation. That is why we do it with the calculator. We should also note that acquiring the skills

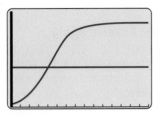

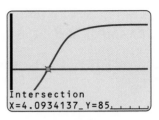

FIGURE 2.77 Adding the line at height 85

FIGURE 2.78 Solution using the crossing-graphs method

necessary to solve such an equation is a significant step forward in your mathematical development.   ■ ■ ■

When we use graphs to solve nonlinear equations in practical settings, it is important for the context to guide us in our choice of the graphing window. The next example illustrates this.

EXAMPLE 2.9   **A Floating Ball**

According to *Archimedes' law*, the weight of water that is displaced by a floating ball is equal to the weight of the ball. A certain wooden ball of diameter 4 feet weighs 436 pounds (see Figure 2.79). If the ball is allowed to float in pure water, Archimedes' law can be used to show that d feet of its diameter will be below the surface of the water, where d is a solution of the equation

$$62.4\pi d^2 \left(2 - \frac{d}{3}\right) = 436.$$

(The left-hand side of this equation represents the weight of the displaced water, and the right-hand side is the weight of the ball.) How much of this ball's diameter is below the surface of the water?

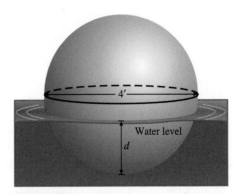

FIGURE 2.79

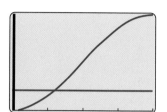

X	Y₁
0	-436
1	-109.3
2	609.52
3	1328.3
4	1655
5	1197.6
6	-436

X=4

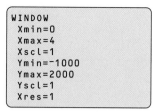

```
WINDOW
 Xmin=0
 Xmax=4
 Xscl=1
 Ymin=-1000
 Ymax=2000
 Yscl=1
 Xres=1
```

FIGURE 2.80 A table for setting vertical screen span

FIGURE 2.81 The properly configured WINDOW menu

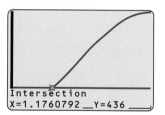

FIGURE 2.82 Preparing to solve using the crossing-graphs method

Solution: To use the crossing-graphs method, we graph the function from the left-hand side of the equation, namely

$$62.4\pi d^2 \left(2 - \frac{d}{3}\right),$$

and we graph the constant function 436 from the right-hand side of the equation. The first step is to enter ⎡2.47⎤ the functions into the calculator and record the appropriate correspondences:

Y_1 = Weight of displaced water on vertical axis

Y_2 = Weight of ball

$X = d$, diameter below surface on horizontal axis .

It is important to remember that the diameter of the ball is 4 feet. Thus we are interested in viewing the graphs only on the span from $d = 0$ to $d = 4$. (See Exercise 3 at the end of this section.) To get the proper vertical span of the viewing screen, we consult the table in Figure 2.80. On the basis of this we set the viewing window as shown in Figure 2.81. Now when we graph, we get the picture shown in Figure 2.82. We use the calculator to find ⎡2.48⎤ where the graphs cross as shown in Figure 2.83. We read the answer $d = 1.18$ feet from the **X=** display in Figure 2.83.   ■ ■ ■

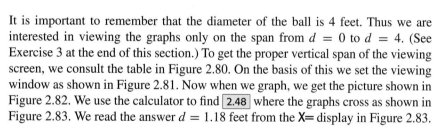

Intersection
X=1.1760792__Y=436

FIGURE 2.83 The part of a floating ball that is below the water line

ANOTHER LOOK

Solving Nonlinear Equations by Factoring

Many nonlinear equations cannot be solved exactly by hand calculation, and some sort of approximation method, such as the crossing-graphs method, may offer the only practical means of solution. But certain special equations can be solved by factoring.

Let's review two factoring formulas that come up often in this context:

$$x^2 + 2ax + a^2 = (x + a)^2$$

$$x^2 + (a + b)x + ab = (x + a)(x + b).$$

The first factoring formula, the perfect-square formula, is really the special case of the second where $a = b$. We can use these formulas to factor such expressions as $x^2 + 5x + 6$. Here we want to use the second formula. Thus we need to find numbers a and b such that $ab = 6$ and $a + b = 5$. The positive-integer possibilities are $a = 6$, $b = 1$, or $a = 3$, $b = 2$. The 6, 1 possibility doesn't work since $6 + 1 = 7$ rather than 5. The 3, 2 possibility does work since $3 + 2 = 5$. Thus

$$x^2 + 5x + 6 = (x + 3)(x + 2).$$

Often the plus and minus signs give important clues about the factors. Here are the four types that occur, along with their expanded forms. We assume that a and b are positive numbers, and we write the resulting forms in terms of positive numbers $k, l, m,$ and n.

$$(x + a)(x + b) = x^2 + kx + l$$
$$(x - a)(x - b) = x^2 - kx + l$$
$$(x + a)(x - b) = x^2 + mx - n \text{ if } a > b$$
$$(x + a)(x - b) = x^2 - mx - n \text{ if } a < b$$

For example, if we consider $x^2 - 5x + 6$, then we know from the signs that we are looking for factors of the form $(x - a)(x - b)$. Now we need, as before, positive integers a and b such that $ab = 6$ and $a + b = 5$. We find that $a = 2$ and $b = 3$ are the required integers. Thus

$$x^2 - 5x + 6 = (x - 3)(x - 2).$$

In the case of $x^2 - x - 6$, we need factors of the form $(x + a)(x - b)$, where $a < b$. We want $ab = 6$ and $b - a = 1$. Thus we should choose $a = 2$ and $b = 3$. Hence

$$x^2 - x - 6 = (x + 2)(x - 3).$$

The key fact that enables us to use factoring to solve equations is that if the product of two numbers is zero, then one factor or the other is zero. That is,

$$\text{if } AB = 0, \text{ then } A = 0 \text{ or } B = 0.$$

Let's see how to use factoring to solve $x^2 + 12 = 7x$. The first step is to bring everything to the left-hand side:

$$x^2 - 7x + 12 = 0.$$

Next we want to factor the left-hand side. This factors in the form $(x - a)(x - b)$, where $ab = 12$ and $a + b = 7$. We find that $a = 4$ and $b = 3$ works. Thus

$$(x - 4)(x - 3) = 0.$$

We now have a product of two numbers that is zero. Thus one factor or the other must be 0:

$$x - 4 = 0 \quad \text{or} \quad x - 3 = 0.$$

Thus we get two solutions, $x = 3$ and $x = 4$.

Note that in this equation the powers on x are all positive integers, and the highest power that occurs is 2. That is to say, the equation is *quadratic*. We found two solutions. That is what we expect in general for a quadratic, although sometimes the two solutions may be the same, or the solutions may be *complex numbers*. For a *cubic* equation (when the highest power is 3), we expect at most three solutions, and for a *quartic* equation (when the highest power is 4), we expect at most four.

We can use factoring to solve cubic equations as well. Let's solve $x^3 + 10x = 7x^2$. As before, we first bring everything to the left-hand side of the equation:

$$x^3 - 7x^2 + 10x = 0.$$

Next we note that each term has a common factor of x, and we factor this out:

$$x(x^2 - 7x + 10) = 0.$$

Now we factor the quadratic term and solve as before:

$$x(x - 5)(x - 2) = 0.$$

Thus we get three solutions: $x = 0$, $x = 5$, and $x = 2$.

Theoretically, any equation involving only positive integral powers of x can be solved by factoring. That is, such an equation can always be factored.[21] But from a practical point of view, the factors may be difficult or impossible to find.

Enrichment Exercises

E-1. **Factoring quadratics:** Factor each of the following.

a. $x^2 - 8x - 9$ b. $x^2 + 6x + 9$

c. $x^2 - 16$ d. $x^2 - 12x + 35$

E-2. **Solving quadratic equations:** Solve each of the following equations.

a. $x^2 + 20 = 9x$ b. $x^2 = 10x - 25$

c. $x^2 = 1$ d. $(x + 1)(x + 2) = 2x^2 + 8x + 6$

E-3. **Solving higher-order equations:** Solve each of the following equations.

a. $x^3 = 3x^2 + 4x$

b. $(2x - 3)(3x + 4)(2x - 5) = 0$

c. $(x - 4)(x^2 + 2x - 15) = 0$

d. $(x^2 - 5x + 6)(x^2 + 7x + 12) = 0$

E-4. **Integral solutions:** A *polynomial* equation is one of the form

$$a_n x^n + a_{n-1} x^{n-1} + \cdots + a_1 x + a_0 = 0.$$

An *integral solution* of such an equation is simply   an integer that satisfies the equation. It is not hard to see that if k is an integral solution of this equation, then k must be a divisor of the constant term a_0. For example, the only divisors

---

[21]The factors may involve complex numbers.

of 4 are 1, -1, 2, -2, 4, and -4. Hence these are the only possible integral solutions of $x^3 + 3x + 4 = 0$. We find by trial and error that -1 is a solution and that the others are not.

Find all the integral solutions of the following polynomial equations.

a. $x^4 + x^3 + x + 1$ b. $x^3 - 6x^2 + 11x - 6$

c. $x^5 + x^4 - x + 7$

E-5. **Rational solutions:** *This is continuation of Exercise E-4.* If p/q is a rational (reduced to lowest terms) solution of the general polynomial equation in Exercise E-4, then it turns out that p is a divisor of a_0 and that q is a divisor of a_n. For example, the divisors of 3 are ± 1 and ± 3, and the divisors of 1 are ± 1. Hence the only possible rational solutions of $3x^3 + x + 1 = 0$ are ± 1 and $\pm\frac{1}{3}$. A check shows that none of these works and hence the equation has no rational solutions. Find the rational solutions of the following equations.

a. $3x^4 - 10x^3 + 6x^2 - 10x + 3 = 0$

b. $2x^3 - x^2 + 2x - 1 = 0$

c. $2x^4 + 3x^2 + 1 = 0$

E-6. **Factoring quadratics using the quadratic formula:** The quadratic formula tells us that the solutions of $ax^2 + bx + c = 0$ are

$$x = \frac{-b \pm \sqrt{b^2 - 4ac}}{2a} .$$

An important theorem known as the *factor theorem* then tells us that $ax^2 + bx + c$ factors as

$$a\left(x - \frac{-b + \sqrt{b^2 - 4ac}}{2a}\right)\left(x - \frac{-b - \sqrt{b^2 - 4ac}}{2a}\right) .$$

This enables us to factor quadratics that do not have integral solutions. Find the factors of the following quadratics.

a. $x^2 + 2x - 2$ b. $x^2 + x - 1$

c. $3x^2 - 6x + 2$

2.4 SKILL BUILDING EXERCISES

S-1. The crossing-graphs method: Solve using the crossing-graphs method: $20/(1 + 2^x) = x$.

S-2. The crossing-graphs method: Solve using the crossing-graphs method: $x^2 - x^3 + 3 = x^5/20$.

S-3. The crossing-graphs method: Solve using the crossing-graphs method: $3^x + x = 2^x + 1$.

S-4. The crossing-graphs method: Solve using the crossing-graphs method: $\sqrt{x^2 + 1} = x^3 + \sqrt{x^4 + 2}$.

S-5. The crossing-graphs method: Solve $5/(x^2 + x + 1) = 1$ using the crossing-graphs method. (*Note:* There are two solutions. Find them both.)

S-6. The single-graph method: Use the single-graph method to solve $20/(1 + 2^x) - x = 0$.

S-7. The single-graph method: Use the single-graph method to solve $5/(x^2 + x + 1) - 1 = 0$. (*Note:* There are two solutions. Find them both.)

S-8. The single-graph method: Use the single-graph method to solve $3^x - 8 = 0$.

S-9. The single-graph method: Use the single-graph method to solve $-x^4/(x^2 + 1) + 1 = 0$. (*Note:* There are two solutions. Find them both.)

S-10. The single-graph method: Use the single-graph method to solve $x^3 - x - 5 = 0$.

2.4 EXERCISES

1. **A population of foxes:** A breeding group of foxes is introduced into a protected area, and the population growth follows a logistic pattern. After t years the population of foxes is given by

$$N = \frac{37.5}{0.25 + 0.76^t} \text{ foxes.}$$

   a. How many foxes were introduced into the protected area?

   b. Make a graph of N versus t and explain in words how the population of foxes increases with time.

   c. When will the fox population reach 100 individuals?

2. **Profit:** The monthly profit P for a widget producer is a function of the number n of widgets sold. The formula is

$$P = -15 + 10n - 0.2n^2.$$

   Here P is measured in thousands of dollars, n is measured in thousands of widgets, and the formula is valid up to a level of 15 thousand widgets sold.

   a. Make a graph of P versus n.

   b. Calculate $P(1)$ and explain in practical terms what your answer means.

   c. Is the graph concave up or concave down? Explain in practical terms what this means.

   d. The break-even point is the sales level at which the profit is 0. Find the break-even point for this widget producer.

3. **Revisiting the floating ball:** In Example 2.9 we solved the equation

$$62.4\pi d^2 \left(2 - \frac{d}{3}\right) = 436$$

   on the interval from $d = 0$ to $d = 4$. If you graph the function $62.4\pi d^2(2 - d/3)$ and the constant function 436 on the span from $d = -2$ to $d = 7$, you will see that this equation has two other solutions. Find these solutions. Is there a physical interpretation of these solutions that makes sense?

4. **The skydiver again:** When a skydiver jumps from an airplane, her downward velocity, in feet per second, before she opens her parachute is given by $v = 176(1 - 0.834^t)$, where t is the number of seconds that have elapsed since she jumped from the airplane. We found earlier that the terminal velocity for the skydiver is 176 feet per second. How long does it take to reach 90% of terminal velocity?

5. **Falling with a parachute:** If an average-size man with a parachute jumps from an airplane, he will fall $12.5(0.2^t - 1) + 20t$ feet in t seconds. How long will it take him to fall 140 feet?

6. **A cup of coffee:** The temperature C of a fresh cup of coffee t minutes after it is poured is given by

$$C = 125e^{-0.03t} + 75 \text{ degrees Fahrenheit}.$$

a. Make a graph of C versus t.

b. The coffee is cool enough to drink when its temperature is 150 degrees. When will the coffee be cool enough to drink?

c. What is the temperature of the coffee in the pot? (*Note:* We are assuming that the coffee pot is being kept hot and is the same temperature as the cup of coffee when it was poured.)

d. What is the temperature in the room where you are drinking the coffee? (*Hint:* If the coffee is left to cool a long time, it will reach room temperature.)

7. **Reaction rates:** In a chemical reaction, the reaction rate R is a function of the concentration x of the product of the reaction. For a certain second-order reaction between two substances, we have the formula

$$R = 0.01x^2 - x + 22.$$

Here x is measured in moles per cubic meter and R is measured in moles per cubic meter per second.

a. Make a graph of R versus x. Include concentrations up to 100 moles per cubic meter.

b. Use functional notation to express the reaction rate when the concentration is 15 moles per cubic meter, and then calculate that value.

c. The reaction is said to be in *equilibrium* when the reaction rate is 0. At what two concentrations is the reaction in equilibrium?

8. **Population growth:** The growth G of a population of lower organisms over a day is a function of the population size n at the beginning of the day. If both n and G are measured in thousands of organisms, the formula is

$$G = -0.03n^2 + n.$$

a. Make a graph of G versus n. Include values of n up to 40 thousand organisms.

b. Calculate $G(35)$ and explain in practical terms what your answer means.

c. For what two population levels will the population grow by 5 thousand over a day?

d. If there is no population to start with, of course there will be no growth. At what other population level will there be no growth?

9. **Van der Waals equation:** In Exercise 14 at the end of Section 2.3, we discussed the *ideal gas law*, which shows the relationship among volume V, pressure p, and temperature T for a fixed amount (1 mole) of a gas. But chemists believe that in many situations, the *van der Waals equation* gives more accurate results. If we measure temperature T in kelvins, volume V in liters, and pressure p in atmospheres (1 atm is the pressure exerted by the atmosphere at sea level), then the relationship for carbon dioxide is given by

$$p = \frac{0.082T}{V - 0.043} - \frac{3.592}{V^2} \text{ atm}.$$

What volume does this equation predict for 1 mole of carbon dioxide at 500 kelvins and 100 atm? (*Suggestion:* Consider volumes ranging from 0.1 to 1 liter.)

10. **Radioactive decay:** The *half-life* of a radioactive substance is the time H that it takes for half of the substance to change form through radioactive decay. This number does not depend on the amount with which you start. For example, carbon 14 is known to have a half-life of $H = 5770$ years. Thus if you begin with 1 gram of carbon 14, then 5770 years later you will have $\frac{1}{2}$ gram of carbon 14. And if you begin with 30 grams of carbon 14, then after 5770 years there will be 15 grams left. In general, radioactive substances decay according to the formula

$$A = A_0 \times 0.5^{t/H},$$

where H is the half-life, t is the elapsed time, A_0 is the amount you start with (the amount when $t = 0$), and A is the amount left at time t.

a. Uranium 228 has a half-life H of 9.3 minutes. Thus the decay function for this isotope of uranium is

$$A = A_0 \times 0.5^{t/9.3},$$

where t is measured in minutes. Suppose we start with 8 grams of uranium 228.

i. How much uranium 228 is left after 2 minutes?

ii. How long will you have to wait until there is only 3 grams left?

b. Uranium 235 is the isotope of uranium that can be used to make nuclear bombs. It has a half-life of 713 million years. Suppose we start with 5 grams of uranium 235.

  i. How much uranium 235 is left after 200 million years?

  ii. How long will you have to wait until there is only 3 grams left?

11. **Radiocarbon dating:** *This is a continuation of Exercise 10.* We rewrite the decay formula as

$$\frac{A}{A_0} = 0.5^{t/H}.$$

This formula shows the fraction of the original amount that is present as a function of time.

   Carbon 14 has a half-life of 5.77 thousand years. It is thought that in recent geological time (the last few million years or so), the amount of C_{14} in the atmosphere has remained constant. As a consequence, all organisms that take in air (trees, people, and so on) maintain the same level of C_{14} so long as they are alive. When a living organism dies, it no longer takes in C_{14}, and the amount present at death begins to decay according to the formula above. This phenomenon can be used to date some archaeological objects.

   Suppose that the amount of carbon 14 in charcoal from an ancient campfire is $\frac{1}{3}$ of the amount in a modern, living tree. In terms of the formula above, this means $A/A_0 = \frac{1}{3}$. When did the tree that was used to make the campfire die? Be sure to explain how you got your answer.

12. **Monthly payment on a loan:** If you borrow $5000 at an APR of r (as a decimal) from a lending institution that compounds interest continuously, and if you wish to pay off the note in 3 years, then your monthly payment M, in dollars, can be calculated using

$$M = \frac{5000(e^{r/12} - 1)}{1 - e^{-3r}}.$$

Your budget will allow a payment of $150 per month, and you are shopping for an interest rate that will give a payment of this size. What interest rate do you need to find?

13. **Grazing kangaroos:** The amount of vegetation eaten in a day by a grazing animal is a function of the amount V of food available (measured as biomass, in units such as pounds per acre).[22] This relationship is called the *functional response*. If there is little vegetation available, the daily intake will be small, since the animal will have difficulty finding and eating the food. As the food biomass increases, so does the daily intake. Clearly, though, there is a limit to the amount the animal will eat, regardless of the amount of food available. This maximum amount eaten is the *satiation level*.

a. For the western grey kangaroo of Australia, the functional response is

$$G = 2.5 - 4.8e^{-0.004V},$$

where $G = G(V)$ is the daily intake (measured in pounds) and V is the vegetation biomass (measured in pounds per acre).

  i. Draw a graph of G against V. Include vegetation biomass levels up to 2000 pounds per acre.

  ii. Is the graph you found in part i concave up or concave down? Explain in practical terms what your answer means about how this kangaroo feeds.

  iii. There is a *minimal* vegetation biomass level below which the western grey kangaroo will eat nothing. (Another way of expressing this is to say that the animal cannot reduce the food biomass below this level.) Find this minimal level.

  iv. Find the satiation level for the western grey kangaroo.

b. For the red kangaroo of Australia, the functional response is

$$R = 1.9 - 1.9e^{-0.033V}, \qquad \longrightarrow$$

---

[22]This exercise and the next are based on the work of J. Short, "Factors affecting food intake of rangelands herbivores," in G. Caughley, N. Shepherd, and J. Short, eds. *Kangaroos* (Cambridge, England: Cambridge University Press, 1987).

where R is the daily intake (measured in pounds) and V is the vegetation biomass (measured in pounds per acre).

   i. Add the graph of R against V to the graph of G you drew in part a.

   ii. A simple measure of the grazing efficiency of an animal involves the minimal vegetation biomass level described above: The lower the minimal level for an animal, the more efficient it is at grazing. Which is more efficient at grazing, the western grey kangaroo or the red kangaroo?

14. **Grazing rabbits and sheep:** *This is a continuation of Exercise 13.* In addition to the kangaroos, the major grazing mammals of Australia include merino sheep and rabbits. For sheep the functional response is

$$S = 2.8 - 2.8e^{-0.01V},$$

and for rabbits it is

$$H = 0.2 - 0.2e^{-0.008V}.$$

Here S and H are the daily intake (measured in pounds), and V is the vegetation biomass (measured in pounds per acre).

   a. Find the satiation level for sheep and that for rabbits.

   b. One concern in the management of rangelands is whether the various species of grazing animals are forced to compete for food. It is thought that competition will not be a problem if the vegetation biomass level provides at least 90% of the satiation level for each species. What biomass level guarantees that competition between sheep and rabbits will not be a problem?

15. **Growth rate:** The per capita growth rate r (on an annual basis) of a population of grazing animals is a function of V, the amount of vegetation available. A positive value of r means that the population is growing, whereas a negative value of r means that the population is declining. For the red kangaroo of

Australia, the relationship has been given[23] as

$$r = 0.4 - 2e^{-0.008V}.$$

Here V is the vegetation biomass, measured in pounds per acre.

   a. Draw a graph of r versus V. Include vegetation biomass levels up to 1000 pounds per acre.

   b. The population size will be stable if the per capita growth rate is zero. At what vegetation level will the population size be stable?

16. **Hosting a convention:** You are hosting a convention for a charitable organization. You pay a rental fee of $25,000 for the convention center, plus you pay the caterer $12 for each person who attends the convention. Suppose you just want to break even.

   a. Use a formula to express the amount you should charge per ticket as a function of the number of people attending. Be sure to explain the meaning of the letters you choose and the units.

   b. Make a graph of the function that gives the amount you should charge per ticket. Include attendance sizes up to 1000.

   c. How many must attend if you are to break even with a ticket price of $50?

17. **Breaking even:** *The background for this exercise can be found in Exercises 11, 12, 13, and 14 in Section 1.4.* A manufacturer of widgets has fixed costs of $700 per month, and the variable cost is $65 per thousand widgets (so it costs $65 to produce 1 thousand widgets). Let N be the number, in thousands, of widgets produced in a month.

   a. Find a formula for the manufacturer's total cost C as a function of N.

   b. The highest price p, in dollars per thousand widgets, at which N can be sold is given by the formula $p = 75 - 0.02N$. Using this, find a formula for the total revenue R as a function of N.

   c. Use your answers to parts a and b to find a formula for the profit P of this manufacturer as a function of N.

---

[23]This formula is based on the work of P. Bayliss and G. Caughley, in G. Caughley, *Ibid.*

d. Use your formula from part c to determine the two break-even points for this manufacturer. Assume that the manufacturer can produce at most 500 thousand widgets in a month.

18. **Tubeworms:** In a study of a marine *tubeworm*, scientists developed a model for the time T (measured in years) required for the tubeworm to reach a length of L meters.[24] On the basis of their model, they estimate that 170 to 250 years are required for the organism to reach a length of 2 meters, making this tubeworm the longest-lived noncolonial marine invertebrate known. The model is

$$T = \frac{20e^{bL} - 28}{b}.$$

Here b is a constant that requires estimation. What is the value of b if the lower estimate of 170 years to reach a length of 2 meters is correct? What is the value of b if the upper estimate of 250 years to reach a length of 2 meters is correct?

19. **Water flea:** F. E. Smith has reported on population growth of the water flea.[25] In one experiment, he found that the time t, in days, required to reach a population of N is given by the relation

$$e^{0.44t} = \frac{N}{N_0}\left(\frac{228 - N_0}{228 - N}\right)^{4.46}.$$

Here N_0 is the initial population size. If the initial population size is 50, how long is required for the population to grow to 125?

20. **Holling's functional response curve:** The total number P of prey taken by a predator depends on the availability of prey. C. S. Holling proposed a function of the form $P = cn/(1 + dn)$ to model the number of prey taken in certain situations.[26] Here n is the density of prey available, and c and d are constants that depend on the organisms involved as well as on other environmental features. Holling took data gathered earlier by T. Burnett on the number of sawfly cocoons found by a small

wasp parasite at given host density. In one such experiment conducted, Holling found the relationship

$$P = \frac{21.96n}{1 + 2.41n},$$

where P is the number of cocoons parasitized and n is the density of cocoons available (measured as number per square inch).

a. Draw a graph of P versus n. Include values of n up to 2 cocoons per square inch.

b. What density of cocoons will ensure that the wasp will find and parasitize 6 of them?

c. There is a limit to the number of cocoons that the wasp is able to parasitize no matter how readily available the prey may be. What is this upper limit?

21. **Radius of a shock wave:** An explosion produces a spherical shock wave whose radius R expands rapidly. The rate of expansion depends on the energy E of the explosion and the elapsed time t since the explosion. For many explosions, the relation is approximated closely by

$$R = 4.16E^{0.2}t^{0.4}.$$

Here R is the radius in centimeters, E is the energy in ergs, and t is the elapsed time in seconds. The relation is valid only for very brief periods of time, perhaps a second or so in duration.

a. An explosion of 50 pounds of TNT produces an energy of about 10^{15} ergs. See Figure 2.84. How long is required for the shock wave to reach a point 40 meters (4000 centimeters) away?

b. A nuclear explosion releases much more energy than conventional explosions. A small nuclear device of yield 1 kiloton releases approximately 9×10^{20} ergs. How long would it take for the shock wave from such an explosion to reach a point 40 meters away? →

---

[24]D. Bergquist, F. Williams, and C. Fisher, "Longevity record for deep-sea invertebrate," *Nature* **403** (2000), 499–500.

[25]See "Population dynamics in *Daphnia magna* and a new model for population growth," *Ecology* **44** (1963), 651–663. The population size is measured per unit volume.

[26]"Some characteristics of simple types of predation and parasitism," *Can. Ent.* **91** (1959), 385–398.

c. The shock wave from a certain explosion reaches a point 50 meters away in 1.2 seconds. How much energy was released by the explosion? The values of E in parts a and b may help you set an appropriate window.

Note: In 1947 the government released film of the first nuclear explosion in 1945, but the *yield* of the explosion remained classified. Sir Geoffrey Taylor used the film to determine the rate of expansion of the shock wave and so was able to publish a scientific paper[27] concluding correctly that the yield was in the 20-kiloton range.

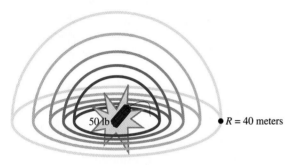

FIGURE 2.84

22. **Friction loss in fire hoses:** When water flows inside a hose, the contact of the water with the wall of the hose causes a drop in pressure from the pumper to the nozzle. This drop is known as *friction loss.*

Although it has come under criticism for lack of accuracy, the most commonly used method for calculating friction loss for flows under 100 gallons per minute uses what is called the *underwriter's formula*:

$$F = \left(2 \left(\frac{Q}{100} \right)^2 + \frac{Q}{200} \right) \left(\frac{L}{100} \right) \left(\frac{2.5}{D} \right)^5 .$$

Here F is the friction loss in pounds per square inch, Q is the flow rate in gallons per minute, L is the length of the hose in feet, and D is the diameter of the hose in inches.

a. In a 500-foot hose of diameter 1.5 inches, the friction loss is 96 pounds per square inch. What is the flow rate?

b. In a 500-foot hose, the friction loss is 80 pounds per square inch when water flows at 65 gallons per minute. What is the diameter of the hose? Round your answer to the nearest $\frac{1}{8}$ inch.

23. **Research project:** Find an equation involving one variable from a textbook in some other subject (not mathematics) or other handy source. Solve that equation using the techniques you've learned in this section. Compare your solution with the solution found in your source. Write a careful description of the original equation, of its use, and of your solution, and add a brief description of the solution found in your source.

---

[27]"The formation of a blast wave by a very intense explosion, II: The atomic explosion of 1945," *Proc. Roy. Soc. A* **201** (1950), 175–186.

2.5 *Optimization*

One key feature of a graph is its *zeros*, the places where it crosses or touches the horizontal axis. We learned how to find these in Section 2.4. Other important features of a graph include its *maxima* and *minima*. The need to locate these will arise, for example, when you want to find the number of items you can produce that will yield a maximum profit or when you wish to lay a pipeline in such a way that the cost is a minimum. Graphs make it easy to find maxima and minima. Thus if we have a function given by a formula, we can *optimize* it (find the optimal values) by first getting the graph.

Optimizing at Peaks and Valleys

In many instances optimal values for a function are located at peaks or valleys of its graph. For example, if a cannon is placed at the origin (X=0, Y=0) and elevated at an angle of 45 degrees (see Figure 2.85), then when it is fired, the cannonball will follow the graph of

$$y = x - g\left(\frac{x}{m}\right)^2.$$

(This simple model ignores air resistance.) Here x represents the number of feet downrange, y is the height in feet, and m is the *muzzle velocity,* in feet per second, of the cannonball. The constant g is the acceleration due to gravity near the surface of the Earth, about 32 feet per second per second. Thus, if the cannon is fired with a muzzle velocity of 250 feet per second, it will follow the graph of

$$y = x - 32\left(\frac{x}{250}\right)^2.$$

Let's use the graph to analyze the flight of the cannonball. The first step is to enter 2.49 the function and record the appropriate correspondences:

$$Y_1 = y, \text{ height on vertical axis}$$

$$X = x, \text{ distance downrange on horizontal axis}.$$

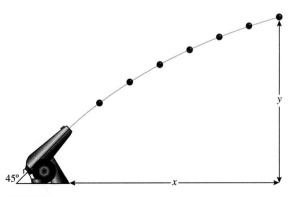

FIGURE 2.85

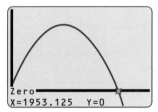

X	Y₁	
0	0	
500	372	
1000	488	
1500	348	
2000	⁻48	
2500	⁻700	
3000	⁻1608	
X=0		

FIGURE 2.86 A table of values for a cannonball flight

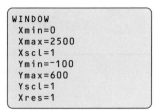

```
WINDOW
 Xmin=0
 Xmax=2500
 Xscl=1
 Ymin=-100
 Ymax=600
 Yscl=1
 Xres=1
```

FIGURE 2.87 Window settings from the table

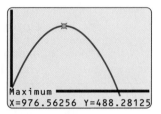

Zero
X=1953.125   Y=0

FIGURE 2.88 Where the cannonball lands

Maximum
X=976.56256 Y=488.28125

FIGURE 2.89 The peak of the cannonball's flight

We need to look at a table of values to help us choose a window. Since we expect the cannonball to travel several hundreds of feet, we made the table ⌐2.50¬ in Figure 2.86 with a starting value of 0 and a table increment of 500. Allowing a little extra room, we set the window as shown in Figure 2.87.

Now graphing produces the cannonball path shown in Figure 2.88. In this picture the horizontal axis shows feet downrange, and the vertical axis shows height above the ground. Thus the horizontal axis is ground level. In Figure 2.88 we have located the place where the graph crosses the horizontal axis ⌐2.51¬ —that is, where the cannonball strikes the ground—and we see that this will happen about 1953 feet downrange. (That is just over a third of a mile.)

Let's find the maximum height of the cannonball, which is at the top of the graph in Figure 2.88. We can get a reasonable estimate for the maximum if we trace the graph and move the cursor as close to the peak as possible. But most graphing calculators have the ability to find maximum values ⌐2.52¬ such as this in a fashion similar to the way they find zeros or where graphs cross. You should consult the *Technology Guide* for the exact keystrokes needed to accomplish this. In Figure 2.89 we have used this feature to locate the peak of the graph, and we see from the prompt at the bottom of the screen that the cannonball reaches a maximum height of 488.28 feet at 976.56 feet downrange.

EXAMPLE 2.10 **Growth of Forest Stands**

In forestry management it is important to know the *growth* and the *yield* of a forest stand.[28] The growth G is the amount by which the volume of wood will increase in a unit of time, and the yield Y is the total volume of wood. A forest manager has determined

---

[28] See Thomas E. Avery and Harold E. Burkhart, *Forest Measurements*, 4th ed. (New York: McGraw-Hill, 1994).

that in a certain stand of age A, the growth $G = G(A)$ is given by the formula

$$G = 32A^{-2}e^{10-32A^{-1}}$$

and that the yield $Y = Y(A)$ is given by the formula

$$Y = e^{10-32A^{-1}}.$$

Here G is measured in cubic feet per acre per year, Y in cubic feet per acre, and A in years.

1. Draw a graph of growth as a function of age that includes ages up to 60 years.

2. At what age is growth maximized?[29]

3. Draw a graph of yield as a function of age. What is the physical meaning of the point on this graph that corresponds to your answer to part 2?

4. To meet market demand, loggers are considering harvesting a relatively young stand of trees. This area was initially clear-cut[30] and left barren. The forest was replanted, with plans to get a new harvest within the next 14 years. At what time in this 14-year period will growth be maximized?

Solution to Part 1: The first step is to enter [2.53] the growth function and record the appropriate correspondences:

$$\mathsf{Y_1} = G, \text{ growth on vertical axis}$$

$$\mathsf{X} = A, \text{ age on horizontal axis}.$$

We need to look at a table of values to help us set the window size. We made the table [2.54] in Figure 2.90 using a starting value of 0 and a table increment of 10. We were told specifically that our graph should include ages up to 60, so we made the graph in Figure 2.91 using a window [2.55] with a horizontal span from $A = 0$ to $A = 60$ and a vertical span from $G = 0$ to $G = 500$.

Solution to Part 2: Growth is maximized at the peak of the graph in Figure 2.91. We have used the calculator [2.56] to locate this point in Figure 2.92. We see from the prompt at the bottom of the screen that a maximum growth of 372.62 cubic feet per acre per year occurs when the stand is 16 years old.

Solution to Part 3: Now we want to enter [2.57] the yield function

$$Y = e^{10-32A^{-1}} \text{ cubic feet per acre}.$$

In Figure 2.93 we see in the $\mathsf{Y_2}$ column values for the yield. It would be good to show graphs of both functions on the same screen, but the relative sizes of function values for growth and yield make this impractical. (Try it to see what happens.)

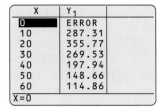

X	Y_1	
0	ERROR	
10	287.31	
20	355.77	
30	269.53	
40	197.94	
50	148.66	
60	114.86	
X=0		

FIGURE 2.90 A table of values for growth

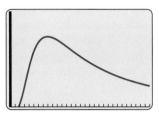

FIGURE 2.91 A graph of forest growth versus age

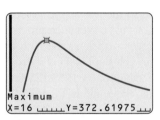

Maximum
X=16 �
Y=372.61975 ⌐

FIGURE 2.92 The age of maximum growth

X	Y_1	Y_2
0	ERROR	ERROR
10	287.31	897.85
20	355.77	4447.1
30	269.53	7580.5
40	197.94	9897.1
50	148.66	11614
60	114.86	12922
X=0		

FIGURE 2.93 A table of values for yield

---

[29] In practice, forest managers are interested in the maximum *mean annual growth* (see Exercise 14). The age of maximum growth can be used to determine the rotation age that will give the largest total wood production over a series of harvests.

[30] Clear-cutting is the practice of harvesting by cutting *all* the trees in an area. The result shows as large swaths of land that are almost barren.

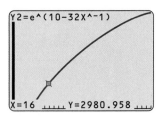

FIGURE 2.94 Yield at maximum growth

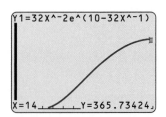

FIGURE 2.95 Maximum growth for a young stand of trees

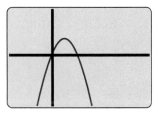

FIGURE 2.96 The formula represents the cannonball path only on a restricted range

Thus we turn off 2.58 the growth function and graph the yield in a window with horizontal span from 0 to 60 and vertical span from 0 to 13,000. The result is shown in Figure 2.94.

In Figure 2.94 we have put the cursor at the age $A = 16$, corresponding to the maximum growth that we found in part 2. We see that if we harvest at the time of maximum growth, we will get a yield of $Y = 2980.96$ cubic feet per acre. Examination of the graph of the yield function shows that it is the steepest at age $A = 16$, which is where the inflection point occurs. The physical meaning of this is that maximum growth corresponds to the fastest increase in yield.

Solution to Part 4: For this part of the problem, we want to look at growth again. Thus we go to the function entry screen and turn Y off and G on 2.59 . We reset the window so that the horizontal span is from 0 to 14 and the vertical span is from 0 to 500. The resulting graph is in Figure 2.95. We see that the graph of growth will be increasing over this span, and so the maximum growth will occur at the right-hand endpoint. That is, we get a maximum growth of $G = 365.73$ cubic feet per acre per year if we wait to the end of the 14-year period to harvest. Do you think it is appropriate to proceed with harvesting at this time? ∎ ∎ ∎

Optimizing at Endpoints

Let's look back at the graph of the flight of the cannonball in Figure 2.88. If we adjust the window, 2.60 we see the view in Figure 2.96. This shows that the graph of $y = x - 32(x/250)^2$ extends far below the horizontal axis and to the left of the vertical axis. This indicates, among other things, that the cannonball would travel underground, which is nonsense. In fact, the function $y = x - 32(x/250)^2$ represents the path of the cannonball only during the period when it is flying through the air. This is from the place where the cannonball was fired ($x = 0$) to the location $x = 1953.125$ feet downrange where it strikes the ground. Beyond these limits, the graph in Figure 2.96 is not a representation of the physical situation. A more accurate picture would show the graph starting at $x = 0$ and ending where the cannonball lands at $x = 1953.125$. That is, an accurate graphical model would have endpoints, and if we wanted to find the *minimum* height of the cannonball, we would not find it at a peak or valley of the graph. Rather, we would find it at ground level: at the endpoints of the graph.

A similar situation occurs in part 4 of Example 2.10, where we are finding the maximum growth for a clear-cut forest. Because of market demand, loggers want to harvest the stand within 14 years from planting, and under these conditions, growth is maximized at the end of the 14-year period. This is at the right endpoint of the graph in Figure 2.95.

For the cannonball, we don't need a graph to tell us that the minimum height will be at ground level, but there are many physical situations where the optimal value is not so obvious but occurs at endpoints of the graph. Let's look, for example, at a simple geometry problem. Suppose we have 100 yards of fence from which we wish to construct two pens. We will use part of the fence to make a square pen and the rest to make a circular pen, as illustrated in Figure 2.97.

If we use s yards of the 100-yard stretch of fence for the square, then the total area $A = A(s)$ enclosed by the two pens turns out to be

$$A = \left(\frac{s}{4}\right)^2 + \frac{(100 - s)^2}{4\pi} \text{ square yards}.$$

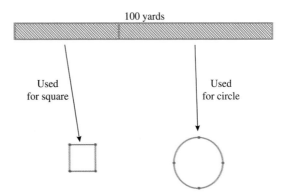

FIGURE 2.97 Cutting fence to make two pens

X	Y_1
0	795.77
20	534.3
40	386.48
60	352.32
80	431.83
100	625
120	931.83
X=0	

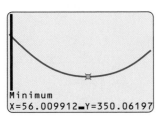

FIGURE 2.98 A table of values for area

FIGURE 2.99 The minimum area enclosed by a square and a circle

We want to investigate how much fence to use for the square, s yards, and how much to use for the circle, $100 - s$ yards, to optimize the total area A. The first step is to enter 2.61 the function and list the variable correspondences:

$$\mathsf{Y}_1 = A, \text{ total area on vertical axis}$$

$$\mathsf{X} = s, \text{ yards used for square on horizontal axis}.$$

Bearing in mind that we have only 100 yards of fence, so that s is between 0 and 100, we made the table of values in Figure 2.98 with a starting value of 0 and an increment of 20. This led us to choose a window with a horizontal span from 0 to 100 and a vertical span from 0 to 1000, which gives the graph shown in Figure 2.99. In Figure 2.99 we have located the minimum value 2.62 for area. We see from the prompt at the bottom of the screen that we enclose the minimum amount of area, $A = 350$ square yards, if we use about $s = 56$ yards of fence to make the square. That leaves $100 - 56 = 44$ yards of fence to make the circle.

Now, let's find how we should use the fence to enclose the maximum area. In Figure 2.99, we found the minimum at a *valley* of the graph, but there is no *peak* that corresponds to the maximum. The critical factor here is that we have only 100 yards of fence to work with, so s is between 0 and 100. The graph shows clearly that the maximum occurs at an endpoint.

We have traced the graph to locate the cursor at the right-hand endpoint in Figure 2.100 and at the left-hand endpoint in Figure 2.101. We see in Figure 2.101 that we get the maximum area of $A = 795.77$ square yards if we use $s = 0$ yards for the square. That is, all the fence goes to make the circle.

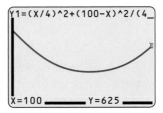

FIGURE 2.100 The right-hand endpoint: the area when all the fence is used for the square

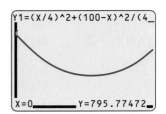

FIGURE 2.101 The maximum area at the left-hand endpoint, when all the fence is used for the circle

Optimizing with Parabolas

In the general case, finding maxima and minima exactly requires the use of calculus. But for quadratic functions the graph is a *parabola*, and there is an easy way to find the maximum or minimum.

In general, a quadratic expression has the form $ax^2 + bx + c$. In the case where a is positive, its graph is a parabola that opens upward and has a minimum but no maximum. An example is shown in Figure 2.102. When a is negative, the graph is a parabola that opens downward and has a maximum but no minimum. A typical example is shown in Figure 2.103. In general, we refer to the maximum or minimum point of a parabola as its *vertex*.

We get the graph of $ax^2 + bx$ from the graph of $ax^2 + bx + c$ by shifting the graph up or down, depending on whether c is positive or negative. Thus the height of the vertex for $ax^2 + bx$ may be different from the height of the vertex for $ax^2 + bx + c$, but the x value, or horizontal location, is the same. Parabolas are symmetric about a vertical line through the vertex. Thus when a parabola crosses the horizontal axis twice, the horizontal location of its vertex is halfway between the crossing points. Since $ax^2 + bx = x(ax + b)$, this parabola crosses the horizontal axis at $x = 0$ and at $x = -b/a$. Its vertex occurs halfway between, at $x = -b/2a$. Since the vertex of $ax^2 + bx$ has the same horizontal location as the vertex of $ax^2 + bx + c$, we have located the vertex of a general quadratic.

Locating the vertex: The vertex of the graph of $ax^2 + bx + c$ is located at $x = -b/2a$.

Thus, for example, the vertex of $2x^2 - 4x + 7$ occurs at

$$x = -\frac{-4}{2 \times 2} = 1.$$

Since the leading coefficient 2 is positive, this is a minimum. The vertical location of the minimum is found by getting the function value at $x = 1$:

$$\text{Vertical location of minimum } = 2 \times 1^2 - 4 \times 1 + 7 = 5.$$

Thus the vertex of the parabola occurs at the point $(1, 5)$. To verify our calculations, we have made the graph of $2x^2 - 4x + 7$ in Figure 2.104.

We can use what we know about the vertex of a parabola to solve certain optimization problems exactly.

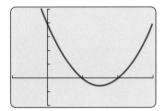

FIGURE 2.102 When the leading coefficient is positive, the parabola has a minimum

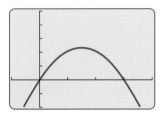

FIGURE 2.103 When the leading coefficient is negative, the parabola has a maximum

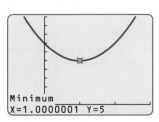

FIGURE 2.104 The graph of $2x^2 - 4x + 7$, showing the vertex at $(1, 5)$

EXAMPLE 2.11   **Maximizing Area**

One hundred yards of fence are to be used to make three sides of a rectangle. The fourth side of the rectangle is the bank of a straight river, as shown in Figure 2.105. The river bank should not be fenced. Let x denote the width of the rectangle, in yards, and assume that the length of the rectangle is along the river.

1. Express the length of the rectangle in terms of x.

2. Express the area of the rectangle in terms of x.

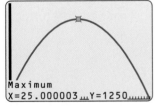

FIGURE 2.105 A rectangle with a river bank on one side

3. What value of x will give the rectangle of maximum area?

4. What is the largest area we can achieve?

Solution to Part 1: We use two sections of x yards of fence to make the left and right sides of the rectangle. That leaves $100 - 2x$ yards for the length of the rectangle.

Solution to Part 2: The area is the width x times the length $100 - 2x$:

$$\text{Area} = x(100 - 2x).$$

Solution to Part 3: To find the maximum of $x(100-2x)$, we first write it in standard form as $-2x^2 + 100x$. Thus the vertex occurs at

$$x = -\frac{100}{2 \times (-2)} = 25 \text{ yards}.$$

Since the leading coefficient is negative, this is a maximum.

Solution to Part 4: To find the largest area, we calculate the value of the area function $x(100 - 2x)$ for $x = 25$:

$$\text{Maximum area} = (25)(100 - 2 \times 25) = 1250 \text{ square yards}.$$

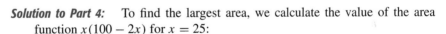

In Figure 2.106 we have plotted the graph of $x(100 - 2x)$ and located the maximum in order to verify our work. ■ ■ ■

Maximum
X=25.000003 Y=1250

FIGURE 2.106 Locating the vertex of $x(100 - 2x)$

Enrichment Exercises

E-1. **Locating the vertex of a parabola:** Find the vertex of each of the following parabolas. Find both the horizontal coordinate and the vertical coordinate, and determine whether the vertex is a maximum or a minimum.

a. $x^2 + 6x - 4$ b. $3x^2 - 30x + 1$

c. $-2x^2 + 12x - 3$

E-2. **More on locating the vertex of a parabola:** Consider the parabola $ax^2 + bx + c$.

a. Show that if a and b have the same sign, then the vertex lies to the left of the vertical axis.

b. Show that if a and b have opposite signs, then the vertex lies to the right of the vertical axis.

E-3. **Determining whether the horizontal axis is crossed:** Use the location of the vertex to show the following.

a. Show that the graph of $x^2 - 2x + 5$ does not cross the horizontal axis.

b. Show that the graph of $x^2 + 4x - 1$ crosses the horizontal axis twice.

E-4. **Finding the least area of a triangle:** One hundred yards of fence are to be used to construct a right triangular pen with a straight river serving as the hypotenuse of the triangle, as shown in Figure 2.107. No fence is placed along the river. Let x denote the height of the triangle, in yards.

a. Express the base (the horizontal side in Figure 2.107) of the triangle in terms of x.

b. Express the area of the triangle in terms of x.

c. What value of x gives the maximum area?

d. What is the greatest area that can be achieved?

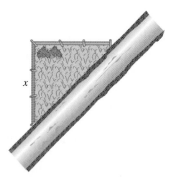

FIGURE 2.107 Making a triangle next to a river

E-5. **Distance from a line:** A typical point on a certain line in the plane has the form $(x, x+1)$. The square of the distance from this point to the origin is given by

$$D = x^2 + (x + 1)^2.$$

Find the x value that gives the point on the line nearest the origin.

E-6. **Size of high schools again:** Solve Exercise 19 below by finding the vertex of a parabola.

2.5 SKILL BUILDING EXERCISES

S-1. Maximum: Find the maximum value of $5x + 4 - x^2$ on the horizontal span of 0 to 5.

S-2. Minimum: Find the minimum value of $2^x - x^2 + 5$ on the horizontal span of 0 to 5.

S-3. Minimum: Find the minimum value of $x + (x + 5)/(x^2 + 1)$ on the horizontal span of 0 to 5.

S-4. Maximum: Find the maximum value of $5x^2 - e^x$ on the horizontal span of 0 to 5.

S-5. Maximum: Find the maximum value of $x^{1/x}$ on the horizontal span of 0 to 10.

S-6. Minimum: Find the minimum value of $4e^{2x} + 3x^2$ on the horizontal span of -2 to 1.

S-7. Maxima and minima: Find all maxima and minima of $f = x^3 - 6x + 1$ with a horizontal span from -2 to 2 and a vertical span from -10 to 10.

S-8. Maxima and minima: Find all maxima and minima of $f = 8x/(1 + x^2)$. Use a horizontal span of -5 to 5.

S-9. Endpoint maximum: Find the maximum value of $x^3 + x$ on the horizontal span of 0 to 5.

S-10. Endpoint minimum: Find the minimum value of $200 - x^3$ on the horizontal span of 0 to 5.

2.5 EXERCISES

1. **The cannon at a different angle:** Suppose a cannon is placed at the origin and elevated at an angle of 60 degrees. If the cannonball is fired with a muzzle velocity of 0.15 mile per second, it will follow the graph of $y = x\sqrt{3} - 160x^2/297$, where distances are measured in miles.

   a. Make a graph that shows the path of the cannonball.

   b. How far downrange does the cannonball travel? Explain how you got your answer.

   c. What is the maximum height of the cannonball, and how far downrange does that height occur?

2. **Profit:** The weekly profit P for a widget producer is a function of the number n of widgets sold. The formula is

$$P = -2 + 2.9n - 0.3n^2.$$

   Here P is measured in thousands of dollars, n is measured in thousands of widgets, and the formula is valid up to a level of 7 thousand widgets sold.

   a. Make a graph of P versus n.

   b. Calculate $P(0)$ and explain in practical terms what your answer means.

   c. At what sales level is the profit as large as possible?

3. **Marine fishery:** One class of models for population growth rates in marine fisheries assumes that the harvest from fishing is proportional to the population size. For one such model, we have

$$G = 0.3n \left(1 - \frac{n}{2}\right) - 0.1n.$$

   Here G is the growth rate of the population, in millions of tons of fish per year, and n is the population size, in millions of tons of fish.

   a. Make a graph of G versus n. Include values of n up to 1.5 million tons.

   b. Use functional notation to express the growth rate if the population size is 0.24 million tons, and then calculate that value.

   c. Calculate $G(1.42)$ and explain in practical terms what your answer means.

   d. At what population size is the growth rate the largest?

4. **Enclosing a field:** You have 16 miles of fence that you will use to enclose a rectangular field.

   a. Draw a picture to show that you can arrange the 16 miles of fence into a rectangle of width 3 miles and length 5 miles. What is the area of this rectangle?

b. Draw a picture to show that you can arrange the 16 miles of fence into a rectangle of width 2 miles and length 6 miles. What is the area of this rectangle?

c. The first two parts of this exercise are designed to show you that you can get different areas for rectangles of the same perimeter, 16 miles. In general, if you arrange the 16 miles of fence into a rectangle of width w miles, then it will enclose an area of $A = w(8 - w)$ square miles.

   i. Make a graph of area enclosed as a function of w, and explain what the graph is showing.

   ii. What width w should you use to enclose the most area?

   iii. What is the length of the maximum-area rectangle that you have made, and what kind of figure do you have?

5. **Forming a pen:** We want to form a rectangular pen of area 100 square feet. One side of the pen is to be formed by an existing building and the other three sides by a fence (see Figure 2.108). Let W be the length, in feet, of the sides of the rectangle perpendicular to the building, and let L be the length, in feet, of the other side.

a. Find a formula for the total amount of fence needed in terms of W and L.

b. Express, as an equation involving W and L, the requirement that the total area formed be 100 square feet.

c. Solve the equation you found in part b for L.

d. Use your answers to parts a and c to find a formula for F, the total amount, in feet, of fence needed, as a function of W alone.

e. Make a graph of F versus W.

f. Determine the dimensions of the rectangle that requires a minimum amount of fence.

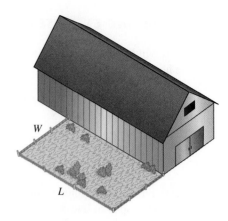

FIGURE 2.108

6. **Sales growth:** In this exercise we develop a model for the growth rate G, in thousands of dollars per year, in sales of a product as a function of the sales level s, in thousands of dollars.[31] The model assumes that there is a limit to the total amount of sales that can be attained. In this situation we use the term *unattained sales* for the difference between this limit and the current sales level. For example, if we expect sales to grow to 3 thousand dollars in the long run, then $3 - s$ gives the unattained sales. The model states that the growth rate G is proportional to the product of the sales level s and the unattained sales. Assume that the constant of proportionality is 0.3 and that the sales grow to 2 thousand dollars in the long run.

a. Find a formula for unattained sales.

b. Write an equation that shows the proportionality relation for G.

c. On the basis of the equation from part b, make a graph of G as a function of s.

d. At what sales level is the growth rate as large as possible?

e. What is the largest possible growth rate?

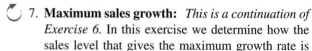 7. **Maximum sales growth:** *This is a continuation of Exercise 6.* In this exercise we determine how the sales level that gives the maximum growth rate is

---

[31]The model we develop for sales growth can be applied in other settings where there is a limit to growth. Examples include the spread of a new technology and population growth under environmental constraints.

related to the limit on sales. Assume, as above, that the constant of proportionality is 0.3, but now suppose that sales grow to a level of 4 thousand dollars in the limit.

a. Write an equation that shows the proportionality relation for G.

b. On the basis of the equation from part a, make a graph of G as a function of s.

c. At what sales level is the growth rate as large as possible?

d. Replace the limit of 4 thousand dollars with another number, and find at what sales level the growth rate is as large as possible. What is the relationship between the limit and the sales level that gives the largest growth rate? Does this relationship change if the proportionality constant is changed?

e. Use your answers in part d to explain how to determine the limit if we are given sales data showing the sales up to a point where the growth rate begins to decrease.

8. **An aluminum can:** The cost of making a can is determined by how much aluminum A, in square inches, is needed to make it. This in turn depends on the radius r and the height h of the can, both measured in inches. You will need some basic facts about cans. See Figure 2.109.

   The surface of a can may be modeled as consisting of three parts: two circles of radius r and the surface of a cylinder of radius r and height h. The area of these circles is πr^2 each, and the area of the surface of the cylinder is $2\pi rh$. The volume of the can is the volume of a cylinder of radius r and height h, which is $\pi r^2 h$.

   In what follows, we assume that the can must hold 15 cubic inches, and we will look at various cans holding the same volume.

a. Explain why the height of any can that holds a volume of 15 cubic inches is given by

$$h = \frac{15}{\pi r^2}.$$

b. Make a graph of the height h as a function of r, and explain what the graph is showing.

c. Is there a value of r that gives the least height h? Explain.

d. If A is the amount of aluminum needed to make the can, explain why

$$A = 2\pi r^2 + 2\pi rh.$$

e. Using the formula for h from part a, explain why we may also write A as

$$A = 2\pi r^2 + \frac{30}{r}.$$

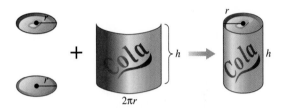

FIGURE 2.109

9. **An aluminum can, continued:** *This is a continuation of Exercise 8.* The cost of making a can is determined by how much aluminum A, in square inches, is needed to make it. As we saw in Exercise 8, we can express both the height h and the amount of aluminum A in terms of the radius r:

$$h = \frac{15}{\pi r^2}$$

$$A = 2\pi r^2 + \frac{30}{r}.$$

a. What is the height, and how much aluminum is needed to make the can, if the radius is 1 inch? (This is a tall, thin can.)

b. What is the height, and how much aluminum is needed to make the can, if the radius is 5 inches? (This is a short, fat can.)

c. The first two parts of this problem are designed to illustrate that for an aluminum can, different surface areas can enclose the same volume of 15 cubic inches.

   i. Make a graph of A versus r and explain what the graph is showing. →

ii. What radius should you use to make the can using the least amount of aluminum?

iii. What is the height of the can that uses the least amount of aluminum?

10. **Cost for a can:** *This is a continuation of Exercises 8 and 9.* Suppose now that we use different materials in making different parts of the can. The material for the side of the can costs $0.10 per square inch, and the material for the top and bottom costs $0.05 per square inch.

   a. Use a formula to express the cost C, in dollars, of the material for the can as a function of the radius r.

   b. What radius should you use to make the least expensive can?

11. **Profit:** *The background for this exercise can be found in Exercises 11, 12, 13, and 14 in Section 1.4.* A manufacturer of widgets has fixed costs of $600 per month, and the variable cost is $60 per thousand widgets (so it costs $60 to produce a thousand widgets). Let N be the number, in thousands, of widgets produced in a month.

   a. Find a formula for the manufacturer's total cost C as a function of N.

   b. The highest price p, in dollars per thousand widgets, at which N can be sold is given by the formula $p = 70 - 0.03N$. Using this, find a formula for the total revenue R as a function of N.

   c. Use your answers to parts a and b to find a formula for the profit P of this manufacturer as a function of N.

   d. Use your formula from part c to determine the production level at which profit is maximized if the manufacturer can produce at most 300 thousand widgets in a month.

12. **Laying phone cable:** City A lies on the north bank of a river that is 1 mile wide. You need to run a phone cable from City A to City B, which lies on the opposite bank 5 miles down the river. You will lay L miles of the cable along the north shore of the river, and from the end of that stretch of cable you will lay W miles of cable running under water directly toward City B. (See Figure 2.110.)

   You will need the following fact about right triangles. A right triangle has two legs, which meet at the right angle, and the hypotenuse, which is the longest side. An ancient and beautiful formula, the *Pythagorean theorem*, relates the lengths of the three sides:

Length of hypotenuse

$$= \sqrt{\text{Length of one leg}^2 + \text{Length of other leg}^2}.$$

   a. Find an appropriate right triangle that shows that $W = \sqrt{1 + (5 - L)^2}$.

   b. Find a formula for the length of the total phone cable P from City A to City B as a function of L.

   c. Make a graph of the total phone cable length P as a function of L, and explain what the graph is showing.

   d. What value of L gives the least length for the total phone cable? Draw a picture showing the least-length total phone cable.

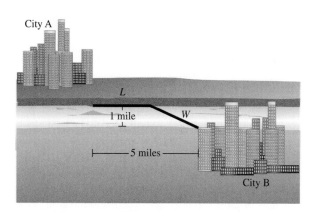

FIGURE 2.110 Laying phone cable

13. **Laying phone cable, continued:** *This is a continuation of Exercise 12.* Suppose it costs $300 per mile to run cable on land but $500 per mile to lay it under water.

   a. Let C be the total cost of the project, in dollars. Explain why $C = 300L + 500W$.

   b. Write a formula for C as a function of L. *Hint:* Use part a of Exercise 12.

   c. Find the amount of cable that runs under water and the total cost of the project if you choose L to be 1 mile long. Draw and properly label a picture that shows this cable plan.

d. Find the amount of cable that runs under water and the total cost of the project if you choose L to be 3 miles long. Draw and properly label a picture that shows this cable plan.

e. Make a graph of the cost C as a function of L, and explain what the graph is showing.

f. What value of L gives the least cost for the project?

g. Find the value of W that corresponds to your answer in part f, and draw a picture showing the least-cost project.

h. Potential legal disputes about easements have caused the cost of laying cable on land to increase to $700 per mile. With this change, find a new formula for the cost C. How does this affect your plans for the least-cost project?

14. **Mean annual growth:** *This is a continuation of Example 2.10.* Forest managers study the *mean annual growth* $M = M(A)$, defined as the yield at age A divided by A.

a. Use the formula for Y in Example 2.10 to find a formula for M as a function of A.

b. Add the graph of M as a function of A to the graph of G.

c. The *rotation age* may be determined as the age of maximum mean annual growth. Determine the rotation age using the graph of M.

d. From your two graphs in part b, can you suggest a different way to find the rotation age?

15. **Growth of fish biomass:** An important model for commercial fisheries is that of Beverton and Holt.[32] It begins with the study of a single *cohort* of fish— that is, all the fish in the study are born at the same time. For a cohort of the North Sea plaice (a type of flatfish), the number $N = N(t)$ of fish in the population is given by

$$N = 1000e^{-0.1t},$$

and the weight $w = w(t)$ of each fish is given by

$$w = 6.32 \left(1 - 0.93e^{-0.095t}\right)^3.$$

Here w is measured in pounds and t in years. The variable t measures the so-called recruitment age, which we refer to simply as the age

The *biomass* $B = B(t)$ of the fish cohort is defined to be the total weight of the cohort, so it is obtained by multiplying the population size by the weight of a fish.

a. If a plaice weighs 3 pounds, how old is it?

b. Use the formulas for N and w given above to find a formula for $B = B(t)$, and then make a graph of B against t. (Include ages through 20 years.)

c. At what age is the biomass the largest?

d. In practice, fish below a certain size can't be caught, so the biomass function becomes relevant only at a certain age.

  i. Suppose we want to harvest the plaice population at the largest biomass possible, but a plaice has to weigh 3 pounds before we can catch it. At what age should we harvest?

  ii. Work part i under the assumption that we can catch plaice weighing at least 2 pounds.

16. **Spawner-recruit model:** In fish management it is important to know the relationship between the abundance of the *spawners* (also called the parent stock) and the abundance of the *recruits*—that is, those hatchlings surviving to maturity.[33] According to the *Ricker model*, the number of recruits R as a function of the number of spawners P has the form

$$R = APe^{-BP}$$

for some positive constants A and B. This model describes well a phenomenon observed in some fisheries: A large spawning group can actually lead to a small group of recruits.[34]   →

---

[32] See R. J. H. Beverton and S. J. Holt, *On the Dynamics of Exploited Fish Populations*, Fishery Investigations, Series 2, Volume 19 (London: Ministry of Agriculture, Fisheries and Food, 1957).

[33] See W. E. Ricker, "Stock and recruitment," *J. Fish. Res. Board Can.* **11** (1954), 559–623.

[34] Biological mechanisms that contribute to this phenomenon are suspected to include competition and cannibalism of the young.

In a study of the sockeye salmon, it was determined that $A = 4$ and $B = 0.7$. Here we measure P and R in thousands of salmon.

a. Make a graph of R against P for the sockeye salmon. (Assume there are at most 3000 spawners.)

b. Find the maximum number of salmon recruits possible.

c. If the number of recruits R is greater than the number of spawners P, then the difference $R - P$ of the recruits can be removed by fishing, and next season there will once again be P spawners surviving to renew the cycle. What value of P gives the maximum value of $R - P$, the number of fish available for removal by fishing?

17. **Rate of growth:** The rate of growth G in the weight of a fish is a function of the weight w of the fish. For the North Sea cod, the relationship is given by

$$G = 2.1w^{2/3} - 0.6w.$$

Here w is measured in pounds and G in pounds per year. The maximum size for a North Sea cod is about 40 pounds.

a. Make a graph of G against w.

b. Find the greatest rate of growth among all cod weighing at least 5 pounds.

c. Find the greatest rate of growth among all cod weighing at least 25 pounds.

18. **Health plan:** The manager of an employee health plan for a firm has studied the balance B, in millions of dollars, in the plan account as a function of t, the number of years since the plan was instituted. He has determined that the account balance is given by the formula $B = 60 + 7t - 50e^{0.1t}$.

a. Make a graph of B versus t over the first 7 years of the plan.

b. At what time is the account balance at its maximum?

c. What is the smallest value of the account balance over the first 7 years of the plan?

19. **Size of high schools:** The farm population has declined dramatically in the years since World War II, and with that decline, rural school districts have been faced with consolidating in order to be economically efficient. One researcher studied data from the early 1960s on expenditures for high schools ranging from 150 to 2400 in enrollment.[35] He considered the cost per pupil as a function of the number of pupils enrolled in the high school, and he found the approximate formula

$$C = 743 - 0.402n + 0.00012n^2,$$

where n is the number of pupils enrolled and C is the cost, in dollars, per pupil.

a. Make a graph of C versus n.

b. What enrollment size gives a minimum per-pupil cost?

c. If a high school had an enrollment of 1200, how much in per-pupil cost would be saved by increasing enrollment to the optimal size found in part b?

20. **Radioactive decay:** Radium 223 is a radioactive substance that itself is a product of the radioactive decay of thorium 227. For one experiment, the amount A of radium 223 present, as a function of the time t since the experiment began, is given by the formula $A = 3(e^{-0.038t} - e^{-0.059t})$, where A is measured in grams and t in days.

a. Make a graph of A versus t covering the first 60 days of the experiment.

b. What was the largest amount of radium 223 present over the first 60 days of the experiment?

c. What was the largest amount of radium 223 present over the first 10 days of the experiment?

d. What was the smallest amount of radium 223 present over the first 60 days of the experiment?

---

[35]This exercise is based on the study by J. Riew, "Economies of scale in high school operation," *Review of Economics and Statistics* **48** (August 1966), 280–287. In our presentation of his results, we have suppressed variables such as teacher salaries.

21. **Water flea:** F. E. Smith has studied population growth for the water flea.[36] Let N denote the population size. In one experiment, Smith found that G, the rate of growth per day in the population, can be modeled by

$$G = \frac{0.44N(228 - N)}{228 + 3.46N}.$$

  a. Draw a graph of G versus N. Include values of N up to 350.

  b. At what population level does the greatest rate of growth occur?

  c. There are two values of N where G is zero. Find these values of N and explain what is occurring at these population levels.

  d. What is the rate of population growth if the population size is 300? Explain what is happening to the population at this level.

22. **An epidemic:** One model for the spread of epidemics gives the number of newly infected individuals t days after the outbreak of the epidemic as

$$\text{New cases} = \frac{\beta n(n + 1)^2 e^{(n+1)\beta t}}{\left(n + e^{(n+1)\beta t}\right)^2}.$$

Here n is the total number of people we expect to be infected over the course of the epidemic, and β depends on the nature of the infection as well as on other environmental factors. For a certain epidemic, the number of new cases is

$$\text{New cases} = \frac{75{,}150 e^{0.3t}}{\left(500 + e^{0.3t}\right)^2}.$$

  a. Make a graph of the number of new cases versus days since the outbreak. Include times up to 30 days.

  b. What is the greatest number of new cases we expect to see in 1 day, and when does that occur?

  c. The local medical facilities can handle no more than 25 new cases per day. During what time period will it be necessary to recruit help from outside sources?

23. **Research project:** Locate a calculus textbook and look in the section entitled optimization, max/min problems, or something similar. Find a worked-out example in the text where a maximum or minimum for a function is found. Solve that same problem using the optimization procedure you've learned in this section. Compare your solution with the solution provided in the calculus text. Write a careful description of the function, its use, and your solution to the problem, and add a very brief description of the solution found in your source.

---

[36] See "Population dynamics in *Daphnia magna* and a new model for population growth," *Ecology* **44** (1963), 651–663. The population size is measured per unit volume.

Summary

Each type of function presentation has its advantages and disadvantages, and one of the most effective methods of mathematical analysis is to take a function presented in one way and express it in another form. It is particularly tedious to get a graph or table of values from a function given by a complicated formula, but doing so can also be very illuminating. The graphing calculator takes the tedium out of this procedure by producing graphs and tables on demand.

Tables and Trends

It is often difficult to determine limiting values or maximum and minimum values from a formula, but as we saw in Chapter 1, this is easily done from a table of values. The calculator produces such tables on demand, and in this respect it is an important problem-solving tool. For example, the probability of getting exactly 3 sixes when you roll N dice is given by the formula

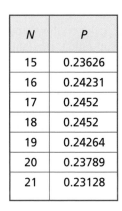

N	P
15	0.23626
16	0.24231
17	0.2452
18	0.2452
19	0.24264
20	0.23789
21	0.23128

$$P = \frac{N(N-1)(N-2)}{750}\left(\frac{5}{6}\right)^N.$$

It is by no means clear from this formula how many dice should be rolled to make the probability of achieving exactly 3 sixes the greatest. A calculator can be used to produce quickly the accompanying partial table of values.

The table shows clearly that the maximum probability is achieved by rolling 17 or 18 dice.

Graphs

As with tables, the graphing calculator can quickly produce graphs from formulas, and effective analysis can proceed. Graphs viewed from different perspectives can differ considerably in appearance. Effective graphical analysis depends on being able to view the graph in an appropriate window. This is done through an intelligent choice of *horizontal span* and *vertical span*. When the function is a model for some physical phenomenon, choosing the horizontal span, or *domain*, may be a matter of common sense. The vertical span, or *range*, can then be chosen by looking at a table of values.

As an example, suppose a leather craftsman has produced 25 belts that he intends to sell at an upcoming art fair for $22.75 each. He has invested a total of $300 in materials. The net profit p, in dollars, from the sale of n belts is given by

$$p = 22.75n - 300.$$

In order to graph this function, we need to choose a horizontal and a vertical span. Since he has only 25 belts, he will sell somewhere between 0 and 25 belts. This is the horizontal span, or domain, for our function. To get an appropriate vertical span, we look at the table of values shown in Figure 2.111. This shows that we should choose a vertical span of about −325 to 300. The resulting graph, which is now available for further analysis, is shown in Figure 2.112.

X	Y₁
0	-300
5	-186.5
10	-73
15	40.5
20	154
25	267.5
30	381

X=0

FIGURE 2.111 A table of values for net profit on belt sales

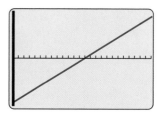

FIGURE 2.112 A graph of net profit on belt sales

Solving Linear Equations

Many times the analysis required to understand a phenomenon involves solving an equation. For many equations, this can be a difficult process, but for *linear equations*, solutions are easy. Linear equations are those that do not involve powers, roots, or other complications of the variable; they are the simplest of all equations. Any linear equation can be solved using two basic rules.

> **Moving terms across the equals sign:** You may add (or subtract) the same thing to (or from) both sides of the equation. This can be thought of as moving a term from one side of the equation to the other, provided that you change its sign.

> **Dividing each term of an equation:** You may divide (or multiply) both sides of an equation by any nonzero number.

As an example, we show how to solve $5x + 7 = 18 - 2x$ using these two rules.

$$5x + 7 = 18 - 2x$$

$$5x = 18 - 7 - 2x \qquad \text{Move 7 across and change its sign.}$$

$$5x + 2x = 11 \qquad \text{Move } 2x \text{ across and change its sign.}$$

$$7x = 11$$

$$x = \frac{11}{7} = 1.57 \qquad \text{Divide each term by 7.}$$

Solving Nonlinear Equations

Equations that are not linear may be difficult or impossible to solve exactly by hand, but the graphing calculator can provide approximate (yet highly accurate) solutions to virtually any equation quickly and easily.

The most useful method for solving equations with a calculator is referred to in this text as the *crossing-graphs method*. We will illustrate it with an example. Suppose the temperature, after t minutes, of an aluminum bar being heated in an oven is given by

$$A = 800 - 730e^{-0.06t} \text{ degrees Fahrenheit}.$$

We want to know when the temperature will reach 600 degrees. In terms of an equation, that means we need to solve

$$800 - 730e^{-0.06t} = 600$$

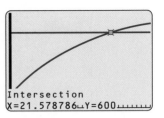

FIGURE 2.113 Finding when
the temperature is 600 degrees

for t. This is done by entering the functions $800 - 730e^{-0.06t}$ and 600 into the calculator. We want to know when these functions are the same—that is, where the graphs cross. We see in Figure 2.113 that this occurs 21.58 minutes after the bar is put in the oven.

Optimization

A strategy similar to that used to solve nonlinear equations can be used to optimize functions given by formulas. We use the calculator to make the graph and then look for peaks or valleys. To illustrate the method, we look for the maximum height of a cannonball following the graph of

$$x - 32\left(\frac{x}{250}\right)^2,$$

where x is the distance downrange, measured in feet. We graph the function as shown in Figure 2.114. That figure shows that the cannonball reaches its maximum height of 488.28 feet when it is 976.56 feet downrange.

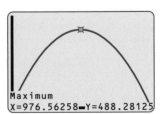

FIGURE 2.114 The maximum
height of a cannonball

Chapter

3

Straight Lines and Linear Functions

Straight lines in the form of city streets, directions, boundaries, and many other things are among the most obvious mathematical objects that we experience in daily life. Historically, mathematicians made extensive studies of the *geometry* of straight lines several centuries before these lines were associated with a *linear formula*. Today it is understood that it is the combination of pictures and formulas that make lines so useful and so easy to handle.

3.1 *The Geometry of Lines*

Characterizations of Straight Lines

One way in which straight lines are often characterized is that they are determined by two points. For example, to describe a straight ramp, it is only necessary to give the locations of the ends of the ramp. In Figure 3.1 we have depicted a ramp with one end atop a 4-foot-high retaining wall and the other on the ground 10 feet away. It is often convenient to represent such lines on coordinate axes. If we choose the horizontal axis to be ground level and let the vertical axis follow the retaining wall, we get the picture in Figure 3.2. The ends of the ramp in Figure 3.1 match the points in Figure 3.2 where the line crosses the horizontal and vertical axes. These crossing points are known as the *horizontal and vertical intercepts*. (Sometimes they are called the *x-intercept* and the *y-intercept*.) Thus in Figure 3.2 the horizontal intercept is 10 and the vertical intercept is 4.

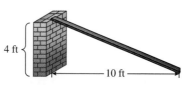

FIGURE 3.1 A ramp on a retaining wall

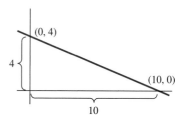

FIGURE 3.2 Representing the ramp line on coordinate axes

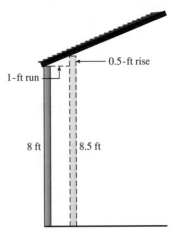

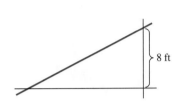

FIGURE 3.3 The roof of a building

FIGURE 3.4 Representing the roof line

To illustrate these ideas further, in Figure 3.3 we have drawn a roof line that is 8 feet high at the outside wall and 8.5 feet high 1 foot toward the interior of the structure. If we want to represent the roof line on coordinate axes, it is natural to let the horizontal axis correspond to ground level and to let the vertical axis follow the outside wall. We have done this in Figure 3.4. We note that the vertical intercept of this line is 8, but in this case it is not immediately apparent what the horizontal intercept is or what its physical significance might be. We will return to this in Example 3.1.

Another way of characterizing a straight line is to say that it rises or falls at the same rate everywhere on the line. (Curved graphs rise or fall at different rates at different places.) If we stand at the outside wall of the house depicted in Figure 3.3 and move 1 foot toward its interior, the roof rises from 8 feet to 8.5 feet. That is, in 1 horizontal foot, the roof rises by 0.5 vertical foot. Since the roof line is straight, we know that if we move 1 more foot toward the interior, the roof will rise by that same amount, 0.5 foot, and the roof at that point will be 9 feet high. Each 1-foot horizontal movement toward the interior results in a 0.5-foot vertical rise in the roof line. In this setting, horizontal change is often referred to as the *run* and vertical change as the *rise*, and we would say that the roof line rises 0.5 foot for each 1-foot run.

EXAMPLE 3.1   **Further Examination of the Roof Line**

In Figure 3.5 we have extended the roof to its peak 14 horizontal feet toward the interior of the structure.

1. How high is the roof line at its peak?

2. Locate the horizontal intercept in Figure 3.6 and explain its physical significance.

Solution to Part 1:   We know that the roof rises 0.5 foot for each 1-foot run. To get to the peak of the roof, we need to make a 14-foot run. That will result in 14 half-foot rises. Thus the roof rises by $14 \times 0.5 = 7$ feet. Since the roof is 8 feet high at the outside wall, that makes it $8 + 7 = 15$ feet high at its peak. This is illustrated in Figure 3.5.

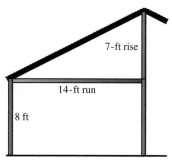

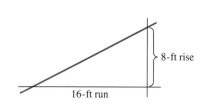

FIGURE 3.5 Extending the roof line to its peak

FIGURE 3.6 Finding the horizontal intercept

Solution to Part 2:   In Figure 3.6, movement 1 foot to the right results in a 0.5-foot rise in the line. Thus a 1-foot movement to the left would give a fall of 0.5 foot. To get down to the horizontal axis, we need to drop a total of 8 units. Since we do that in 0.5-foot steps, we need to move 16 units to the left of the vertical axis to get to the place in Figure 3.6 where the line crosses the horizontal axis. Thus the horizontal intercept is −16 (negative, since it's to the left of zero). This is where an extended roof line would meet the ground. It might, for example, be important if we wanted to build an "A-frame" structure. The minus sign here means that an extended roof line would meet the ground 16 feet in the exterior direction from the wall.

■ ■ ■

The Slope of a Line

The number 0.5 associated with the roof line in Example 3.1 is the *rate of change* in height with respect to horizontal distance. For straight lines, this is universally termed the *slope* and is commonly denoted by the letter m. The slope is one of the most important and useful features of a line. It tells the vertical change along the line when there is a horizontal change of 1 unit. In other words, a line of slope m exhibits m units of rise for each unit of run. Another way to express this is to say that the rise is proportional to the run, and the constant of proportionality is called the slope. Horizontal change is sometimes denoted by Δx and vertical change by Δy. For a line of slope m, the following three equations say exactly the same thing using these different terms:

$$\text{Rise} = m \times \text{Run} \tag{3.1}$$

$$\text{Vertical change} = m \times \text{Horizontal change} \tag{3.2}$$

$$\Delta y = m \times \Delta x. \tag{3.3}$$

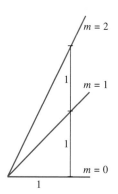

FIGURE 3.7 Some lines with positive slope

Figure 3.7 shows lines of slopes 0, 1, and 2. Note that the horizontal line has slope $m = 0$. This makes sense because a horizontal change results in no vertical change at all. The line with slope $m = 1$ rises 1 unit for each unit of run. Thus it slants upward at a 45-degree angle. The line of slope $m = 2$ is much steeper, showing a rise of 2 units for each unit of run. In general, larger positive slopes mean steeper lines that point upward from left to right.

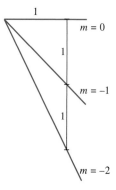

FIGURE 3.8 Some lines with negative slope

If a line has negative slope, then a run of 1 unit will result in a *drop* rather than a rise. The line in Figure 3.8 with $m = -1$ drops 1 unit for each unit of run. Similarly, the line with $m = -2$ drops 2 units for each unit of run. In general, large negative slopes mean steep lines that point downward.

Getting Slope from Points

Equations (3.1), (3.2), and (3.3) show us how to use the slope of a line to determine vertical change. These are all linear equations, so we can use division to solve each of them for *m*. This will give us equations that tell how to find the slope *m* of a line. Once again, all three equations say the same thing using different notations:

$$\text{Rate of change} = \text{slope} = m = \frac{\text{Rise}}{\text{Run}} \tag{3.4}$$

$$\text{Rate of change} = \text{slope} = m = \frac{\text{Vertical change}}{\text{Horizontal change}} \tag{3.5}$$

$$\text{Rate of change} = \text{slope} = m = \frac{\Delta y}{\Delta x}. \tag{3.6}$$

KEY IDEA 3.1

The Slope of a Line

- The slope, or rate of change, *m*, of a line shows how steeply it is increasing or decreasing. It tells the vertical change along the line when there is a horizontal change of 1 unit.

  If *m* is positive, the line is rising from left to right. Larger positive values of *m* mean steeper lines.

  If $m = 0$, the line is horizontal.

  If *m* is negative, the line is falling from left to right. Negative values of *m* that are larger in size correspond to lines that fall more steeply.

- The slope *m* of a line can be calculated using

$$m = \frac{\text{Vertical change}}{\text{Horizontal change}}.$$

- The slope *m* can be used to calculate vertical change:

$$\text{Vertical change} = m \times \text{Horizontal change}.$$

EXAMPLE 3.2   A Circus Tent

The outside wall of the circus tent depicted in Figure 3.9 is 10 feet high. Five feet toward the center pole, the tent is 12 feet high. The center pole of the tent is 60 feet from the outside wall.

1. What is the slope of the line that follows the roof of the circus tent?

2. Use the slope to find the height of the tent at its center pole.

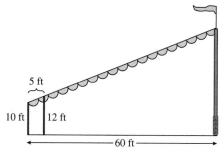

FIGURE 3.9 A circus tent

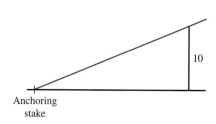

FIGURE 3.10 An anchor rope

3. A rope is attached to the roof of the tent at the outside wall and anchors the tent to a stake in the ground. The rope follows the line of the tent roof as shown in Figure 3.10. How far away from the tent wall is the anchoring stake?

Solution to Part 1:   To calculate the slope we use Equation (3.4). At the outside wall, the circus tent is 10 feet high. If we run 5 feet toward the center pole, the height increases to 12 feet; that is a rise of 2 feet. Thus

$$m = \frac{\text{Rise}}{\text{Run}} = \frac{2}{5} = 0.4 \text{ foot per foot}.$$

Solution to Part 2:   To find the height at the center pole, we first find the rise from the top of the outside wall to the center pole. We do this by using Equation (3.1) and the value of the slope we found in part 1. From the outside wall to the center pole is a run of 60 feet, so

$$\text{Rise} = m \times \text{Run} = 0.4 \times 60 = 24 \text{ feet}.$$

Thus from the top of the outside wall to the center pole, the height increases by 24 feet. To find the height at the center pole, we need to add the height of the outside wall. We find that the height of the center pole is $24 + 10 = 34$ feet.

Solution to Part 3:   Since the slope of the line is 0.4, if we stand at the anchoring stake and move toward the tent, each horizontal foot that we move results in a 0.4-foot rise in the rope. We need the total rise to be 10 feet:

$$\text{Rise} = m \times \text{Run}$$

$$10 = 0.4 \times \text{Run}$$

$$\frac{10}{0.4} = \text{Run}$$

$$25 = \text{Run}.$$

Thus the anchoring stake is located 25 feet outside the wall of the tent. If the vertical axis corresponds to the outside wall of the tent, then the horizontal intercept of the roof line is −25.

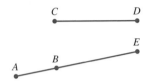

FIGURE 3.11 Extending a segment

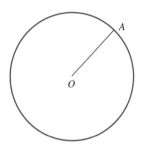

FIGURE 3.12 Constructing a circle with given radius and given center

FIGURE 3.13 A given line and a given point not on the line

FIGURE 3.14 The unique parallel through P

Plane Geometry, Parallel, and Perpendicular Lines

Lines were extensively studied long before anything like algebra was invented. One of the most important such studies was Euclid's *Elements*, which examined plane geometry from an axiomatic point of view. Euclid began with the following five *axioms* or *postulates*.

Postulate 1: Given points P and Q with $P \neq Q$, there is a unique line l that passes through both P and Q.

Postulate 2: For every segment AB and for every segment CD, there is a unique point E such that B is between A and E and segment CD is congruent to BE.

Intuitively this postulate says we can always extend a given segment by a given length. See Figure 3.11.

Postulate 3: Given points O and A with $O \neq A$, there exists a circle with center O and radius OA.

Here Euclid had in mind the notion of construction. The postulate tells us that we can construct a circle with given center and given radius. See Figure 3.12.

Postulate 4: All right angles are congruent to one another.

Postulate 5, the parallel postulate: Given a line l and a point P not on l, there is a unique line n through P that is parallel to l. See Figures 3.13 and 3.14.

Euclid deduced from these five simple axioms 465 propositions that contained all the geometric knowledge of his time. Euclid's *Elements* is one of the most important mathematical texts in the history of mathematics. Modern mathematicians recognize many flaws in the work, but even with its flaws it represented a milestone in mathematical development. And indeed, the flaws contained in Euclid's *Elements* stimulated a wealth of crucially important mathematical development.

Early mathematicians had no difficulty accepting the first four of Euclid's axioms. But the fifth was thought not to be so obvious as the other four, and mathematicians tried into the 19th century to derive the fifth postulate from the first four. All such attempts were doomed to failure, but in making the effort, investigators developed a great deal of modern mathematics. It was finally discovered in the 19th century by Bolyai, Gauss, Lobachevsky, Beltrami, and others that the fifth postulate is in fact independent of the first four and that one gets a perfectly legitimate geometry by discarding Euclid's parallel axiom and replacing it with a different one:

The hyperbolic parallel axiom: Given a line l and a point P not on l, there are infinitely many lines through P that are parallel to l.

Euclid's first four axioms, together with the hyperbolic parallel axiom, are the basis for *hyperbolic geometry*. The world of hyperbolic geometry may seem unfamiliar. The following are a few theorems that are true in hyperbolic geometry.

Hyperbolic geometry theorem 1: There are no rectangles.

Hyperbolic geometry theorem 2: Similar triangles (triangles with congruent angles) are congruent.

Hyperbolic geometry theorem 3: The angle sum of any triangle is strictly less than 180 degrees.

The difference (180 − Angle sum) is known as the *defect* of the triangle.

Hyperbolic geometry theorem 4: The area of a triangle is proportional to its defect.

This has the curious consequence that areas of triangles are bounded. Triangles of arbitrarily large area do not exist.

Hyperbolic geometry theorem 5: There is a natural unit of length.

In both Euclidean and hyperbolic geometry there is a natural unit of angle measure known as the *radian*, but in Euclidean geometry there is no natural unit of length.

There are many other seemingly odd theorems in hyperbolic geometry. You may be surprised to know that hyperbolic geometry fits remarkably well with Einstein's theory of special relativity and is an important tool in modern physics for describing the universe. Perhaps it isn't so odd after all.

One important result of Euclidean geometry is the theory of similar triangles— that is, triangles with congruent angles. It is similarity theory that makes trigonometry possible.

Euclidean geometry theorem 1: The angle sum of any triangle is 180 degrees.

Euclidean geometry theorem 2: If $\triangle ABC$ is similar to $\triangle A'B'C'$, then

$$\frac{|AB|}{|A'B'|} = \frac{|AC|}{|A'C'|} = \frac{|BC|}{|B'C'|}.$$

We are using $|AB|$ here to denote the length of the segment AB. See Figure 3.15.

A typical situation where we find similar triangles is shown in Figure 3.16. Here DE is parallel to AB. An important theorem in Euclidean geometry tells us that since DE is parallel to AB, $\angle CAB$ is congruent to $\angle CDE$ and $\angle CBA$ is congruent to $\angle CED$. Since the third angle is shared by both triangles, it follows that $\triangle ABC$ is similar to $\triangle DEC$.

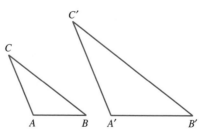

FIGURE 3.15 Similar triangles

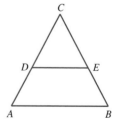

FIGURE 3.16 Similarity resulting from *AB* being parallel to *DE*

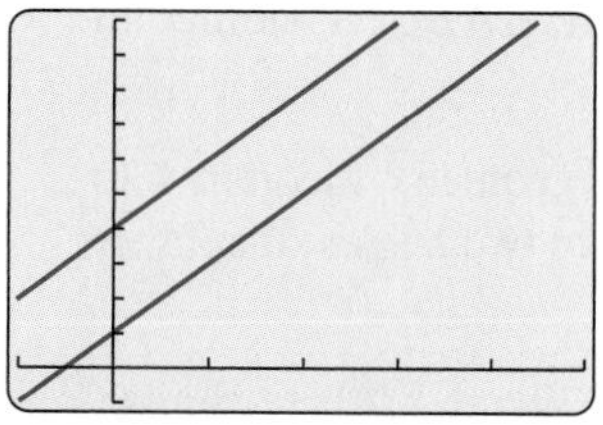

Now we apply this discussion to parallel and perpendicular lines. As we noted earlier in this section, the slope of a line determines the direction in which it points. It is then not surprising to note that lines with the same slope are parallel or coincident. In Figure 3.17 we show the graphs of $y = 2x + 1$ and $y = 2x + 4$. They have the

FIGURE 3.17 Parallel lines resulting from the same slope

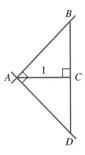

FIGURE 3.18
Perpendicular
lines

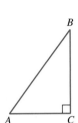

FIGURE 3.19 Reorienting similar
triangles

same slope, 2, and indeed appear to be parallel. Note that we get the upper graph by shifting the lower graph up 3 units.

We want to see how the slopes of lines that are perpendicular compare. In Figure 3.18 the line through AB is perpendicular to the line through AD. We take the line through AC as the horizontal axis. Since the length of AC is 1, the slope m_1 of the line through AB is $|BC|$, and the slope m_2 of the line through AD is $-|DC|$. Since $\angle BAD$ is a right angle, we have $\angle DAC° = 90 - \angle BAC°$. But also, since the angle sum of $\triangle CAB$ is 180 degrees, we have that $\angle ABC° = 90 - \angle BAC°$. We conclude that $\angle DAC$ is congruent to $\angle ABC$. A similar argument shows that $\angle CDA$ is congruent to $\angle CAB$. Hence the top triangle in Figure 3.18 is similar to the bottom triangle. In Figure 3.19 we have oriented the triangles to show the proper correspondence of angles.

Since the triangles are similar, we have

$$\frac{|BC|}{|AC|} = \frac{|AC|}{|DC|}.$$

Now, using the facts that $|AC| = 1$, $|BC| = m_1$, and $|DC| = -m_2$, we have

$$m_1 = -\frac{1}{m_2}.$$

Thus if two lines are perpendicular, then the slope of one is the negative reciprocal of the slope of the other.

Enrichment Exercises

E-1. **An alternative version of similarity:** Show that if $\triangle ABC$ is similar to $\triangle A'B'C'$, then $\frac{|AB|}{|AC|} = \frac{|A'B'|}{|A'C'|}$.

E-2. **Two-angle criterion for similarity:** Use Euclidean geometry theorem 1 to show that if two angles of one triangle are congruent to two angles of another triangle, then the two triangles are similar.

E-3. **Calculating with similarity:** The triangles shown in Figure 3.20 are similar. The lengths of some of the sides are given. Find a and b.

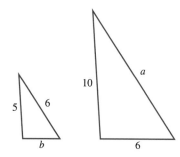

FIGURE 3.20 Similar triangles with some lengths given

FIGURE 3.21 An angle *A*

E-4. **Defining the sine function:** For an angle *A* shown in Figure 3.21, we define the sine function as follows. Add the segment BC such that BC is perpendicular to AB. This makes a right triangle as shown in Figure 3.22. Then we define

$$\sin A = \frac{|BC|}{|AC|}.$$

The difficulty is that the definition appears to depend on the choice of the segment we added. Use similarity theory to show that if we use another segment DE (perpendicular to AD) as shown in Figure 3.23, then

$$\frac{|BC|}{|AC|} = \frac{|DE|}{|AE|}.$$

FIGURE 3.22 Adding a segment to make a right triangle

FIGURE 3.23 Adding a different segment

E-5. **The Pythagorean theorem:** Similarity theory can be used to establish the famous Pythagorean theorem. Consider the right triangle in Figure 3.24. In Figure 3.25 we have added a perpendicular to make two smaller right triangles.

a. Use the two-angle criterion for similarity (see Exercise E-2) to show that $\triangle ABC$ is similar to both $\triangle DAC$ and $\triangle DBA$. The proper orientation to show the similarity is displayed in Figure 3.26 and Figure 3.27.

b. Use the fact that $\triangle DAC$ is similar to $\triangle ABC$ to show that $cd = a^2$.

c. Use the fact that $\triangle DBA$ is similar to $\triangle ABC$ to show that $ce = b^2$.

d. Use the information from parts b and c above to show that $a^2 + b^2 = c^2$.

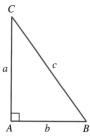

FIGURE 3.24 A right triangle

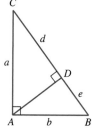

FIGURE 3.25 Two smaller right triangles

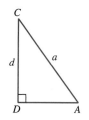

FIGURE 3.26 $\triangle DAC$ is similar to $\triangle ABC$

FIGURE 3.27 $\triangle DBA$ is similar to $\triangle ABC$

E-6. **Another proof of the Pythagorean theorem:** This is a proof by cut and paste. The square in Figure 3.28 clearly has area c^2. Trace a copy of the square, and cut out the two triangles indicated. Rearrange the two triangles and the remainder of the square to make the picture in Figure 3.29. Explain how this provides a proof of the Pythagorean theorem.

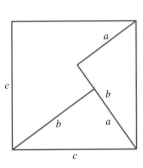

FIGURE 3.28 A square with two triangles marked for cutting

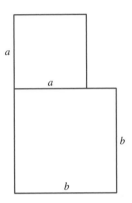

FIGURE 3.29 The remainder and two triangles arranged into a new figure

E-7. **A right triangle:** Show that the points $(1, 1)$, $(3, 4)$, and $(4, -1)$ form the vertices of a right triangle.

E-8. **Parallel and perpendicular lines:** For each of the following pairs of lines, determine whether the lines are parallel, perpendicular, or neither.

a. $3x + 2y = 4$ and $6x + 4y = 9$

b. $5x - 7y = 3$ and $4x - 3y = 8$

c. $5x - 7y = 15$ and $15y - 21x = 7$

d. $ax + by = c$ and $akx + aky = d$ $(a \neq 0)$

E-9. **A system of equations with no solution:** Give a geometric explanation of why the following system of equations has no solution.

$$2x + 7y = 9$$
$$8x + 28y = 10.$$

3.1   *SKILL BUILDING EXERCISES*

S-1. Slope from rise and run: One end of a ladder is on the ground. The top of the ladder rests at the top of an 8-foot wall. The wall is 2 horizontal feet from the base of the ladder. What is the slope of the line made by the ladder? (Assume that the positive direction points from the base of the ladder toward the wall.)

S-2. Slope from rise and run: One end of a ladder is on the ground. The top of the ladder rests at the top of a 15-foot wall. The wall is 3 horizontal feet from the base of the ladder. What is the slope of the line made by the ladder? (Assume that the positive direction points from the base of the ladder toward the wall.)

S-3. Height from slope and horizontal distance: The base of a ladder is 3 horizontal feet from the wall where its top rests (see Figure 3.30). The slope of the line made by the ladder is 2.5. What is the vertical height of the top of the ladder? (Assume that the positive direction points from the base of the ladder toward the wall.)

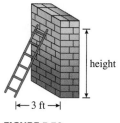

height

|← 3 ft →|

FIGURE 3.30

S-4. Height from slope and horizontal distance: The base of a ladder is 4 horizontal feet from the wall where its top rests. The slope of the line made by the ladder is 1.7. What is the vertical height of the top of the ladder? (Assume that the positive direction points from the base of the ladder toward the wall.)

S-5. Horizontal distance from height and slope: A ladder leans against a wall so that its slope is 1.75. The top of the ladder is 9 vertical feet above the ground. What is the horizontal distance from the base of the ladder to the wall? (Assume that the positive direction points from the base of the ladder toward the wall.)

S-6. Horizontal distance from height and slope: A ladder leans against a wall so that its slope is 2.1. The top of the ladder is 12 vertical feet above the ground. What is the horizontal distance from the base of the ladder to the wall? (Assume that the positive direction points from the base of the ladder toward the wall.)

S-7. Slope from two points: Take west to be the positive direction. The height of a sloped roof above the place where I stand is 12 feet (see Figure 3.31). If I move 3 feet west, the height is 10 feet. What is the slope?

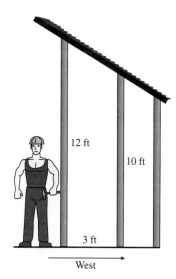

12 ft

10 ft

3 ft

West

FIGURE 3.31

S-8. Continuation of Exercise S-7: If I move 5 additional feet west, what is the height of the roof above the place where I stand?

S-9. A circus tent: I am at the center of a circus tent, where the height is 22 feet (see Figure 3.32). I am facing due west, which I take to be the positive direction. The slope of the tent line is −0.8. If I walk 7 feet west, how high is the tent?   →

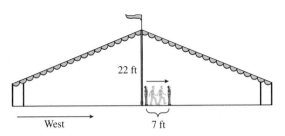

22 ft

West

7 ft

FIGURE 3.32

S-10. **More on the circus tent:** Assume that the roof of the circus tent in Exercise S-9 extends in a straight line to the ground. How far from the center of the tent does the roof meet the ground?

3.1 EXERCISES

1. **A line with given intercepts:** On coordinate axes, draw a line with vertical intercept 4 and horizontal intercept 3. Do you expect its slope to be positive or negative? Calculate the slope.

2. **A line with given vertical intercept and slope:** On coordinate axes, draw a line with vertical intercept 3 and slope 1. What is its horizontal intercept?

3. **Another line with given vertical intercept and slope:** A line has vertical intercept 8 and slope -2. What is its horizontal intercept?

4. **A line with given horizontal intercept and slope:** A line has horizontal intercept 6 and slope 3. What is its vertical intercept?

5. **Lines with the same slope:** On the same coordinate axes, draw two lines, each of slope 2. The first line has vertical intercept 1, and the second has vertical intercept 3. Do the lines cross? In general, what can you say about lines with the same slope?

6. **Where lines with different slopes meet:** On the same coordinate axes, draw one line with vertical intercept 2 and slope 3 and another with vertical intercept 4 and slope 1. Do these lines cross? If so, do they cross to the right or left of the vertical axis? In general, if one line has its vertical intercept below the vertical intercept of another, what conditions on the slope will ensure that the lines cross to the right of the vertical axis?

7. **A ramp to a building:** The base of a ramp sits on the ground (see Figure 3.33). Its slope is 0.4, and it extends to the top of the front steps of a building 15 horizontal feet away.

   a. How high is the ramp 1 horizontal foot toward the building from the base of the ramp?

   b. How high is the top of the steps relative to the ground?

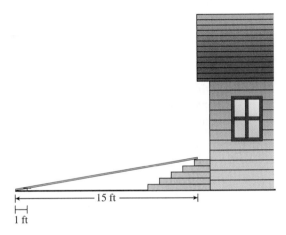

15 ft

1 ft

FIGURE 3.33

8. **A wheelchair service ramp:** The *Americans with Disabilities Act* (ADA) requires, among other things, that wheelchair service ramps have a slope not exceeding $\frac{1}{12}$.

   a. Suppose the front steps of a building are 2 feet high. You want to make a ramp conforming to ADA standards that reaches from the ground to the top of the steps. How far away from the building is the base of the ramp?

   b. Another way to give specifications on a ramp is to give allowable inches of rise per foot of run. In these terms, how many inches of rise does the ADA requirement allow in 1 foot of run?

9. **A cathedral ceiling:** A cathedral ceiling shown in Figure 3.34 is 8 feet high at the west wall of a room.

As you go from the west wall toward the east wall, the ceiling slants upward. Three feet from the west wall, the ceiling is 10.5 feet high.

a. What is the slope of the ceiling?

b. The width of the room (the distance from the west wall to the east wall) is 17 feet. How high is the ceiling at the east wall?

c. You want to install a light in the ceiling as far away from the west wall as possible. You intend to change the bulb, when required, by standing at the top of your small stepladder. If you stand on your stepladder, you can reach 12 feet high. How far from the west wall should you install the light?

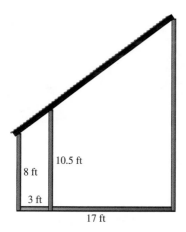

FIGURE 3.34 A cathedral ceiling

10. **Roof trusses:** Trusses as shown in Figure 3.35 are to be constructed to support the roof of a building. The truss is to have a 16-foot horizontal base (joist) that spans from one wall to the opposite wall. The vertical center strut is 4 feet long.

a. Vertical struts are located 3 horizontal feet inside each wall. How long are these vertical struts?

b. The rafter extends 1.5 horizontal feet outside the wall. If the top of the wall is 8 feet above the floor, how high above the floor is the outside tip of the rafter?

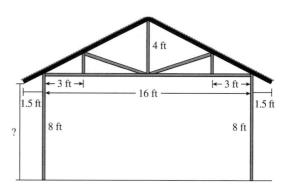

FIGURE 3.35 A roof truss

11. **Cutting plywood siding:** Plywood siding is to be used to cover the exterior wall of a house. Plywood siding comes in sheets 4 feet wide and 8 feet high. On the exterior wall shown in Figure 3.36, three pieces of siding are shown cut to conform to the roof line. The piece on the far right is 1 foot high on the shorter side and 2 feet 6 inches high on the longer side (toward the peak of the roof). To make proper cuts on the next two sheets of plywood, we need to know the lengths h and k shown in Figure 3.36. Find these lengths.

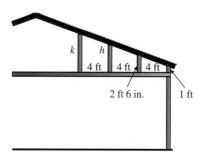

FIGURE 3.36 Plywood siding for a house

12. **An overflow pipeline:** An overflow pipeline for a pond is to run in a straight line from the pond at maximum water level a distance of 96 horizontal feet to a drainage area that is 5 vertical feet below the maximum water level (see Figure 3.37). How much lower is the pipe at the end of each 12-foot horizontal stretch? →

FIGURE 3.37

FIGURE 3.37

13. **Looking over a wall:** Twenty horizontal feet north of a 50-foot building is a 35-foot wall (see Figure 3.38). A man 6 feet tall wishes to view the top of the building from the north side of the wall. How far north of the wall must he stand in order to view the top of the building?

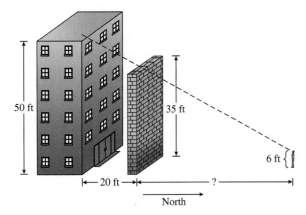

FIGURE 3.38

14. **The Mississippi River:** For purposes of this exercise, we will think of the Mississippi River as a straight line beginning at its headwaters, Lake Itasca, Minnesota, at an elevation of 1475 feet above sea level, and sloping downward to the Gulf of Mexico 2340 miles to the south.

   a. Think of the southern direction as pointing to the right along the horizontal axis. What is the slope of the line representing the Mississippi River? Be sure to indicate what units you are using.

   b. Memphis, Tennessee, sits on the Mississippi River 1982 miles south of Lake Itasca. What is the elevation of the river as it passes Memphis?

   c. How many miles south of Lake Itasca would you find the elevation of the Mississippi to be 200 feet?

15. **A road up a mountain:** You are driving on a straight road that is inclined upward toward the peak of a

mountain (see Figure 3.39). You pass a sign that reads "Elevation 4130 Feet." Three horizontal miles farther you pass a sign that reads "Elevation 4960 Feet."

   a. Think of the direction in which you are driving as the positive direction. What is the slope of the road? (*Note:* Units are important here.)

   b. What is the elevation of the road 5 horizontal miles from the first sign?

   c. You know that the peak of the mountain is 10,300 feet above sea level. How far in horizontal distance is the first sign from the peak of the mountain?

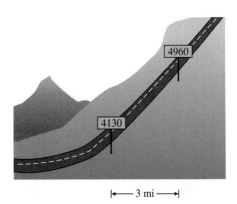

FIGURE 3.39

16. **An underground water source:** An underground aquifer near Seiling, Oklahoma, sits on an impermeable layer of limestone. West of Seiling the limestone layer is thought to slope downward in a straight line. In order to map the limestone layer, hydrologists started at Seiling heading west and drilled *sample wells.* Two miles west of Seiling the limestone layer was found at a depth of 220 feet. Three miles west of Seiling the limestone layer was found at a depth of 270 feet.

   a. What would you expect to be the depth of the limestone layer 5 miles west of Seiling?

   b. You want to drill a well down to the limestone layer as far west from Seiling as you can. Your budget will allow you to drill a well that is 290 feet deep. How far west of Seiling can you go to drill the well?

   c. Four miles west of Seiling someone drilled a well and found the limestone layer at 273 feet. Were the hydrologists right in saying that the limestone layer slopes downward in a straight line west of Seiling? Explain your reasoning.

17. **Earth's umbra:** Earth has a shadow in space, just as people do on a sunny day. The darkest part[1] of that shadow is a conical region in space known as the *umbra*. A representation of Earth's umbra is shown in Figure 3.40. Earth has a radius of about 3960 miles, and the umbra ends at a point about 860,000 miles from Earth. The moon is about 239,000 miles from Earth and has a radius of about 1100 miles. Consider a point on the opposite side of Earth from the sun and at a distance from Earth equal to the moon's distance from Earth. What is the radius of the umbra at that point? Can the moon fit inside Earth's umbra? What celestial event occurs when this happens?

FIGURE 3.40 The Earth's umbra

18. **Earth's penumbra:** *This is a continuation of Exercise 17.* Earth also has a partial shadow known as the *penumbra*. The penumbra is represented in Figure 3.41. The penumbra has a radius of about 10,000 miles at a point on the opposite side of Earth from the sun and at a distance from Earth equal to the moon's distance from Earth. The penumbra has a conical shape and, if extended toward the sun, would reach an apex shown in Figure 3.41. How far away from Earth is this apex?

FIGURE 3.41 The Earth's penumbra

19. **The umbra of the moon:** *This is a continuation of Exercise 17.* A total eclipse of the sun occurs when we are in the umbra of the moon. The size of the moon's umbra on Earth's surface is so small[2] that we can consider that the umbra reaches its apex at Earth's center. The sun is 93,498,600 miles away from Earth. What is the radius of the sun? (The actual radius is 434,994 miles. Because of our simplified assumptions, you will get a slightly different, though relatively close, answer.)

---

[1] In this part of the shadow the sun is not visible, and it would be totally dark but for reflected light from the moon and light bent by Earth's atmosphere. Blue light is scattered by Earth's atmosphere more easily than red light. This is why the moon in total eclipse has a coppery color.

[2] At its largest, it has a radius of about 84 miles. If the moon were just a little farther away from Earth, total solar eclipses would never occur.

3.2   *Linear Functions*

Constant Rates of Change

We now turn to some special types of functions, *linear functions*, that are intimately related to straight lines. A linear function is one whose *rate of change,* or *slope,* is always the same. Let's look at an example to show what we mean. Suppose the CEO of a company wants to have a dinner catered for employees. He finds that he must pay a dining hall rental fee of $275 and an additional $28 for each meal served. Then the cost $C = C(n)$, in dollars, is a function of the number n of meals served. If an unexpected guest arrives for dinner, then the CEO will have to pay an additional $28. This is the rate of change in C, and it is always the same no matter how many dinner guests are already seated. In more formal terms, when the variable n increases by 1, the function C changes by 28. This means that C is a linear function of n with slope or rate of change 28.

The rate of change, or slope, of a linear function can be used in ways that remind us of how slope is used for straight lines. Suppose, for example, that the caterer, having anticipated that 35 people would attend the dinner, sent the CEO a bill for $1255. But 48 people actually attended the dinner. What should the total price of the dinner be under these circumstances? Since 13 additional people attended the dinner, and the price per meal is $28, we can calculate the additional charge over $1255:

$$\text{Additional charge} = 28 \times \text{Additional people}$$

$$= 28 \times 13$$

$$= 364 \text{ dollars} .$$

Thus the total cost for 48 people is $1255 + 364 = 1619$ dollars.

In general, the slope or rate of change of a linear function is the amount the function changes when the variable increases by 1 unit. Furthermore, we can use the slope m just as we did above to calculate the change in function value that corresponds to a given change in the variable. We state the relationship in two equivalent ways:

$$\text{Change in function} = m \times \text{Change in variable}$$

$$\text{Additional function value} = m \times \text{Additional variable value} .$$

We can use division to restate these so that they show how to calculate m from changes in function value corresponding to changes in the variable. Once again, the following two equations say exactly the same thing using different words:

$$m = \frac{\text{Change in function}}{\text{Change in variable}}$$

$$m = \frac{\text{Additional function value}}{\text{Additional variable value}} .$$

Linear Functions

A function is linear if it has a constant slope or rate of change. This slope is then the amount the function changes when the variable increases by 1 unit.

Suppose $y = y(x)$ is a linear function of x. Then

- The slope m, or rate of change in y with respect to x, can be calculated from the change in y corresponding to a given change in x:

$$m = \frac{\text{Change in function}}{\text{Change in variable}} = \frac{\text{Change in } y}{\text{Change in } x}.$$

- The slope m can be used to calculate the change in y resulting from a given change in x:

$$\text{Change in function} = m \times \text{Change in variable}.$$

Or, using letters,

$$\text{Change in } y = m \times \text{Change in } x.$$

EXAMPLE 3.3   **Oklahoma Income Tax**

The amount of income tax $T = T(I)$, in dollars, owed to the State of Oklahoma is a linear function of the taxable income I, in dollars, at least over a suitably restricted range of incomes. According to one of the year-2001 Oklahoma Income Tax tables, a single Oklahoma resident taxpayer with a taxable income of $15,000 owes $780 in Oklahoma income tax. In functional notation this is $T(15,000) = 780$. If the taxable income is $15,500, then the tables show a tax liability of $825.

1. Calculate the rate of change in T with respect to I, and explain in practical terms what it means.

2. How much does the taxpayer owe if the taxable income is $15,350?

Solution to Part 1:   Since we are thinking of the tax T as a function of the variable I representing income, we calculate the slope as follows:

$$m = \frac{\text{Change in } T}{\text{Change in } I} = \frac{\text{Change in tax}}{\text{Change in income}} = \frac{825 - 780}{15,500 - 15,000} = \frac{45}{500} = 0.09.$$

This means that for each additional dollar earned, the taxpayer can expect to pay 9 cents more in Oklahoma income tax. In economics this is known as the *marginal tax rate*, because it is the rate at which new money that you earn is taxed. It is worth noting that the marginal tax rate normally does not appear directly in tax tables but must be calculated as we did here. It can provide crucial information for financial planning.

Solution to Part 2:   An income of $15,350 is an additional $350 income over the $15,000 level. Now we can use the marginal tax rate m calculated in part 1 to get the additional tax:

$$\text{Additional tax} = m \times \text{Additional income} = 0.09 \times 350 = 31.50 .$$

Thus we would owe the tax on $15,000—that is, $780—plus an additional tax of $31.50, for a total tax liability of $811.50.   ■ ■ ■

Linear Functions and Straight Lines

If we look more carefully at the catered dinner, we can write a formula for the total cost $C = C(n)$ when there are n dinner guests:

$$\text{Cost} = \text{Cost of food} + \text{Rent}$$

$$\text{Cost} = \text{Price per meal} \times \text{Number of guests} + \text{Rent}$$

$$C = 28n + 275 .$$

If we graph this function as we did in Figure 3.42, we see the fundamental relationship between linear functions and straight lines: The graph of a linear function is a straight line. (We used a horizontal span of -1 to 3 and a vertical span of 250 to 400.)

But there is still more to find out. In Figure 3.42 we placed the cursor at X=0, and we see that the vertical intercept of the graph is 275. That is the rental fee for the dining hall, or, more to the point, it is the value of the linear function C when the variable n is zero. This value $C(0)$ is often referred to as the *initial value* of the function. It is true in general that the vertical intercept of the graph corresponds to the initial value of the linear function and is denoted b.

In Figure 3.43 we have put the cursor at X=1, and we see that a horizontal change of 1 unit causes the graph to rise from 275 to 303. That means the slope of the straight-line graph is $303 - 275 = 28$, and that is the same as the rate of change of the linear function. Once again, this is true in general: The slope of a linear function is the same as the slope of its graph.

Finally, we note that the horizontal intercept of the graph (not shown in Figure 3.42) is the solution of the equation $C(n) = 0$. The form of C and the relationships we have observed here between linear functions and straight-line graphs are typical.

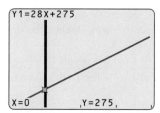

FIGURE 3.42 The graph of the linear function $C = 28n + 275$ is a straight line

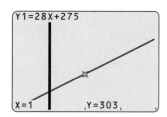

FIGURE 3.43 The slope of the graph is the same as the rate of change of the function

KEY IDEA 3.3

The Relationships Between Linear Functions and Straight Lines

1. The formula for a linear function is

$$y = \text{slope} \times x + \text{Initial value}$$

$$y = mx + b.$$

2. The graph of a linear function is a straight line.

3. The slope of a linear function is the same as the slope of its graph.

4. The vertical intercept of the graph corresponds to the initial value b of the linear function—that is, the value of the function when the variable is 0.

5. The horizontal intercept of the graph corresponds to the value of the variable when the linear function is zero. This is the solution of the equation

$$\text{Linear function} = 0.$$

EXAMPLE 3.4 **Selling Jewelry at an Art Fair**

Suppose you pay $192 to rent a booth for selling necklaces at an art fair. The necklaces sell for $32 each.

1. Explain why the function that shows your net income (revenue from sales minus rental fee) as a function of the number of necklaces sold is a linear function.

2. Write a formula for this function.

3. Use functional notation to show your net income if you sell 25 necklaces, and then calculate the value.

4. Make the graph of the net income function.

5. Identify the vertical intercept and slope of the graph, and explain in practical terms what they mean.

6. Find the horizontal intercept and explain its meaning in practical terms.

Solution to Part 1:   We choose variable and function names: Let n be the number of necklaces sold, and let $P = P(n)$ be the net income in dollars. Each time n increases by 1—that is, when 1 additional necklace is sold—the value of P, the net income, increases by the same amount, $32. Thus the rate of change for P is always the same, and hence it is a linear function.

Solution to Part 2: You pay $192 to rent the booth and take in $32 for each necklace sold.

$$\text{Net income} = \text{Income from sales} - \text{Rent}$$

$$\text{Net income} = \text{Price} \times \text{Number sold} - \text{Rent}$$

$$P = 32n - 192 \text{ dollars}.$$

Solution to Part 3: If 25 necklaces are sold, then in functional notation the net income is $P(25)$. To calculate that, we put 25 into the formula in place of n.

$$P(25) = 32 \times 25 - 192 = 608 \text{ dollars}.$$

Solution to Part 4: First we enter $\boxed{3.1}$ the function and record appropriate correspondences:

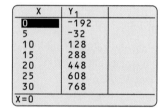

FIGURE 3.44 A table for setting the window

$$\mathbf{Y_1} = P, \text{ net income on vertical axis}$$

$$\mathbf{X} = n, \text{ necklaces sold on horizontal axis}.$$

To choose a window size, we made the table in Figure 3.44 using a starting value of 0 and an increment of 5. This led us to choose a horizontal span of $n = 0$ to $n = 30$ and a vertical span of $P = -200$ to $P = 800$. This makes the graph in Figure 3.45, and we note that, as expected, the graph of our linear function P is a straight line.

Solution to Part 5: We know in general that the vertical intercept of the graph of $y = mx + b$ is b and the slope is m. Thus for $P = 32n - 192$, the vertical intercept is -192, and the slope is 32. The vertical intercept -192 is the net income if no necklaces are sold. That is, you lost $192 because you had to pay rent for the booth but sold no necklaces. The slope of 32 dollars per necklace indicates that, for each additional necklace sold, your net income increases by $32. Thus the slope represents the price of each necklace. We should note that economists refer to the slope here as the *marginal income*. It shows the additional income taken when 1 additional item is sold.

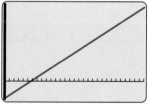

FIGURE 3.45 Net income for selling necklaces

Solution to Part 6: The horizontal intercept occurs where the graph crosses the horizontal axis. That is where $P(n) = 0$. We will show how to find that by direct calculation.

$$P(n) = 0$$

$$32n - 192 = 0$$

$$32n = 192 \qquad \text{Add 192 to both sides.}$$

$$n = \frac{192}{32} = 6 \qquad \text{Divide by 32.}$$

Thus the horizontal intercept occurs at $n = 6$ necklaces. That is where $P(n) = 0$, and in this case it is the "break-even" point. You need to sell 6 necklaces to avoid losing money on this venture.

We found the horizontal intercept here by hand calculation, but we should note that it could also be found with the calculator by looking at the graph or a table.

■ ■ ■

Linear Equations from Data

There are a number of ways in which information about a linear function may be given, but the following three situations are quite common. We will illustrate each case using variations on Example 3.4.

Getting a linear equation if you know the slope and initial value: In Example 3.4, we were effectively told the slope, 32, of the linear function P and its initial value, -192. As we have already noted, this enables us immediately to write a formula:

$$P = \text{Slope} \times n + \text{Initial value}$$

$$P = 32n - 192.$$

This works in general. If we know the slope m of a linear function and its initial value b, we can immediately write down the formula as $y = mx + b$.

Getting a linear equation if you know the slope and one data point: Suppose that we were given the information for Example 3.4 in a different way: We are told that the price for each necklace is $32 and that if we sell $n = 8$ necklaces, we will have a net income of $P = 64$ dollars. Now we know the slope, 32, but we don't know the initial value. We can get it by solving an equation. The information tells us that the formula for P is

$$P = 32n + \text{Initial value}. \tag{3.7}$$

We also know that when $n = 8$, $P = 64$. We put these values into Equation (3.7) and proceed to solve for the initial value:

$$64 = 32 \times 8 + \text{Initial value}$$

$$64 = 256 + \text{Initial value}$$

$$-192 = \text{Initial value}.$$

Now we have the slope and initial value, and we can write the formula for P as $P = 32n - 192$. This works in general. If we know the slope and one data point for a linear function, we can solve an equation as we did above to find the initial value and then write down the formula.

Getting a linear equation from two data points: Suppose the information in Example 3.4 were given as two data points. For example, suppose that we make a net income $P = 64$ dollars when we sell $n = 8$ necklaces and a net income of $P = 160$ dollars when $n = 11$ necklaces are sold. In this case, we know neither the slope nor the initial value. Since the slope is so important, we get it first, using a familiar formula:

$$\text{Slope} = \frac{\text{Change in function}}{\text{Change in variable}} = \frac{160 - 64}{11 - 8} = 32 \text{ dollars per necklace}.$$

Now we know the slope, 32, and the data point $P = 64$ when $n = 8$. With this information, we can proceed exactly as we did above to get the initial value and then the formula for P.

We also know that $P = 160$ when $n = 11$. What would happen if we used this data point with the slope rather than $n = 8$ and $P = 64$? Would the equation we find be different? Let's see:

$$P = \text{slope} \times n + \text{Initial value}$$

$$P = 32n + \text{Initial value}$$

$$160 = 32 \times 11 + \text{Initial value}$$

$$160 = 352 + \text{Initial value}$$

$$-192 = \text{Initial value}.$$

We got the same initial value as we did before, and hence we will get the same formula for P, namely $P = 32n - 192$.

In general, if you are given two data points and need to find the linear function that they determine, you proceed in two steps. First, use the formula

$$\text{Slope} = \frac{\text{Change in function}}{\text{Change in variable}}$$

to compute the slope. Next, use the slope you found and put *either* of the given data points into the equation

$$y = \text{Slope} \times x + \text{Initial value}$$

to solve for the initial value.

How to Get the Equation of a Linear Function

1. **If you know the slope and the initial value,** use

$$y = \text{Slope} \times x + \text{Initial value}.$$

2. **If you know the slope and one data point,** put the given slope and data point into the equation

$$y = \text{Slope} \times x + \text{Initial value}$$

and solve it for the initial value. You can now get the formula as in part 1 above.

3. **If you know two data points,** first use the formula

$$\text{Slope} = \frac{\text{Change in function}}{\text{Change in variable}}$$

to get the slope. Now use either of the data points and proceed as in part 2 above.

EXAMPLE 3.5 Changing Celsius to Fahrenheit

The temperature $F = F(C)$ in Fahrenheit is a linear function of the temperature C in Celsius. A lab assistant placed a Fahrenheit thermometer beside a Celsius thermometer and observed the following: When the Celsius thermometer reads 30 degrees ($C = 30$), the Fahrenheit thermometer reads 86 degrees ($F = 86$). When the Celsius thermometer reads 40 degrees, the Fahrenheit thermometer reads 104 degrees.

1. Use a formula to express F as a linear function of C.

2. At sea level, water boils at 212 degrees Fahrenheit. What temperature in degrees Celsius makes water boil?

3. Explain in practical terms what the slope means in this setting.

Solution to Part 1:   Since we know two points, the first step is to calculate the slope. When C changes from 30 to 40, F changes from 86 to 104. Since we are thinking of F as the function and of C as the variable, we use

$$\text{Slope} = \frac{\text{Change in function}}{\text{Change in variable}} = \frac{\text{Change in } F}{\text{Change in } C} = \frac{104 - 86}{40 - 30} = 1.8.$$

Thus the slope is 1.8 degrees Fahrenheit per degree Celsius. Now we use the slope $m = 1.8$, and the data point $F = 86$ when $C = 30$, to make an equation that we solve for the initial value:

$$F = \text{Slope} \times C + \text{Initial value}$$
$$86 = 1.8 \times 30 + \text{Initial value}$$
$$86 = 54 + \text{Initial value}$$
$$32 = \text{Initial value}.$$

Now we know the slope, 1.8, and the initial value, 32, so $F = 1.8C + 32$. It is worth noting the significance of the initial value we found. The temperature 32 degrees Fahrenheit is the freezing point for water, and this corresponds to a temperature of 0 degrees Celsius. That is, on the Celsius scale, a temperature of 0 is the freezing point for water.

Solution to Part 2:   We want to know the value of C when $F = 212$. Thus we put in 212 for F in the equation we found in part 1 and then solve for C:

$$F = 1.8C + 32$$
$$212 = 1.8C + 32$$
$$180 = 1.8C$$
$$\frac{180}{1.8} = C$$
$$100 = C.$$

Thus water boils at 100 degrees Celsius.

Solution to Part 3: As we found in part 1, the slope is 1.8. Remember that the slope tells how much F changes when C changes by 1. Thus a change of 1 degree Celsius results in a change of 1.8 degrees Fahrenheit. ■ ■ ■

ANOTHER LOOK

Forms of Equations

There are special forms of the equation of a line that enable us to get equations of lines quickly. The first we will look at is the *point–slope* form of a line. This is useful if we know the slope m of a line and a point (a, b) through which it passes. If (x, y) is any point on the line, then we can calculate the slope of the line using (a, b) and (x, y):

$$m = \frac{y - b}{x - a}.$$

This equation can be written as

$$y - b = m(x - a).$$

This is the point–slope form of the line. For example, suppose we know that a line with slope 2 goes through the point $(3, 4)$. Using the point–slope form, we have

$$y - 4 = 2(x - 3)$$
$$y - 4 = 2x - 6$$
$$y = 2x - 2.$$

The following example uses the point–slope form, along with the facts about parallel and perpendicular lines derived in the "Another Look" subsection of Section 3.1.

EXAMPLE 3.6 Finding Equations of Lines

1. Find the equation of a line through $(2, 3)$ that is parallel to $y = 3x + 1$.

2. Find the equation of a line through $(2, 3)$ that is perpendicular to $y = 3x + 1$.

Solution to Part 1: The line we want is parallel to $y = 3x + 1$, so its slope is 3. Thus we want to find the equation of a line that has slope 3 and passes through $(2, 3)$. Using the point–slope form, we have

$$y - 3 = 3(x - 2)$$
$$y - 3 = 3x - 6$$
$$y = 3x - 3.$$

Solution to Part 2:   The line we want is perpendicular to $y = 3x + 1$, so its slope is $-\frac{1}{3}$. Thus we want to find the equation of a line that has slope $-\frac{1}{3}$ and passes through $(2, 3)$. Using the point–slope form, we have

$$y - 3 = -\frac{1}{3}(x - 2)$$

$$y - 3 = -\frac{1}{3}x + \frac{2}{3}$$

$$y = -\frac{1}{3}x + \frac{11}{3}.$$

Another important special form is the *two-point* form. If we know that a line passes through two points (a, b) and (c, d), we can use these two points to calculate the slope:

$$m = \frac{d - b}{c - a}.$$

On the other hand, if (x, y) is any point on the line, we can use (x, y) and (a, b) to find the slope:

$$m = \frac{y - b}{x - a}.$$

Equating these two representations of the slope, we get the two-point form of the equation of a line:

$$\frac{y - b}{x - a} = \frac{d - b}{c - a},$$

which can be written as

$$y - b = \left(\frac{d - b}{c - a}\right)(x - a).$$

For example, if we know that a line passes through the points $(1, 3)$ and $(5, 7)$, we get the equation of the line immediately as follows:

$$y - 3 = \left(\frac{7 - 3}{5 - 1}\right)(x - 1)$$

$$y - 3 = 1(x - 1)$$

$$y = x + 2.$$

Enrichment Exercises

E-1. **Parallel lines:** Find the equation of the line parallel to $y = 3x - 2$ that passes through the point $(3, 3)$.

E-2. **Perpendicular lines:** Find the equation of the line perpendicular to $y = 4x + 1$ that passes through $(8, 2)$.

E-3. **Finding equations of lines:** Find equations for each of the following lines.

a. Passes through (2, 1) with slope 3.

b. Passes through (1, 1) with slope −4.

c. Passes through (1, 2) and (5, 10).

d. Passes through (3, 1) and (−2, 2).

E-4. **Finding the equation of a line:** Find the equation of the line that is parallel to $3x + 2y = 4$ and has the same horizontal intercept as $2x - y = 8$.

3.2   SKILL BUILDING EXERCISES

S-1. Slope from two values: Suppose f is a linear function such that $f(2) = 7$ and $f(5) = 19$. What is the slope of f?

S-2. Slope from two values: Suppose f is a linear function such that $f(3) = 9$ and $f(8) = 5$. What is the slope of f?

S-3. Function value from slope and run: Suppose f is a linear function such that $f(3) = 7$. If the slope of f is 2.7, then what is $f(5)$?

S-4. Function value from slope and run: Suppose f is a linear function such that $f(5) = 2$. If the slope of f is 3.1, then what is $f(12)$?

S-5. Run from slope and rise: Suppose that f is a linear function with slope -3.4 and that $f(1) = 6$. What value of x gives $f(x) = 0$?

S-6. Run from slope and rise: Suppose that f is a linear function with slope 2.6 and that $f(5) = -3$. What value of x gives $f(x) = 0$?

S-7. Linear equation from slope and point: Suppose that f is a linear function with slope 4 and that $f(3) = 5$. Find the equation for f.

S-8. Linear equation from slope and point: Suppose that f is a linear function with slope -3 and that $f(2) = 8$. Find the equation for f.

S-9. Linear equation from two points: Suppose f is a linear function such that $f(4) = 8$ and $f(9) = 2$. Find the equation for f.

S-10. Linear equation from two points: Suppose f is a linear function such that $f(3) = 5$ and $f(7) = -4$. Find the equation for f.

3.2   EXERCISES

1. **Getting Celsius from Fahrenheit:** Water freezes at 0 degrees Celsius, which is the same as 32 degrees Fahrenheit. Also water boils at 100 degrees Celsius, which is the same as 212 degrees Fahrenheit.

   a. Use the freezing and boiling points of water to find a formula expressing Celsius temperature C as a linear function of the Fahrenheit temperature F.

   b. What is the slope of the function you found in part a? Explain its meaning in practical terms.

   c. In Example 3.5 we showed that $F = 1.8C + 32$. Solve this equation for C and compare the answer with that obtained in part a.

2. **A trip to a science fair:** An elementary school is taking a busload of children to a science fair. It costs $130.00 to drive the bus to the fair and back, and the school pays each student's $2.00 admission fee.

   a. Use a formula to express the total cost C, in dollars, of the science fair trip as a linear function of the number n of children who make the trip.

   b. Identify the slope and initial value of C, and explain in practical terms what they mean.

   c. Explain in practical terms what $C(5)$ means, and then calculate that value.

   d. Solve the equation $C(n) = 146$ for n. Explain what the answer you get represents.

3. **Digitized pictures on a disk drive:** The hard disk drive on a computer holds 2 gigabytes of information. That is 2000 megabytes. The formatting information, operating system, and applications software take up 781 megabytes of disk space. The operator wants to store on his computer a collection of digitized pictures, each of which requires 2.3 megabytes of storage space.

   a. We think of the total amount of storage space used on the disk drive as a function of the number of pictures that are stored on the drive. Explain why this function is linear.

   b. Find a formula to express the total amount of storage space used on the disk drive as a linear function of the number of pictures that are stored on the drive. (Be sure to identify what the letters you use mean.) Explain in practical terms what the slope of this function is.

   c. Express using functional notation the total amount of storage space used on the disk drive if there are 350 pictures stored on the drive, and then calculate that value.   →

d. After putting a number of pictures on the disk drive, the operator executes a *directory* command, and at the end of the list the computer displays the message *598,000,000 bytes free*. This message means that there are 598 megabytes of storage space left on the computer. How many pictures are stored on the disk drive? How many additional pictures can be added before the disk drive is filled?

4. **Speed of sound:** The speed of sound in air changes with the temperature. When the temperature T is 32 degrees Fahrenheit, the speed S of sound is 1087.5 feet per second. For each degree increase in temperature, the speed of sound increases by 1.1 feet per second.

   a. Explain why speed S is a linear function of temperature T. Identify the slope of the function.

   b. Use a formula to express S as a linear function of T.

   c. Solve for T in the equation from part b to obtain a formula for temperature T as a linear function of speed S.

   d. Explain in practical terms the meaning of the slope of the function you found in part c.

5. **Total cost:** The *total cost C* for a manufacturer during a given time period is a function of the number N of items produced during that period. To determine a formula for the total cost, we need to know two things. The first is the manufacturer's *fixed costs*. This amount covers expenses such as plant maintenance and insurance, and it is the same no matter how many items are produced. The second thing we need to know is the cost for each unit produced, which is called the *variable cost*.

   Suppose that a manufacturer of widgets has fixed costs of $1500 per month and that the variable cost is $20 per widget (so it costs $20 to produce 1 widget).

   a. Explain why the function giving the total monthly cost C, in dollars, of this widget manufacturer in terms of the number N of widgets produced in a month is linear. Identify the slope and initial value of this function, and write down a formula.

   b. Another widget manufacturer has a variable cost of $12 per widget, and the total cost is $3100 when 150 widgets are produced in a month. What are the fixed costs for this manufacturer?

   c. Yet another widget manufacturer has determined the following: The total cost is $2700 when 100 widgets are produced in a month, and the total cost is $3500 when 150 widgets are produced in a month. What are the fixed costs and variable cost for this manufacturer?

6. **Total revenue and profit:** *This is a continuation of Exercise 5.* The *total revenue R* for a manufacturer during a given time period is a function of the number N of items produced during that period. In this exercise we assume that the selling price per unit of the item is a constant, so it does not depend on the number of items produced. The *profit P* for a manufacturer is the total revenue minus the total cost. If the profit is zero, then the manufacturer is at a *break-even point*.

   We consider again the manufacturer of widgets in Exercise 5 with fixed costs of $1500 per month and a variable cost of $20 per widget. Suppose the manufacturer sells 100 widgets for $2300 total.

   a. Use a formula to express the total monthly revenue R, in dollars, of this manufacturer in a month as a function of the number N of widgets produced in a month.

   b. Use a formula to express the monthly profit P, in dollars, of this manufacturer as a function of the number of widgets produced in a month. Explain how the slope and initial value of P are derived from the fixed costs, variable cost, and price per widget.

   c. What is the break-even point for this manufacturer?

   d. Make graphs of total monthly cost and total monthly revenue. Include monthly production levels up to 1200 widgets. What is the significance of the point where the graphs cross?

7. **Slowing down in a curve:** A study of average driver speed on rural highways by A. Taragin[3] found a linear relationship between average speed S, in miles

[3] "Driver Performance on Horizontal Curves," *Proceedings* **33** (Washington, D.C.: Highway Research Board, 1954), 446–466.

per hour, and the amount of curvature D, in degrees, of the road. On a straight road ($D = 0$), the average speed was found to be 46.26 miles per hour. This was found to decrease by 0.746 mile per hour for each additional degree of curvature.

a. Find a linear formula relating speed to curvature.

b. Express using functional notation the speed for a road with a curvature of 10 degrees, and then calculate that value.

8. **Real estate sales:** A real estate agency has fixed monthly costs associated with rent, staff salaries, utilities, and supplies. It earns its money by taking a percentage commission on total real estate sales. During the month of July, the agency had total sales of $832,000 and showed a net income (after paying fixed costs) of $15,704. In August total sales were $326,000 with a net income of only $523.

a. Use a formula to express net income as a linear function of total sales. Be sure to identify what the letters that you use mean.

b. Plot the graph of net income and identify the slope and vertical intercept.

c. What are the real estate agency's fixed monthly costs?

d. What percentage commission does the agency take on the sale of a home?

e. Find the horizontal intercept and explain what this number means to the real estate agency.

9. **Currency conversion:** The number P of British pounds you can get from a bank is a linear function of the number D of American dollars you pay. An American tourist arriving at Heathrow airport in England went to a banking window at the airport and gave the teller 83 American dollars. She received 31 British pounds in exchange. In this exercise, assume there is no service charge for exchanging currency.

a. What is the rate of change, or slope, of P with respect to D? Explain in practical terms what this number means. (*Note:* You need two values to calculate a slope, but you were given only one. If you think about it, you know one other value. How many British pounds can you get for zero American dollars?)

b. A few days later, the American tourist went to a bank in Plymouth and exchanged 130 American

dollars for British pounds. How many pounds did she receive?

c. Upon returning to the airport, she found that she still had £12.32 in British currency in her purse. In preparation for the trip home, she exchanged that for American dollars. How much money, in American dollars, did she get?

10. **Growth in height:** Between the ages of 7 and 11 years, a certain boy grows 2 inches taller each year. At age 9 he is 48 inches tall.

a. Explain why, during this period, the function giving the height of the boy in terms of his age is linear. Identify the slope of this function.

b. Use a formula to express the height of the boy as a linear function of his age during this period. Be sure to identify what the letters that you use mean.

c. What is the initial value of the function you found in part b?

d. Studying a graph of the boy's height as a function of his age from birth to age 7 reveals that the graph is increasing and concave down. Does this indicate that his actual height (or length) at birth was larger or smaller than your answer to part c? Be sure to explain your reasoning.

11. **Adult male height and weight:** Here is a rule of thumb relating weight to height among adult males. If a man is 1 inch taller than another, then we expect him to be heavier by 5 pounds.

a. Explain why, according to this rule of thumb, among typical adult males the weight is a *linear* function of the height. Identify the slope of this function.

b. A related rule of thumb is that a typical man who is 70 inches tall weighs 170 pounds. On the basis of these two rules of thumb, use a formula to express the trend giving weight as a linear function of height. (Be sure to identify the meaning of the letters that you use.)

c. If a man weighs 152 pounds, how tall would you expect him to be?

d. An atypical man is 75 inches tall and weighs 190 pounds. In terms of the trend formula you found in part b, is he heavy or light for his height?

12. **Lean body weight in males:** Your *lean body weight* L is the amount you would weigh if all the fat in your body were to disappear. One text gives the following estimate of lean body weight L (in pounds) for young adult males:

$$L = 98.42 + 1.08W - 4.14A \,,$$

where W is total weight in pounds and A is abdominal circumference in inches.[4]

a. Consider a group of young adult males who have the same abdominal circumference. If their weight increases but their abdominal circumference remains the same, how does their lean body weight change?

b. Consider a group of young adult males who have the same weight. If their abdominal circumference decreases but their weight stays the same, how does their lean body weight change?

c. Suppose a young adult male has a lean body weight of 144 pounds. Over a period of time, he gains 15 pounds in total weight, and his abdominal circumference increases by 2 inches. What is his lean body weight now?

13. **Lean body weight in females:** *This is a continuation of Exercise 12.* The text cited in Exercise 12 gives a more complex method of calculating lean body weight for young adult females:

$$L = 19.81 + 0.73W + 21.2R - 0.88A$$
$$- 1.39H + 2.43F.$$

Here L is lean body weight in pounds, W is weight in pounds, R is wrist diameter in inches, A is abdominal circumference in inches, H is hip circumference in inches, and F is forearm circumference in inches. Assuming the validity of the formulas given here and in Exercise 12, compare increase in lean body weight of young adult males and of young adult females if their weight increases but all other factors remain the same.

14. **Horizontal reach of straight streams:** If a fire hose is held horizontally, then the distance the stream will

travel depends on the water pressure and on the *horizontal factor* for the nozzle. The horizontal factor H depends on the diameter of the nozzle. For a 0.5-inch nozzle, the horizontal factor is 56. For each $\frac{1}{8}$-inch increase in nozzle diameter, the horizontal factor increases by 6.

a. Explain why the function giving the horizontal factor H in terms of the nozzle diameter d (measured in inches) is linear.

b. Use a formula to express H as a linear function of d.

c. Once the horizontal factor H is known, we can calculate the distance S in feet that a horizontal stream of water can travel by using

$$S = \sqrt{Hp} \,.$$

Here p is pressure in pounds per square inch. How far will a horizontal stream travel if the pressure is 50 pounds per square inch and the nozzle diameter is 1.75 inches?

d. Firefighters have a nozzle with a diameter of 1.25 inches. The pumper generates a pressure of 70 pounds per square inch. The hose nozzle is 75 feet from a fire. Can a horizontal stream of water reach the fire?

15. **Vertical reach of fire hoses:** If a fire hose is held vertically, then the height the stream will travel depends on water pressure and on the *vertical factor* for the nozzle. The vertical factor V depends on the diameter of the nozzle. For a 0.5-inch nozzle, the vertical factor is 85. For each $\frac{1}{8}$-inch increase in nozzle diameter, the vertical factor increases by 5.

a. Explain why the function giving the vertical factor V in terms of the nozzle diameter d is linear.

b. Use a formula to express V as a linear function of d (measured in inches).

c. Once the vertical factor is known, we can calculate the height S in feet that a vertical stream of water can travel by using

$$S = \sqrt{Vp} \,.$$

---

[4] D. Kirkendall, J. Gruber, and R. Johnson, *Measurement and Evaluation for Physical Educators*, 2nd ed. (Champaign, IL: Human Kinetics Publishers, 1987).

Here p is pressure in pounds per square inch. How high will a vertical stream travel if the pressure is 50 pounds per square inch and the nozzle diameter is 1.75 inches?

d. Firemen have a nozzle with a diameter of 1.25 inches. The pumper generates a pressure of 70 pounds per square inch. From street level, they need to get water on a fire 60 feet overhead. Can they reach the fire with a vertical stream of water?

16. **Budget constraints:** Your family likes to eat fruit, but because of budget constraints, you spend only $5 each week on fruit. Your two choices are apples and grapes. Apples cost $0.50 per pound, and grapes cost $1 per pound. Let a denote the number of pounds of apples you buy and g the number of pounds of grapes. Because of your budget, it is possible to express g as a linear function of the variable a. To find the linear formula, we need to find its slope and initial value.

   a. If you buy one more pound of apples, how much less money do you have available to spend on grapes? Then how many fewer pounds of grapes can you buy?

   b. Use your answer to part a to find the slope of g as a linear function of a. (*Hint:* Remember that the slope is the change in the function that results from increasing the variable by 1. Should the slope of g be positive or negative?)

   c. To find the initial value of g, determine how many pounds of grapes you can buy if you buy no apples.

   d. Use your answers to parts b and c to find a formula for g as a linear function of a.

17. **More on budget constraints:** *This is a continuation of Exercise 16.* Another way to find a linear formula giving the number of pounds g of grapes that you can buy in terms of a, the number pounds of apples, is to write the budget constraint as an equation and solve it for g.

   a. If you buy 5 pounds of apples, then of course you spend $0.50 \times 5 = 2.50$ dollars on apples. In general, if you buy a pounds of apples, how much money do you spend on apples? Your answer should be a formula involving a.

   b. Write a formula involving g to express how much money you spend on grapes if you buy g pounds of grapes.

   c. Use your answers to parts a and b to write an equation expressing the budget constraint of $5 in terms of a and g.

   d. Solve the equation you found in part c for g, and thus find a formula expressing g as a linear function of a. Compare this with your answer to part d of Exercise 16.

18. **Traffic signals:** The number of seconds n for the yellow light is critical to safety at a traffic signal. One study[5] recommends the following formula for setting the time:

$$n = t + 0.5\frac{v}{a} + \frac{w+l}{v}.$$

Here t is the perception-reaction time of a driver in seconds, v is the average approach speed in feet per second, a is the deceleration rate in feet per second per second, w is the crossing-street width in feet, and l is the average vehicle length in feet. If we assume that the average perception-reaction time is 1 second, the approach velocity is 40 miles per hour (58.7 feet per second), the deceleration rate is 15 feet per second per second, and the average vehicle length is 20 feet, then n can be expressed as the following linear function of crossing-street width:

$$n = 3.3 + 0.017w.$$

   a. Under the given assumptions, what is the minimum time the yellow light should be on, no matter what the width of the crossing street?

   b. If the crossing street for one signal is 10 feet wider than the crossing street for another signal, how should the lengths of the yellow light times compare?

   c. Calculate $n(70)$ and explain in practical terms what your answer means.

   d. What crossing-street width would warrant a 5-second yellow light?

---

[5]P. L. Olson and R. W. Rothery, "Driver response to the amber phase of traffic signals," *Traffic Engineering* **XXXII (5)** (1962), 17–20, 29.

19. **Sleeping longer:** A certain man observed that each night he was sleeping 15 minutes longer than he had the night before, and he used this observation to predict the day of his death.[6] If he made his observation right after sleeping 8 hours, how long would it be until he slept 24 hours (and so would never again wake)?

20. **The saros cycle:** Total solar eclipses occur on a regular and predictable basis. One that could be viewed in parts of the United States occurred on March 7, 1970. The *saros cycle* is a period of 18 years, $11\frac{1}{3}$ days, and solar eclipses occur according to this cycle. For the purposes of this exercise, we ignore leap years and assume that a year is exactly 365 days. Because the saros cycle is not a whole number of days, but instead is one-third of a day longer than a day, at the end of the saros cycle Earth will have rotated one-third revolution beyond its location at the beginning of the cycle. Thus if a solar eclipse is viewable at a certain location, the next solar eclipse will not be. How long after March 7, 1970, will a solar eclipse again be viewable from the United States?[7]

21. **Life on other planets:** The number N of civilizations per galaxy that might be able to communicate with us can be expressed in the formula

$$N = SPZBIL.$$

Here S is the number of stars per galaxy, about 2×10^{11}. The remaining variables, together with their pessimistic and optimistic estimates,[8] are given in the table that follows.

Variable	Meaning	Pessimistic value	Optimistic value
P	Fraction of stars with planets	0.01	0.5
Z	Planets per star in life zone for at least 4 billion years	0.01	1
B	Fraction of suitable planets on which life begins	0.01	1
I	Fraction of life forms that evolve to intelligence	0.01	1
L	Fraction of star's life during which a technological society survives	10^{-8}	10^{-4}

a. Calculate the most pessimistic and most optimistic values for N.

b. The single variable with the greatest range of values is L, the fraction of a star's life during which a technological society, once established, can survive. The pessimistic estimate corresponds to 100 years, the optimistic estimate to 1,000,000 years. Using pessimistic estimates for other variables, what value of L will give 1 communicating civilization per galaxy?

c. Is this formula helpful in determining whether there is life on other planets? Explain your reasoning.

---

[6]This is a legend about the mathematician De Moivre, who died in 1754.

[7]The actual date is April 8, 2024, and the eclipse should be viewable through much of the central United States.

[8]Data taken from Michael A. Seeds, *Foundations of Astronomy* (Belmont, CA: Wadsworth, 1994).

Modeling Data with Linear Functions

Information about physical and social phenomena is frequently obtained by gathering data or sampling. For example, collecting data is the basic tool that census takers use to get information about the population of the United States. Once data are gathered, an important key to further analysis is to produce a *mathematical model* describing the data. A model is a function that (1) represents the data either exactly or approximately and (2) incorporates patterns in the data. In many cases, such a model takes the form of a linear function.

Testing Data for Linearity

Let's look at a hypothetical example. One of the most important events in the development of modern physics was Galileo's description of how objects fall. In about 1590 he conducted experiments in which he dropped objects and attempted to measure their downward velocities $V = V(t)$ as they fell. Here we measure velocity V in feet per second and time t as the number of seconds after release. If Galileo had been able to nullify air resistance, and if he had been able to measure velocity without any experimental error at all, he might have recorded the following table of values.

t = seconds	0	1	2	3	4	5
V = feet per second	0	32	64	96	128	160

What conclusions can be drawn from this table of data? In this case, the key is to look at how velocity changes as the rock falls. From $t = 0$ to $t = 1$, the velocity changes from 0 feet per second to 32 feet per second. That is a change of $32 - 0 = 32$ feet per second. From $t = 1$ to $t = 2$, the velocity changes from 32 feet per second to 64 feet per second, a change in velocity of $64 - 32 = 32$ feet per second. Thus the change in velocity is the same, 32 feet per second, during the first and second seconds of the fall. If we continue this, we obtain the following table.

Change in t	From 0 to 1	From 1 to 2	From 2 to 3	From 3 to 4	From 4 to 5
Change in V	32	32	32	32	32

We see that the rate of change in velocity, the *acceleration*, is always the same, 32 feet per second per second. We know this is characteristic of linear functions; their rate of change is always the same. Thus it is reasonable to describe these data with a linear formula. That is, we want a formula of the form $V = mt + b$, where m is the slope and b is the initial value. Since V changes by 32 when t changes by 1, the slope of the linear function is $m = 32$ feet per second per second. We also know that $b = 0$ because the initial velocity is zero. We conclude that the velocity of the rock t seconds after it is dropped is given by

$$V = 32t + 0 = 32t.$$

Linear Models

This linear function serves as a *mathematical model* for the experimentally gathered data, and it gives us more information than is apparent from the data table alone. For example, we can use the linear formula to calculate the velocity of the rock 10.8 seconds after it is released, information that is not given by the table:

$$V(10.8) = 10.8 \times 32$$

$$= 345.6 \text{ feet per second}.$$

The key physical observation that Galileo made was that falling objects have constant acceleration, 32 feet per second per second. He did additional experiments to show that if air resistance is ignored, then this acceleration does not depend on the weight or size of the object. The same table of values would result, and the same acceleration would be calculated, whether the experiment was done with a pebble or with a cannonball. According to tradition, Galileo conducted some of his experiments in public, dropping objects from the top of the leaning tower of Pisa. His observations got him into serious trouble with the authorities because they conflicted with the accepted premise of Aristotle that heavier objects would fall faster than lighter ones.

The feature of this data table that allowed us to construct a linear model was that the change in the function, velocity, was always the same. If a data table (with evenly spaced values for the variable) does not exhibit this behavior, then the data cannot be modeled with a linear function, and some more complicated model may be sought. For example, if instead of measuring velocity, we had measured the distance $D = D(t)$ that the rock fell, we would have obtained the following data.

t = seconds	0	1	2	3	4	5
D = feet	0	16	64	144	256	400

The corresponding table of differences is

Change in t	From 0 to 1	From 1 to 2	From 2 to 3	From 3 to 4	From 4 to 5
Change in D	16	48	80	112	144

We see that the change in D is not constant, so D is not a linear function. A deeper analysis can show that $D(t) = 16t^2$.

When Data Are Linear

A data table for $y = y(x)$ with evenly spaced values for x can be modeled with a linear function if it shows a constant change in y.

If the change in y is not constant, then the data cannot be modeled exactly by a linear function.

EXAMPLE 3.7 **Sampling Voter Registration**

In this hypothetical experiment, a political analyst compiled data on the number of registered voters in Payne County, Oklahoma, each year from 1995 through 2000. For each of the following possible data tables that the analyst might have obtained, determine whether the data can be modeled with a linear function. If so, find such a formula and predict the number of registered voters in Payne County in the presidential election year 2004.

Hypothetical Data Table 1:

Date	1995	1996	1997	1998	1999	2000
Registered voters	28,321	28,542	29,466	30,381	30,397	31,144

Hypothetical Data Table 2:

Date	1995	1996	1997	1998	1999	2000
Registered voters	28,321	28,783	29,245	29,707	30,169	30,631

Solution for Data Table 1: First we choose variable and function names: Let d be the number of years since 1995, and let N be the number of registered voters. The table for N as a function of d is then

d	0	1	2	3	4	5
N	28,321	28,542	29,466	30,381	30,397	31,144

To determine whether the data can be modeled by a linear function, we need to look at the change in the number of registered voters from year to year.

Change in d	0 to 1	1 to 2	2 to 3	3 to 4	4 to 5
Change in N	221	924	915	16	747

This table shows clearly that the change in registered voters is not constant from year to year, and we conclude that it is not appropriate to model these data with a linear function.

Solution for Data Table 2: Again, we let d be the number of years since 1995 and N the number of registered voters. The table for N as a function of d is then

d	0	1	2	3	4	5
N	28,321	28,783	29,245	29,707	30,169	30,631

We construct a table of differences using Data Table 2.

Change in d	0 to 1	1 to 2	2 to 3	3 to 4	4 to 5
Change in N	462	462	462	462	462

Here we see that the number of registered voters increases by 462 each year, so we know that we can model the data with a linear function. To determine the formula, we need to know the slope and the initial value. Since 462 is the change in N corresponding to a change in d of 1, the slope of our linear function is 462 registered voters per year. The initial value is given by the first entry in the table for N: $N(0) = 28,321$. Thus the formula is $N = 462d + 28,321$. To predict the number of registered voters in the year 2004, we use this formula, putting in 9 for d:

$$N(9) = 462 \times 9 + 28,321 = 32,479\,.$$

We can get this same prediction using only the slope rather than the formula. The change in d from 5 (the year 2000) to 9 (the year 2004) is 4 years, so we find

$$\text{Change in } N = m \times \text{Change in } d$$

$$\text{Change in } N = 462 \times 4$$

$$\text{Change in } N = 1848\,.$$

Thus from $d = 5$ (the year 2000) to $d = 9$ (the year 2004), we expect an increase of 1848 registered voters. That gives $30,631 + 1848 = 32,479$ registered voters in the year 2004, the same answer we got using the formula. ▪ ▪ ▪

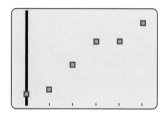

FIGURE 3.46 Voter registration data entered in a calculator table

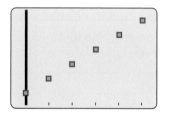

FIGURE 3.47 A view of nonlinear data

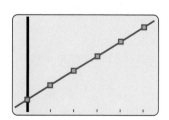

FIGURE 3.48 Linear data points from Data Table 2

We note that the choice of the variable d as the number of years since 1995 greatly simplifies finding the linear formula, since the initial value can then be found as the first entry in the table. We could choose the variable to be the actual date. This would not affect the slope, but we would have to calculate the initial value using the slope and one of the data points in the original table. The resulting initial value would be different from the one we found above, since it would represent the number of registered voters present in A.D. 0! That is of course a nonsensical figure, and for this reason, as well as for the purpose of simplifying the calculation, it is better to choose the variable as we did in the example. This is an important point to remember for subsequent examples and exercises in which the variable is time.

We also note that Data Table 2 in Example 3.7 is not very realistic. It almost never happens that statistics of this sort turn out to be exactly linear. But many times such data can be closely approximated by a linear function. We will see how to deal with that in the next section.

Graphing Discrete Data

Calculating differences can always tell you whether data are linear, but many times it is advantageous to view such data graphically. Most graphing calculators will allow you to enter data tables and to display them graphically. In Figure 3.46 we have entered 3.2 the data for Data Table 1 from Example 3.7. They are graphed 3.3 in Figure 3.47. This picture clearly verifies our earlier contention that these are not linear data; the points do not fall on a straight line.

Figure 3.47 contrasts nicely with the plot of linear data from Data Table 2 that is shown in Figure 3.48. We see that the points do indeed fall on a straight line, clearly showing the linear nature of the data. Finally, we can add 3.4 the graph of the linear function $N = 462d + 28,321$ that we found in part 2 of Example 3.7 to the screen as shown in Figure 3.49.

Figure 3.48 and Figure 3.49 illustrate clearly what it means to model linear data. The data are displayed in Figure 3.48. The model is the line in Figure 3.49 that passes through the data points but also fills in the gaps and extends beyond the data points.

EXAMPLE 3.8   **Newton's Second Law of Motion**

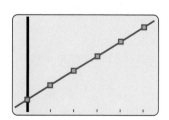

FIGURE 3.49 Adding the linear model

Newton's second law of motion shows how force on an object, measured in *newtons*,[9] is related to acceleration of the object, measured in meters per second per second. The following experiment might be conducted in order to discover Newton's second law. Objects of various masses, measured in kilograms, were given an acceleration of 5 meters per second, and the associated forces were measured and recorded in the table below.

Mass	1	1.3	1.6	1.9	2.2
Force	5	6.5	8	9.5	11

---

[9]One newton is about a quarter of a pound.

1. Check differences to show that these are linear data.

2. Find the slope of a linear model for the data, and explain in practical terms what the slope means.

3. Construct a linear model for the data.

4. What force does your model show for an object of mass 1.43 kilograms that is accelerating at 5 meters per second?

5. Make a graph showing the data, and overlay it with the graph of the linear model you made in part 3.

6. Newton's second law of motion says that

$$\text{Force} = \text{Mass} \times \text{Acceleration}.$$

Do the results of our hypothetical experiment provide support for the validity of the second law? What additional experiments might be appropriate to provide further verification?

Solution to Part 1:   First we choose variable and function names: Let m be the mass and F the force. Now we make a table of differences.

Change in m	1 to 1.3	1.3 to 1.6	1.6 to 1.9	1.9 to 2.2
Change in F	1.5	1.5	1.5	1.5

Since the change in F is always the same, 1.5 newtons, we conclude that the data are linear.

Solution to Part 2:   Part 1 shows that it is appropriate to model the data with a linear function. In finding the slope we must be careful. For Data Table 2 of Example 3.7, the common difference in N, the number of registered voters, and the slope were the same. This will occur exactly when the data for the variable are given in steps of 1 unit, as was the case for the variable d in Example 3.7. Here the change in variable is not 1; rather, it is 0.3. Now we get the slope with a familiar formula:

$$\text{Slope} = \frac{\text{Change in function}}{\text{Change in variable}} = \frac{\text{Change in } F}{\text{Change in } m} = \frac{1.5}{0.3} = 5 \text{ newtons per kilogram}.$$

To explain the slope in practical terms, we recall that the slope is the change in the function that results from a unit increase in the variable. Thus the slope of 5 newtons per kilogram means that when the mass increases by 1 kilogram, the associated force increases by 5 newtons. (For future reference, it is worth noting that the slope 5 turned out to be exactly the same as the acceleration. This is not an accident!)

Solution to Part 3:   Since we know the slope, we need only get the initial value b. We cannot find this directly from the table. Instead, we use the slope together with any

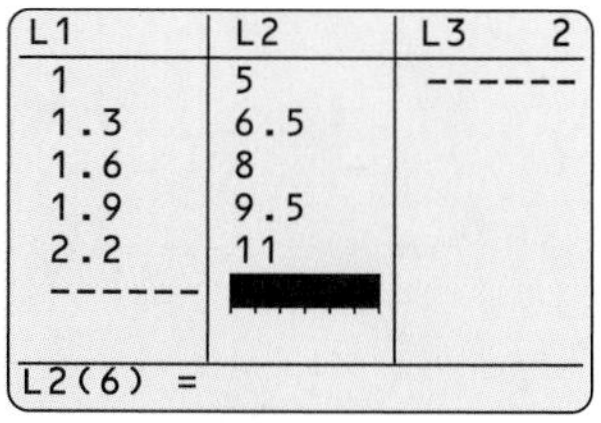

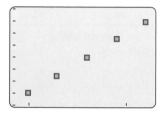

FIGURE 3.50 Data entered for force versus mass

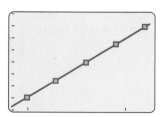

FIGURE 3.51 A plot of data for force versus mass

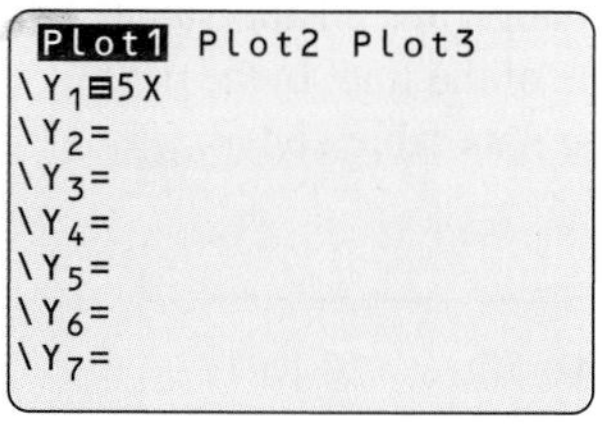

```
Plot1  Plot2  Plot3
\Y₁≣5X
\Y₂=
\Y₃=
\Y₄=
\Y₅=
\Y₆=
\Y₇=
```

FIGURE 3.52 Entering the linear model

FIGURE 3.53 The linear model overlaying the data

of the data points we wish. We will make the calculation with the first data point, $F(1) = 5$:

$$F = 5m + b$$
$$5 = 5 \times 1 + b$$
$$0 = b.$$

Thus the initial value b is 0, and we arrive at the linear model $F = 5m$.

In this case we could also have found the initial value by reasoning as follows: If the object has a mass of zero kilograms, the force will surely be 0, so the initial value is 0. Although this method gives the correct answer here, in general we should be cautious about going outside of the table to get data points, since the linear model may be appropriate only over the range of values in the table. See Exercise 11.

Solution to Part 4:   To get the force for a mass of 1.43 kilograms, we use this in place of the mass m in our linear model:

$$F = 5m = 5 \times 1.43 = 7.15 \text{ newtons}.$$

Solution to Part 5:   The first step is to enter ⎡3.5⎤ the data as shown in Figure 3.50. Once the data are entered, we can plot ⎡3.6⎤ them to get Figure 3.51. As expected, the points in Figure 3.51 fall on a straight line, giving striking visual verification for our conclusion in part 1 that it is appropriate to make a linear model for the data. To overlay the data with the model we made, we enter the function ⎡3.7⎤ as shown in Figure 3.52 and then make the graph ⎡3.8⎤ .We see in Figure 3.53 that our model overlays the data exactly. This provides a good way for us to check our work. If the line had missed some of the points, we would know that we had made an error somewhere.

Solution to Part 6:   Our model verifies Newton's second law for objects with an acceleration of 5 meters per second per second. There are many other experiments that might be appropriate to give further evidence. One important experiment would be to take an object of fixed mass and measure what happens to the force when acceleration is varied. See Exercise 8 below.   ■ ■ ■

ANOTHER LOOK

Unevenly Spaced Data

When data are evenly spaced, we need only calculate differences to see whether the data are linear. If the data are not evenly spaced, we must work a bit harder. Consider the data in the following table.

x	2	5	6	10	12
y	7	13	15	23	27

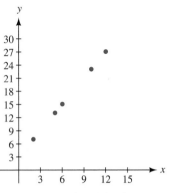

FIGURE 3.54 A plot of the data points

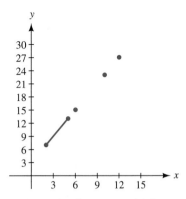

FIGURE 3.55 The segment joining the first two data points

In Figure 3.54 we have plotted the data points. In Figure 3.55 we have added the line segment joining the first two data points. If these data are linear, then when we extend the segment, it will pass through the next data point. That is, the segments joining each data point to the next all have the same slope. But the slope of the segment joining the data points is the average rate of change from one data point to the next. Thus data are linear if the average rates of change are all the same. Furthermore, the common average rate of change is the slope of the line. In the next table, we have calculated the average rates of change for the data table above.

Interval	2 to 5	5 to 6	6 to 10	10 to 12
Average rate of change	2	2	2	2

The average rates of change are all the same, so the data are linear with slope 2. Since we know the line passes through $(2, 7)$ we can use the point–slope form to get the equation of the line:

$$y - 7 = 2(x - 2)$$
$$y - 7 = 2x - 4$$
$$y = 2x + 3.$$

Enrichment Exercises

E-1. **Testing for linearity:** Test each of the following data sets to see whether they are linear. For those that are linear, find a linear model.

a.

x	2	5	7	8	11
y	5	14	20	23	32

b.

x	3	4	6	9	10
y	4	6	10	18	20

c.

x	1	3	7	9	12
y	−1	7	23	31	43

E-2. **Linear and angular diameter:** If a spherical object is some distance away, then one measure of its actual diameter d is the angle through which our eye moves in traveling from one side of the object to the other. This angle is the *angular diameter a* of the object, and, when this angle is very small (no more than a few degrees), an approximate linear relation holds between d and a. The following table gives angular diameter in minutes of arc (one-sixtieth of a degree) and diameter in kilometers for spherical objects 1 kilometer away.

Angular diameter a	6	11	23	40
Diameter d	0.00174	0.00319	0.00667	0.01160

a. Show that the data in the table are linear.

b. Express d as a linear function of a.

c. For small angles such as those we are considering, the diameter of an object that is k kilometers away and shows an angular diameter of a minutes of arc equals k times the diameter of an object that is 1 kilometer away and shows an angular diameter of a minutes of arc. For small angles, what is the relationship between distance, diameter, and angular diameter?

d. The moon is about 384,000 kilometers away and shows an angular diameter of about 31.17 minutes of arc. What is the diameter of the moon?

E-3. **Finding data points:** The following table contains linear data, but some data points are missing. Find the missing data points.

x	2	5		8	
y	5		17	23	29

E-4. **Testing for linearity:** Test the following data to see whether they are linear. If they are linear, find a linear model. Assume that $a \neq 0$.

x	a	2a	5a
y	b + a	2b + a	5b + a

3.3 *SKILL BUILDING EXERCISES*

S-1. **Testing data for linearity:** Test the following data to see whether they are linear.

x	2	4	6	8
y	12	17	22	27

S-2. **Testing data for linearity:** Test the following data to see whether they are linear.

x	2	4	6	8
y	12	17	21	25

S-3. **Making a linear model:** Make a linear model for the data in Exercise S-1.

S-4. **Making a linear model:** The data in Exercise S-2 are not linear, but if we omit the first point, the remaining data are linear. Make a linear model for these last three points.

S-5. **Graphing discrete data:** Plot the data from the table in Exercise S-2 above.

S-6. **Adding a graph to a data plot:** Add the graph of $y = 2.5x + 7$ to the data plot from Exercise S-5.

S-7. **Entering and graphing data:** Enter the following data and plot f against x.

x	1	2	3	4	5
f	8	6	5	3	1

S-8. **Adding a graph:** Add to the picture in Exercise S-7 the graph of $y = -1.7x + 9.7$.

S-9. **Editing data:** Plot the squares of the data points in the table from Exercise S-7.

S-10. **Data that are linear:** Plot the data from Exercise S-1 along with the linear model from Exercise S-3.

3.3 *EXERCISES*

Note: If no variable and function names are given in an exercise, you are of course expected to choose them and give the appropriate units. In your choice, when the variable is time, you should keep in mind the discussion following Example 3.7. Also, some of the data tables in this exercise set have been altered to allow for exact rather than approximate linear modeling.

Date	Value
1988	4.9
1989	7.1
1990	9.3
1991	11.5
1992	13.7

1. **Auto parts production:** The accompanying table shows the value, in billions of dollars, of auto parts produced in the United States on the given date.

   a. By calculating differences, show that these data can be modeled using a linear function.

   b. Plot the data points.

   c. Find a linear formula that models these data.

   d. Add the graph of the function you found in part c to your graph of data points.

   e. Use the formula you found in part c to estimate the value of auto parts made in the United States in 1996.

2. **Tuition at American private universities:** The following table shows the average yearly tuition and required fees, in dollars, charged by American private universities in the school year beginning in the given year.

Date	Average tuition
1994	$13,821
1995	$14,687
1996	$15,553
1997	$16,419
1998	$17,285

a. Show that these data can be modeled by a linear function, and find its formula.

b. Plot the data points and add the graph of the linear formula you found in part a.

c. What prediction does this formula give for average tuition and fees at American private universities for the academic year beginning in 2003?

3. **Tuition at American public universities:** *This is a continuation of Exercise 2.* The following table shows the average yearly in-state tuition and required fees, in dollars, charged by American public universities in the school year beginning in the given year.

Date	Average tuition
1994	$2816
1995	$2984
1996	$3152
1997	$3320
1998	$3488

a. Show that these data can be modeled by a linear function, and find its formula.

b. What is the slope for the linear function modeling tuition and required fees for public universities?

c. What is the slope of the linear function modeling tuition and required fees for private universities? (*Note:* See Exercise 2 above.)

d. Explain what the information in parts b and c tells you about the rate of increase in tuition in public versus private institutions.

e. Which shows the larger percentage increase from 1997 to 1998?

4. **Total cost:** *The background for this exercise can be found in Exercises 5 and 6 in Section 3.2.* The following table gives the total cost C, in dollars, for a widget manufacturer as a function of the number N of widgets produced during a month.

Number N	Total cost C
200	7900
250	9650
300	11,400
350	13,150

a. What are the fixed costs and variable cost for this manufacturer?

b. The manufacturer wants to reduce the fixed costs so that the total cost at a monthly production level of 350 will be $12,975. What will the new fixed costs be?

c. Instead of reducing the fixed costs as in part b, the manufacturer wants to reduce the variable cost so that the total cost at a monthly production level of 350 will be $12,975. What will the new variable cost be?

5. **Total revenue and profit:** *This is a continuation of Exercise 4.* In general, the highest price p per unit of an item at which a manufacturer can sell N items is not constant but is rather a function of N. Suppose the manufacturer of widgets in Exercise 4 has developed the following table showing the highest price p, in dollars, of a widget at which N widgets can be sold.

Number N	Price p
200	43.00
250	42.50
300	42.00
350	41.50

→

a. Find a formula for p in terms of N modeling the data in the table.

b. Use a formula to express the total monthly revenue R, in dollars, of this manufacturer in a month as a function of the number N of widgets produced in a month. Is R a linear function of N?

c. On the basis of the tables in this exercise and the preceding one, use a formula to express the monthly profit P, in dollars, of this manufacturer as a function of the number of widgets produced in a month. Is P a linear function of N?

6. **Dropping rocks on Mars:** The behavior of objects falling near Earth's surface depends on the mass of Earth. On Mars, a much smaller planet than Earth, things are different. If Galileo had performed his experiment on Mars, he would have obtained the following table of data.

t = seconds	V = feet per second
0	0
1	12.16
2	24.32
3	36.48
4	48.64
5	60.8

a. Show that these data can be modeled by a linear function, and find a formula for the function.

b. Calculate $V(10)$ and explain in practical terms what your answer means.

c. Galileo found that the acceleration due to gravity of an object falling near Earth's surface was 32 feet per second per second. Physicists normally denote this number by the letter g. If Galileo had lived on Mars, what value would he have found for g?

7. **The Kelvin temperature scale:** Physicists and chemists often use the *Kelvin* temperature scale. In order to determine the relationship between the Fahrenheit and Kelvin temperature scales, a lab assistant put Fahrenheit and Kelvin thermometers side by side and took readings at various temperatures. The following data were recorded.

K = kelvins	F = degrees Fahrenheit
200	−99.67
220	−63.67
240	−27.67
260	8.33
280	44.33
300	80.33

a. Show that the temperature F in degrees Fahrenheit is a linear function of the temperature K in kelvins.

b. What is the slope of this linear function? (*Note:* Be sure to take into account that the table lists kelvins in jumps of 20 rather than in jumps of 1.)

c. Find a formula for the linear function.

d. Normal body temperature is 98.6 degrees Fahrenheit. What is that temperature in kelvins?

e. If temperature increases by 1 kelvin, by how many degrees Fahrenheit does it increase? If temperature increases by 1 degree Fahrenheit, by how many kelvins does it increase?

f. The temperature of 0 kelvins is known as *absolute zero*. It is not quite accurate to say that all molecular motion ceases at absolute zero, but at that temperature the system has its minimum possible total energy. It is thought that absolute zero cannot be attained experimentally, although temperatures lower than 0.0000001 kelvin have been attained. Find the temperature of absolute zero in degrees Fahrenheit.

8. **Further verification of Newton's second law:** This exercise represents a hypothetical implementation of the experiment suggested in the solution of part 6 of Example 3.8. A mass of 15 kilograms was subjected to varying accelerations, and the resulting force was measured. In the following table, acceleration is in meters per second per second, and force is in newtons.

Acceleration	Force
8	120
11	165
14	210
17	255
20	300

a. Construct a table of differences and explain how it shows that these data are linear.

b. Find a linear model for the data.

c. Explain in practical terms what the slope of this linear model is.

d. Express, using functional notation, the force resulting from an acceleration of 15 meters per second per second, and then calculate that value.

e. Explain how this experiment provides further evidence for Newton's second law of motion.

9. **Market supply:** The following table shows the quantity S of wheat, in billions of bushels, that wheat suppliers are willing to produce in a year and offer for sale at a price P, in dollars per bushel.

S = quantity of wheat	P = price
1.0	$1.35
1.5	$2.40
2.0	$3.45
2.5	$4.50

In economics, it is customary to plot S on the horizontal axis and P on the vertical axis, so we will think of S as a variable and of P as a function of S.

a. Show that these data can be modeled by a linear function, and find its formula.

b. Make a graph of the linear formula you found in part a. This is called the *market supply curve*.

c. Explain why the market supply curve should be increasing. (*Hint:* Think about what should happen when the price increases.)

d. How much wheat would suppliers be willing to produce in a year and offer for sale at a price of $3.90 per bushel?

10. **Market demand:**   *This is a continuation of Exercise 9.* The following table shows the quantity D of wheat, in billions of bushels, that wheat consumers are willing to purchase in a year at a price P, in dollars per bushel.

D = quantity of wheat	P = price
1.0	$2.05
1.5	$1.75
2.0	$1.45
2.5	$1.15

In economics, it is customary to plot D on the horizontal axis and P on the vertical axis, so we will think of D as a variable and of P as a function of D.

a. Show that these data can be modeled by a linear function, and find its formula.

b. Add the graph of the linear formula you found in part a, which is called the *market demand curve*, to your graph of the market supply curve from Exercise 9.

c. Explain why the market demand curve should be decreasing.

d. The *equilibrium price* is the price determined by the intersection of the market demand curve and the market supply curve. Find the equilibrium price determined by your graph in part b.

11. **Sports car:** An automotive engineer is testing how quickly a newly designed sports car accelerates from rest. He has collected data giving the velocity, in miles per hour, as a function of the time, in seconds, since the car was at rest. Here is a table giving a portion of his data:

Time	Velocity
2.0	27.9
2.5	33.8
3.0	39.7
3.5	45.6

→

a. By calculating differences, show that the data in this table can be modeled by a linear function.

b. What is the slope for the linear function modeling velocity as a function of time? Explain in practical terms the meaning of the slope.

c. Use the data in the table to find the formula for velocity as a linear function of time that is valid over the time period covered in the table.

d. What would your formula from part c give for the velocity of the car at time 0? What does this say about the validity of the linear formula over the initial segment of the experiment? Explain your answer in practical terms.

e. Assume that the linear formula you found in part c is valid from 2 seconds through 5 seconds. The marketing department wants to know from the engineer how to complete the following statement: "This car goes from 0 to 60 mph in _____ seconds." How should they fill in the blank?

12. **High school graduates:** The following table shows the number, in millions, graduating from high school in the United States in the given year.[10]

Year	Number graduating (in millions)
1985	2.83
1987	2.65
1989	2.47
1991	2.29

a. By calculating differences, show that these data can be modeled using a linear function.

b. What is the slope for the linear function modeling high school graduations? Explain in practical terms the meaning of the slope.

c. Find a formula for a linear function that models these data.

d. Express, using functional notation, the number graduating from high school in 1994, and then calculate that value.

13. **Later high school graduates:** *This is a continuation of Exercise 12.* The following table shows the number, in millions, graduating from high school in the United States in the given year.

Year	Number graduating (in millions)
1995	2.59
1997	2.75
1999	2.91
2001	3.07

a. Find the slope of the linear function modeling high school graduations, and explain in practical terms the meaning of the slope.

b. Find a formula for a linear function that models these data.

c. Express, using functional notation, the number graduating from high school in 2002, and then calculate that value.

d. The actual number graduating from high school in 1994 was about 2.52 million. Compare this with the value given by the formula in part b and with your answer to part d of Exercise 12. Which is closer to the actual value? In general terms, what was the trend in high school graduations from 1985 to 2001?

14. **Tax table:** Here are selected entries from the 1999 tax table that shows the federal income tax owed by those married and filing jointly. The taxable income and the tax are both in dollars.

Taxable income	Tax	Taxable income	Tax
42,600	6394	43,200	6507
42,700	6409	43,300	6535
42,800	6424	43,400	6563
42,900	6439	43,500	6591
43,000	6454	43,600	6619
43,100	6479	43,700	6647

---

[10]The tables in this exercise and the next are adapted from the *1999 Statistical Abstract of the United States.*

Over what two parts of this table is the tax a linear function of the taxable income? Find formulas for both linear functions, and explain in practical terms what the slopes mean.

15. **Sound speed in oceans:** Marine scientists use a linear model for the speed c of sound in the oceans as a function of the salinity S at a fixed depth and temperature.[11] In the following table, one scientist recorded data for c and S at a depth of 500 meters and a temperature of 15 degrees Celsius. Here c is measured in meters per second and S in parts per thousand.

Salinity S	Sound speed c
35.0	1515.36
35.6	1516.08
36.2	1516.80
36.8	1517.62
37.4	1518.24

a. Looking over the table, the scientist realizes that one of the entries for sound speed is in error. Which entry is it, and what is the correct speed?

b. Explain in practical terms the meaning of the slope of the linear model for c in terms of S at the given depth and temperature.

c. Calculate $c(39)$ and explain in practical terms what it means.

16. **Focal length:** A refracting telescope has a main lens, or *objective lens,* and a second lens, the *eyepiece* (see Figure 3.56). For a given magnification M of the telescope, the focal length F_o of the objective lens is a linear function of the focal length F_e of the eyepiece. For example, a telescope with magnification $M = 80$ times can be constructed using various combinations of lenses. The following table gives some samples of focal lengths for telescopes with magnification $M = 80$. Here focal lengths are in centimeters.

F_e	F_o
0.3	24
0.5	40
0.7	56
0.9	72

a. Construct a linear model for the data.

b. In this example the magnification M is 80. In general F_o is proportional to F_e, and the constant of proportionality is M. Use this relation to write a formula for F_o in terms of F_e and M.

c. Solve the equation you obtained in part c for M and thus obtain a formula for magnification as a function of objective lens focal length and eyepiece focal length.

d. To achieve a large magnification, how should objective and eyepiece lenses be selected?

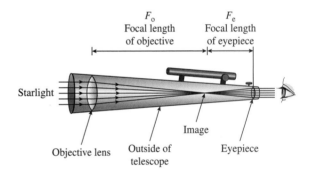

FIGURE 3.56

17. **Measuring the circumference of the Earth:** Eratosthenes, who lived in Alexandria around 200 B.C., learned that at noon on the summer solstice the sun shined vertically into a well in Syene (modern Aswan, due south of Alexandria). He found that on the same day in Alexandria the sun was about 7 degrees short of being directly overhead. Since 7 degrees is about $\frac{1}{50}$ of a full circle of 360 degrees, he concluded that the distance from Syene to Alexandria was about $\frac{1}{50}$ of the Earth's circumference. He knew from travelers that it was a 50-day trip and that

---

[11]The model here is based on the formula of Kenneth V. Mackenzie, "Nine-term equation for sound speed in the oceans," *J. Acoust. Soc. Am.* **70** (1981), 807–812.

camels could travel 100 *stades*[12] per day. What was Eratosthenes' measure, in stades, of the circumference of the Earth? One estimate is that the stade is 0.104 mile. Using this estimate, what was Eratosthenes' measurement of the circumference of the Earth in miles?

18. **A research project on tax tables:** Here are selected entries from the 2000 tax table that shows the federal income tax owed by those married and filing jointly. The taxable income and the tax are both in dollars.

Taxable income	Tax	Taxable income	Tax
43,100	6469	43,700	6559
43,200	6484	43,800	6574
43,300	6499	43,900	6599
43,400	6514	44,000	6627
43,500	6529	44,100	6655
43,600	6544	44,200	6683

a. Over what two parts of this table is the tax a linear function of the taxable income? Find formulas for both linear functions.

b. Compare your answer to part a with your answer to Exercise 14. Consider especially the income level where the transition from one linear function to the next takes place.

c. As a research project, examine tax tables from earlier years and investigate tax law to discover how the transition level is determined. Determine what the phrase *bracket creep* means.

---

[12]There is some uncertainty about just how long the stade was, so we only know that his estimate is too large by between 4 and 14 percent. In any case, it is considerably better than Columbus's estimate (which was about 40% too small) some 1700 years later. The mathematicians of the day knew full well that Columbus's estimate was nowhere near correct, and that is one of the reasons why he had so much trouble getting financing for his expedition to Cathay.

3.4 *Linear Regression*

In real life, rarely is information gathered that perfectly fits any simple formula. In cases such as government spending, many factors influence the budget, including the political make-up of the legislature. In the case of scientific experiments, variations may be due to *experimental error*, the inability of the data gatherer to obtain exact measurements; there may also be elements of chance involved. Under these circumstances, it may be necessary to obtain an approximate rather than an exact mathematical model.

The Regression Line

To illustrate this idea, let's look at total federal education expenditures, in billions of dollars, by the United States as reported by the *1995 Statistical Abstract of the United States* and recorded in the following table.

Date	1985	1986	1987	1988	1989
Expenditures in billions	27.0	29.0	30.4	30.9	32.1

To study these data, we assign variable and function names: Let t be the number of years since 1985, and let E be the expenditures, in billions of dollars. The table for E as a function of t is then

t	0	1	2	3	4
E	27.0	29.0	30.4	30.9	32.1

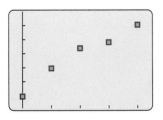

FIGURE 3.57 Data points that are almost on a straight line

It is difficult to discern, by looking at the table, the pattern of spending on public education. We can check to see whether these data can be modeled by a linear function by making a table of changes.

Change in t	From 0 to 1	From 1 to 2	From 2 to 3	From 3 to 4
Change in E	2.0	1.4	0.5	1.2

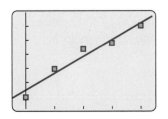

FIGURE 3.58 A line that almost fits the data

The change in E is not constant, so we know that these data cannot be modeled exactly by a linear function, but let's explore further. If we plot the data points as we have done in Figure 3.57, we see that they *almost* fall on a straight line. To show this better, we have added, in Figure 3.58, a straight line that passes *close* to all the data points. This line is known as the *least-squares fit* or *linear regression* line, and it appears that the data follow this line fairly closely. Thus, although we cannot model the data *exactly* with a linear function, it seems reasonable to use the line in Figure 3.58 as an *approximate* model for federal education spending.

There are many ways of approximating data that are nearly linear, but the regression line is the one that is most often used.[13] It is the line that gives the least possible total of the squares of the vertical distances from the line to the data points. This is why the regression line is sometimes referred to as the least-squares fit. The mathematical concepts needed to derive the formula for regression lines are beyond the scope of this course. Furthermore, although it is possible to calculate a formula for the regression line by hand, the procedure is cumbersome. Instead, this calculation is nearly always done with a calculator or computer, and that is how we will do it here. You should refer to Chapter 4 of the *Technology Guide* to see the exact keystrokes needed to execute this important procedure.

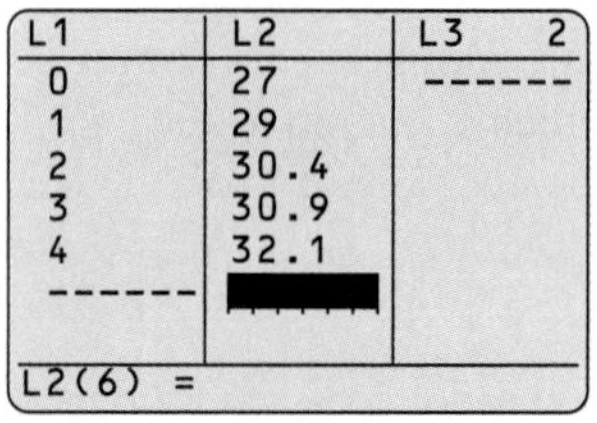

With the data properly entered 3.9 in the calculator as shown in Figure 3.59, we can get 3.10 the regression line information shown in Figure 3.60. As usual, the calculator chooses its own letter[14] names, and we need to record the appropriate correspondences:

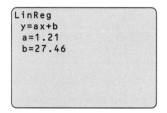

FIGURE 3.59 Data for federal education spending

$$X = t, \text{ the variable}$$

$$Y = E, \text{ the function name}$$

$$a{=}1.21 = \text{slope of the regression line}$$

$$b{=}27.46 = \text{vertical intercept of the regression line}.$$

We use this information to make the regression line model for E as a function of t:

$$E = 1.21t + 27.46. \tag{3.8}$$

FIGURE 3.60 The regression line parameters

This is the line we added 3.11 to the data plot in Figure 3.57 to get the picture in Figure 3.58.

It is important to remember that even though we have written an equality, Equation (3.8) in fact is a model that only approximates the relationship between t and E given by the data table. For example, the initial value according to the linear model in Equation (3.8) is 27.46, whereas the entry in the table for $t = 0$ is $E = 27$. We will use the equals sign as above, but you should be aware that in this setting many would prefer to replace it by an approximation symbol, $\approx$, or to use different letters for the regression equation. If you do use different letters for function and variable, it is crucial that you state clearly what they represent.

Uses of the Regression Line: Slope and Trends

The most useful feature of the regression line is its slope, which in many cases provides the key to understanding data. For education spending, the slope, 1.21 billion dollars per year, of the regression line tells us that during the period from 1985 to 1989, education spending grew by about 1.21 billion dollars per year. It tells how the data are changing.

A plot that shows the regression line with the data can be useful in analyzing trends. For example, in Figure 3.58 the third data point (corresponding to 1987) lies above the regression line, whereas the fourth data point (corresponding to 1988) lies below it.

---

[13]Those who are interested in experimenting with different linear approximations or want more information about the regression equation should see the discussion at the end of this section.

[14]An additional number, often denoted r, is shown at the bottom of the screen display for some calculators. This is a statistical measure of how closely the line fits the data, and we will not make use of it.

Certainly spending was higher in 1988 than in 1987, but it could be argued, on the basis of the position of the data points relative to the line, that in 1987 spending was ahead of the trend, whereas in 1988 it was slightly behind the trend.

It is tempting to use the regression line to predict the future. We will show how to do this and then discuss the pitfalls in such practice. What level of spending does the regression line in Equation (3.8) predict for 1990? Since t is the number of years since 1985, we want to get the value of E when $t = 5$. This information is not shown in the table, so we approximate the value of $E(5)$ using the regression line with $t = 5$:

$$\text{Projected spending in 1990} = 1.21 \times 5 + 27.46 = 33.51 \text{ billion dollars}.$$

Thus we project 1990 spending to be about 33.5 billion dollars. Consulting the source for our data, the *1995 Statistical Abstract of the United States*, we find that federal education expenditures in 1990 were in fact $E = 33.4$ billion dollars. In this case, the projection given by the regression line was pretty accurate.

Let's try to use the regression line to go the other way—that is, to estimate money spent before 1985. In particular, we want to determine the level of federal education expenditures in 1970. That is 15 years before 1985, so we want the value of E when $t = -15$. We use the regression line with $t = -15$:

$$\text{Estimated spending in 1970} = 1.21 \times (-15) + 27.46 = 9.31 \text{ billion dollars}.$$

Thus we estimate 1970 spending to have been about 9.3 billion dollars. Once again, we refer to the *1995 Statistical Abstract of the United States* and see that the actual federal expenditures for education in 1970 were 26.6 billion dollars. In this case, the value given by the regression line is a very bad estimate of the real value.

The regression line is a powerful tool for analysis of certain kinds of data, but caution in its use is essential. We saw above that the regression line gave a good estimate for 1990 expenditures but a very bad one for 1970. In general, using the regression line to extrapolate beyond the limits of the data is risky, and the risk increases dramatically for long-range extrapolations. What the regression line really shows is the *linear trend* established by the data. Thus, for our 1990 projection, it would be appropriate to say, "If the trend established in the late 1980s had persisted, then federal education spending in 1990 would have been about 33.51 billion dollars." A check of the 1990 data showed that this trend did indeed persist into 1990. An appropriate statement for our 1970 analysis might be "If the trend established in the late 1980s had been valid since 1970, then federal expenditures in 1970 would have been about 9.31 billion dollars." Since the actual expenditure was much more, we might proceed by gathering more data to see exactly how spending changed and then conduct historical, political, and economic investigations into why it changed.

Newspaper and magazine articles frequently use data gathered today to make predictions about what might be expected in the future, and in many cases this is done with the regression line. When such data are handled by professionals who have sophisticated tools and insights available to help them, valuable information can be gained. But sometimes predictions based on gathered data are made by people who know a great deal less about handling data than you will by the time you complete this course. As a citizen, it is important that you be able to bring your own insight to bear on such analyses. It is risky to use the regression line to make projections unless you have good reason to believe that the data you are looking at are nearly linear and that the linear nature of the data will persist into the future. On the other hand, the regression line can clearly show the trend of almost linear data and can appropriately be used to determine whether trends persist. If information is available that indicates that linear trends are persisting, then the regression line can be used to make forecasts.

As an example, suppose a legislator who took office in 1989 ran on a platform calling for a "significant" increase in federal education spending in 1990 and 1991. Does the record show that the legislator was able to fulfill his campaign promise? If so, then we should be able to detect a "significant" increase in 1990 and 1991 over the trend established from 1985 through 1989. We have already seen that the trend of the late 1980s leads us to project an expenditure of about 33.51 billion dollars. A similar analysis gives projected spending in 1991:

$$\text{Projected spending in } 1991 = 1.21 \times 6 + 27.46 = 34.72 \text{ billion dollars},$$

so the projected spending is about 34.7 billion dollars. Checking the record, we find that actual federal spending, 33.4 billion in 1990 and 34.4 billion in 1991, was slightly less than the trend of the late 1980s would suggest. That is what we might expect to happen if no additional impetus for funding increase were applied. It is easy to conceive of this legislator taking credit for a 1.3 billion dollar increase in 1990 and a 1 billion dollar increase in 1991. We leave it to the reader to decide whether a re-election vote would be merited.

EXAMPLE 3.9   Military Expenditures

The following table shows the amount $M = M(t)$ of money, in billions of dollars, spent by the United States on national defense.[15] In the time row, $t = 0$ corresponds to 1985, $t = 1$ refers to 1986, and so on.

$t =$ years since 1985	0	1	2	3	4
$M =$ billions of dollars	252.7	263.3	282.0	290.4	303.6

1. Plot the data points. Does it appear that it is appropriate to approximate these data with a straight line?

2. Find the equation of the regression line for M as a function of t, and add the graph of this line to your data plot.

3. Explain in practical terms the meaning of the slope of the regression line model we found in part 2.

4. Compare the slope of the regression line for defense spending with the slope of the regression line for spending on education that we made earlier. What conclusions do you draw from this comparison?

5. Use the regression equation to estimate military spending by the United States in 1990 and 1995. The actual military expenditures in 1990 and 1995 were $299.3 billion and $271.6 billion, respectively. Did the trend established in the late 1980s persist until 1990? Did it persist through 1995?

---

[15] From the *1995 Statistical Abstract of the United States.*

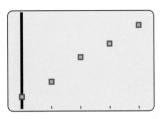

L1	L2	L3	2
0	252.7	------	
1	263.3		
2	282		
3	290.4		
4	303.6		
------	▬▬▬▬		

L2(6) =

FIGURE 3.61 National defense spending data

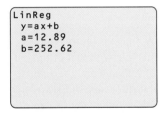

FIGURE 3.62 A plot of national defense spending

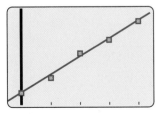

```
LinReg
 y=ax+b
 a=12.89
 b=252.62
```

FIGURE 3.63 Regression line parameters for defense spending

FIGURE 3.64 The regression line added to the data plot

Solution to Part 1:   First we enter ⬚3.12⬚ the data as shown in Figure 3.61. Next we plot ⬚3.13⬚ the data as shown in Figure 3.62. Note that the horizontal axis corresponds to years since 1985, the vertical axis to billions of dollars of military expenditures. From Figure 3.62, it is clear that the data points do not lie exactly on a line, but they do almost line up, and it is not unreasonable to model these data with a straight line.

Solution to Part 2:   We use the calculator to get ⬚3.14⬚ the regression line parameters shown in Figure 3.63. We record the appropriate associations:

$$X = t, \text{ the variable, years since 1985}$$

$$Y = M, \text{ the function name, billions spent}$$

$$a=12.89 = \text{slope of the regression line}$$

$$b=252.62 = \text{vertical intercept of the regression line}.$$

Thus the regression line model we want is $M = 12.89t + 252.62$. We emphasize once more that even though we have written an equality, this does not establish an exact relationship between t and M. Rather, it is the approximation provided by the regression model. We add ⬚3.15⬚ this equation to the function list and get the graph in Figure 3.64. It is important to note here that the regression line we found passes near the data points. If it didn't, then we would look for a mistake in our work; if we found that no error had been made, we would conclude that it might not be appropriate to model the data using the regression line.

Solution to Part 3:   The slope of the regression line shows the rate at which military expenditures were growing during the late 1980s. We can conclude that in this period, defense spending was growing at a rate of about 12.89 billion dollars per year.

Solution to Part 4:   In Part 3 we noted that during the late 1980s, defense spending was growing at a rate of 12.89 billion dollars per year. Earlier we found that the slope of the regression line for federal education spending was 1.21 billion dollars per year. Thus not only was military spending much greater during the late 1980s; it was growing over ten times as fast as federal education spending. (There are other comparisons that may be made that do not show such a sharp distinction. You may note, for example, that the percentage increases are nearly the same.)

Solution to Part 5:   The year 1990 corresponds to $t = 5$, and 1995 corresponds to $t = 10$. We use these values in the regression equation to make the projections:

Projected 1990 spending $= 12.89 \times 5 + 252.62 = 317.07$ billion dollars

Projected 1995 spending $= 12.89 \times 10 + 252.62 = 381.52$ billion dollars.

Since the actual military expenditures in 1990 and 1995 were $299.3 billion and $271.6 billion, respectively, we see that the trend of the late 1980s did not persist even until 1990. By 1995 the departure from the trend is dramatic.

We emphasize that the failure of the regression model to predict accurately spending in 1990 and 1991 is not a shortcoming of the model. Rather, it provides a convincing argument that something happened to change the trend of military spending in the late 1980s. It shows in a striking way the influence of the end of the Cold War on military spending.   ▨ ▨ ▨

ANOTHER LOOK

Linear Regression Formula, the Error

Before giving the formula for linear regression, let's consider how we might tell how well a given line fits a data set. Let's look at the data set from the beginning of this section on federal education expenditures. Here x is years since 1985, and y is federal education expenditures in billions of dollars.

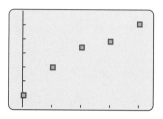

FIGURE 3.65 A plot of the data

x	0	1	2	3	4
y	27.0	29.0	30.4	30.9	32.1

One way to find an equation of a line to fit the data would be to use the initial value of 27 from the table and to take for the slope the average rate of change from the beginning to the end of this period, which is $\frac{32.1-27.0}{4-0} = 1.275$. This gives us a linear function that we'll call Est:

$$\text{Est} = 1.275x + 27 .$$

This is not the regression line, but it is not an unreasonable way to fit the data, and we see from Figure 3.65 and Figure 3.66 that, at least visually, the fit is pretty good.

We want to get a quantitative comparison of the fit of this line with that of the regression line. To do this, we introduce the idea of *errors of estimate* and the *sum of squares of errors of estimate*. In the table below, we calculate the value of Est for each data point, then the error between Est and the actual data, and then the square of the error (to three decimal places).

FIGURE 3.66 A proposed linear fit

Year x	Data y	Estimate Est	Error of estimate y − Est	Square of error of estimate $(y - \text{Est})^2$
0	27.0	27.0	0	0
1	29.0	28.275	0.725	0.526
2	30.4	29.55	0.85	0.723
3	30.9	30.825	0.075	0.006
4	32.1	32.1	0	0

The sum of the squares of the errors of the estimate is simply the sum of the last column, which is 1.255. Through squaring and computing the sum, each error (representing a less-than-perfect fit of the line) is accounted for in this one number 1.255. This number is one way of measuring the fit of the line from Est to the data points.

Let's calculate the sum of the squares of errors for the regression line Reg = $1.21x + 27.46$ that we found earlier for these data. We use the regression line to make a table just like the one above.

Year x	Data y	Estimate Reg	Error of estimate $y - \text{Reg}$	Square of error of estimate $(y - \text{Reg})^2$
0	27.0	27.46	−0.46	0.212
1	29.0	28.67	0.33	0.109
2	30.4	29.88	0.52	0.270
3	30.9	31.09	−0.19	0.036
4	32.1	32.3	−0.2	0.04

Totaling the last column, we get a sum-of-squares error for the regression line of 0.667. This is smaller than the sum-of-squares error for the Est line, and we conclude that by this measure, the regression line is a better fit of the data. In fact, that is what the regression line is. It is the line that, among all lines, has the smallest sum-of-squares error.

To learn how to calculate the regression line, we introduce the idea of *deviations from the average*. To begin, the average of the x's in the data table above is the sum of the x's divided by the number of items. Thus in this case the average is

$$\bar{x} = \frac{0 + 1 + 2 + 3 + 4}{5} = 2.$$

The *x-deviations* are the differences between the various values of x and the average $\bar{x}$. In a similar way, we calculate that the average $\bar{y}$ of the y values is 29.88, and we define the y-deviations in terms of this average. For reasons that will become clear later, certain other calculations are also made.

x	y	x-Deviation $x - \bar{x}$	y-Deviation $y - \bar{y}$	x-Deviation squared $(x - \bar{x})^2$	Product of deviations $(x - \bar{x}) \times (y - \bar{y})$
0	27.0	−2	−2.88	4	5.76
1	29.0	−1	−0.88	1	0.88
2	30.4	0	0.52	0	0
3	30.9	1	1.02	1	1.02
4	32.1	2	2.22	4	4.44

Let's use Sx^2 for the sum of the x-deviations squared and Sxy for the sum of the product of the deviations. In this example we total the fifth column to get $Sx^2 = 10$, and we total the last column to get $Sxy = 12.10$.

In this notation, the regression line is defined by

$$\text{Reg} = \bar{y} + \frac{Sxy}{Sx^2}(x - \bar{x}).$$

Thus for this data table, the regression line is

$$\text{Reg} = 29.88 + \frac{12.10}{10}(x - 2),$$

which simplifies to

$$\text{Reg} = 1.21x + 27.46.$$

That is the formula the calculator gives. Your calculator goes through the same computations to find the equation of the regression line as we did here, albeit much faster. The derivation of this formula and the proof that it is actually the line with the smallest sum of squares of errors are beyond the scope of this text.

Enrichment Exercises

E-1. **Fitting lines:** Consider again the data on federal education expenditures. Find an equation for another estimate line that you think is a better fit than the estimate Est used above. Calculate the sum of the squares of the errors for it.

E-2. **Calculating regression lines:** Calculate by hand the regression line for each of the following data sets.

a.

x	0	2	4	6
y	1	6	13	20

b.

x	0	3	5	9
y	−1	6	8	16

c.

x	1	2	6	7
y	6	9	23	26

E-3. **Calculating the error:** For each data set in Exercise E-2, calculate the sum-of-squares error.

E-4. **Military expenditures:** In this exercise we use the data table on military expenditures M from Example 3.9.

a. Calculate (by hand) the regression line L for military expenditures M as a function of t.

b. Evaluate the formula for L from part a at $t = \bar{t}$. How does this value compare with $\overline{M}$?

E-5. **The meaning of errors:**

  a. Suppose you had an estimate line such that the sum of the squares of the errors of the estimate was 0. What would that say about how well the line fits the data points?

  b. Calculate the sum of the errors (not the squares) for the regression lines you found in Exercise E-2.

E-6. **Correlation coefficient:** The *correlation coefficient r* is a measure of how well a regression line fits data. The value of r is always between -1 and 1. If the fit is exact, the value of r is 1 or -1. Values of r near zero indicate a poor fit. Most calculators can be set to display the correlation coefficient.[16] For each of the following data sets, get the regression line, plot the data and the regression line on the same screen, and report the value of the correlation coefficient.

  a.

x	2	4	6	8	10
y	5	9	13	17	21

  b.

x	2	4	6	8	10
y	5	9	15	17	21

  c.

x	2	4	6	8	10
y	5	9	20	17	21

----

[16] See the *Technology Guide*.

S-1. Slope of regression line: The slope of the regression line for a certain data set is a positive number. Do you expect the data values to be increasing or decreasing?

S-2. Meaning of slope of regression line: For a certain school district, the slope of the regression line for money spent on education as a function of the year is $2300 per year. What does this mean in practical terms?

S-3. Meaning of slope of regression line: The slope of the regression line for federal agricultural spending is larger than the slope of the regression line for federal spending on research and development. Explain in practical terms what this relationship means.

S-4. Meaning of slope of regression line: For many animals, speed running, in miles per hour, is approximately a linear function of the length, in inches, of the animal. The slope of the regression line for speed running as a function of length is 2.03. Explain in practical terms what this means.

S-5. Plotting data and regression lines: For the following data set: (a) Plot the data. (b) Find the equation of the regression line. (c) Add the graph of the regression line to the plot of the data points.

x	1	2	3	4	5
y	2.3	2	1.8	1.4	1.3

S-6. Plotting data and regression lines: For the following data set: (a) Plot the data. (b) Find the equation of the regression line. (c) Add the graph of the regression line to the plot of the data points.

x	1.3	2.5	3.3	4.2	5.1
y	2.6	2.6	2	1.8	1.5

S-7. Plotting data and regression lines: For the following data set: (a) Plot the data. (b) Find the equation of the regression line. (c) Add the graph of the regression line to the plot of the data points.

x	2.3	3.7	5.1	6.4	8.2
y	4.8	5.3	7.2	9.6	10.3

S-8. Plotting data and regression lines: For the following data set: (a) Plot the data. (b) Find the equation of the regression line. (c) Add the graph of the regression line to the plot of the data points.

x	4.1	5.7	7.3	8.9	10.5
y	7.7	8.3	8.4	8.9	9.1

S-9. Plotting data and regression lines: For the following data set: (a) Plot the data. (b) Find the equation of the regression line. (c) Add the graph of the regression line to the plot of the data points.

x	5.2	8.9	12.6	16.3	20
y	−3.1	−4.8	−5.3	−7.1	−7.9

S-10. Plotting data and regression lines: For the following data set: (a) Plot the data. (b) Find the equation of the regression line. (c) Add the graph of the regression line to the plot of the data points.

x	16.3	20	23.7	27.4	31.1
y	51.1	68.8	80.3	86.2	99.6

3.4 EXERCISES

1. **Is a linear model appropriate?** The number, in thousands, of bacteria in a petri dish is given by the table below. Time is measured in hours.

Time in hours since experiment began	Number of bacteria in thousands
0	1.2
1	2.4
2	4.8
3	9.6
4	19.2
5	38.4
6	76.8

The growth of bacteria

The table below shows enrollment,[17] in millions of people, in public colleges in the United States during the years from 1996 through 2000.

Date	Enrollment in millions
1996	11.1
1997	11.2
1998	11.4
1999	11.5
2000	11.6

Enrollment in public colleges

a. Plot the data points for number of bacteria. Does it look reasonable to approximate these data with a straight line?

b. Plot the data points for college enrollment. Does it look reasonable to approximate these data with a straight line?

2. **College enrollment:** *This is a continuation of Exercise 1. We use the data in the college enrollment table that appears there.*

a. Find the equation of the regression line model for college enrollment as a function of time, and add its graph to the data plot made in Exercise 1.

b. Explain the meaning of the slope of the line you found in part a.

c. Express, using functional notation, the enrollment in American public colleges in 2002, and then estimate that value.

d. Enrollment in American public colleges in 1993 was 11.2 million. Does it appear that the trend established in the late 1990s was valid as early as 1993?

3. **Tourism:** The number, in millions, of tourists who visited the United States from overseas is given in the following table.[18]

Date	Millions of tourists
1994	18.46
1995	20.64
1996	22.66
1997	24.19

a. Plot the data.

b. Find the equation of the regression line and add its graph to your data plot. (Round the regression line parameters to two decimal places.)

c. Explain in practical terms the meaning of the slope.

d. Express, using functional notation, the number of tourists who visited the United States in 1992, and then estimate that value.

---

[17]This is taken from the *Statistical Abstract of the United States* and includes projections.

[18]These data are from the *1999 Statistical Abstract of the United States*.

4. **Cable TV:** The following table gives the percentage of American homes with cable TV.

Date	Percent of homes with cable
1985	46.2
1986	48.1
1987	50.5
1988	53.8

a. Plot the data points.

b. Find the equation of the regression line and add its graph to the plotted data.

c. In 1990, 58.6% of American homes had cable TV. If you had been a marketing strategist in 1988 with only the data in the table above available, what would have been your prediction for the percentage of American homes that would have cable TV in 1990?

5. **Long jump:** The following table shows the length, in meters, of the winning long jump in the Olympic Games for the indicated year. (One meter is 39.37 inches.)

Year	1900	1904	1908	1912
Length	7.19	7.34	7.48	7.60

a. Find the equation of the regression line that gives the length as a function of time. (Round the regression line parameters to three decimal places.)

b. Explain in practical terms the meaning of the slope of the regression line.

c. Plot the data points and the regression line.

d. Would you expect the regression line formula to be a good model of the winning length over a long period of time? Be sure to explain your reasoning.

e. There were no Olympic Games in 1916 because of World War I, but the winning long jump in the 1920 Olympic Games was 7.15 meters. Compare this with the value that the regression line model

gives. Is the result consistent with your answer to part d?

6. **Driving:** You are driving on a highway. The following table gives your speed S, in miles per hour, as a function of the time t, in seconds, since you started making your observations.

Time t	Speed S
0	54
15	59
30	63
45	66
60	68

a. Find the equation of the regression line that expresses S as a linear function of t.

b. Explain in practical terms the meaning of the slope of the regression line.

c. On the basis of the regression line model, when do you predict that your speed will reach 70 miles per hour? (Round your answer to the nearest second.)

d. Plot the data points and the regression line.

e. Use your plot in part d to answer the following: Is your prediction in part c likely to give a time earlier or later than the actual time when your speed reaches 70 miles per hour?

7. **The effect of sampling error on linear regression:** A stream that feeds a lake is flooding, and during this flooding period the depth of water in the lake is increasing. The actual depth of the water at a certain point in the lake is given by the linear function $D = 0.8t + 52$ feet, where t is measured in hours since the flooding began. A hydrologist does not have this function available and is trying to determine experimentally how the water level is rising. She sits in a boat and, each half-hour, drops a weighted line into the water to measure the depth to the bottom. The motion of the boat and the waves at the surface make exact measurement impossible. Her compiled data are given in the following table.

t = hours since flooding began	D = measured depth in feet
0	51.9
0.5	52.5
1	52.9
1.5	53.3
2	53.7

a. Plot the data points.

b. Find the equation of the regression line for D as a function of t, and explain in practical terms the meaning of the slope.

c. Add the graph of the regression line to the plot of the data points.

d. Add the graph of the depth function $D = 0.8t + 52$ to the picture. Does it appear that the hydrologist was able to use her data to make a close approximation of the depth function?

e. What was the actual depth of the water at $t = 3$ hours?

f. What prediction would the hydrologist's regression line give for the depth of the water at $t = 3$?

8. **Gross national product:** The United States gross national product, in trillions of dollars, is given in the table below.

Date	Gross national product
1985	4.02
1986	4.23
1987	4.52
1988	4.88
1989	5.20

a. Find the equation of the regression line, and explain the meaning of its slope. (Round regression line parameters to two decimal places.)

b. Plot the data points and the regression line.

c. In 1989 a prominent economist predicted that by 1993, the gross national product would reach 6.6 trillion dollars. Does your information from part a support that conclusion? If not, when would you predict a gross national product of 6.6 trillion dollars?

9. **Japanese auto sales in the United States:** For 1992 through 1996, the table below shows the total U.S. sales, in millions, of Japanese automobiles (excluding light trucks).[19]

Date	Japanese cars sold
1992	2.48
1993	2.38
1994	2.39
1995	2.32
1996	2.29

a. Plot the data points. In your opinion, does it appear appropriate to approximate these data with a linear function?

b. Get the equation of the regression line (rounding parameters to three decimal places), and explain in practical terms the meaning of the slope. In particular, comment on the meaning of the sign of the slope.

c. Add the graph of the regression line to the data plot in part a. In your opinion, does this picture make the use of the regression line here appear to be more or less appropriate?

d. The U.S. Department of Commerce, International Trade Administration, forecasted that there would be 2.11 million Japanese cars sold in the United States in 1997 and 2.04 million in 1998. How does the forecast obtained from the regression line compare with these figures?

_____

[19]From the 1995 *U.S. Industrial Outlook*.

10. **Running speed versus length:** The following table gives the length L, in inches, of an animal and its maximum speed R, in feet per second, when it runs.[20] (For comparison, 10 feet per second is about 6.8 miles per hour.)

Animal	Length L	Speed R
Deermouse	3.5	8.2
Chipmunk	6.3	15.7
Desert crested lizard	9.4	24.0
Grey squirrel	9.8	24.9
Red fox	24.0	65.6
Cheetah	47.0	95.1

a. Does this table support the generalization that larger animals run faster?

b. Plot the data points. Does it appear that running speed is approximately a linear function of length?

c. Find the equation of the regression line for R as a function of L, and explain in practical terms the meaning of its slope. (Round regression line parameters to two decimal places.) Add the plot of the regression line to the data plot in part b.

d. Judging on the basis of the plot in part c, which is faster *for its size*, the red fox or the cheetah?

11. **Antimasonic voting:** In *A Guide to Quantitative History*,[21] R. Darcy and Richard C. Rohrs use mathematics to investigate the influence of religious zeal, as evidenced by the number C of churches in a given township, on the percent M of voting that was antimasonic in Genesee County, New York, from 1828 to 1832. The data used there are partially reproduced in the accompanying table.

Township	C = Number of church buildings	M = Percent antimasonic voting
Alabama	2	60.0
Attica	4	65.0
Stafford	5	74.0
Covington	6	81.7
Elbe	7	88.3

Solely on the basis of the data above, analyze the premise that "The percent of antimasonic voting in Genesee County around 1830 had a direct dependence on the number of churches in the township." Your analysis should include a statement of why you believe there is a relationship and how that relationship might appropriately be described. You are encouraged to consult the book cited for a deeper and more authoritative analysis.

12. **Running ants:** A scientist collected the following data on the speed, in centimeters per second, at which ants ran at the given ambient temperature, in degrees Celsius.[22]

Temperature	Speed
25.6	2.62
27.5	3.03
30.3	3.57
30.4	3.56
32.2	4.03
33.0	4.17
33.8	4.32

---

[20] The table is adapted from J. T. Bonner, *Size and Cycle* (Princeton, NJ: Princeton University Press, 1965).

[21] Published by Praeger Publishers (Westport, CT) in 1995.

[22] The table is taken from the data of H. Shapley, "Note on the thermokinetics of Dolichoderine ants," *Proc. Nat. Acad. Sci.* **10** (1924), 436–439.

a. Find the equation of the regression line, giving the speed as a function of the temperature.

b. Explain in practical terms the meaning of the slope of the regression line.

c. Express, using functional notation, the speed at which the ants run when the ambient temperature is 29 degrees Celsius, and then estimate that value.

d. The scientist observes the ants running at a speed of 2.5 centimeters per second. What is the ambient temperature?

13. **Expansion of steam:** When water changes to steam, its volume increases rapidly. At a normal atmospheric pressure of 14.7 pounds per square inch, water boils at 212 degrees Fahrenheit and expands in volume by a factor of 1700 to 1. But when water is sprayed into hotter areas, the expansion ratio is much greater. This principle can be applied to good effect in fire fighting. The steam can occupy such a large volume that oxygen is expelled from the area and the fire may be smothered. The table below shows the approximate volume, in cubic feet, of 50 gallons of water converted to steam at the given temperatures, in degrees Fahrenheit.

T = Temperature	V = cubic feet of steam
212	10,000
400	12,500
500	14,100
800	17,500
1000	20,000

a. Make a linear model of volume V as a function of T.

b. If one fire is 100 degrees hotter than another, what is the increase in the volume of steam produced by 50 gallons of water?

c. Calculate $V(420)$ and explain in practical terms what your answer means.

d. At a certain fire, 50 gallons of water expanded to 14,200 cubic feet of steam. What was the temperature of the fire?

14. **Technological maturity versus use maturity:** There are a number of processes used industrially for separation (of salt from water, for example). Such a process has a *technological maturity*, which is the percentage of perfection that the process has reached. A separation process with a high technological maturity is a very well-developed process. It also has a *use maturity*, which is the percentage of total reasonable use. High use maturity indicates that the process is being used to near capacity. In 1987 Keller collected from a number of experts their estimation of technological and use maturities. The table below is adapted from those data.

Process	Technological maturity (%)	Use maturity (%)
Distillation	87	87
Gas absorption	81	76
Ion exchange	60	60
Crystallization	64	62
Electrical separation	24	13

a. Construct a linear model for use maturity as a function of technological maturity.

b. Explain in practical terms the meaning of the slope of the regression line.

c. Express, using functional notation, the use maturity of a process that has a technological maturity of 89%, and then estimate that value.

d. Solvent extraction has a technological maturity of 73% and a use maturity of 61%. Is solvent extraction being used more or less than would be expected from its technological development? How might this information affect an entrepreneur's decision whether to get into the business of selling solvent equipment to industry?

e. Construct a linear model for technological maturity as a function of use maturity.

15. **Whole crop weight versus rice weight:** A study by Horie[23] compared the dry weight B of brown rice with the dry weight W of the whole crop (including stems and roots). These data, in tons per hectare,[24] for the variety *Nipponbare* grown in various environmental conditions are partially presented in the table below.

Whole crop weight W	Rice weight B
6	1.8
11.1	3.2
13.7	3.7
14.9	4.3
17.6	5.2

a. Find an approximate linear model for B as a function of W.

b. Which sample might be considered to have a lower rice weight than expected from the whole crop weight?

c. How much additional rice weight can be expected from 1 ton per hectare of additional whole crop weight?

16. **Rice production in Asia:** Improved agricultural practices, including better utilization of fertilizers and use of improved plant varieties, have resulted in increased rice yields in Asia. The accompanying table shows the average yield[25] Y, in tons per hectare, as a function of the number of years t since 1950.

t = Years since 1950	Y = Average yield
5	1.9
10	1.8
20	2.4
25	2.5
35	3.3

a. One study used a linear function to approximate yield as a function of time. Find an approximate linear model for Y as a function of t.

b. Explain what the slope of the linear model tells you.

c. Use the model to calculate $Y(30)$, and explain in practical terms what your answer means.

d. What yield would the model estimate for 1990? (The actual yield was 3.6 tons per hectare.)

e. The 1993 report of the International Rice Research Institute predicted that rice requirements in Asia would grow exponentially. If the rice requirements in 2005 are predicted to be 4.92 tons per hectare, will rice production in 2005 meet those requirements?

17. **Energy cost of running:** Physiologists have studied the steady-state oxygen consumption (measured per unit of mass) in a running animal as a function of its velocity (i.e., its speed). They have determined that the relationship is approximately linear, at least over an appropriate range of velocities. The following table gives the velocity v, in kilometers per hour, and the oxygen consumption E, in milliliters of oxygen per gram per hour, for the rhea, a large, flightless South American bird.[26] (For comparison, 10 kilometers per hour is about 6.2 miles per hour.)

---

[23] T. Horie, "The effects of climatic variations on agriculture in Japan. 5: The effects on rice yields in Hokkaido." In M. L. Parry, T. R. Carter, and N. T. Konijn, eds. *The Impact of Climatic Variations on Agriculture, Vol. 1: Assessment in Cool Temperate and Cold Regions* (Dordrecht, The Netherlands: Kluwer Academic Publishers, 1988), 809–826.

[24] A hectare is about $2\frac{1}{2}$ acres.

[25] IRRI, 1991. *World Rice Statistics 1990* (Los Banos, The Philippines: International Rice Research Institute).

[26] The table is based on C. R. Taylor, R. Dmi'el, M. Fedak, and K. Schmidt-Nielsen, "Energetic cost of running and heat balance in a large bird, the rhea," *Am. J. Physiol.* **221** (1971), 597–601.

Velocity *v*	Oxygen consumption *E*
2	1.0
5	2.1
10	4.0
12	4.3

a. Find the equation of the regression line for *E* in terms of *v*.

b. The slope of the linear function giving oxygen consumption in terms of velocity is called the *cost of transport* for the animal, since it measures the energy required to move a unit mass by 1 unit distance. What is the cost of transport for the rhea?

c. Physiologists have determined the general approximate formula $C = 8.5W^{-0.40}$ for the cost of transport *C* of an animal weighing *W* grams. If the rhea weighs 22,000 grams, is its cost of transport from part b higher or lower than what the general formula would predict? Is the rhea a more or a less efficient runner than a typical animal its size?

d. What would your equation from part a lead you to estimate for the oxygen consumption of a rhea at rest? Would you expect that estimate to be higher or lower than the actual level of oxygen consumption of a rhea at rest?

 18. **Laboratory experiment:** This lab uses a motion detector and a calculator-based laboratory (CBL^{TM}) unit. The motion detector measures the distance from the detector to an object in front of it, while the CBL records the data. After collecting the data, the CBL unit sends the data to the calculator, which displays a graph of the distance recorded with respect to time. In this lab, we drop a basketball toward the motion detector and measure the height and speed of the ball during the drop. For a detailed description, go to

http://math.college.hmco.com/students

and then, from the book's website, go to the Ball Drop experiment.

 19. **Laboratory experiment:** This lab uses a pressure sensor, a temperature probe, and a calculator-based laboratory (CBL^{TM}) unit. In this experiment we will test the *pressure–temperature law* by varying the temperature of an air sample and measuring the resulting changes in pressure. For a detailed description, go to

http://math.college.hmco.com/students

and then, from the book's website, go to the Pressure and Temperature experiment.

 20. **Laboratory experiment:** This lab uses a motion detector and a calculator-based laboratory (CBL^{TM}) unit. The motion detector measures the distance from the detector to an object in front of it, while the CBL records the data. After collecting the data, the CBL unit sends the data to the calculator, which displays a graph of the distance recorded with respect to time. In this lab, we roll a basketball toward the motion detector and measure the position of the ball as a function of time. For a detailed description, go to

http://math.college.hmco.com/students

and then, from the book's website, go to the Rolling Basketball experiment.

21. **Research project:** For this project, you should collect the height and weight of at least five adult males. Combine these data with those of others in your class, and analyze the combined data by finding the slope of the regression line for weight as a function of height. Compare that answer with the rule of thumb from Exercise 11 of Section 3.2. For a detailed description, go to

http://math.college.hmco.com/students

and then, from the book's website, go to the Weight versus Height experiment.

3.5 *Systems of Equations*

Many physical problems can be described by a system of two equations in two unknowns, and often the desired information is found by *solving the system of equations*. As we shall see, this involves nothing more than finding the intersection of two lines, and we already know how to do that since we can find the intersection of *any* two graphs.

Graphical Solutions of Systems of Equations

To show the method, we look at a simple example. We have $900 to spend on the repair of a gravel drive. We want to make the repairs using a mix of coarse gravel priced at $28 per ton and fine gravel priced at $32 per ton. To make a good driving surface, we need 3 times as much fine gravel as coarse gravel. How much of each will our budget allow us to buy?

As in any *story problem*, the first step is to convert the words and sentences into symbols and equations. Let's begin by writing equations using the words *tons of coarse gravel* and *tons of fine gravel*. We have a $900 budget, and we spend $28 per ton for coarse gravel and $32 per ton for fine gravel:

$$28 \times \boxed{\text{Tons of coarse gravel}} + 32 \times \boxed{\text{Tons of fine gravel}} = 900. \quad (3.9)$$

We also know that we must use 3 times as much fine gravel as coarse. In other words, we multiply the amount of coarse gravel by 3 to get the amount of fine gravel:

$$\boxed{\text{Tons of fine gravel}} = 3 \times \boxed{\text{Tons of coarse gravel}}. \quad (3.10)$$

Now let's choose variable names:

$$c = \text{Amount of coarse gravel}$$

$$f = \text{Amount of fine gravel}.$$

We complete the process by putting these letter names into Equation (3.9) and Equation (3.10), and we arrive at a *system of equations* that we need to solve:

$$28c + 32f = 900$$

$$f = 3c.$$

To *solve the system* just means to find values of c and f that make *both equations true at the same time*. There are many ways to do this, but we can easily change this into a problem that we already know how to do with the calculator. The key step is first to solve each of the equations for one of the variables. Since the second equation is already solved for f, we do the same for the first equation, $28c + 32f = 900$:

$$28c + 32f = 900$$

$$32f = 900 - 28c$$

$$f = \frac{900 - 28c}{32}.$$

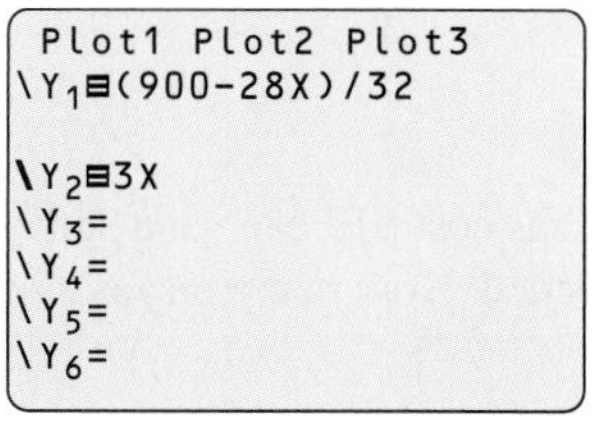

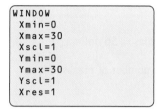

FIGURE 3.67 Entering functions for the purchase of gravel

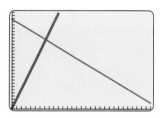

FIGURE 3.68 Configuring the graphing window

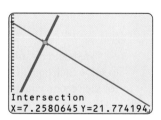

FIGURE 3.69 Graphing a system of equations for gravel

Thus we can replace the original system of equations by

$$f = \frac{900 - 28c}{32} = 3c.$$

We want to find where *both* of these equations are true. That is, we want to find where their graphs cross. Thus we can proceed using the crossing-graphs method for solving equations that we learned in Chapter 2.

Let's recall the procedure. The first step is to enter $\boxed{3.16}$ both equations as shown in Figure 3.67. We record the appropriate variable correspondences:

$$Y_1 = \frac{900 - 28c}{32}, \text{ first expression for } f \text{ on vertical axis}$$

$$Y_1 = 3c, \text{ second expression for } f \text{ on vertical axis}$$

$$X = c, \text{ coarse gravel on horizontal axis}.$$

A common-sense estimate can help us choose our graphing range. We are buying $900 worth of gravel that costs about $30 per ton, so we will certainly buy no more than about 30 tons of either type. Thus we choose a viewing window $\boxed{3.17}$ with a horizontal span from $c = 0$ to $c = 30$ and a vertical span from $f = 0$ to $f = 30$. The properly configured window is in Figure 3.68.

Now when we graph, we get the display $\boxed{3.18}$ in Figure 3.69, where the thin line is the graph of $f = (900 - 28c)/32$ and the thick line is the graph of $f = 3c$. The solution we seek is the crossing point, which we find $\boxed{3.19}$ as we did in Chapter 2. We see from Figure 3.70 that, rounded to two decimal places, we should buy $c = 7.26$ tons of coarse gravel and $f = 21.77$ tons of fine gravel.

FIGURE 3.70 The solution of the system

KEY IDEA 3.6

How to Solve a System of Two Equations in Two Unknowns

Step 1:   Solve both equations for one of the variables.

Step 2:   The solution of the system of equations is the intersection point of the two graphs. This is found using the crossing-graphs method from Chapter 2.

EXAMPLE 3.10   A Picnic

We have \$56 to spend on pizzas and sodas for a picnic. Pizzas cost \$12 each and sodas cost \$0.50 each. Four times as many sodas as pizzas are needed. How many pizzas and how many sodas will our budget allow us to buy?

Solution: The cost of the picnic can be written as

$$12 \times \text{Number of pizzas} + 0.50 \times \text{Number of sodas} = \text{Cost of picnic}.$$

Since the picnic is to cost \$56, we can rewrite this as

$$12 \times \text{Number of pizzas} + 0.50 \times \text{Number of sodas} = 56 \text{ dollars}.$$

If we use S for the number of sodas and P for the number of pizzas, this is

$$12P + 0.5S = 56.$$

We also know that we need four times as many sodas as pizzas. In other words, to get the number of sodas, we multiply the number of pizzas by 4:

$$\text{Number of sodas} = 4 \times \text{Number of pizzas}$$

In terms of the letters S and P, this is

$$S = 4P.$$

Thus we need to solve the system of equations

$$12P + 0.5S = 56$$
$$S = 4P.$$

We begin by solving the first equation for S. (Alternatively, it would be correct to solve each equation for P. We chose to solve for S since the work is already done for the second equation.) We have

$$0.5S = 56 - 12P$$
$$S = \frac{56 - 12P}{0.5}.$$

Now we need to use the crossing-graphs method to solve the system of equations

$$S = \frac{56 - 12P}{0.5} = 4P.$$

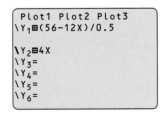

Plot1 Plot2 Plot3
\Y₁⊟(56−12X)/0.5

\Y₂⊟4X
\Y₃=
\Y₄=
\Y₅=
\Y₆=

FIGURE 3.71 Entering a system of equations for a picnic

We enter 3.20 both functions as shown in Figure 3.71 and record the appropriate variable correspondences:

$$\mathbf{Y_1} = S, \text{ sodas on vertical axis}$$

$$\mathbf{X} = P, \text{ pizzas on horizontal axis}.$$

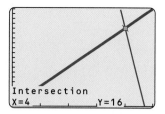

FIGURE 3.72 How many pizzas and sodas to buy

Since pizzas cost $12 each, we can't buy more than 5 of them. And that means we will buy no more than 20 sodas. Thus, to make the graph, we set [3.21] the horizontal span from $P = 0$ to $P = 5$ and the vertical span from $S = 0$ to $S = 20$. This configuration produces the graphs shown in Figure 3.72, where the thin line is the graph of $S = (56 - 12P)/0.5$, and the thick line is the graph of $S = 4P$.

We find the intersection point [3.22] using the calculator, and we see from Figure 3.72 that the solution is X=4 and Y=16. Since X corresponds to P and Y corresponds to S, we conclude that we can buy 4 pizzas and 16 sodas.   ■ ■ ■

An alternative algebraic solution of this system of equations is provided below.

Algebraic Solutions

We can, if we wish, solve systems of equations without using the calculator. Let's see how to do that with the system of equations for the picnic in Example 3.10:

$$12P + 0.5S = 56$$
$$S = 4P.$$

The first step is to solve one equation for one of the variables. In general, we can use either equation and either variable. Sometimes, as in this case, a wise choice can save work. Accordingly, we avoid unnecessary work by choosing to solve the second equation for S:

$$S = 4P.$$

Now we put the expression $4P$ in place of S in the first equation:

$$12P + 0.5(4P) = 56.$$

The result is that the variable S has been *eliminated*, and we are left with a linear equation that involves only the single variable P. We learned in Section 2.3 how to solve such equations. We get

$$12P + 0.5(4P) = 56$$
$$12P + 2P = 56$$
$$14P = 56$$
$$P = \frac{56}{14}$$
$$P = 4.$$

This gives us the solution for P. We put this value for P back into the equation $S = 4P$ to get the value for S:

$$S = 4P = 4 \times 4 = 16.$$

We get the solution $P = 4$ and $S = 16$, and this agrees with our earlier solution using the calculator.

How to Solve a System of Two Equations in Two Unknowns Algebraically

Step 1:   Solve one of the equations for one of the variables.

Step 2:   Put the expression you got in step 1 into the other equation.

Step 3:   Solve the resulting linear equation in one variable.

Step 4:   To get the other variable, put the value you got from step 3 into the expression you found in step 1.

ANOTHER LOOK

Solution by Elimination, Matrices

There is another method for solving systems of linear equations that is often useful. It is called *elimination* or *Gaussian elimination*, and we illustrate it first by looking at the following system of equations:

$$3x + 2y = 14$$
$$5x - 2y = 18.$$

Note that if we add the two equations together, the y's cancel out. We keep the first equation and replace the second by the sum of the two:

$$3x + 2y = 14$$
$$8x = 32.$$

Now, since $8x = 32$, we have that $x = 4$. We plug this value into the first equation to get y:

$$3 \times 4 + 2y = 14$$
$$2y = 2$$
$$y = 1.$$

Hence the solution is $x = 4$ and $y = 1$.

Sometimes it may be necessary to adjust one equation or the other before we can eliminate a variable. Consider, for example, the following system of equations:

$$2x + 4y = 16$$
$$3x - y = 3.$$

We cannot eliminate a variable by adding (or subtracting) one equation to (or from) another. But note that if we multiply the top equation by 3 and the bottom equation by 2, then $6x$ will appear as the first term in each equation:

$$6x + 12y = 48 \tag{3.11}$$

$$6x - 2y = 6. \tag{3.12}$$

Now we can subtract the equations to eliminate x. As before, we keep the first equation and replace the second by Equation (3.12) minus Equation (3.11):

$$6x + 12y = 48$$

$$-14y = -42.$$

This last equation gives immediately that $y = 3$. If we put this value back into the first equation, we get $x = 2$.

One of the most important things about the elimination method is that it allows us to handle systems of equations with many unknowns. Let's look at the following system of three equations in three unknowns:

$$x + y + z = 6 \tag{3.13}$$

$$2x + 3y - z = 5 \tag{3.14}$$

$$3x - y + z = 4. \tag{3.15}$$

The first step is to use Equation (3.13) to eliminate x from Equations (3.14) and (3.15). In abbreviated form, this is what we will do:

Replace Equation (3.14) by Equation (3.14) $- 2$ Equation (3.13).

Replace Equation (3.15) by Equation (3.15) $- 3$ Equation (3.13).

The result is

$$x + y + z = 6 \tag{3.16}$$

$$y - 3z = -7 \tag{3.17}$$

$$-4y - 2z = -14. \tag{3.18}$$

Next we use Equation (3.17) to eliminate y from Equation (3.18). The required step is

Replace Equation (3.18) by Equation (3.18) $+ 4$ Equation (3.17).

The result is

$$x + y + z = 6 \tag{3.19}$$

$$y - 3z = -7 \tag{3.20}$$

$$-14z = -42. \tag{3.21}$$

Now Equation (3.21) tells us that $z = 3$. We put this value into Equation (3.20) and find that $y = 2$. Finally, we put these two values into Equation (3.19) to get $x = 1$.

The method of elimination can be further refined by the use of an *augmented matrix*. We will illustrate what we have in mind by looking once more at the system of equations we solved above:

$$x + y + z = 6 \tag{3.22}$$

$$2x + 3y - z = 5 \tag{3.23}$$

$$3x - y + z = 4. \tag{3.24}$$

To get the augmented matrix of this system, we pick off the coefficients and arrange them in an array:

$$\begin{pmatrix} 1 & 1 & 1 & 6 \\ 2 & 3 & -1 & 5 \\ 3 & -1 & 1 & 4 \end{pmatrix}.$$

We allow the following *row operations* on the matrix, which correspond to operations on the system of equations:

Operation 1: Swap two rows. This corresponds to writing the system of equations in a different order.

Operation 2: Multiply a row by a nonzero number. This corresponds to multiplying an equation by that number. We will indicate the operation of multiplying the ith row by a with the notation aR_i.

Operation 3: Add a multiple of one row to another row. This corresponds to adding a multiple of one equation to another. We will indicate the operation of replacing the ith row by the ith row plus a times the jth row with the notation $R_i + aR_j$.

Eliminating variables from a system of equations corresponds to using row operations on the augmented matrix to make certain entries zero. We start in the upper left-hand corner of the matrix and clear out all entries beneath it. We illustrate this using our earlier example and indicate the required operations:

$$\begin{pmatrix} 1 & 1 & 1 & 6 \\ 0 & 1 & -3 & -7 \\ 0 & -4 & -2 & -14 \end{pmatrix} \quad \begin{matrix} \\ R_2 - 2R_1 \\ R_3 - 3R_1 \end{matrix}$$

Compare this with the elimination of the x variable from two of the equations above. Next we move to the next entry on the diagonal (in this case, the second row and second column) and clear out all entries below it:

$$\begin{pmatrix} 1 & 1 & 1 & 6 \\ 0 & 1 & -3 & -7 \\ 0 & 0 & -14 & -42 \end{pmatrix} \quad \begin{matrix} \\ \\ R_3 + 4R_2 \end{matrix}$$

This matrix corresponds to the system of equations

$$x + y + z = 6$$

$$y - 3z = -7$$

$$-14z = -42.$$

These are the same as Equations (3.19) through (3.21) above, and we can solve them immediately.

The use of the augmented matrix to solve a system of equations may at first seem artificial. But it allows us to program a computer to solve large systems of equations involving hundreds or even thousands of variables. Furthermore, a deeper study of matrices leads us to a central area of mathematics known as *linear algebra*.

Enrichment Exercises

E-1. **Using elimination to solve systems of equations:** Use the method of elimination to solve the following systems of equations.

a. $x + y = 5$
   $x - y = 1$

b. $2x - y = 0$
   $3x + 2y = 14$

c. $3x + 2y = 6$
   $4x - 3y = 8$

E-2. **Solving larger systems of equations:** Solve the following systems of equations.

a. $x + y + z = 3$
   $2x - y + 2z = 3$
   $3x + 3y - z = 9$

b. $x - y - z = -2$
   $3x - y + 3z = 8$
   $2x + 2y + z = 6$

c. $x + y + z + w = 8$
   $2x - y + z - w = -1$
   $3x + y - z + 2w = 9$
   $x + 2y + 2z + w = 12$

E-3. **Solving using the augmented matrix:** Solve each of the systems in Exercise E-2 by using the augmented matrix.

E-4. **Systems of equations with many solutions:** Consider the following system of equations:

$$x + y + z = 6$$
$$2x + y - z = 1$$
$$4x + 3y + z = 13.$$

a. Show that solution of the system of equations by use of the augmented matrix leads to the matrix

$$\begin{pmatrix} 1 & 1 & 1 & 6 \\ 0 & 1 & 3 & 11 \\ 0 & 0 & 0 & 0 \end{pmatrix}.$$

b. The augmented matrix in part a corresponds to the following system of equations:

$$x + y + z = 6$$
$$y + 3z = 11.$$

i. Solve the second equation for y to get a formula for y in terms of z. →

ii. Solve the first equation for x, and then use the formula for y from part i to get a formula for x in terms of z.

c. Show (by direct substitution) that for any value of z, the values of x and y found in part b give a solution to the original system of equations.

E-5. **A system of equations with no solution:** Show that the following system of equations has no solution:

$$x + y + z = 1$$
$$x + y - z = 2$$
$$x + y = 5.$$

E-6. **Determinants:** The *determinant* of the 2-by-2 matrix $\begin{pmatrix} a & b \\ c & d \end{pmatrix}$ is denoted by $\begin{vmatrix} a & b \\ c & d \end{vmatrix}$ and is defined by

$$\begin{vmatrix} a & b \\ c & d \end{vmatrix} = ad - bc.$$

Calculate the following determinants.

a. $\begin{vmatrix} 1 & 2 \\ 3 & 4 \end{vmatrix}$
b. $\begin{vmatrix} 2 & 2 \\ 6 & 5 \end{vmatrix}$

c. $\begin{vmatrix} 3 & 1 \\ 3 & 4 \end{vmatrix}$
d. $\begin{vmatrix} 1 & 2 \\ 1 & 2 \end{vmatrix}$

E-7. **Cramer's rule:** *This is a continuation of Exercise E-6. Cramer's rule gives a simple formula for the solution of a system of two equations in two unknowns. If*

$$\begin{vmatrix} a & b \\ c & d \end{vmatrix} \neq 0,$$

then the solution of

$$ax + by = e$$
$$cx + dy = f$$

is given by

$$x = \frac{\begin{vmatrix} e & b \\ f & d \end{vmatrix}}{\begin{vmatrix} a & b \\ c & d \end{vmatrix}} \quad \text{and} \quad y = \frac{\begin{vmatrix} a & e \\ c & f \end{vmatrix}}{\begin{vmatrix} a & b \\ c & d \end{vmatrix}}.$$

Use Cramer's rule to solve the following systems of equations.

a. $x + 2y = 5$
   $x - y = -1$

b. $x + y = 3$
   $x - y = 0$

c. $5x + 3y = 3$
   $2x + 3y = 4$

d. $6x - y = 5$
   $3x + 4y = 2$

3.5 SKILL BUILDING EXERCISES

S-1. An explanation: Explain why the solution of a system of two equations in two unknowns is the intersection point of the two graphs.

S-2. What is the solution? For a certain system of two linear equations in two unknowns, the graphs are distinct parallel lines. What can you conclude about the solution of the system of equations?

S-3. Crossing graphs: Solve using crossing graphs.

$$3x + 4y = 6$$
$$2x - 6y = 5$$

S-4. Crossing graphs: Solve using crossing graphs.

$$-7x + 21y = 79$$
$$13x + 17y = 6$$

S-5. Crossing graphs: Solve using crossing graphs.

$$0.7x + 5.3y = 6.6$$
$$5.2x + 2.2y = 1.7$$

S-6. Crossing graphs: Solve using crossing graphs.

$$-6.6x - 26.5y = 17.1$$
$$6.9x + 5.5y = 8.4$$

S-7. Hand calculation: Solve the system of equations in Exercise S-3 by hand calculation.

S-8. Hand calculation: Solve the system of equations in Exercise S-4 by hand calculation.

S-9. Hand calculation: Solve the system of equations in Exercise S-5 by hand calculation.

S-10. Hand calculation: Solve the system of equations in Exercise S-6 by hand calculation.

3.5 EXERCISES

1. **A party:** You have $36 to spend on refreshments for a party. Large bags of chips cost $2.00 and sodas cost $0.50. You need to buy five times as many sodas as bags of chips. How many bags of chips and how many sodas can you buy?

2. **Mixing feed:** A milling company wants to mix alfalfa, which contain 20% protein, and *wheat mids,*[27] which contain 15% protein, to make cattle feed.

   a. If you make a mixture of 30 pounds of alfalfa and 40 pounds of wheat mids, how many pounds of protein are in the mixture? (*Hint:* In the 30 pounds of alfalfa there are $30 \times 0.2 = 6$ pounds of protein.)

   b. Write a formula that gives the amount of protein in a mixture of a pounds of alfalfa and w pounds of wheat mids.

   c. Suppose the milling company wants to make 1000 pounds of cattle feed that contains 17% protein. How many pounds of alfalfa and how many pounds of wheat mids must be used?

3. **An order for bulbs:** You have space in your garden for 55 small flowering bulbs. Crocus bulbs cost $0.35 each and daffodil bulbs cost $0.75 each. Your budget allows you to spend $25.65 on bulbs. How many crocus bulbs and how many daffodil bulbs can you buy?

4. **American dollars and British pounds:** Assume that at the current exchange rate, the British pound is worth $2.66 in American dollars. In your wallet are some dollar bills and several British pound notes. There are 17 bills altogether, which have a total value of $30.28 in American dollars. How many American dollars and how many British pound notes do you have in your wallet?

5. **Population growth:** There are originally 255 foxes and 104 rabbits on a particular game reserve. The fox population grows at a rate of 33 foxes per year, and the rabbits increase at a rate of 53 rabbits per year. Under these conditions, how long does it take for the number of rabbits to catch up with the number

---

[27]When the wheat kernel is removed to produce flour, some parts of the whole wheat, including the bran, are often pressed into pellets known as *wheat mids* and sold as a feed ingredient.

of foxes? How many of each animal will be present at that time?

6. **Teacher salaries:** The table below shows the average salaries, in thousands of dollars, of elementary and secondary classroom teachers in public schools in the given year.[28]

Year	Elementary	Secondary
1994	35.2	36.6
1995	36.2	37.5
1996	37.3	38.4
1997	38.2	39.1
1998	39.1	39.9

On the basis of the regression lines, and assuming that the trends continue, when would you expect that the average salary for elementary teachers in public schools will be the same as that for secondary teachers?

7. **Competition between populations:** In this exercise we consider the problem of competition between two populations that vie for resources but do not prey on each other. Let m be the size of the first population, let n be the size of the second (both measured in thousands of animals), and assume that the populations coexist eventually. An example of one common model for the interaction is

Per capita growth rate for m is $3(1 - m - n)$

Per capita growth rate for n is $2(1 - 0.7m - 1.1n)$.

At an *equilibrium point* the per capita growth rates for m and for n are both zero. If the populations reach such a point, then they will continue at that size indefinitely. Find the equilibrium point in the example above.

8. **Market supply and demand:** The quantity of wheat, in billions of bushels, that wheat suppliers are willing to produce in a year and offer for sale is called the *quantity supplied* and is denoted by S. The quantity supplied is determined by the price P

of wheat, in dollars per bushel, and the relation is $P = 2.13S - 0.75$.

The quantity of wheat, in billions of bushels, that wheat consumers are willing to purchase in a year is called the *quantity demanded* and is denoted by D. The quantity demanded is also determined by the price P of wheat, and the relation is $P = 2.65 - 0.55D$.

At the *equilibrium price,* the quantity supplied and the quantity demanded are the same. Find the equilibrium price for wheat.

9. **Boron uptake:** Many factors influence a plant's uptake of boron from the soil, but one key factor is soil type. One experiment[29] compared plant content C of boron, in parts per million, with the amount B, in parts per million, of water-soluble boron in the soil. In *Decatur silty clay* the relation is given by $C = 33.78 + 37.5B$.

In *Hartsells fine sandy loam* the relation is given by $C = 31.22 + 71.17B$.

a. What amount of water-soluble boron available will result in the same plant content of boron for Decatur silty clay and Hartsells fine sandy loam? (If you choose to solve this problem graphically, we suggest a horizontal span of 0 to 0.5 for B.)

b. For available boron amounts larger than that found in part a, which of the two soil types results in the larger plant content of boron?

10. **Male and female high school graduates:** The table below shows the percentage of male and female high school graduates who enrolled in college within 12 months of graduation.[30]

Year	1960	1965	1970	1975
Males	54%	57.3%	55.2%	52.6%
Females	37.9%	45.3%	48.5%	49%

a. Find the equation of the regression line for percentage of male high school graduates entering college as a function of time.

---

[28] From the *1999 Statistical Abstract of the United States.*

[29] J. I. Wear and R. M. Patterson, "Effect of soil pH and texture on the availability of water-soluble boron in the soil," *SSSA Proc.* **26** (1962), 334–335.

[30] From the *1995 Statistical Abstract of the United States.*

b. Find the equation of the regression line for percentage of female high school graduates entering college as a function of time.

c. Assume that the regression lines you found in part a and part b represent trends in the data. If the trends persisted, when would you expect first to have seen the same percentage of female and male graduates entering college? (You may be interested to know that this actually occurred for the first time in 1980. The percentages fluctuated but remained very close during the 1980s. In the 1990s significantly more female graduates entered college than did males. In 1992, for example, the rate for males was 59.6% compared with 63.8% for females.)

11. **Fahrenheit and Celsius:** If you know the temperature C in degrees Celsius, you can find the temperature F in degrees Fahrenheit from the formula

$$F = \frac{9}{5}C + 32 .$$

At what temperature will a Fahrenheit thermometer read exactly twice as much as a Celsius thermometer?

12. **A bag of coins:** A bag contains 30 coins, some dimes and some quarters. The total amount of money in the bag is $3.45. How many dimes and how many quarters are in the bag?

13. **Parabolic mirrors:** Reflector telescopes use *parabolic mirrors* because they have the shape that will reflect incoming light to a single point, the *focal point* (see Figure 3.73). The graph of $y = x^2$ has the shape of a parabola, for example. Light traveling vertically down (parallel to the y-axis) encounters the parabola and is reflected according to the law that the angle of incidence equals the angle of reflection.

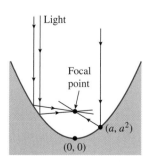

FIGURE 3.73

This law implies that vertical light rays encountering the parabola at a point (a, a^2) will be reflected along the line

$$4ay + (1 - 4a^2)x = a .$$

a. Find where the light rays reflected from the points $(2, 4)$ and $(3, 9)$ meet.

b. The point you found in part a is the focal point. Show that every vertical light ray that is reflected from the mirror passes through this point.

14. **An interesting system of equations:** What happens when you try to solve the following system of equations? Can you explain what is going on?

$$x + y = 1$$
$$x + y = 2$$

15. **Another interesting system of equations:** What happens when you try to solve the following system of equations? Can you explain what is going on?

$$x + 2y = 3$$
$$-2x - 4y = -6$$

16. **A system of three equations in three unknowns:** Consider the following system of three equations in three unknowns.

$$2x - y + z = 3$$
$$x + y + 2z = 9$$
$$3x + 2y - z = 4$$

a. Solve the first equation for z.

b. Put the solution you got in part a for z into *both* the second and third equations.

c. Solve the system of two equations in two unknowns that you found in part b.

d. Write the solution of the original system of three equations in three unknowns.

17. **An application of three equations in three unknowns:** A bag of coins contains nickels, dimes, and quarters. There are a total of 21 coins in the bag, and the total amount of money in the bag is $3.35. There is one more dime than there are nickels. How many dimes, nickels, and quarters are in the bag?

Summary

One of the simplest and most important geometric objects is the straight line, and the line is intimately tied to the idea of a *linear function*. This is a function with a constant rate of change, and the constant is known as the *slope*.

The Geometry of Lines

An important characterization of a straight line such as that determined by a roof is that it rises or falls at the same rate everywhere. This rate is known as the *slope* of the line and is commonly denoted by the letter m. Critical relationships involving the slope are summarized below.

Calculating the slope: The slope m can be calculated using

$$m = \frac{\text{Vertical change}}{\text{Horizontal change}}.$$

Fundamental property of the slope: The slope of a line tells how steeply it is increasing or decreasing.

- If m is positive, the line is rising.
- If $m = 0$, the line is horizontal.
- If m is negative, the line is falling.

Using the slope: The slope m can be used to calculate vertical change.

$$\text{Vertical change} = m \times \text{Horizontal change}.$$

Linear Functions

A linear function is one with a constant rate of change. As with lines, this is known as the slope and is denoted by m. Linear functions can be written in a special form:

$$y = mx + b,$$

where m is the slope and b is the *initial value*. The graph of a linear function is a straight line. The slope of the linear function is the same as the slope of the line, and the initial value of the function is the same as the vertical intercept of the line.

The slope of a linear function is used much like the slope of a line.

- The slope m can be calculated from the change in y corresponding to a given change in x:

$$m = \frac{\text{Change in function}}{\text{Change in variable}} = \frac{\text{Change in } y}{\text{Change in } x}.$$

- The slope can be used to calculate the change in y resulting from a given change in x:

$$\text{Change in } y = m \times \text{Change in } x.$$

For example, the income tax liability T is, over a suitably restricted range of incomes, a linear function of the adjusted gross income I. Suppose the tax table shows a tax liability of $T = 780$ dollars when the adjusted gross income is $I = 15,000$ dollars and shows a tax liability of $825 for an adjusted gross income of $15,500.

- We can calculate the slope of the linear function $T = T(I)$ as follows:

$$m = \frac{\text{Change in tax}}{\text{Change in income}} = \frac{825 - 780}{15,500 - 15,000} = \frac{45}{500} = 0.09 \,.$$

This means that a taxpayer in this range could expect to pay 9 cents tax on each additional dollar earned. Economists would term this the *marginal tax rate*.

- Once the slope, or marginal tax rate, is known, one can calculate the tax liability for other incomes. If the adjusted gross income is $15,350, the additional tax over that owed on $15,000 is given by

$$\text{Additional tax} = m \times \text{Additional income} = 0.09 \times 350 = 31.50 \text{ dollars}\,.$$

Thus the total tax liability is $780 + 3150 = 811.50$ dollars.

Modeling Data with Linear Functions

It is appropriate to model observed data with a linear function, provided that the data show a constant rate of change. When the data for the variable are evenly spaced, this is always evidenced by a constant change in function values. For example, if a rock is dropped and its velocity is recorded each second, one might collect the following data:

$t =$ seconds	0	1	2	3	4	5
$V =$ feet per second	0	32	64	96	128	160

To see whether it is appropriate to model velocity with a linear function, we look at successive differences:

Change in t	From 0 to 1	From 1 to 2	From 2 to 3	From 3 to 4	From 4 to 5
Change in V	32	32	32	32	32

Since the change in V is always the same, we conclude that V is a linear function of t. We can get a formula for this linear function by first calculating the slope:

$$m = \frac{\text{Change in } V}{\text{Change in } t} = \frac{32}{1} = 32\,,$$

so then

$$V = \text{Slope} \times t + \text{Initial value} = 32t + 0 = 32t.$$

Linear Regression

Experimentally gathered data are always subject to error, and rarely will data points show an exactly constant rate of change. When appropriate, we can still model the data with a linear function by using *linear regression*. Linear regression gives the line with the least possible total of the squares of the vertical distances from the line to the data points. It is by far the most commonly used method of approximating linear data, and it is nearly always accomplished with a computer or calculator.

The education expenditures $E = E(t)$ of the federal government in billions of dollars are given by the following table.

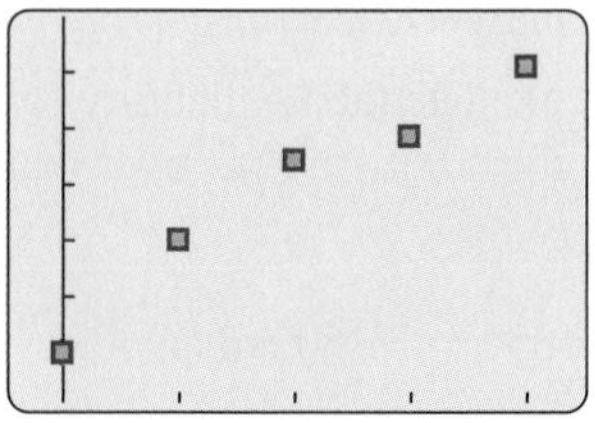

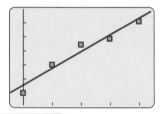

FIGURE 3.74 Educational spending

t = years since 1985	0	1	2	3	4
E = billions spent	27.0	29.0	30.4	30.9	32.1

The data table does not show a constant rate of change, but the plot of the data in Figure 3.74 shows that the data points nearly fall on a straight line. If we wish to model this with a linear function, we use the calculator to get the regression line $E = 1.21t + 27.46$. In Figure 3.75 we have added the regression line to the data points to show how the approximate model works.

The regression line shows the linear trend, if any exists, exhibited by observed data. Extreme caution should be exercised in any use of the regression line beyond the range of observed data points.

FIGURE 3.75 Regression line model for educational spending

Systems of Equations

Many physical problems are described by a system of two linear equations in two unknowns, and often the needed information is obtained by *solving the system of equations*. This can be accomplished either graphically or by hand calculation. The graphical solution is the easier way and fits nicely with the spirit of this course. The procedure is as follows.

Step 1: Solve both equations for one of the variables.

Step 2: The solution of the system is the intersection point of the two graphs. This is found using the crossing-graphs method.

For example, the purchase of c tons of coarse gravel and f tons of fine gravel might be described by the following system of equations:

$$28c + 32f = 900$$

$$f = 3c.$$

We solve the first equation for f to obtain the equivalent system

$$f = \frac{900 - 28c}{32}$$

$$f = 3c.$$

If we graph each of these functions and find the crossing point, we discover that we should buy $c = 7.26$ tons of coarse gravel and $f = 21.77$ tons of fine gravel.

4

Exponential Functions

Exponential functions occur almost as often as linear functions in nature, science, and mathematics. We have already encountered examples of exponential functions in looking at population growth, radioactive decay, free fall subject to air resistance, and interest on bank loans. We will now look more closely at the structure of such functions and show where the formulas we presented in those applications originated.

4.1 *Exponential Growth and Decay*

Recall that a linear function $y = y(x)$ with slope m is one that has a constant rate of change m. Another way of saying this is that a linear function changes by constant sums of m. That is, when x is increased by 1, we get the new value of y by adding m to the old y value. In contrast, an *exponential function* $N = N(t)$ with *base a* is one that changes by constant *multiples* of a. That is, when t is increased by 1, we get the new value of N by multiplying the current N value by a.

Exponential Growth

Exponential functions occur naturally in some population growth analyses, and we look there to provide an illustration. The simplest population studies involve organisms such as bacteria that reproduce by cell division. Consider, for example, an experiment where there are initially 3000 bacteria in a petri dish, and the population doubles each hour. We let $N = N(t)$ denote the number of bacteria present after t hours. The number of bacteria present when the experiment began is $N(0) = 3000$. This is the *initial value* of the function. It is the starting value of N, or the value of N when $t = 0$. During the first hour of the experiment the number of bacteria doubles, so after 1 hour there are $N(1) = 2 \times 3000 = 6000$ bacteria present. During the next hour the population doubles again, so there are $N(2) = 2 \times 6000 = 12,000$ bacteria present after 2 hours. In general we can get the population N at 1 hour in the future by multiplying the current N value by 2. This means that N is an exponential function with base 2. When the base of an exponential function is greater than 1, we will sometimes refer to the base as the *growth factor*. In this case we would say that the bacteria exhibit *exponential growth* with an hourly growth factor of 2 and an initial value of 3000.

If we calculate several values of the function N for bacteria that was described above and write the answer in a special form, we can see how to get a formula for N:

$$N(0) = 3000 = 3000 \times 2^0$$
$$N(1) = 3000 \times 2 = 3000 \times 2^1$$
$$N(2) = 3000 \times 2 \times 2 = 3000 \times 2^2$$
$$N(3) = 3000 \times 2 \times 2 \times 2 = 3000 \times 2^3$$
$$N(4) = 3000 \times 2 \times 2 \times 2 \times 2 = 3000 \times 2^4.$$

The pattern should be evident. To get the population N after t hours, we multiply the initial amount by the hourly growth factor 2 a total of t times, and that is the same as multiplying the initial population by 2^t:

$$N = 3000 \times 2^t.$$

This formula makes it easy to calculate the number of bacteria present at any time t. For example, after 6 hours there are $N(6) = 3000 \times 2^6 = 192,000$ bacteria present. Or after $1\frac{1}{2}$ hours there are $N(1.5) = 3000 \times 2^{1.5} = 8485$ bacteria present. (Here we have rounded our answer to the nearest integer because we don't expect to see parts of bacteria in the petri dish.)

The formula $N = 3000 \times 2^t$ that we obtained is typical of exponential functions in general. The formula for an exponential function $N = N(t)$ with base a and initial value P is

$$N = Pa^t.$$

Exponential Decay

Let's look at the bacteria experiment again, but this time let's suppose that an antibiotic has been introduced into the petri dish, so that rather than growing in number, each hour half of the bacteria die. Under these conditions there will be only 1500 bacteria left after 1 hour. After another hour, there will be 750 bacteria left. In general, we can get the number of bacteria present 1 hour in the future by multiplying the present population by $\frac{1}{2}$. Thus, under these conditions, N is an exponential function with base $\frac{1}{2}$. Because the population is decreasing rather than increasing, we would say that the population exhibits *exponential decay* with an hourly *decay factor* of $\frac{1}{2}$. Just as we did in the case where the population was growing, we can show a pattern that will lead us to a formula for $N = N(t)$:

$$N(0) = 3000 = 3000 \times \left(\frac{1}{2}\right)^0$$

$$N(1) = 3000 \times \frac{1}{2} = 3000 \times \left(\frac{1}{2}\right)^1$$

$$N(2) = 3000 \times \frac{1}{2} \times \frac{1}{2} = 3000 \times \left(\frac{1}{2}\right)^2$$

$$N(3) = 3000 \times \frac{1}{2} \times \frac{1}{2} \times \frac{1}{2} = 3000 \times \left(\frac{1}{2}\right)^3$$

$$N(4) = 3000 \times \frac{1}{2} \times \frac{1}{2} \times \frac{1}{2} \times \frac{1}{2} = 3000 \times \left(\frac{1}{2}\right)^4.$$

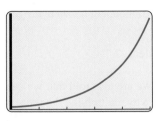

FIGURE 4.1 Exponential growth

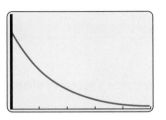

FIGURE 4.2 Exponential decay

In general, we see that $N = 3000 \times \left(\frac{1}{2}\right)^t$. This is an exponential function with initial value 3000 and base, or hourly decay factor, $\frac{1}{2}$.

In Figure 4.1 we have made the graph of the exponential function 3000×2^t. A table of values led us to choose a viewing window 4.1 with a horizontal span of $t = 0$ to $t = 5$ and a vertical span of $N = 0$ to $N = 100{,}000$. The graph is increasing and concave up, and this is typical of exponential growth. Growth is slow at first, but it later becomes more rapid. In Figure 4.2 we have graphed $3000 \times \left(\frac{1}{2}\right)^t$ using 4.2 a horizontal span of $t = 0$ to $t = 5$ and a vertical viewing span of $N = 0$ to $N = 3500$. This graph is decreasing and concave up, and it shows typical behavior for exponential decay. Decay is initially rapid, but the decay rate later slows.

KEY IDEA 4.1

Exponential Functions

A function $N = N(t)$ is exponential with base a if N changes in constant multiples of a. That is, if t is increased by 1, the new value of N is found by multiplying by a.

1. The formula for an exponential function with base a and initial value P is

$$N = Pa^t.$$

2. If $a > 1$, then N shows exponential growth with *growth factor a*. The graph of N will be similar in shape to that in Figure 4.1.

3. If $a < 1$, then N shows exponential decay with *decay factor a*. The graph of N will be similar in shape to that in Figure 4.2. The limiting value of such a function is 0.

EXAMPLE 4.1   Radioactive Decay

Radioactive substances decay over time, and the rate of decay depends on the element. If, for example, there are G grams of heavy hydrogen H_3 in a container, then as a result of radioactive decay, 1 year later there will be $0.783G$ grams of heavy hydrogen left. Suppose we begin with 50 grams of heavy hydrogen.

1. Use a formula to express the number G of grams left after t years as an exponential function of t. Identify the base or yearly decay factor and the initial value.

2. How much heavy hydrogen is left after 5 years?

3. Plot the graph of $G = G(t)$ and describe in words how the amount of heavy hydrogen that is present changes with time.

4. How long will it take for half of the heavy hydrogen to decay?

Solution to Part 1:   The initial value, 50 grams, is given. We are also told that we can get the amount G of heavy hydrogen 1 year in the future by multiplying the present amount by 0.783. Thus $G = G(t)$ is an exponential function with initial value 50 grams and base, or yearly decay factor, 0.783. We conclude that the formula for G is given by

$$G = \text{Initial value} \times (\text{Yearly decay factor})^t$$

$$G = 50 \times 0.783^t.$$

Solution to Part 2:   To get the amount of heavy hydrogen left after 5 years, we use the formula above to calculate $G(5)$:

$$G(5) = 50 \times 0.783^5 = 14.72 \text{ grams}.$$

Solution to Part 3:   We made the graph shown in Figure 4.3 using a horizontal span of $t = 0$ to $t = 10$ and a vertical span of $G = 0$ to $G = 60$. You may wish to look at a table of values to see why we chose these settings. We note that in this graph the correspondences are

$$X = t, \text{ years on horizontal axis}$$

$$Y = G, \text{ grams left on vertical axis}.$$

The features of this graph are typical of exponential decay. The amount of heavy hydrogen decreases rapidly over the first few years, but after that the rate of decay slows.

Solution to Part 4:   We want to know when there will be 25 grams of heavy hydrogen left. That is, we want to solve the equation

$$50 \times 0.783^t = 25$$

for t. We can do this using the crossing-graphs method. In Figure 4.4 we have added the graph of $G = 25$ and then used $\boxed{4.3}$ the calculator to get the intersection point

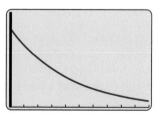

FIGURE 4.3 The decay of heavy hydrogen

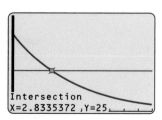

Intersection
X=2.8335372 ,Y=25

FIGURE 4.4 The half-life of heavy hydrogen

shown in Figure 4.4. We see that there will be 25 grams of heavy hydrogen left after about 2.83 years. This number is known as the *half-life* of heavy hydrogen because it tells how long it takes for half of it to decay. It is a fact that the half-life of a radioactive substance does not depend on the initial amount. Thus we would have gotten the same answer, 2.83 years, if we had started with 1000 grams of heavy hydrogen and asked how long it took until only 500 grams were left. See Exercise 7 at the end of this section. ■ ■ ■

Constant Proportional Change

The growth or decay factor is a crucial bit of information for an exponential function, but in practice its value is rarely given directly. One way in which exponential functions are commonly described involves percentage, or proportional, growth or decay. For example, the first census of the United States in 1790 showed a resident population of 3.93 million people. For certain populations, it is reasonable to expect that each year the population will grow by some fixed percentage. At least from 1790 through 1860, when the pattern was disrupted by the Civil War, this occurred in the United States, and the population grew by about 3% each year. That means that by the end of each year the population had grown to 103% of its value at the beginning of the year. Or if $U = U(t)$ represents the U.S. population, in millions, t years since 1790, the value of U at 1 year in the future can be found by multiplying the current value of U by 1.03. Thus U is an exponential function with initial value 3.93 and yearly growth factor 1.03. This observation enables us to write the formula for U:

$$U = \text{ Initial value} \times (\text{Yearly growth factor})^t$$
$$U = 3.93 \times 1.03^t.$$

In this example of an exponential function, the population is always changing by the same percentage. This is in fact an alternative characterization of exponential functions: The percentage or proportional growth (or decay) rate is always the same. For populations, percentage rate of change is often called *per capita rate of change*. Note in this example that we get the yearly growth factor 1.03 by adding 1 to the decimal form 0.03 of the yearly percentage growth rate: $1.03 = 1 + 0.03$. This makes sense: The number 0.03 represents the growth that occurs over a year, and the number 1 represents the population already in place at the start of the year.

Let's examine this in the case of exponential decay. Suppose that because of disease or famine, the population in a mythical medieval land decreased at a rate of 3% per year. That means that the population at the end of each year would be 97% of its value at the beginning of the year. Thus this is an exponential function with yearly decay factor 0.97 and yearly percentage decay rate 0.03. In this case we get the decay factor 0.97 by subtracting the percentage decay rate 0.03 from 1, that is $0.97 = 1 - 0.03$.

If we also know that the initial size of this medieval population was 250,000, and if we let $M = M(t)$ be the population after t years, we have collected the information we need to write a formula for M:

$$M = \text{ Initial value} \times (\text{Yearly decay factor})^t$$
$$M = 250{,}000 \times 0.97^t.$$

The relationships between growth or decay factors and percentage rates of change that we found in the two examples above are typical.

Alternative Characterization of Exponential Functions

A function is exponential if it shows constant percentage (or proportional) growth or decay.

1. For an exponential function with discrete (yearly, monthly, etc.) percentage growth rate r as a decimal, the growth factor is $a = 1 + r$.

2. For an exponential function with discrete percentage decay rate r as a decimal, the decay factor is $a = 1 - r$.

EXAMPLE 4.2   **Compound Interest**

A credit card holder begins the year owing $395.00 to a bank card. The bank card charges 1.2% interest on the outstanding balance each month. For purposes of this exercise, assume that no additional payments or charges are made and that no additional service charges are levied. Let $B = B(t)$ denote the balance of the account t months after January 1.

1. Explain why B is an exponential function of t. Identify the monthly percentage growth rate, the monthly growth factor, and the initial value. Write a formula for $B = B(t)$.

2. How much is owed after 7 months?

3. Assume that there is a limit of $450 on the card and that the bank will demand a payment the first month this limit is exceeded. How long is it before a payment is required?

Solution to Part 1:   For each dollar that is owed at the beginning of the month, an additional 0.012 dollar will be owed at the end of the month. Thus for each dollar owed at the beginning of the month, $1.012 will be owed at the end of the month. That means the value of B at the end of the month can be found by multiplying the balance at the beginning of the month by 1.012. Therefore, B is an exponential function with monthly growth factor 1.012. The monthly percentage growth rate is the monthly interest rate, 1.2%. The initial value is $395. Thus B is given by the formula

$$B = \text{Initial value} \times (\text{Monthly growth factor})^t$$
$$B = 395 \times 1.012^t.$$

Solution to Part 2: To calculate the balance after 7 months, we put $t = 7$ into the formula we found in part 1:

$$B(7) = 395 \times 1.012^7 = 429.40 \text{ dollars}.$$

Solution to Part 3: To find when the credit limit is reached, we need to solve

$$395 \times 1.012^t = 450$$

X	Y$_1$	
6	424.31	
7	429.4	
8	434.55	
9	439.77	
10	445.04	
11	**450.38**	
12	455.79	

Y$_1$=450.383771297

FIGURE 4.5 A table of values for a bank card

for t. We could solve this using the crossing-graphs method, but in this case it is easier to make a table of values for $B(t)$ as we have done in Figure 4.5. The table shows that the credit limit will be exceeded in 11 months. ■ ■ ■

Unit Conversion

Sometimes it is important to represent exponential functions in units that are not the same as those that are given. For example, we noted earlier that from 1790 to 1860, the U.S. population grew exponentially with yearly growth factor 1.03. (That is, it grew by 3% per year.) If we are interested in modeling the population as it changed each decade, we want to know the *decade* growth factor. Since the yearly growth factor is 1.03, in 10 years we would multiply by 1.03 a total of 10 times—that is, by 1.03^{10}:

Decade growth factor = (Yearly growth factor)^{10}

Decade growth factor = $1.03^{10} = 1.344$, rounded to three decimal places.

Thus, if we want the population at the end of a decade, we multiply the population at the beginning of the decade by 1.344. This is the decade growth factor, and if we use d to denote decades since 1790, we have the information we need to write a formula for our exponential function:

U = Initial value × (Decade growth factor)^d

$U = 3.93 \times 1.344^d.$

Let's turn the problem around. Census data are normally collected each decade, and rather than being given information on a yearly basis, we might have been told that from 1790 to 1860 the U.S. population grew by 34.4% per decade. Then the *decade* growth factor would be 1.344. If we wanted to express the exponential function in terms of years, we would need to recover the yearly growth factor from the decade growth factor. Since the population grows by a factor of 1.344 each decade, in d decades it will grow by a factor of 1.344^d. In particular, in 1 year, which is one-tenth of a decade, it will grow by a factor of

Yearly growth factor = (Decade growth factor)$^{1/10}$

Yearly growth factor = $1.344^{1/10} = 1.03$, rounded to two decimal places.

Thus we have recovered the yearly growth factor of 1.03 and can conclude that the population grew by 3% per year.

The reasoning we have used above applies in general.

Unit Conversion for Growth Factors

If the growth or decay factor for 1 period of time is a, then the growth or decay factor A for k periods of time is given by

$$A = a^k.$$

This relationship may be illustrated with the following diagram.

GROWTH FACTOR FOR ONE PERIOD OF TIME	Raise to Kth Power ⟶ ⟵ Raise to 1/Kth Power	GROWTH FACTOR FOR K PERIODS OF TIME

EXAMPLE 4.3 Getting the Decay Factor from the Half-life

It is standard practice to give the rate at which a radioactive substance decays in terms of its *half-life*. That is the amount of time it takes for half of the substance to decay. The half-life of carbon 14 is 5770 years.

1. If you start with 1 gram of carbon 14, how long will it take for only $\frac{1}{4}$ gram to remain?

2. What is the yearly decay factor rounded to five decimal places?

3. Find the yearly percentage decay rate and explain what it means in practical terms.

4. Assuming that we start with 25 grams of carbon 14, find a formula that gives the amount of carbon 14 left after t years.

Solution to Part 1: To say that the half-life is 5770 years means that, no matter how much carbon 14 is originally in place, at the end of 5770 years, half of that amount will remain. If we start with 1 gram, in 5770 years $\frac{1}{2}$ gram will remain. In 5770 more years, there will be half of that amount, $\frac{1}{4}$ gram, remaining. It takes two half-lives, or $5770 + 5770 = 11{,}540$ years, for 1 gram of carbon 14 to decay to $\frac{1}{4}$ gram of carbon 14.

Solution to Part 2: In one half-life, carbon 14 decays by a factor of $\frac{1}{2}$. Thus in h half-lives, it will decay by a factor of

$$\left(\frac{1}{2}\right)^h.$$

Since 1 year is $\frac{1}{5770}$ half-life, in 1 year carbon 14 will decay by a factor of

$$\text{Yearly decay factor} = (\text{Half-life decay factor})^{1/5770}$$

$$\text{Yearly decay factor} = \left(\frac{1}{2}\right)^{1/5770} = 0.99988, \text{ rounded to five decimal places}.$$

Solution to Part 3: The yearly decay factor is 0.99988. Thus the yearly percentage decay rate is $1 - 0.99988 = 0.00012$. This means that each year the amount of carbon 14 is reduced by 0.012%.

Solution to Part 4: The initial value is 25, and from part 2 we know that the yearly decay factor is 0.99988. Thus if we let $C = C(t)$ be the amount, in grams, of carbon 14 left after t years, then

$$C = \text{Initial value} \times (\text{Yearly decay factor})^t$$

$$C = 25 \times 0.99988^t.$$

Exponential Functions and Daily Experience

Exponential functions are common in daily news reports, and it is important that citizens spot them and understand their significance. Economic reports are often driven by exponential functions, and inflation is a prime example. When it is reported that the inflation rate is 3% per year, the meaning is that prices are increasing at a rate of 3% each year, and therefore prices are growing exponentially. It might be reported as good economic news that "Last year's 5% inflation rate is now down to 4%." Although this should indeed be encouraging, the lower rate does not change the fact that we are still dealing with an exponential function whose graph will eventually get very steep. The reduction from 5% to 4% changes the growth factor from 1.05 to 1.04 and so delays our reaching the steep part of the curve, but, as is shown in Figure 4.6, if the price function remains exponential in nature, then over the long term, catastrophic price increases will occur.

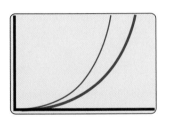

FIGURE 4.6 5% inflation (thin graph) versus 4% inflation (thick graph)

Short-term population growth is another phenomenon that is, in many settings, exponential in nature. It is the shape of the exponential curve, not necessarily the exact exponential formula, that causes concerns about eventual overcrowding. Many environmental issues related to human and animal populations are inextricably tied to exponential functions. It may be relatively inexpensive to dispose properly of large amounts of materials at a waste cleanup site, but further cleanup may be much more expensive. That is, as we begin the cleanup process, the amount of objectionable material remaining to be dealt with may be a decreasing exponential function. Since certain toxic substances are dangerous even in minute quantities, it can be very expensive to reduce them to safe levels. Once again, it is the nature of exponential decay, not the exact formula, that contributes to the astronomical expense of environmental cleanup.

Whenever phenomena are described in terms of percentage change, they may be modeled by exponential functions, and understanding how exponential functions behave is the key to understanding their true behavior. One of the most important features of exponential phenomena is that they may change at one rate over a period of time but change at a much different rate later on. Exponential growth may be modest for a time, but eventually the function will increase at a dramatic rate. Similarly, exponential decay may show encouragingly rapid progress at first, but such rates of decrease cannot continue indefinitely.

Elementary Properties of Exponents

We will look at exponential functions from a more algebraic point of view here. First we recall the following basic laws of exponents, where we assume that a is positive.

Product law: $a^b a^c = a^{b+c}$

Quotient law: $\dfrac{a^b}{a^c} = a^{b-c}$

Power law: $\left(a^b\right)^c = a^{bc}$

These rules can be used to verify the basic properties of exponential functions. For example, let's verify the fundamental property that exponential functions change by constant multiples. When t is increased by 1, we may consider $N = Pa^t$ as the old value of an exponential function and Pa^{t+1} as the new value. Now

$$\text{New value} = Pa^{t+1} = \left(Pa^t\right)a^1 = \text{Old value} \times a.$$

Similarly, we can use these rules to verify how unit conversion works. Let's look first at a specific example. Suppose that in the formula $N = Pa^t$, the variable t represents time measured in years. Then we could more properly write the formula as

$$N = Pa^{t \text{ years}}.$$

Suppose now that we wish to change units to decades. Since d decades is $10d$ years, we have

$$Pa^{10d \text{ years}} = P\left(a^{10}\right)^{d \text{ decades}}.$$

Thus the decade growth factor is the yearly growth factor raised to the tenth power.

The same idea works in general. Suppose we are measuring t in some unit—let's call it the old unit—for which the growth factor is a. Then

$$N = Pa^{t \text{ old units}}.$$

Suppose we wish to measure using a new unit that is k old units. Then

$$n \text{ new units} = nk \text{ old units}.$$

Hence

$$Pa^{nk \text{ old units}} = P\left(a^k\right)^{n \text{ new units}}.$$

Thus we get the new unit growth factor by raising the old unit growth factor to the kth power.

It is easy to write down the formula for an exponential function if we know its initial value and growth factor. But we can also find the formula from different information. For example, suppose we know that an exponential function N has growth factor 1.4 and that $N(2) = 5$. Then we know that $N = P \times 1.4^t$ and we want to find P. Now since $N(2) = 5$, we have

$$5 = P \times 1.4^2.$$

We can easily solve this equation for P:

$$P = \frac{5}{1.4^2} = 2.55 \,.$$

Thus $N = 2.55 \times 1.4^t$.

Before looking at how we find exponential functions if we are given two points, we want to note that the power law for exponents enables us to solve certain types of exponential equations. For example, if we know that $a^3 = 7$, then we can find the value of a by raising each side of the equation to the $\frac{1}{3}$ power:

$$a^3 = 7$$
$$\left(a^3\right)^{1/3} = 7^{1/3}$$
$$a^{3(1/3)} = 1.91$$
$$a = 1.91 \,.$$

Now suppose we know that N is an exponential function and that $N(2) = 5$ and $N(6) = 12$. Since N is exponential, it has the form Pa^t. The two bits of information we have tell us that

$$5 = Pa^2$$
$$12 = Pa^6.$$

If we divide the bottom equation by the top, note that the P cancels out:

$$\frac{12}{5} = \frac{Pa^6}{Pa^2} = a^{6-2} = a^4.$$

Solving as above, we have

$$a = \left(\frac{12}{5}\right)^{1/4} = 1.24 \,.$$

We put this value of a into the equation $5 = Pa^2$:

$$5 = P \times 1.24^2$$
$$P = \frac{5}{1.24^2}$$
$$P = 3.25 \,.$$

Thus the formula for N is 3.25×1.24^t.

Enrichment Exercises

E-1. **Practice with exponents:** Use the laws of exponents to simplify the following expressions.

a. $\dfrac{a^3 b^2}{a^2 b^3}$

b. $\left(\left(a^2 \right)^3 \right)^4$

c. $a^3 b^2 a^4 b^{-1}$

E-2. **Solving exponential equations:** Solve the following equations for the variable a, assuming $a \neq 0$.

a. $5a^3 = 7$

b. $6a^4 = 2a^2$

c. $a^5 b^2 = ab$ (assuming $b \neq 0$)

d. $a^t = ab$ (assuming $t \neq 1$)

E-3. **Finding exponential functions:**

a. Find an exponential function N with growth factor 6 if $N(2) = 7$.

b. Find an exponential function N such that $N(3) = 4$ and $N(7) = 8$.

c. Find an exponential function N with growth factor a if $N(2) = 3$.

d. Find an exponential function N if $N(2) = m$ and $N(4) = n$. Assume that m and n are positive.

E-4. **Substitution:** Solve the following equations by first making the suggested substitution.

a. $x^4 - 3x^2 + 2 = 0$. Substitute $y = x^2$. You should get a quadratic you can solve by factoring.

b. $x^6 - 5x^3 + 6 = 0$. Substitute $y = x^3$.

c. $x^{10} - 7x^5 + 12 = 0$. Substitute $y = x^5$.

E-5. **Finding initial value:** Suppose that $N = N(t)$ is an exponential function with growth factor 1.94 and $N(5) = 6$. Find a formula for N.

E-6. **Exponential function from two points:** Suppose that $N = N(t)$ is an exponential function such that $N(4) = 9$ and $N(7) = 22$. Find a formula for N.

4.1 SKILL BUILDING EXERCISES

S-1. **Function value from initial value and growth factor:** Suppose that f is an exponential function with growth factor 2.4 and that $f(0) = 3$. Find $f(2)$. Find a formula for $f(x)$.

S-2. **Function value from initial value and decay factor:** Suppose that f is an exponential function with decay factor 0.094 and that $f(0) = 400$. Find $f(2)$. Find a formula for $f(x)$.

S-3. **Finding the growth factor:** Suppose that f is an exponential function with $f(4) = 8$ and $f(5) = 10$. What is the growth factor for f?

S-4. **Exponential decay:** Is the graph of exponential decay versus time increasing or decreasing?

S-5. **Rate of change:** What can be said about the rate of change of an exponential function?

S-6. **Percentage growth:** A certain phenomenon has an initial value of 10 and grows at a rate of 7% per year. Give an exponential function that describes this phenomenon.

S-7. **Percentage decay:** A certain phenomenon has initial value 10 and decays by 4% each year. Give an exponential function that describes this phenomenon.

S-8. **Changing units:** A certain phenomenon has a yearly growth factor of 1.17. What is its monthly growth factor? What is its decade growth factor?

S-9. **Percentage change:** A bank account grows by 9% each year. By what percentage does it grow each month?

S-10. **Percentage change:** A certain radioactive substance decays at a rate of 17% each year. By what percentage does it decay each month?

4.1 EXERCISES

1. **Exponential growth with given initial value and growth factor:** Write the formula for an exponential function with initial value 23 and growth factor 1.4. Plot its graph.

2. **Exponential decay with given initial value and decay factor:** Write the formula for an exponential function with initial value 200 and decay factor 0.73. Plot its graph.

3. **A population with given per capita growth rate:** A certain population has a yearly per capita growth rate of 2.3%, and the initial value is 3 million.

   a. Use a formula to express the population as an exponential function.

   b. Express using functional notation the population after 4 years, and then calculate that value.

4. **Unit conversion with exponential growth:** The exponential function $N = 3500 \times 1.77^d$, where d is measured in decades, gives the number of individuals in a certain population.

   a. Calculate $N(1.5)$ and explain what your answer means.

   b. What is the percentage growth rate per decade?

   c. What is the yearly growth factor rounded to three decimal places? What is the yearly percentage growth rate?

   d. What is the growth factor (rounded to two decimal places) for a century? What is the percentage growth rate per century?

5. **Unit conversion with exponential decay:** The exponential function $N = 500 \times 0.68^t$, where t is measured in years, shows the amount, in grams, of a certain radioactive substance present.

   a. Calculate $N(2)$ and explain what your answer means.

   b. What is the yearly percentage decay rate?

   c. What is the monthly decay factor rounded to three decimal places? What is the monthly percentage decay rate?

   d. What is the percentage decay rate per second? (*Note:* For this calculation, you will need to use all the decimal places that your calculator can show.)

6. **A savings account:** You initially invest $500 in a savings account that pays a yearly interest rate of 4%.

 a. Write a formula for an exponential function giving the balance in your account as a function of the time since your initial investment.

 b. What monthly interest rate best represents this account? Round your answer to three decimal places.

 c. Calculate the decade growth factor.

 d. Use the formula you found in part a to determine how long it will take for the account to reach $740. Explain how this is consistent with your answer to part c.

7. **Half-life of heavy hydrogen:** We stated in Example 4.1 that the half-life of a radioactive substance does not depend on the initial amount. Using the information from Example 4.1, show that it takes the same amount of time for 100 grams of heavy hydrogen to decay to 50 grams as for 50 grams to decay to 25 grams. How long will it take for 100 grams of heavy hydrogen to decay to 6.25 grams? (*Note:* You can do this without your calculator!)

8. **How fast do exponential functions grow?** At age 25 you start to work for a company and are offered two rather fanciful retirement options.

 Retirement option 1: When you retire, you will be paid a lump sum of $25,000 for each year of service.

 Retirement option 2: When you start to work, the company will deposit $10,000 into an account that pays a monthly interest rate of 1%. When you retire, the account will be closed and the balance given to you.

 Which retirement option is more favorable to you if you retire at age 65? What if you retire at age 55?

9. **Inflation:** The yearly *inflation rate* tells the percentage by which prices increase. For example, from 1990 through 2000 the inflation rate in the United States remained stable at about 3% each year. In 1990 an individual retired on a fixed income of $36,000 per year. Assuming that the inflation rate remains at 3%, determine how long it will take for the retirement income to deflate to half its 1990 value. (*Note:* To say that retirement income has deflated to half its 1990 value means that prices have doubled.)

10. **Rice production in Asia:** In Exercise 16 of Section 3.4, it was estimated that annual rice yield in Asia was increasing by about 0.05 ton per hectare each year. The International Rice Research Institute[1] predicts an increase in demand for Asian rice of about 2.1% per year. Discuss the future of rice production in Asia if these estimates remain valid.

11. **The MacArthur–Wilson theory of biogeography:** Consider an island separated from the mainland, which contains a pool of potential colonizer species. The MacArthur–Wilson theory of biogeography[2] hypothesizes that some species from the mainland will migrate to the island but that increasing competition on the island will lead to species extinction. They further hypothesize that both the rate of migration and the rate of extinction of species are exponential functions, and that an *equilibrium* occurs when the rate of extinction matches the rate of immigration. This equilibrium point is thought to be the point at which immigration and extinction stabilize. Suppose that, for a certain island near the mainland, the rate of immigration of new species is given by

$$I = 4.2 \times 0.93^t \text{ species per year}$$

and that the rate of species extinction on the island is given by

$$E = 1.5 \times 1.1^t \text{ species per year}.$$

According to the MacArthur–Wilson theory, how long will be required for stabilization to occur, and what are the immigration and extinction rates at that time?

12. **Long-term population growth:** Although exponential growth can often be used to model population

---

[1] IRRI, *IRRI Rice Almanac 1993–1995* (Los Banos, the Philippines: International Rice Research Institute, 1993).

[2] R. H. MacArthur and E. O. Wilson, *The Theory of Island Biogeography* (Princeton, NJ: Princeton University Press, 1967). By the same authors: "An Equilibrium Theory of Insular Zoogeography," *Evolution* **17** (1963), 373–387.

growth accurately for some periods of time, there are inevitably, in the long term, limiting factors that make purely exponential models inaccurate. If the U.S. population had continued to grow by 3% each year from 1790, when it was 3.93 million, until today, what would the population of the United States have been in 2000? For comparison, according to census data, the population of the United States in 2000 was 281,421,906. The population of the world was just over 6 billion people.

13. **The population of Mexico:**[3] In 1980 the population of Mexico was about 67.38 million. For the years 1980 through 1985, the population grew at a rate of about 2.6% per year.

    a. Find a formula for an exponential function that gives the population of Mexico.

    b. Express using functional notation the population of Mexico in 1983, and then calculate that value.

    c. Use the function you found in part a to predict when the population of Mexico will reach 90 million.

14. **Cleaning contaminated water:** A tank of water is contaminated with 60 pounds of salt. In order to bring the salt concentration down to a level consistent with EPA standards, clean water is being piped into the tank, and the well-mixed overflow is being collected for removal to a toxic-waste site. The result is that at the end of each hour there is 22% less salt in the tank than at the beginning of the hour. Let $S = S(t)$ denote the number of pounds of salt in the tank t hours after the flushing process begins.

    a. Explain why S is an exponential function and find its hourly decay factor.

    b. Give a formula for S.

    c. Make a graph of S that shows the flushing process during the first 15 hours, and describe in words how the salt removal process progresses.

    d. In order to meet EPA standards, there can be no more than 3 pounds of salt in the tank. How long must the process continue before EPA standards are met?

    e. Suppose this cleanup procedure costs $8000 per hour to operate. How much does it cost to reduce the amount of salt from 60 pounds to 3 pounds? How much does it cost to reduce the amount of salt from 3 pounds to 0.1 pound?

15. **Tsunami waves in Crescent City:** Crescent City, California, has historically been at risk for tsunami waves.[4] The probability P (as a decimal) of no tsunami wave of height 15 feet or more striking Crescent City over a period of Y years decreases as the time interval increases. Increasing the time interval by 1 year decreases the probability[5] by about 2%.

    a. Explain why P is an exponential function of Y.

    b. What is the decay factor for P?

    c. Recall that the probability of a certainty is 1. What is the initial value of P?

    d. Find a formula for P as a function of Y.

    e. What is the probability of no tsunami waves 15 feet or higher striking Crescent City over a 10-year period? Over a 100-year period?

    f. Find a formula for the probability Q that at least one wave 15 feet or higher will strike Crescent City over a period of Y years. *Suggestion:* It should be helpful to note that the probability of the occurrence of an event plus the probability that it will not occur is the probability of a certainty, 1.

16. **Growth of bacteria:** The organism *E. coli* is a common bacterium. Under certain conditions it undergoes cell division approximately each 20 minutes. During cell division, each cell divides into two cells.

    a. Explain why the number of *E. coli* cells present is an exponential function of time.

    b. What is the hourly growth factor for *E. coli*?

    c. Express the population N of *E. coli* as an exponential function of time t measured in hours. (Use N_0 to denote the initial population.)

    d. How long will it take a population of *E. coli* to triple in size?

17. **Grains of wheat on a chess board:** A children's fairy tale tells of a clever elf who extracted from

---

[3]Adapted from D. Hughes-Hallett, A. M. Gleason, et al., *Calculus,* 3rd ed. (New York: John Wiley, 2002).

[4]On March 28, 1964, a tsunami wave approximately 25 feet high struck Crescent City.

[5]This value is taken from data provided on page 294 of Robert L. Wiegel, ed., *Earthquake Engineering* (Englewood Cliffs, NJ: Prentice-Hall, 1970).

a king the promise to give him one grain of wheat on a chess board square today, two grains on an adjacent square tomorrow, four grains on an adjacent square the next day, and so on, doubling the number of grains each day until all 64 squares of the chess board were used. How many grains of wheat did the hapless king contract to place on the 64th square? There are about 1.1 million grains of wheat in a bushel. Assume that a bushel of wheat sells for $4.25. What was the value of the wheat on the 64th square?

18. **The Beer–Lambert–Bouguer law:** When light strikes the surface of a medium such as water or glass, its intensity decreases with depth. The *Beer–Lambert–Bouguer law* states that the percentage of decrease is the same for each additional unit of depth. In a certain lake, intensity decreases about 75% for each additional meter of depth.

    a. Explain why intensity I is an exponential function of depth d in meters.

    b. Use a formula to express intensity I as an exponential function of d. (Use I_0 to denote the initial intensity.)

    c. Explain in practical terms the meaning of I_0.

    d. At what depth will the intensity of light be one-tenth of the intensity of light striking the surface?

19. **The photic zone:** *This is a continuation of Exercise 18.* In the ocean, the *photic zone* is the region where there is sufficient light for photosynthesis to occur. (See Figure 4.7.) For marine phytoplankton, the photic zone extends from the surface of the ocean to a depth where the light intensity is about 1% of surface light. Near Cape Cod, Massachusetts, the depth of the photic zone is about 16 meters. In waters near Cape Cod, by what percentage does light intensity decrease for each additional meter of depth?

20. **Decibels:** Sound exerts a pressure P on the human ear. This pressure increases as the loudness of the sound increases. It is convenient to measure the loudness D in decibels and the pressure P in dynes per square centimeter. It has been found that each increase of 1 decibel in loudness causes a 12.2%

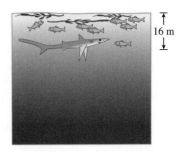

FIGURE 4.7

increase in pressure. Furthermore, a sound of loudness 97 decibels produces a pressure of 15 dynes per square centimeter.

    a. Explain why P is an exponential function of D and find the growth factor.

    b. Find $P(0)$ and explain in practical terms what your answer means.

    c. Find an exponential model for P as a function of D.

    d. When pressure on the ear reaches a level of about 200 dynes per square centimeter, physical damage can occur. What decibel level should be considered dangerous?

21. **Headway on four-lane highways:** When traffic is flowing on a highway, the *headway* is the average time between vehicles. On four-lane highways, the probability P that the headway is at least t seconds is given to a good degree of accuracy[6] by

$$P = e^{-qt},$$

where q is the average number of vehicles per second traveling one way on the highway.

    a. On a four-lane highway carrying an average of 500 vehicles per hour in one direction, what is the probability that the headway is at least 15 seconds? *Note:* 500 vehicles per hour is $\frac{500}{3600} = 0.14$ vehicle per second.

    b. On a four-lane highway carrying an average of 500 vehicles per hour, what is the decay factor for

---

[6]See Institute of Traffic Engineers, *Transportation and Traffic Engineering Handbook,* ed. John E. Baerwald (Englewood Cliffs, NJ: Prentice-Hall, 1976), page 102.

the probability that headways are at least t seconds? *Reminder:* An important law of exponents tells us that $a^{bc} = \left(a^b\right)^c$.

22. **APR and APY:** Recall that financial institutions sometimes report the annual interest rate that they offer on investments as the APR, often called the *nominal* interest rate. To indicate how an investment will actually grow, they advertise the *annual percentage yield*, or APY.[7] In mathematical terms, this is the yearly percentage growth rate for the exponential function that models the account balance. In this exercise and the next, we study the relationship between the APR and the APY. We assume that the APR is 10%, or 0.1 as a decimal.

   To determine the APY when we know the APR, we need to know how often interest is compounded. For example, suppose for the moment that interest is compounded twice a year. Then to say that the APR is 10% means that, in half a year, the balance grows by $\frac{10}{2}$ % (or 5%). In other words, the $\frac{1}{2}$-year percentage growth rate is $\frac{0.1}{2}$ (as a decimal). Thus the $\frac{1}{2}$-year growth factor is $1 + \frac{0.1}{2}$. To find the *yearly* growth factor, we need to perform a unit conversion: One year is 2 half-year periods, so the yearly growth factor is $\left(1 + \frac{0.1}{2}\right)^2$, or 1.1025.

   a. What is the yearly growth factor if interest is compounded four times a year?

   b. Assume that interest is compounded n times each year. Explain why the formula for the yearly growth factor is

$$\left(1 + \frac{0.1}{n}\right)^n.$$

   c. What is the yearly growth factor if interest is compounded daily? Give your answer to four decimal places.

23. **Continuous compounding:** *This is a continuation of Exercise 22.* In this exercise we examine the relationship between the APR and the APY when interest is compounded continuously—in other words, at every instant. We will see by means of an example that the relationship is

$$\text{Yearly growth factor} = e^{\text{APR}}, \qquad (4.1)$$

   and so

$$\text{APY} = e^{\text{APR}} - 1 \qquad (4.2)$$

   if both the APR and the APY are in decimal form and interest is compounded continuously. Assume that the APR is 10%, or 0.1 as a decimal.

   a. The yearly growth factor for continuous compounding is just the limiting value of the function given by the formula in part b of Exercise 22. Find that limiting value to four decimal places.

   b. Compute e^{APR} with an APR of 0.1 (as a decimal).

   c. Use your answers to parts a and b to verify that Equation (4.1) holds in the case where the APR is 10%.

   Note: On the basis of part a, one conclusion is that there is a limit to the increase in the yearly growth factor (and hence in the APY) as the number of compounding periods increases. We might have expected the APY to increase without limit for more and more frequent compounding.

---

[7]For loans, financial institutions often refer to this as the effective annual rate, or EAR.

4.2 *Modeling Exponential Data*

Just as we did with linear data, we will show how to recognize exponential data and develop the appropriate tools for constructing exponential models.

Recognizing Exponential Data

The table below shows how the balance in a savings account grows over time since the initial investment.

Time in months	0	1	2	3	4
Savings balance	$3500.00	$3542.00	$3585.50	$3627.52	$3671.04

Let t be the time in months since the initial investment, and let B be the balance in dollars. Certainly a linear model for the data is not appropriate, since the balance is growing more and more quickly over time. For example, over the first month the balance grows by $3542.00 - 3500.00 = 42.00$ dollars, whereas over the second month it grows by $3585.50 - 3542.00 = 43.50$ dollars. In fact, we expect that the balance will grow as an exponential function if the interest rate is constant. How can we test the data to see whether they really are exponential in nature? If the data are growing exponentially, then each month we should get the new balance by multiplying the old balance by the monthly growth factor:

$$\text{New } B = \text{Monthly growth factor} \times \text{Old } B.$$

Using division to rewrite this yields

$$\frac{\text{New } B}{\text{Old } B} = \text{Monthly growth factor}. \tag{4.3}$$

Thus a data table representing an exponential function should show common quotients if we divide each function entry by the one preceding it, assuming that we have evenly spaced values for the variable. In the following table, we have calculated this quotient for each of the data entries, rounding our answers to three decimal places.

Time increment	From 0 to 1	From 1 to 2	From 2 to 3	From 3 to 4
Ratios of B	$\dfrac{3542.00}{3500.00} = 1.012$	$\dfrac{3585.50}{3542.00} = 1.012$	$\dfrac{3627.52}{3585.50} = 1.012$	$\dfrac{3671.04}{3627.52} = 1.012$

According to Equation (4.3), this table tells us two things. First, the ratios are always the same, so the data are exponential. Second, the common quotient 1.012 is the monthly growth factor. This means that the monthly interest rate is $1.012 - 1 = 0.012$ as a decimal, or 1.2%.

Comparing this with how we handled linear data in Chapter 3, we notice an important analogy. Linear functions are functions that change by *constant sums*, so we detect linear data by looking for a common *difference* in function values when we use successive data points, assuming evenly spaced values for the variable. When our data set is given in variable increments of 1, this common difference is the slope of the linear model. Exponential functions are functions that change by *constant multiples*, so we detect exponential data by looking for a common *quotient* when we use successive data points, again assuming evenly spaced values for the variable. When our data set is given in increments of 1, this common quotient is the growth (or decay) factor of the exponential model.

Constructing an Exponential Model

In our discussion of the balance in a savings account, we saw that the data were exponential. Since we know the growth factor, we need only one further bit of information to make an exponential model: the initial value. But that also appears in the table as $B(0) = 3500.00$ dollars. Thus the formula, which will serve as our exponential model for B as a function of time t in months, is

$$B = \text{Initial value} \times (\text{Monthly growth factor})^t$$

$$B = 3500.00 \times 1.012^t \text{ dollars}.$$

In some situations the data themselves are not exponential, but the difference from a limiting value can be modeled by an exponential function.

EXAMPLE 4.4   **Contaminated Well**

Underground water seepage, such as that from a toxic waste site, can contaminate water wells many miles away. Suppose that water seeping from a toxic waste site is polluted with a certain contaminant at a level of 64 milligrams per liter. Several miles away, monthly tests are made on a water well to monitor the level of this contaminant in the drinking water. In the table below, we record the difference between the contaminant level of the waste site (64 milligrams per liter) and the contaminant level of the water well (in milligrams per liter).

Time in months	0	1	2	3	4	5
Contaminant level difference	64	45.44	32.26	22.91	16.26	11.55

1. Explain why the nature of the data suggests that an exponential model may be appropriate.

2. Test to see that the data are exponential.

3. Find an exponential model for the difference in contaminant level.

4. Use your answer in part 3 to make a model for the contaminant level as a function of time.

5. This contaminant is considered dangerous in drinking water when it reaches a level of 57 milligrams per liter. When will this dangerous level be reached?

Solution to Part 1:   Let t be the time, in months, since the leakage began, and let D be the difference, in milligrams per liter, between the contaminant level of the waste site and that of the water well. The table shows that the difference D is narrowing, so the water well is becoming more heavily polluted. We note further that D decreases quickly at first but that the rate of decrease slows later on. This is a feature of exponential decay, and it leads us to suspect that these data may be appropriately modeled by an exponential function.

Solution to Part 2:   In the following table, we have calculated successive quotients for the data entries, rounding our answers to two decimal places.

Time increment	From 0 to 1	From 1 to 2	From 2 to 3	From 3 to 4	From 4 to 5
Ratios of D	$\dfrac{45.44}{64} = 0.71$	$\dfrac{32.26}{45.44} = 0.71$	$\dfrac{22.91}{32.26} = 0.71$	$\dfrac{16.26}{22.91} = 0.71$	$\dfrac{11.55}{16.26} = 0.71$

Since the common quotient is 0.71, the data can be modeled by an exponential function.

Solution to Part 3:   The decay factor for D is the common quotient 0.71 that we found in part 2. The initial value of D is in the table: $D(0) = 64$ milligrams per liter. Thus we have

$$D = \text{Initial value} \times (\text{Decay factor})^t$$
$$D = 64 \times 0.71^t.$$

Solution to Part 4:   Let $C = C(t)$ be the contamination level of the water well in milligrams per liter. Since D represents the contaminant level of the waste site, 64 milligrams per liter, minus the contaminant level C of the water well, we find a formula for C as follows:

$$D = 64 - C$$
$$C + D = 64$$
$$C = 64 - D$$
$$C = 64 - 64 \times 0.71^t \text{ milligrams per liter}.$$

Solution to Part 5:   We want to solve the equation

$$64 - 64 \times 0.71^t = 57$$

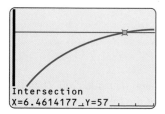

X	Y₁	Y₂
3	41.094	57
4	47.737	57
5	52.453	57
6	55.802	57
7	58.179	57
8	59.867	57
9	61.066	57

X=7

FIGURE 4.8 A table of values for contaminant levels

Intersection
X=6.4614177 Y=57

FIGURE 4.9 When the contaminant level reaches 57 milligrams per liter

for t. We use the crossing-graphs method to do that. We entered 4.4 the contaminant function $C = 64 - 64 \times 0.71^t$ and the target level, 57 milligrams per liter, into the calculator. We record the appropriate variable correspondences:

$$Y_1 = C, \text{ contaminant level on vertical axis}$$

$$X = t, \text{ months on horizontal axis}.$$

We made the table of values in Figure 4.8 to find out how to set up the graphing window. This table shows that the critical level of contamination will occur sometime before 7 months. This led us to choose a horizontal span of $t = 0$ to $t = 8$. Since the contaminant level starts at 0 and will never be more than 64, we add a little extra room, using a vertical span of $C = 0$ to $C = 75$. The resulting graphs are in Figure 4.9. We have used the calculator to get the crossing point 4.5, and from the prompt at the bottom of Figure 4.9, we see that the contamination level of the water well will reach 57 milligrams per liter 6.46 months after the leakage began.

■ ■ ■

The graph in Figure 4.9 not only answers for us the question we asked but also provides a display that shows the classic exponential behavior for this phenomenon. As we noted earlier, the difference D between the contaminant level of the waste site and that of the water well decreases quickly at first, but the rate of decrease slows later on. As Figure 4.9 shows, this means that the contamination itself increases rapidly at first, but as the level nears that of the polluted source, the rate of increase slows.

EXAMPLE 4.5   **Making Ice**

A freezer maintains a constant temperature of 6 degrees Fahrenheit. An ice tray is filled with tap water and placed in the refrigerator to make ice. The difference between the temperature of the water and that of the freezer was sampled each minute and recorded in the table below.

Time in minutes	0	1	2	3	4	5
Temperature difference	69.0	66.3	63.7	61.2	58.8	56.5

1. Test to see that the data are exponential.

2. Find an exponential model for temperature difference.

3. Use your answer in part 2 to make a model for the temperature of the cooling water as a function of time.

4. When will the temperature of the water reach 32 degrees?

Solution to Part 1:   Let t be the time in minutes and D the temperature difference. To test whether the data are exponential, we make a table of successive quotients. (We rounded our answers to two decimal places.)

Time increment	From 0 to 1	From 1 to 2	From 2 to 3	From 3 to 4	From 4 to 5
Ratios of D	$\dfrac{66.3}{69.0} = 0.96$	$\dfrac{63.7}{66.3} = 0.96$	$\dfrac{61.2}{63.7} = 0.96$	$\dfrac{58.8}{61.2} = 0.96$	$\dfrac{56.5}{58.8} = 0.96$

The table of values shows a common quotient of 0.96, and we conclude that the data can be modeled by an exponential function.

Solution to Part 2: We know that the decay factor for D is the common ratio 0.96 that we calculated in part 1. We also know that when $t = 0$, $D = 69.0$, and that is the initial value of D. Thus we have

$$D = \text{Initial value} \times (\text{Decay factor})^t$$

$$D = 69 \times 0.96^t.$$

Solution to Part 3: Let W be the temperature of the water. The function D is the temperature W of the water minus 6, the temperature of the freezer:

$$W - 6 = D$$

$$W = 6 + D$$

$$W = 6 + 69 \times 0.96^t \text{ degrees Fahrenheit}.$$

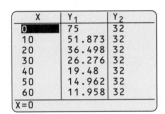

X	Y₁	Y₂
0	75	32
10	51.873	32
20	36.498	32
30	26.276	32
40	19.48	32
50	14.962	32
60	11.958	32

X=0

FIGURE 4.10 A table of values for water temperature

Solution to Part 4: We want to solve the equation

$$6 + 69 \times 0.96^t = 32.$$

We enter 4.6 the water temperature function W and the target temperature, 32 degrees Fahrenheit. We record the variable correspondences:

$$\mathsf{Y_1} = W, \text{ water temperature, on vertical axis}$$

$$\mathsf{X} = t, \text{ minutes, on horizontal axis}.$$

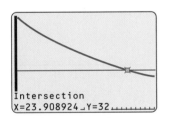

Intersection
X=23.908924 ⌐Y=32

FIGURE 4.11 When the temperature is 32 degrees

We made the table of values in Figure 4.10 using an increment value of 10 minutes to get an estimate of when the temperature will reach 32 degrees. The table shows that this will happen within 30 minutes. Thus we set up our graphing window using a horizontal span of $t = 0$ to $t = 30$ and a vertical span of $W = 0$ to $W = 75$. The graphs with the crossing point that was calculated 4.7 are in Figure 4.11. We see that the temperature will reach 32 degrees in 23.91 minutes. ▪ ▪ ▪

Growth and Decay Factor Units in Exponential Modeling

In the contaminated well example, the decay factor was the common quotient we calculated. This will always occur when time measurements are given in 1-unit increments. But when data are measured in different increments, adjustments to account for units must be made to get the right decay factor. We can show what we mean by looking at the contaminated water well example again, but this time we suppose that measurements were taken every 3 months. Under those conditions, the data table would have been as follows:

t = months	0	3	6	9	12	15
D = difference in contaminant level	64	22.91	8.2	2.93	1.05	0.38

We test to see whether the data are exponential by calculating successive ratios, rounding to two decimal places.

Time increment	From 0 to 3	From 3 to 6	From 6 to 9	From 9 to 12	From 12 to 15
Ratios of D	$\dfrac{22.91}{64} = 0.36$	$\dfrac{8.2}{22.91} = 0.36$	$\dfrac{2.93}{8.2} = 0.36$	$\dfrac{1.05}{2.93} = 0.36$	$\dfrac{0.38}{1.05} = 0.36$

Since the successive ratios are the same, 0.36, we conclude once again that the data are exponential. But the common ratio 0.36 is not the monthly decay factor. Rather, it is the *3-month decay factor*. To get the monthly decay factor, we need to change the units. Since 1 month is one-third of the 3-month recording time, we get

Monthly decay factor = (3-month decay factor)$^{1/3}$

Monthly decay factor = $0.36^{1/3} = 0.71$ rounded to two decimal places .

The initial value is 64, so we arrive at the same exponential model, 64×0.71^t, as we did when we modeled the data given in 1-month intervals.

KEY IDEA 4.4

Modeling Exponential Data

1. To test whether data with evenly spaced values for the variable show exponential growth or decay, calculate successive quotients. If the quotients are all the same, it is appropriate to model the data with an exponential function. If the quotients are not all the same, some other model will be needed.

2. To model exponential data, use the common successive quotient for the growth (or decay) factor, but if the data are not measured in 1-unit increments, it will be necessary to make an adjustment for units. One of the data points can be used to determine the initial value.

EXAMPLE 4.6 **Finding the Time of Death**

One important topic of forensic medicine is the determination of the time of death. A method that is sometimes used involves temperature. Suppose at 6:00 P.M. a body is discovered in a basement of a building where the ambient air temperature is maintained at 72 degrees.[8] At the moment of death, the body temperature was 98.6 degrees, but after death the body cools, and eventually its temperature matches the ambient air temperature. Beginning at 6:00 P.M., the body temperature is measured and the difference $D = D(t)$ between body temperature and ambient air temperature is recorded. The measurement is repeated every 2 hours.

t = hours since 6:00 P.M.	0	2	4	6	8
D = temperature difference	12.02	8.08	5.44	3.65	2.45

1. Show that the data can be modeled by an exponential function.

2. Find an exponential model for the data that shows temperature difference as a function of hours.

3. Find a formula for a function $T = T(t)$ that gives the temperature of the body at time t.

4. What was the time of death?

Solution to Part 1: To see whether the data are exponential, we calculate successive quotients, rounding to two decimal places.

Time increment	From 0 to 2	From 2 to 4	From 4 to 6	From 6 to 8
Ratios of D	$\dfrac{8.08}{12.02} = 0.67$	$\dfrac{5.44}{8.08} = 0.67$	$\dfrac{3.65}{5.44} = 0.67$	$\dfrac{2.45}{3.65} = 0.67$

Since the successive quotients are the same, we conclude that the data show exponential decay.

Solution to Part 2: To give an exponential model, we need to know the initial value and the hourly decay factor. From the data table, the initial value is $D(0) = 12.02$. Our calculation of successive quotients in part 1 gave us the 2-hour decay factor, 0.67.

---

[8]If the air temperature fluctuates, establishing time of death using body temperature is more difficult. It is noted in the book *Helter Skelter*, by Vincent Bugliosi, that body temperature was used to help establish the time of death in the Charles Manson group murders of Sharon Tate and others.

We want the 1-hour decay factor. Since 1 hour is half of 2 hours, we get

Hourly decay factor = (2-hour decay factor)$^{1/2}$

Hourly decay factor = $0.67^{1/2} = 0.82$, rounded to two decimal places.

Thus we have

$$D = 12.02 \times 0.82^t.$$

Solution to Part 3: The function $D = 12.02 \times 0.82^t$ gives the difference between the temperature of the body and that of the air. Since we are assuming that the air has a temperature of 72 degrees, we add that to D to get the temperature of the body:

$$T = 72 + 12.02 \times 0.82^t.$$

Solution to Part 4: To find the time of death, we need to know when the temperature was that of a living person, 98.6 degrees. Thus we need to solve the equation

$$72 + 12.02 \times 0.82^t = 98.6$$

for t. Before we begin, we note that since $t = 0$ corresponds to some time several hours after death, the value of t that we are looking for is surely negative. We solve the equation by using the crossing-graphs method. In Figure 4.12 we have made the graph of T versus t using a horizontal span of $t = -5$ to $t = 5$ hours and a vertical span of $T = 70$ to $T = 120$ degrees. You may wish to consult a table of values to see why we chose these window settings. We then added the graph of the target temperature $T = 98.6$ and found the intersection point. From the bottom of Figure 4.12, we see that death occurred about 4 hours before the body was found. Since $t = 0$ corresponds to 6:00 P.M., we set the time of death at about 2:00 P.M.

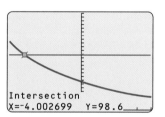

FIGURE 4.12 Finding the time of death

Testing Unevenly Spaced Data

All of the data sets we have looked at thus far have involved evenly spaced data. If the data are unevenly spaced, then we must work a bit harder to test whether they are exponential. Consider the following data set.

x	2	4	7	9
y	18	162	4374	39,366

If the data are exponential, then a formula of the form $y = Pa^x$ must be valid. Now from the first data point we have $18 = Pa^2$, and from the second data point we have

$162 = Pa^4$. Dividing the second equation by the first gives (since $\frac{162}{18} = 9$)

$$9 = \frac{Pa^4}{Pa^2} = a^2.$$

Thus $a^2 = 9$. What happens is that the ratio $\frac{\text{New } y}{\text{Old } y}$ is not a but rather a raised to a power, and that power is the difference in x values. When we do this for each successive ratio, we get

$$a^2 = 9$$

$$a^3 = \frac{4374}{162} = 27$$

$$a^2 = \frac{39{,}366}{4374} = 9.$$

From the first equation we find $a = 3$. You should check to see that this does indeed satisfy the remaining equations. Thus the data are exponential with growth factor $a = 3$. We use the first data point to get the initial value:

$$P \times a^2 = 18$$

$$P \times 3^2 = 18$$

$$P = \frac{18}{9} = 2.$$

Thus the exponential model for the data is $y = 2 \times 3^x$.

Let's look at a second data set.

x	2	4	7	9
y	12	48	388	1552

Calculating ratios gives the following equations:

$$a^2 = \frac{48}{12} = 4$$

$$a^3 = \frac{388}{48} = 8.08$$

$$a^2 = \frac{1552}{388} = 4.$$

From the first equation we have that $a = 2$. But that does not satisfy the second equation. We conclude that the data are not exponential.

The test in general is as follows: Calculate ratios to get a system of exponential equations. Solve the first equation for a. Then test this value in the remaining equations. If it satisfies each equation, the data are exponential. If it fails to satisfy even one equation, then the data are not exactly exponential.

Enrichment Exercises

E-1. **Testing for exponential data:** Test each of the following data sets to see whether it is exponential. For those that are exponential, find an exponential model.

a.

x	2	5	7	11
y	48	3072	49,152	12,582,912

b.

x	2	4	8	10
y	4	1	0.0625	0.015625

c.

x	2	5	7	9
y	36	972	8748	78,740

E-2. **Finding missing points:** The following data set is exponential. Fill in the missing data.

x	1	3		6
y		54	162	1458

E-3. **Testing exponential data:** Show that the following data set is exponential and find an exponential model.

x	a	$a + 3$	$a + 5$	$a + 6$
y	1	b^3	b^5	b^6

4.2 *SKILL BUILDING EXERCISES*

S-1. **Finding the growth factor:** Suppose that $N = N(t)$ is an exponential function with initial value 7. If t is increased by 1, N is multiplied by 8. Find a formula for N.

S-2. **Finding the growth factor:** Suppose that $N = N(t)$ is an exponential function with initial value 6. If t is increased by 1, N is divided by 13. Find a formula for N.

S-3. **Finding the growth factor:** Suppose $N = N(t)$ is an exponential function with initial value 12. If t is increased by 7, the effect is to multiply N by 62. Find a formula for N.

S-4. **Testing exponential data:** Determine whether the following table shows exponential data.

x	0	1	2	3
y	2.6	7.8	23.4	70.2

S-5. **Testing exponential data:** Determine whether the following table shows exponential data.

x	0	2	4	6
y	5	10	20	40

S-6. **Testing exponential data:** Determine whether the following table shows exponential data.

x	0	2	4	6
y	5	9	21	43

S-7. **Modeling exponential data:** Make an exponential model for the data from Exercise S-4.

S-8. **Modeling exponential data:** Make an exponential model for the data from Exercise S-5.

S-9. **Testing exponential data:** Determine whether the following table shows exponential data.

x	0	2	4	6
y	6	18	54	143

S-10. **Modeling exponential data:** Make an exponential model for the data from Exercise S-9.

4.2 *EXERCISES*

1. **Making an exponential model:** Show that the accompanying data are exponential and find a formula for an exponential model.

t	$f(t)$	t	$f(t)$
0	3.80	3	4.27
1	3.95	4	4.45
2	4.11	5	4.62

2. **An exponential model with unit adjustment:** Show that the following data are exponential and find a formula for an exponential model. (*Note:* It will be necessary to make a unit adjustment. For this problem, round your answers to three decimal places.)

t	g(t)	t	g(t)
0	38.30	12	16.04
4	28.65	16	11.99
8	21.43	20	8.97

3. **Data that are not exponential:** Show that the following data are not exponential.

t	h(t)	t	h(t)
0	4.9	3	200.2
1	26.6	4	352.1
2	91.7	5	547.4

4. **Linear and exponential data:** One of the two tables below shows linear data, and the other shows exponential data. Identify which is which, and find models for both.

t	f(t)	t	f(t)
0	6.70	3	10.46
1	7.77	4	12.13
2	9.02	5	14.07

Table A

t	g(t)	t	g(t)
0	5.80	3	10.99
1	7.53	4	12.72
2	9.26	5	14.45

Table B

5. **Magazine sales:** The following table shows the income from sales of a certain magazine, measured in thousands of dollars, at the start of the given year.

Year	Income
1994	7.76
1995	8.82
1996	9.88
1997	10.94
1998	12.00
1999	13.08
2000	14.26
2001	15.54

Over an initial period the sales grew at a constant rate, and over the rest of the time the sales grew at a constant percentage rate. Calculate differences and ratios to determine what these time periods are, and find the growth rate or percentage growth rate, as appropriate.

6. **Population growth:** A population of animals is growing exponentially, and an ecologist has made the following table of the population size, in thousands, at the start of the given year.

Year	Population in thousands
1996	5.25
1997	5.51
1998	5.79
1999	6.04
2000	6.38
2001	6.70

Looking over the table, the ecologist realizes that one of the entries for population size is in error.

→

Which entry is it, and what is the correct population? (Round the ratios to two decimal places.)

7. **An investment:** You have invested money in a savings account that pays a fixed monthly interest on the account balance. The following table shows the account balance over the first 5 months.

Time in months	Savings balance
0	$1750.00
1	$1771.00
2	$1792.25
3	$1813.76
4	$1835.52
5	$1857.55

a. How much money was originally invested?

b. Show that the data are exponential and find an exponential model for the account balance.

c. What is the monthly interest rate?

d. What is the yearly interest rate?

e. Suppose that you made this investment on the occasion of the birth of your daughter. Your plan is to leave the money in the account until she starts college at age 18. How large a college fund will she have?

f. How long does it take your money to double in value? How much longer does it take it to double in value again?

8. **A bald eagle murder mystery:** At 3:00 P.M. a park ranger discovered a dead bald eagle that had been impaled by an arrow. Only two archers were found in the region. The first archer is able to establish that, between 11:00 A.M. and 1:00 P.M., he was in a nearby diner having lunch. The second archer can show that he was in camp with friends between 9:00 A.M. and 11:00 A.M. The air temperature in the park has remained at a constant 62 degrees. Beginning at 3:00 P.M., the difference $D = D(t)$ between the temperature of the dead eagle and that of the air was measured and recorded in the table below. (Here t is the time in hours since 3:00 P.M.)

t = hours since 3:00 P.M.	D = temperature difference
0	26.83
1	24.42
2	22.22
3	20.22
4	18.40
5	16.74

This table, together with the fact that the body temperature of a living bald eagle is 105 degrees, exonerates one of the archers, but the other may remain suspect. Which archer's innocence is established?

9. **A skydiver:** When a skydiver jumps from an airplane, his downward velocity increases until the force of gravity matches air resistance. The velocity at which this occurs is known as the *terminal velocity*. It is the upper limit on the velocity a skydiver in free fall will attain (in a stable, spread position), and for a man of average size, its value is about 176 feet per second (or 120 miles per hour). A skydiver jumped from an airplane, and the difference $D = D(t)$ between the terminal velocity and his downward velocity in feet per second was measured at 5-second intervals and recorded in the following table.

t = seconds into free fall	D = velocity difference
0	176.00
5	73.61
10	30.78
15	12.87
20	5.38
25	2.25

a. Show that the data are exponential and find an exponential model for D. (Round all your answers to two decimal places.)

b. What is the percentage decay rate per second for the velocity difference of the skydiver? Explain in practical terms what this number means.

c. Let $V = V(t)$ be the skydiver's velocity t seconds into free fall. Find a formula for V.

d. How long would it take the skydiver to reach 99% of terminal velocity?

10. **The half-life of U^{239}:** Uranium 239 is an unstable isotope of uranium that decays rapidly. In order to determine the rate of decay, 1 gram of U^{239} was placed in a container, and the amount remaining was measured at 1-minute intervals and recorded in the table below.

Time in minutes	Grams remaining
0	1
1	0.971
2	0.943
3	0.916
4	0.889
5	0.863

a. Show that these are exponential data and find an exponential model. (For this problem, round all your answers to three decimal places.)

b. What is the percentage decay rate each minute? What does this number mean in practical terms?

c. Use functional notation to express the amount remaining after 10 minutes and then calculate that value.

d. What is the half-life of U^{239}?

11. **An inappropriate linear model for radioactive decay:** *This is a continuation of Exercise 10.* Physicists have established that radioactive substances display constant percentage decay, and thus radioac-

tive decay is appropriately modeled exponentially. This exercise is designed to show how using data without an understanding of the phenomenon that generated them can lead to inaccurate conclusions.

a. Plot the data points from Exercise 10. Do they appear almost to fall on a straight line?

b. Find the equation of the regression line and add its graph to the one you made in part a.

c. If you used the regression line as a model for decay of U^{239}, how long would it take for the initial 1 gram to decay to half that amount? Compare this with your answer to part d of Exercise 10.

d. The linear model represented by the regression line makes an absurd prediction concerning the amount of uranium 239 remaining after 1 hour. What is this prediction?

12. **Account growth:** The table below shows the balance B in a savings account, in dollars, in terms of time t, measured as the number of years since the initial deposit was made.

Time t	Balance B
0	125.00
1	131.25
2	137.81
3	144.70
4	151.94

a. Was the yearly interest rate constant over the first 4 years? If so, explain why and find that rate. If not, explain why not. (Round the ratios to two decimal places.)

b. Estimate $B(2.75)$ and explain in practical terms what your answer means. (Assume that interest is compounded and deposited continuously.)

13. **Rates vary:** The table below shows the balance in a savings account, in dollars, in terms of time, measured as the number of years since the initial deposit was made.   →

Time	Balance
0	250.00
1	262.50
2	275.63
3	289.41
4	302.43
5	316.04
6	330.26

Explain in terms of interest rates how the account grew. (Round the ratios to three decimal places.)

14. **Wages:** A worker is reviewing his pay increases over the past several years. The table below shows the hourly wage W, in dollars, that he earned as a function of time t, measured in years since the beginning of 1990.

Time t	Wage W
1	15.30
2	15.60
3	15.90
4	16.25

a. By calculating ratios, show that the data in this table are exponential. (Round the quotients to two decimal places.)

b. What is the yearly growth factor for the data?

c. The worker can't remember what hourly wage he earned at the beginning of 1990. Assuming that W is indeed an exponential function, determine what that hourly wage was.

d. Find a formula giving an exponential model for W as a function of t.

e. What percentage raise did the worker receive each year?

f. Given that prices increased by 34% over the decade of the 1990s, use your model to deter-

mine whether the worker's wage increases kept pace with inflation.

15. **Stochastic population growth:** Many populations are appropriately modeled by an exponential function, at least for a limited period of time. But there are many factors contributing to the growth of any population, and many of them depend on chance. There are a number of ways to produce stochastic models. We consider one that is an illustration of a simple *Monte Carlo method*. We want to model a population that is initially 500 and grows at an average rate of 2% per year. To do this we make a table of values for population according to the following procedure. The first entry in the table is for time $t = 0$, and it records the initial value 500. To get the entry corresponding to $t = 1$, we roll a die and change the population according to the face that appears, using the following rule.

Face appearing	Population change
1	Down 2%
2	Down 1%
3	No change
4	Up 2%
5	Up 4%
6	Up 9%

To get the entry corresponding to $t = 2$, we roll a die and change the population from $t = 1$, again using the above rule. This procedure is then followed for $t = 3$, and so on.

a. Using this procedure, make a table recording the population values for years 0 through 10.

b. Plot the data points from your table and the exponential model on the same screen. Comment on the level of agreement.

16. **Growth rate of a tubeworm:** A recent study of the growth rate and life span of a marine tubeworm concludes that it is the longest-lived non-colonial

marine invertebrate known.[9] Since tubeworms live on the ocean floor and have a long life span, scientists do not measure their age directly. Instead, scientists measure their growth rate at various lengths and then construct a model for growth rate in terms of length. On the basis of that model, scientists can find a relationship between *age* and length. This is a good example of how rates of change can be used to determine a relationship when direct measurement is difficult or impossible. The table below shows for a tubeworm the rate of growth in length, measured in meters per year, at the given length, in meters.

Length in meters	Growth rate in meters per year
0	0.0510
0.5	0.0255
1.0	0.0128
1.5	0.0064
2.0	0.0032

a. Often in biology the growth rate is modeled as a decreasing linear function of length. For some organisms, however, it may be appropriate to model the growth rate as a decreasing exponential function of length. Use the data in the table to decide which model is more appropriate for the tubeworm, and find that model. Give a practical explanation of the slope or percentage decay rate, whichever is applicable.

b. Use functional notation to express the growth rate at a length of 0.64 meter, and then calculate that value using your model from part a.

---

[9]D. Bergquist, F. Williams, and C. Fisher, "Longevity record for deep-sea invertebrate," *Nature* **403** (2000), 499–500. Their conservative estimate for the life span is between 170 and 250 years.

4.3 *Modeling Nearly Exponential Data*

As in the case of linear data, rarely can experimentally gathered data be modeled exactly by an exponential function, but such data can in many cases be closely approximated by an exponential model. A key step in analyzing exponential data is to establish their relationship with linear data. This is done using the *logarithm* function.

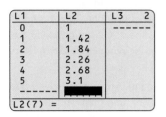

FIGURE 4.13 Linear data entered in the calculator

From Linear to Exponential Data

To establish the link between linear and exponential data, we look at an example. You should verify that the data in Table 4.1 are linear by showing that successive differences give a common value of 0.42. This is the slope for the linear data.

We have entered 4.8 these data in Figure 4.13 and then plotted 4.9 them in Figure 4.14. As expected, the points fall on a straight line.

Let's see what happens if we replace each *y* value in the table by e^y. This process is often referred to as *exponentiating* the data. Table 4.2 is the resulting table (with calculations rounded to four decimal places). (We have chosen to round to four decimal places on this occasion so that our hand-generated work will match what happens on the calculator later.)

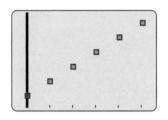

FIGURE 4.14 A plot of linear data

We have added 4.10 the data from Table 4.2 to the calculator data entry screen in Figure 4.15 and then plotted 4.11 them in Figure 4.16. The plot in Figure 4.16 has a familiar and not surprising shape; it looks like exponential data. You should verify that the data in Table 4.2 are indeed exponential by showing that successive quotients give a common value of 1.5219. This is the growth factor for the exponential data.

We should also take note of the relationship between the slope 0.42 of the linear data and the growth factor 1.5219 of the exponentiated data. Since we used e^x to generate the data in Table 4.2, the following calculation is not surprising:

$$e^{\text{slope of linear data}} = e^{0.42} = 1.5219 = \text{growth factor of exponential data}.$$

This same relationship holds for the initial value 1 for the linear data and the initial value 2.7183 for the exponential data, and you should verify the following calculation:

$$e^{\text{linear initial value}} = e^1 = 2.7183 = \text{exponential initial value}.$$

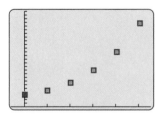

FIGURE 4.15 Entering exponentiated data

x	0	1	2	3	4	5
$y = y(x)$	1	1.42	1.84	2.26	2.68	3.1

TABLE 4.1 A table of linear data

x	0	1	2	3	4	5
e^y	2.7183	4.1371	6.2965	9.5831	14.5851	22.1980

TABLE 4.2 Exponentiating linear data

FIGURE 4.16 Plot of exponentiated data

The relationships we saw here are true in general, and the following schematic is presented to summarize these relationships and make them easy to remember.

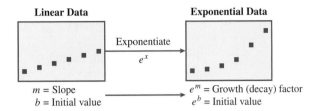

From Exponential to Linear Data: The Logarithm

The key tool needed to reverse the link—that is, to convert exponential data into linear data—is a special function known as the *natural logarithm*. This is denoted by $\ln x$, and it is closely related to the exponential function e^x. In formal mathematical terms, $\ln x$ and e^x are *inverse functions*. If we apply the exponential function to a number x, that number is put as an exponent for e, giving e^x. The logarithm does just the opposite; the logarithm applied to e^x removes the e and leaves just the exponent x. That is, $\ln e^x = x$. Thus, for example, $\ln e^2 = 2$. Informally speaking, the logarithm undoes exponentiation. More formally, $\ln x$ is the power of e that gives x. The values of such expressions as $\ln 7$ are in general very difficult to calculate by hand, and it is appropriate to get the evaluation from the calculator 4.12 as we have done in Figure 4.17. This new function, the natural logarithm, is no more complicated than the exponential function e^x, and a little experience with it will make it seem more familiar.

Let's see how the logarithm works to straighten out exponential data. This time we start with the exponential data in Table 4.2. In Figure 4.18 we have entered the exponential data in the third column. We know already that if we plot this data, we get the picture in Figure 4.16. We want to see what happens if we take the logarithm of each entry in the third column, as we have done 4.13 in Figure 4.19. In this figure, each entry in the second column is the logarithm of the corresponding entry in the third column; for example, 1 is the logarithm of 2.7183. The key thing to note here is that as a result, we recover in the second column exactly the linear data from Table 4.1. This shows clearly how the logarithm reverses the effects of exponentiation. That is, the logarithm of exponential data is linear. Thus, if we plot it, we will get the picture of linear data in Figure 4.14. When we go in this direction, from exponential to linear data using the

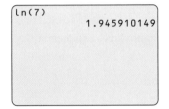

FIGURE 4.17 Calculating ln 7

L1	L2	L3	2
0	█████	2.7183	
1		4.1371	
2		6.2965	
3		9.5831	
4		14.585	
5		22.198	
------	------	------	
L2(1)=			

FIGURE 4.18 Exponential data entered in the calculator

L1	L2	L3	3
0	1	2.7183	
1	1.42	4.1371	
2	1.84	6.2965	
3	2.26	9.5831	
4	2.68	14.585	
5	3.1	22.198	
------	------	------	
L2(1)=1			

FIGURE 4.19 Using ln x to recover linear data from exponential data

logarithm, we get expected relationships among initial values, slope, and growth (or decay) factor:

$$\ln(\text{growth factor}) = \ln 1.5219 = 0.42 = \text{slope for linear data}$$

$$\ln(\text{exponential initial value}) = \ln 2.7183 = 1 = \text{linear initial value}.$$

As before, we provide the following schematic, which is intended to summarize our findings and suggest an easy way of remembering them.

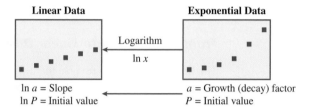

KEY IDEA **4.5**

The Connection Between Exponential and Linear Data

From linear to exponential: Exponentiating linear data produces exponential data.

$$e^{\text{slope}} = \text{growth (decay) factor}.$$

$$e^{\text{linear initial value}} = \text{exponential initial value}.$$

From exponential to linear: The logarithm of exponential data is linear.

$$\ln(\text{growth (decay) factor}) = \text{slope}.$$

$$\ln(\text{exponential initial value}) = \text{linear initial value}.$$

Exponential Regression

We want to see how the link we have established between linear and exponential data can be used to make important analyses. The following data, taken from the *1996 Information Please Almanac*, show the U.S. population from 1800 to 1860, just prior to the Civil War.

Date	1800	1810	1820	1830	1840	1850	1860
Population in millions	5.31	7.24	9.64	12.87	17.07	23.19	31.44

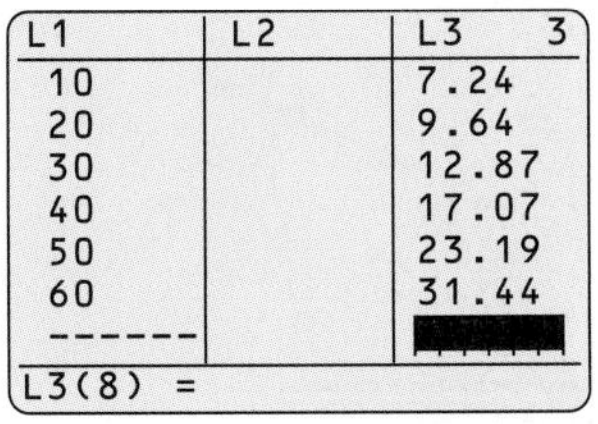

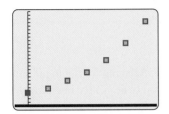

FIGURE 4.20 Entering pre–Civil War population data

Let t be the time in years since 1800 and N the population in millions. The table for N as a function of t is then

t	0	10	20	30	40	50	60
N	5.31	7.24	9.64	12.87	17.07	23.19	31.44

FIGURE 4.21 A plot of pre–Civil War data

Since the data table shows population growth, it is not unreasonable to suspect that the data might be exponential in nature. If you calculate successive quotients of N, you will find that the data are not exactly exponential, but as with linear regression, it may still be appropriate to make an exponential model that approximates the data. That is the first question we want to answer: "Is it reasonable to approximate the data with an exponential model?"

First we enter 4.14 the data in the calculator as we have done in Figure 4.20. Next we plot 4.15 them to get a look at the overall trend. Figure 4.21 shows the classic shape of exponential growth, and the idea that the data may be exponential is reinforced.

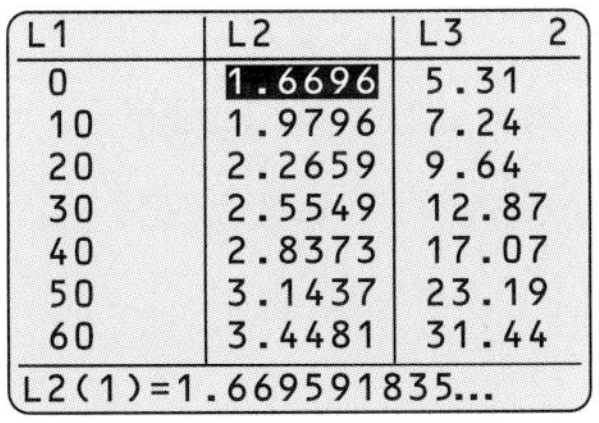

However, a caution is in order here. It is very difficult in general to be sure that data are exponential by looking at their plot, because there are other types of data that will give a similar appearance. The key to determining whether it is appropriate to model data by an exponential function is the logarithm/exponential link we have established between exponential and linear data. If these data are indeed exponential, then if we take the logarithm, the result should be linear data, and the plot of these data should show the points falling (approximately) on a straight line.[10] This is how we want to proceed. In Figure 4.22 we have entered the logarithm 4.16 of the data in the second column, and in Figure 4.23 we have plotted 4.17 them.

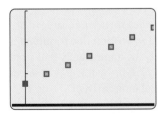

FIGURE 4.22 Automatic entry of the logarithm of the data

Figure 4.23 gives striking evidence that the logarithm of population data is nearly linear—and hence that the original population data are nearly exponential.

Now that we have convincing evidence that the original data set is exponential, we want to find an exponential function to model it. We find that there is another advantage to having "straightened out" the data with the logarithm; it enables us to apply what we know about linear data and make 4.18 a regression line. The regression line parameters for ln N are shown in Figure 4.24. Let's write down what all the letters there mean. We have rounded to four decimal places.[11]

FIGURE 4.23 The logarithm of the data appears to be linear

$$Y_1 = \ln N, \text{ the logarithm of population}$$

$$X = t, \text{ years since } 1800$$

$$a = 0.0294 = \text{slope of the regression line}$$

$$b = 1.6747 = \text{initial value of the regression line}.$$

---

[10]Traditionally this has been accomplished by plotting the data points on a semilogarithmic scale or on *semilogarithmic graphing paper*.

[11]In exploiting the logarithm/exponential link, it is often necessary to use more digits than our standard convention of two.

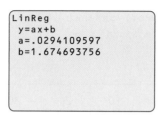

FIGURE 4.24 Regression line data for ln N

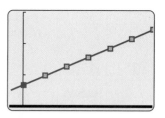

FIGURE 4.25 Regression line added to logarithm of data

Thus we can write the formula for the regression line as

$$\ln N = 0.0294t + 1.6747.$$

In Figure 4.25 we have added 4.19 the regression line to the plot of the logarithm of the data, and this reinforces our perception that these points lie nearly in a straight line.

There is one final step to making an exponential model for N. We used the logarithm to linearize the data and get a regression line for ln N. To go back to N and the original data, we should exponentiate. In performing this step, it is crucial to remember how slope, initial value, and growth factors are affected by this process.

Slope of $\ln N = 0.0294 \longrightarrow$ Growth factor of $N = e^{0.0294} = 1.0298$

Initial value of $\ln N = 1.6747 \longrightarrow$ Initial value of $N = e^{1.6747} = 5.3372$.

Thus we can approximate N with the exponential function

$$N = 5.3372 \times 1.0298^t.$$

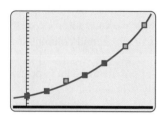

FIGURE 4.26 Exponential model for U.S. population data

In Figure 4.26 we have added the graph 4.20 of this function to the original data plot of N in Figure 4.21. The level of agreement between the exponential curve and the data points is striking.[12]

The process we have followed here is known as *exponential regression*, and we use the resulting formula in the same ways in which we used the regression line formula for linear data. For example, the growth factor 1.0298 for N tells us immediately that from 1800 to 1860, U.S. population grew at a rate of 2.98% per year, or by about 34.1% per decade. We might also use the formula to estimate the population in 1870. To do this, we would put 70 in for t:

Exponential regression estimate for $1870 = 5.3372 \times 1.0298^{70} = 41.69$ million.

There is some uncertainty about the actual U.S. population in 1870. The 1870 census reported the number as 38.56 million. This figure was later revised to 39.82 million because it was thought that the southern population had been undercounted. Whether we use the original or the revised 1870 census estimate, it is clear that U.S. population grew a good deal less than we would have expected from the exponential trend established from 1800 to 1860. Such a discrepancy leads us to seek a historical explanation, and the most

---

[12]Although exponential models are often used to study population growth, such close agreement between the model and real data is unusual.

obvious culprit is the Civil War. In fact the death and disruption of the Civil War may have had long-lasting effects on U.S. population growth. From 1790 to 1860, population grew at a steady rate of around 34.1% per decade. From 1860 to 1870, population grew by only 27%, and this rate steadily declined to its historical low of 7.2% from 1930 to 1940, the decade of the Great Depression.

Exponential Regression

When there is reason to believe that data for a function $N = N(t)$ might be approximately modeled by an exponential function, exponential regression can be performed as follows:

Step 1: Take the logarithm of the data for N and plot it. If N represents approximately exponential data, these points will lie nearly in a straight line.

Step 2: Get the regression line for $\ln N$.

Step 3: Get the exponential regression formula for N from the regression line for $\ln N$ by exponentiating. The key relationships are

$$e^{(\text{slope of regression line})} = \text{growth (decay) factor}.$$

$$e^{(\text{initial value of regression line})} = \text{initial value for } N.$$

EXAMPLE 4.7 **Colonial Population Growth**

The following data table shows colonial population from 1610 to 1670, where t is the number of years since 1610 and $C = C(t)$ is the population in thousands.[13] The source for the data, *The 1996 Information Please Almanac*, cautions that records from this period are spotty, so the numbers should be considered estimates.

t = years since 1610	0	10	20	30	40	50	60
C = population in thousands	0.35	2.3	4.6	26.6	50.4	75.1	111.9

1. Plot the data to get an overall view of their nature.

2. Plot the logarithm of the data. Does the plot indicate that the original data might be modeled by an exponential function?

---

[13] Historical records from these periods ignored Native Americans.

3. Use exponential regression to construct a model for C. (Round regression line parameters to three decimal places.) Add the graph of the exponential model to the plot of population data.

4. Compare the population growth rate as given by exponential regression during colonial times with that in the 60 years preceding the Civil War.

L1	L2	L3 3
0	------	.35
10		2.3
20		4.6
30		26.6
40		50.4
50		75.1
60		111.9

L3(7)=111.9

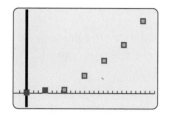

FIGURE 4.27 Entering colonial data

Solution to Part 1: We enter 4.21 the data in the calculator as shown in Figure 4.27 and then plot 4.22 them as in Figure 4.28. The points in Figure 4.28 show some features of exponential growth; it is slow at first and faster later on, but the picture certainly leaves room for doubt.

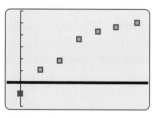

FIGURE 4.28 Plot of colonial data

Solution to Part 2: We get the logarithm 4.23 of the data as shown in Figure 4.29. When we plot the logarithm of the data as we did in Figure 4.30, we see that the points may generally line up, but they show some erratic behavior. The widely dispersed distribution of the points should cause us to be somewhat skeptical of these data being modeled by a linear function—and hence of the original data being modeled by an exponential function.

L1	L2	L3 2
0	⁻1.05	.35
10	.83291	2.3
20	1.5261	4.6
30	3.2809	26.6
40	3.92	50.4
50	4.3188	75.1
60	4.7176	111.9

L2(1)=⁻1.04982212...

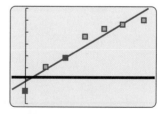

FIGURE 4.29 Getting the logarithm of the data

Solution to Part 3: We get the regression line parameters for ln C as shown in Figure 4.31. Let's record the appropriate correspondences:

$$\mathsf{Y}_1 = \ln C, \text{ logarithm of population}$$
$$\mathsf{X} = t, \text{ years since 1610}$$
$$\mathsf{a} = 0.095, \text{ slope of regression line}$$
$$\mathsf{b} = -0.351, \text{ initial value of regression line}.$$

Thus the regression line for ln C is given by

$$\ln C = 0.095t - 0.351\,.$$

In Figure 4.32 we have added 4.24 the graph of the regression line for ln C to the plot of the logarithm of the data. The points are scattered about the line, but agreement remains dubious.

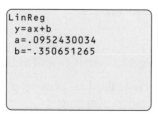

FIGURE 4.30 Plotting ln C: The distribution is erratic

```
LinReg
 y=ax+b
 a=.0952430034
 b=⁻.350651265
```

FIGURE 4.31 Regression line parameters for ln C

FIGURE 4.32 Regression line added to plot of ln C versus t

We exponentiate to get the model for C:

$$\text{Growth factor} = e^{0.095} = 1.1$$

$$\text{Initial value} = e^{-0.351} = 0.704 \,.$$

Using these values, we get the exponential model for C:

$$C = 0.704 \times 1.1^{t} \,.$$

Finally, in Figure 4.33, we add $\boxed{4.25}$ the graph of the exponential model to the plot of data points for C.

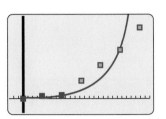

FIGURE 4.33 Exponential model added to plot of C versus t

Solution to Part 4: According to the exponential model we have constructed, colonial population grew at a rate of about 10% per year as compared with 2.98% per year from 1800 to 1860. But since the colonial model is so inaccurate, the estimate of 10% is unreliable. ∎ ∎ ∎

ANOTHER LOOK

Laws of Logarithms

We want to give an algebraic explanation of how exponential regression works. To do this, we need first to look a little more closely at logarithms. We introduced the natural logarithm $\ln x$ as the power of e that gives x. Thus $\ln x = p$ if and only if $x = e^{p}$. We use this to obtain the basic laws of logarithms, which are nothing more than the basic laws of exponents in reverse:

Product law: $\ln(xy) = \ln x + \ln y$ if $x, y > 0$

Quotient law: $\ln\left(\dfrac{x}{y}\right) = \ln x - \ln y$ if $x, y > 0$

Power law: $\ln\left(x^{y}\right) = y \ln x$ if $x > 0$

Let's verify the product law. Suppose $\ln x = p$ and $\ln y = q$. Then $x = e^{p}$ and $y = e^{q}$. Hence

$$xy = e^{p} e^{q} = e^{p+q} \,.$$

Thus $p + q$ is the power of e that gives xy. We conclude that

$$\ln(xy) = p + q = \ln x + \ln y \,.$$

The formal demonstration of the quotient law is very similar, and you will have the opportunity to try it yourself in the exercises. We proceed to the power law. Once again we let $\ln x = p$ so that $e^{p} = x$. Raising each side of this equation to the y power gives

$$x^{y} = \left(e^{p}\right)^{y} = e^{py} \,.$$

Thus py is the power of e that gives x^y. We conclude that

$$\ln\left(x^y\right) = yp = y \ln x .$$

We can use the basic laws of logarithms to provide a formal justification for some of the most important properties of logarithms. For example, in this section we relied heavily on the statement that the logarithm transforms exponential data into linear data—that is, the logarithm of any exponential function is a linear function. To verify this, suppose $N = Pa^t$ is an exponential function. We need to show that $\ln N$ is a linear function of t. We compute that

$$\ln N = \ln\left(Pa^t\right) = \ln P + \ln\left(a^t\right) = \ln P + t \ln a .$$

If we let $K = \ln P$ and $L = \ln a$, then we can write $\ln N = K + Lt$, which is clearly a linear function of t. We have shown a little more: We have shown that if we take the logarithm of the exponential function $N = Pa^t$, then we get a linear function with initial value $\ln P$ and slope $\ln a$.

We are now in a position to verify that our program for exponential regression works. Suppose we begin with data that come from some (unknown) exponential function $N = Pa^t$. We take the logarithm of the data and get a regression line of the form $\ln N = K + Lt$, where $K = \ln P$ and $L = \ln a$. Since $K = \ln P$, we have $P = e^K$. That is, we get the initial value of the exponential function by exponentiating the initial value of the regression line. Similarly, since $L = \ln a$, we have $a = e^L$. Thus we get the growth factor by exponentiating the slope of the regression line.

The laws of logarithms enable us to do some important calculations. Suppose, for example, that $\ln x = 3$ and $\ln y = 4$. We can use the laws of logarithms to calculate $\ln(x^2/y^3)$ as follows:

$$\ln\left(\frac{x^2}{y^3}\right) = \ln x^2 - \ln y^3$$

$$= 2 \ln x - 3 \ln y$$

$$= 2 \times 3 - 3 \times 4$$

$$= -6 .$$

Enrichment Exercises

E-1. **Verifying the quotient law:** Verify that the quotient law holds for logarithms.

E-2. **Exponentiating linear functions:** Suppose $y = mx + b$ is a linear function. Show that if we exponentiate, we get an exponential function with initial value e^b and growth or decay factor e^m.

E-3. **An important fact about logarithms:** Show that $e^{\ln x} = x$.

E-4. **Calculating using the laws of logarithms:** Suppose that $\ln x = 4$ and $\ln y = 5$. Calculate each of the following:

a. $\ln\left(\dfrac{x^4}{y^2}\right)$

b. $\ln x^3 y^5$

c. $\ln\left(\dfrac{1}{x}\right)$

d. $\ln \sqrt{xy}$

E-5. **More calculations using laws of logarithms:** Use the laws of logarithms and the fact that $\ln e = 1$ to calculate the following:

a. $\ln e^3$

b. $\ln\left(\dfrac{\sqrt{e}}{e^4}\right)$

c. $\ln\dfrac{1}{e}$

d. $\ln\ln\left(e^{(e^4)}\right)$

E-6. **Logarithms to other bases:** In general, we define the logarithm to the base $a > 0$ as follows: $\log_a x$ is the power of a that gives x. (Thus $\ln x = \log_e x$.) For example, $\log_2 8 = 3$ since $2^3 = 8$. Calculate the following:

a. $\log_2 64$

b. $\log_2 \dfrac{1}{4}$

c. $\log_{27} 3$

d. $\log_8 8$

e. $\log_a 1$

f. $a^{\log_a x}$

g. $\log_a a^x$

E-7. **Change of base:** *This is a continuation of Exercise E-6.* Show that

$$a^{(\log_a b)(\log_b x)} = x.$$

Explain why this gives the *change-of-base* formula

$$\log_b x = \frac{\log_a x}{\log_a b}.$$

E-8. **Using the change-of-base formula:** *This exercise uses the change-of-base formula from Exercise E-7.* Use the change-of-base formula to get the special case

$$\log_b x = \frac{\ln x}{\ln b}.$$

Use this formula and your calculator to evaluate the following:

a. $\log_7 12$

b. $\log_{3.2} 5.8$

c. $\log_5 e$

4.3   *SKILL BUILDING EXERCISES*

S-1. **Exponential transformation:** If the table of values

Input value	a	b	...
Function value	z	w	...

consists of linear data, then what kind of data would appear in the following table?

Input value	a	b	...
Function value	e^z	e^w	...

S-2. **Logarithmic conversion:** If the table of values

Input value	a	b	...
Function value	z	w	...

consists of exponential data, then what kind of data would appear in the following table?

Input value	a	b	...
Function value	$\ln z$	$\ln w$	...

S-3. **Slope and growth factor:** The slope of the regression line for the natural logarithm of exponential data is −0.77. Find the decay factor for the exponential function.

S-4. **Regression line to exponential model:** The regression line for the natural logarithm of exponential data is $y = 3x + 4$. Find the exponential model for the data.

S-5. **Exponential regression:** Use exponential regression to fit the following data set. Give the ex-

ponential model, and plot the data along with the model.

x	1	2	3	4	5
y	4.1	8.7	19.2	28.6	64.7

S-6. **Exponential regression:** Use exponential regression to fit the following data set. Give the exponential model, and plot the data along with the model.

x	1	2	3	4	5
y	0.7	0.3	0.1	0.05	0.01

S-7. **Exponential regression:** Use exponential regression to fit the following data set. Give the exponential model, and plot the data along with the model.

x	4.2	7.9	10.8	15.5	20.2
y	7.5	8.1	8.5	10.2	12.3

S-8. **Exponential regression:** Use exponential regression to fit the following data set. Give the exponential model, and plot the data along with the model.

x	y
22.4	0.053
27.3	0.025
29.4	0.011
34.1	0.005
38.6	0.002

S-9. **Exponential regression:** Use exponential regression to fit the following data set. Give the exponential model, and plot the data along with the model.

x	1	2	3	4	5
y	3.7	4.3	6.1	9.1	13.6

S-10. **Exponential regression:** Use exponential regression to fit the following data set. Give the exponential model, and plot the data along with the model.

x	y
3	33.5
7	988.8
9	5470.8
10	12,830
15	893,442

4.3 EXERCISES

Special rounding instructions: For this exercise set, round all regression line parameters to three decimal places.

1. **Using the definition of natural logarithm:** Recall that ln x is the power of e that gives x.

   a. Without using your calculator, find the value of ln e^4.

   b. Use your calculator to show that $e^6 = 403.43$, rounded to two decimal places. Use this calculation to give the value of ln 403.43. Verify your answer by calculating the logarithm with the calculator.

2. **The common logarithm:** The *common logarithm,* or *logarithm to the base 10,* works just like the natural logarithm[14] except that it uses 10 in place of e. Scientists and mathematicians normally use log x to denote the common logarithm. Formally, log x is the power of 10 that gives x.

   a. What is log 10? (*Hint:* What power of 10 gives 10?)

   b. What is log 1000?

3. **Population growth:** The following table shows the size, in thousands, of an animal population at the start of the given year.

Year	Population (thousands)
1997	1.56
1998	1.62
1999	1.71
2000	1.77
2001	1.84

   a. Plot the natural logarithm of the data points. Does this plot make it look reasonable to approximate the original data with an exponential function?

   b. Find the regression line for the natural logarithm of the data and add its graph to the plot of the logarithm.

---

[14]Historically, the common logarithm was used before the natural logarithm. It can be used to do all the things we did in this chapter if e is replaced by 10 at each step. Its properties are developed in the next section.

4. **Magazine circulation:** The following table shows the circulation, in thousands, of a magazine at the start of the given year.

Year	Circulation (thousands)
1996	2.64
1997	2.77
1998	2.94
1999	3.08
2000	3.25
2001	3.42

a. Plot the natural logarithm of the data points. Does this plot make it look reasonable to approximate the original data with an exponential function?

b. Find the regression line for the natural logarithm of the data and add its graph to the plot of the logarithm.

5. **Sales growth:** The total sales S, in thousands of dollars, of a small firm is growing exponentially with time t (measured in years since the start of 2001). Analysis of the sales growth has given the following linear model for the natural logarithm of sales:

$$\ln S = 0.049t + 2.230.$$

a. Find an exponential model for sales.

b. By what percentage do sales grow each year?

c. Calculate $S(6)$ and explain in practical terms what your answer means.

d. When would you expect sales to reach a level of 12 thousand dollars?

6. **Population decline:** The population N, in thousands, of a city is decreasing exponentially with time t (measured in years since the start of 2000). City analysts have given the following linear model for the natural logarithm of population:

$$\ln N = -0.051t + 1.513.$$

a. Find an exponential model for population.

b. By what percentage is the population decreasing each year?

c. Express using functional notation the population at the start of 2003 and then calculate that value.

d. When will the population fall to a level of 3 thousand?

7. **Cable TV:** The following table shows the percentage of American households with cable TV for the years from 1976 through 1984.

Date	Percent with cable TV
1976	15.1
1977	16.6
1978	17.9
1979	19.4
1980	22.6
1981	28.3
1982	35
1983	40.5
1984	43.7

a. Plot the natural logarithm of the data points. Does this plot make it look reasonable to approximate the original data with an exponential function?

b. Find the regression line for the natural logarithm of the data and add its graph to the plot of the logarithm.

c. Use exponential regression to construct an exponential model for the original cable TV data.

d. Plot the original data points and the exponential model.

e. What was the yearly percentage growth rate from 1976 through 1984 for the percentage of homes with cable TV?

f. In 1984 an executive had a plan that could make money for the company, provided that at least 65% of American homes could be expected to have cable TV by 1987. Solely on the basis of an exponential model for the data in the table, would it be reasonable for the executive to implement the plan?

8. **Auto parts production workers:** The following table, taken from the *1994 U.S. Industrial Outlook*, shows the average hourly wages for American auto parts production workers from 1987 through 1994.

Date	Hourly wage
1987	$13.79
1988	$14.72
1989	$14.99
1990	$15.35
1991	$15.70
1992	$16.15
1993	$16.50
1994	$16.85

a. Plot the natural logarithm of the data. Does it appear that it is reasonable to model auto parts worker wages using an exponential function?

b. Find the equation of the regression line for the natural logarithm of the data.

c. Make an exponential model for auto parts worker wages.

d. What was the yearly percentage growth rate in average hourly wages for auto parts producers during this period?

e. From 1987 through 1994, inflation was about 3.8% per year. If hourly wages beginning at $13.79 in 1987 had kept pace with inflation, what would be the average hourly wage in 1994?

f. What percentage raise should a worker receiving a wage of $16.85 in 1994 get in order to bring wages in line with inflation?

9. **National health care spending:** The following table shows national health care costs, measured in billions of dollars.

Date	Costs in billions
1950	12.7
1960	26.9
1970	75
1980	248
1990	600

a. Plot the natural logarithm of the data. Does it appear that the data on health care spending can be appropriately modeled by an exponential function?

b. Find the equation of the regression line for the logarithm of the data and add its graph to the plot in part a.

c. By what percent per year were national health care costs increasing during the period from 1950 through 1990?

d. Find an exponential function that approximates the original data for health care costs.

e. Use functional notation to express how much money was spent on health care in the year 2000, and then estimate that value.

10. **A bad data point:** A scientist sampled data for a natural phenomenon that she has good reason to believe is appropriately modeled by an exponential function $N = N(t)$. A laboratory assistant reported that there may have been an error in recording one of the data points but is not certain which one. Which is the suspect data point? Explain your reasoning.

t	N
0	21.3
1	37.5
2	66.0
3	102.3
4	204.4
5	359.7
6	663.1

11. **Grazing rabbits:** The amount A of vegetation (measured in pounds) eaten in a day by a grazing animal is a function of the amount V of food available (measured in pounds per acre).[15] Even if vegetation is abundant, there is a limit, called the *satiation level*, to the amount the animal will eat. The following table shows, for rabbits, the difference D between the satiation level and the amount A of food eaten for a variety of values of V.

V = vegetation level	D = satiation level $- A$
27	0.16
36	0.12
89	0.07
134	0.05
245	0.01

a. Draw a plot of ln D against V. Does it appear that D is approximately an exponential function of V?

b. Find the equation of the regression line for ln D against V and add its graph to the plot in part a.

c. Find an exponential function that approximates D.

d. The satiation level of a rabbit is 0.18 pound per day. Use this, together with your work in part c, to find a formula for A.

e. Find the vegetation level V for which the amount of food eaten by the rabbit will be 90% of its satiation level of 0.18 pound per day.

12. **Growth in length:** In the fishery sciences it is important to determine the length of a fish as a function of its age. One common approach, the von Bertalanffy model, uses a decreasing exponential function of age to describe the growth in length yet to be attained; in other words, the difference between the maximum length and the current length is supposed to decay exponentially with age. The following table shows the length L, in inches, at age t, in years, of the North Sea sole.[16]

t = age	L = length
1	3.7
2	7.5
3	10.0
4	11.5
5	12.7
6	13.5
7	14.0
8	14.4

The maximum length attained by the sole is 14.8 inches.

a. Make a table showing, for each age, the difference D between the maximum length and the actual length L of the sole.

b. Find the exponential function that approximates D.

c. Find a formula expressing the length L of a sole as a function of its age t.

d. Draw a graph of L against t.

e. If a sole is 11 inches long, how old is it?

f. Calculate $L(9)$ and explain in practical terms what your answer means.

13. **Nearly linear or exponential data:** One of the two tables below shows data that are better approximated with a linear function, and the other shows data that are better approximated with an exponential function. Make plots to identify which is which, and then use the appropriate regression to find models for both.

---

[15]This exercise is based on the work of J. Short, "The functional response of kangaroos, sheep and rabbits in an arid grazing system," *Journal of Applied Ecology* **22** (1985), 435–447.

[16]The table is from the work of A. Bückmann, as described by R. J. H. Beverton and S. J. Holt, *On the Dynamics of Exploited Fish Populations*, Fishery Investigations, Series 2, Volume 19 (London: Ministry of Agriculture, Fisheries and Food, 1957).

t	f(t)
1	3.62
2	23.01
3	44.26
4	62.17
5	83.25

Table A

t	g(t)
1	3.62
2	5.63
3	8.83
4	13.62
5	21.22

Table B

14. **Atmospheric pressure:** The following table gives a measurement of atmospheric pressure, in grams per square centimeter, at the given altitude, in kilometers.[17]

Altitude	Atmospheric pressure
5	569
10	313
15	172
20	95
25	52

(For comparison, 1 kilometer is about 0.6 mile, and 1 gram per square centimeter is about 2 pounds per square foot.)

a. Plot the natural logarithm of the data, and find the equation of the regression line for the natural logarithm of the data.

b. Make an exponential model for the data on atmospheric pressure.

c. What is the atmospheric pressure at an altitude of 30 kilometers?

d. Find the atmospheric pressure on Earth's surface. This is termed *standard atmospheric pressure*.

e. At what altitude is the atmospheric pressure equal to 25% of standard atmospheric pressure?

15. **Sound pressure:** Sound exerts pressure on the human ear. Increasing loudness corresponds to greater pressure. The table below shows the pressure P, in dynes per square centimeter, exerted on the ear by sound with loudness D, measured in decibels.

Loudness D	Pressure P
65	0.36
85	3.6
90	6.4
105	30
110	50

a. Plot the natural logarithm of the data. Does it appear reasonable to model pressure as an exponential function of loudness?

b. Find an exponential model of P as a function of D.

c. How is pressure on the ear affected when loudness is increased by 1 decibel?

16. **Magnitude and distance:** Astronomers measure brightness of stars using both the *absolute magnitude*, a measure of the true brightness of the star, and the *apparent magnitude*, a measure of the brightness of a star as it appears from Earth.[18] The difference between apparent and absolute magnitude should yield information about the distance to the star. The table below gives magnitude difference m and distance d, measured in light-years, for several stars.

→

Star	Magnitude difference m	Distance d
Algol	2.56	105
Aldebaran	1.56	68
Capella	0.66	45
Canopus	2.38	98
Pollux	0.13	35
Regulus	2.05	84

[17]This exercise is based on *Space Mathematics* by B. Kastner, published in 1985 by NASA.

[18]Larger apparent magnitudes correspond to dimmer stars.

a. Plot $\ln d$ against m and determine whether it is reasonable to model distance as an exponential function of magnitude difference.

b. Give an exponential model for the data.

c. If one star shows a magnitude difference 1 greater than the magnitude difference that a second star shows, how do their distances from Earth compare?

d. Alphecca shows a magnitude difference of 1.83. How far is Alphecca from Earth?

e. Alderamin is 52 light-years from Earth and has an apparent magnitude of 2.47. Find the absolute magnitude of Alderamin.

17. **Growth in length of haddock:** A study by Riatt showed that the maximum length a haddock could be expected to grow is about 53 centimeters. Let $D = D(t)$ denote the difference between 53 centimeters and the length at age t years. The table below gives experimentally collected values for D.

Age t	Difference D
2	28.2
5	16.1
7	9.5
13	3.3
19	1.0

a. Find an exponential model of D as a function of t.

b. Let $L = L(t)$ denote the length in centimeters of a haddock at age t years. Find a model for L as a function of t.

c. Plot the graph of the experimentally gathered data for the length L at ages 2, 5, 7, 13, and 19 years along with the graph of the model you made for L. Does this graph show that the 5-year-old haddock is a bit shorter or a bit longer than would be expected?

d. A fisherman has caught a haddock that measures 41 centimeters. What is the approximate age of the haddock?

18. **Caloric content versus shell length:** In 1965 Robert T. Paine[19] gathered data on the length L, in millimeters, of the shell and the caloric content C, in calories, for a certain mollusk. The table below is adapted from those data.

L = length	C = calories
7.5	92
13	210
20	625
24	1035
31	1480

a. Find an exponential model of calories as a function of length.

b. Plot the graph of the data and the exponential model. Which of the data points show a good deal less caloric content than the model would predict for the given length?

c. If length is increased by 1 millimeter, how is caloric content affected?

19. **Injury versus speed:** The following data are adapted from a report[20] by D. Solomon that relates the number N of persons injured per 100 accident-involved vehicles to the travel speed s in miles per hour.

s	20	30	40	50	60
N	25	32	38	43	60

a. Find an exponential model for N as a function of s.

[19] See "Natural history, limiting factors and energetics of the opisthobranch *Navanax inermis*," *Ecology* **46** (1965), 603–619.

[20] *Accidents on Main Rural Highways Related to Speed, Driver, and Vehicle* (Washington, DC: Federal Highway Administration, July 1964), p. 11.

b. Calculate $N(70)$ and explain in practical terms what your answer means.

c. How does an increase in 1 mile per hour of speed affect the number of people injured per 100 accident-involved vehicles?

20. **Gray wolves in Wisconsin:** Gray wolves were among the first mammals protected under the Endangered Species Act in the 1970s. Wolves recolonized in Wisconsin beginning in 1980. Their population has grown reliably since 1985 as follows:[21]

Year	Wolves	Year	Wolves
1985	15	1993	40
1986	16	1994	57
1987	18	1995	83
1988	28	1996	99
1989	31	1997	145
1990	34	1998	178
1991	40	1999	197
1992	45	2000	266

a. Explain why an exponential model may be appropriate.

b. Are these data exactly exponential? Explain.

c. Find an exponential model for these data.

d. Plot the data and the exponential model.

e. Comment on your graph in part d. Which data points are below or above the number predicted by the exponential model?

21. **Walking in Seattle:** It is common in large cities for people to travel to the center city and then walk to their final destination. In Seattle the percent P of pedestrians who walk at least D feet from parking facilities in the center city is given partially in the accompanying table[22].

Distance D	Percent P walking at least D feet
300	60
500	40
1000	18
1500	9
2000	5

a. Make a model of P as an exponential function of D. (Round the decay factor to three decimal places.)

b. What percentage of pedestrians walk at least 200 feet from parking facilities?

c. When models of this sort are made, it is important to remember that many times they are only rough indicators of reality. Often they apply only to parts of the data and may tell very little about extremes of the data. Explain why it is not appropriate to view this as an accurate model for very short distances walked.

22. **Traffic in the Lincoln Tunnel:** Characteristics of traffic flow include density D, which is the number of cars per mile, and average speed s in miles per hour. Traffic system engineers have investigated several methods for relating density to average speed. One study[23] considered traffic flow in the north tube of the Lincoln Tunnel and fitted an exponential function to observed data. Those data are partially presented in the table below.

Speed s	32	25	20	17	13
Density D	34	53	74	88	102

a. Make an approximate exponential model of D as a function of s. $\rightarrow$

[21] Data courtesy of the International Wolf Center, Ely, Minnesota. For more information about wolves and their recovery, go to **http://www.wolf.org**.

[22] Adapted from Institute of Traffic Engineers, *Transportation and Traffic Engineering Handbook*, ed. John E. Baerwald (Englewood Cliffs, NJ: Prentice-Hall, 1976, 180).

[23] H. Greenberg, *A Mathematical Analysis of Traffic Flow*, Tunnel Traffic Capacity Study, the Port of New York Authority, New York, 1958.

b. Express using functional notation the density of traffic flow when the average speed is 28 miles per hour, and then calculate that density.

c. If average speed increases by 1 mile per hour, what can be said about density?

23. **Frequency of earthquakes:** The table below gives the average number N of earthquakes[24] of magnitude at least M that occur each year worldwide.

Magnitude M	Number N with magnitude at least M
6	95.8
6.1	77.8
6.6	27.6
7	12.1
7.3	6.5
8	1.5

a. Gutenberg and Richter fitted these data with an exponential function. Find an approximate exponential model for the data. (Round the decay factor to three decimal places.)

b. How many earthquakes per year of magnitude at least 5.5 can be expected?

c. Gutenberg and Richter found that the model fit observed data well up to a magnitude of about 8, but for magnitudes above 8 there were considerably fewer quakes than predicted by the model. How many quakes per year of magnitude 8.5 or greater does the model predict?

d. What is the limiting value for N? Explain in practical terms what this means.

e. How does the number of earthquakes per year of magnitude M or greater compare with the number of earthquakes of magnitude $M + 1$ or greater?

24. **Laboratory experiment:** This lab uses a motion detector and a calculator-based laboratory (CBL™) unit. The motion detector measures the distance from the detector to an object in front of it, and the CBL records the data. After collecting the data, the CBL unit sends the data to the calculator, which displays a graph of the distance recorded with respect to time. In this lab, we drop a basketball away from the motion detector and measure the rebound heights as it bounces. For a detailed description, go to

http://math.college.hmco.com/students

and then, from the book's website, go to the Bouncing Basketball experiment.

25. **Laboratory experiment:** This lab uses a temperature probe and a calculator-based laboratory (CBL™) unit. In this experiment we will investigate how objects cool by heating the probe in hot water and allowing the probe to cool in the air. For a detailed description, go to

http://math.college.hmco.com/students

and then, from the book's website, go to the Newton's Law of Cooling experiment.

26. **Laboratory experiment:** This lab uses a temperature probe and a calculator-based laboratory (CBL™) unit. In this experiment we will investigate how objects heat by cooling the probe in ice water and allowing the probe to heat in the air. For a detailed description, go to

http://math.college.hmco.com/students

and then, from the book's website, go to the Newton's Law of Heating experiment.

27. **Research project:** For this project, you should collect and analyze data for a population of M&M's™. Start with four candies, toss them on a plate, and add one for each candy that has the M side up; record the data. Repeat this seven times and see how close the data are to being exponential. For a detailed description, go to

http://math.college.hmco.com/students

and then, from the book's website, go to the Population Growth experiment.

---

[24]Data adapted from B. Gutenberg and C. F. Richter, "Earthquake magnitude, intensity, energy, and acceleration," *Bull. Seism. Soc. Am.* **46** (1956).

4.4   *Logarithmic Functions*

Up to now, we have used logarithms as an aid to working with exponential functions. But logarithms also have important direct applications to the world around us. The most common applications, such as the Richter scale for measuring seismic disturbances, make use of the common logarithm rather than the natural logarithm, so we will emphasize the common, or base-10, logarithm in this section.

The Richter Scale

The properties of the logarithm are easier to understand if we first look at a familiar scale that is logarithmic in nature. The seismograph measures ground movement during an earthquake, and that movement is normally reported using the *Richter scale.*

The Richter scale was developed in 1935 by Charles F. Richter as a device for comparing the sizes of earthquakes. A moderate earthquake may measure 5.3 on the Richter scale, and a strong earthquake may measure 6.3. The largest shocks ever recorded have a magnitude of 8.9. Thus a small change in the Richter number indicates a large change in the severity of the earthquake. In fact, an earthquake 6.3 in magnitude is 10 times more powerful than a 5.3 earthquake, and a 7.3 earthquake is 10 times 10, or 100, times more powerful than a 5.3 earthquake. In general, an increase of t units on the Richter scale indicates an earthquake 10^t times as strong. This is the key factor in understanding how the Richter scale works.

EXAMPLE 4.8   **Comparing Some Famous Earthquakes**

On December 16, 1811, an earthquake occurred near New Madrid, Missouri, that temporarily reversed the course of the Mississippi River. This was actually one of a series of earthquakes in the area, one of which is estimated to have had a Richter magnitude of 8.8. The area was sparsely populated at the time, and there were thought to be few fatalities. On October 17, 1989, a calamitous earthquake measuring 7.1 on the Richter scale occurred in the San Francisco Bay area. The earthquake killed 67 and injured over 3000.

1. How much more powerful was the New Madrid quake than the 1989 San Francisco quake?

2. If an earthquake 1000 times as powerful as the 1989 San Francisco earthquake occurred, what would its Richter scale measurement be?

3. On June 21, 1990, an earthquake 4 times as powerful as the 1989 San Francisco quake struck northwestern Iran. What was the Richter scale reading for the Iran quake?

Solution to Part 1:   The New Madrid quake registered $8.8 - 7.1 = 1.7$ points higher on the Richter scale than the San Francisco quake. Thus the New Madrid quake was $10^{1.7} = 50.12$ times stronger than the San Francisco quake.

Solution to Part 2: A quake 10^t times more powerful registers t points higher on the Richter scale. Since $1000 = 10^3$, the proposed quake would register 3 points higher, or $7.1 + 3 = 10.1$, on the Richter scale.

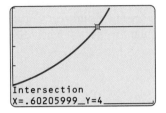

FIGURE 4.34 Solving $10^t = 4$

Solution to Part 3: This problem is similar to part 2, except that it is easier to write 1000 as a power of 10 than it is to write 4 as a power of 10. We need to solve the equation

$$10^t = 4.$$

The crossing-graphs method can be used to do this. (We will discover a more direct method shortly.) In Figure 4.34 we have plotted 10^t, and we see from the prompt at the bottom of the screen that $10^t = 4$ when $t = 0.60$. That means that the Iran quake was $10^{0.6}$ times more powerful than the San Francisco quake. Thus the Iran quake registered $7.1 + 0.6 = 7.7$ on the Richter scale. ▧ ▧ ▧

How the Common Logarithm Works

Recall that $\log x$ is the power of 10 that gives x. Alternatively, $\log x$ is the solution for t of the equation $10^t = x$. This makes common logarithms of integral powers of 10 easy to calculate. Note in the table below that the value recorded for the logarithm in the second row is indeed the power on 10 given in the first row.

x	$1 = 10^0$	$10 = 10^1$	$100 = 10^2$	$1000 = 10^3$	$10,000 = 10^4$
$\log x$	0	1	2	3	4

This table suggests two important features of the logarithm. The first is that, although the logarithm function increases without bound, it increases very slowly. One million is 10 to the sixth power, so $\log 1,000,000 = 6$. Similarly, the logarithm of 1 billion is only 9. We can see how the logarithm function grows slowly by looking at its graph. In Figure 4.35 we have graphed $\log x$ with a horizontal span of 0 to 10 and a vertical span of -1 to 2. Note that the graph crosses the horizontal axis at $x = 1$ since $\log 1 = 0$. In Figure 4.36 we use a horizontal span of 10 to 1,000,000 and a vertical span of 0 to 7. Note that the graph is indeed increasing slowly and that it is concave down. That is, it increases at a decreasing rate.

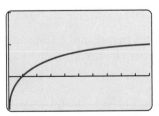

FIGURE 4.35 $\log x$ from 0 to 10

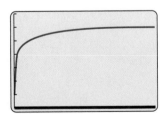

FIGURE 4.36 $\log x$ up to 1 million

The second feature of the logarithm that is apparent from the table is that increasing x by a factor of 10 increases the logarithm by 1 unit. This is precisely what happens with the Richter scale, which indicates that indeed it is a logarithmic scale.

KEY IDEA 4.7

The Common Logarithm

- $\log x$ is the power of 10 that gives x.
  Alternatively, $\log x$ is the solution for t of the equation $10^t = x$.
- If we multiply a number by 10^t, the logarithm is increased by t units.
- The function $\log x$ increases very slowly, and its graph is concave down.

Note that the second property in the Key Idea above is precisely what happens with the Richter scale. This property is critical when we are working with the Richter scale and with the logarithm in general.

The alternative definition of the logarithm also gives us an easy way to solve problems such as part 3 of Example 4.8. There we solved the equation $10^t = 4$ using the crossing-graphs method. The alternative definition of the logarithm says that the solution of this equation is $t = \log 4 = 0.60$. One of the reasons why the logarithm is so useful is that it enables us to solve such equations directly.

There are other important and familiar scales which are logarithmic in nature. Here is an example.

EXAMPLE 4.9   Decibel as a Measure of Loudness

The intensity of sound is measured in watts per square meter. Normally this is given as a *relative intensity*.[25] Loudness of sound is measured in *decibels*, and the number of decibels is 10 times the logarithm of the relative intensity:

$$\text{Decibels} = 10 \log(\text{relative intensity}).$$

1. If one sound has a relative intensity 100 times that of another sound, how do their decibel readings compare?

2. A stereo speaker is producing music at 60 decibels. If a second stereo speaker producing music at 60 decibels is placed next to the first, the relative intensity doubles. What is the decibel reading of the combined speakers?

---

[25]Physicists have adopted a base intensity of 10^{-12} watts per square meter. The relative intensity is intensity divided by the base intensity.

Solution to Part 1:   Relative intensity is increased by a factor of $100 = 10^2$. Thus the logarithm is increased by 2 units. Since the decibel is 10 times the logarithm, the decibel level is increased by $10 \times 2 = 20$ units.

Solution to Part 2:   We know that if relative intensity is increased by a factor of 10^t, then t is added to the logarithm. The relative intensity in this case is doubled. Thus we need first to solve $10^t = 2$. The solution is $t = \log 2 = 0.30$. Thus, doubling the relative intensity is the same as increasing the relative intensity by a factor of $10^{0.3}$. This means that the logarithm is increased by 0.3 unit. Since the decibel is 10 times the logarithm, we need to add $10 \times 0.3 = 3$ units on the decibel scale. The pair of speakers produces a reading of 63 decibels.   ■ ■ ■

It is important to note that the decibel scale is logarithmic in nature because that is how we actually hear. As sound increases in intensity, perceived loudness increases logarithmically. The same principle applies to vision. Increasing intensity of light produces, to the human eye, a logarithmic increase in perceived brightness. Fechner's law, proposed in 1859, states that *if stimulus is increased in geometrical progression, its resulting sensation increases in an arithmetic progression.* This means that sensation is a logarithmic function of stimulus. Although it should be considered a rule of thumb rather than a physical law, Fechner's law is at least correct for human vision and hearing.

The Logarithm as the Inverse of the Exponential Function

We can think of the Richter scale in two different ways. We might record how powerful the quake was and then calculate the Richter magnitude. To examine this, let R be the Richter magnitude for an earthquake of relative strength P.[26] We have already noted that the Richter magnitude is a logarithmic function of P, and in fact

$$R = \log P.$$

But we can also look at things in the opposite or *inverse* way, thinking of the relative strength P as a function of the Richter magnitude R. In this setting, we look at the Richter reading and determine how powerful the quake was. How do we calculate P from R? We know that increasing P by a factor of 10 adds 1 unit to R. Thus, adding 1 unit to R causes P to be multiplied by 10. This means that P is an exponential function of R with growth factor 10, and

$$P = 10^R.$$

In general, there is a close relationship between the exponential and logarithmic functions: One is the inverse of the other.

---

[26]By relative strength we mean the energy released at the source of the quake divided by the energy released at the source of a quake with magnitude 0 on the Richter scale.

KEY IDEA 4.8

The Logarithm as an Inverse

- The function $\log x$ is the inverse of the function 10^t:

$$t = \log x \text{ means that } x = 10^t.$$

- For the logarithmic function $\log x$, increasing x by a factor of 10^t adds t units to the logarithm.
- For the exponential function 10^x, increasing x by t units increases the function by a factor of 10^t.

EXAMPLE 4.10 **Apparent Magnitude and Brightness**

The apparent brightness of a star observed from Earth is a logarithmic function of the intensity of the light arriving from the star. Astronomers use the *magnitude* scale to measure apparent brightness. The *relative intensity* I is calculated by forming the ratio of the intensity of light from the star Vega to the intensity of light from the star of interest. The magnitude m is then given by

$$m = 2.5 \log I.$$

On the magnitude scale, higher magnitudes indicate dimmer stars.

1. Since Vega is used as a base scale, the relative intensity of light coming from Vega is 1. What is the magnitude of Vega?

2. Light arriving at Earth from the star Phecda has a relative intensity of 9.46. That is, light from Vega is 9.46 times as intense as light from Phecda. What is the magnitude of Phecda?

3. Since magnitude is a logarithmic function of relative intensity, relative intensity is an exponential function of magnitude. Use a formula to express relative intensity as an exponential function of magnitude.

4. Higher magnitudes indicate dimmer stars, and very bright stars can even have negative magnitudes. Sirius is the star that appears brightest of all in the night sky. It has an apparent magnitude of -1.45. How does the intensity of light arriving from Sirius compare with that of light arriving from Vega?

Solution to Part 1: For Vega, $I = 1$. Thus its magnitude is given by $m = 2.5 \log 1$. Since $10^0 = 1$, we have $\log 1 = 0$. Thus the magnitude of Vega is 0.

Solution to Part 2: The magnitude of Phecda is calculated using $m = 2.5 \log 9.46 = 2.44$.

Solution to Part 3:   We first rearrange the formula for magnitude by dividing both sides by 2.5:

$$m = 2.5 \log I$$

$$\frac{m}{2.5} = \log I .$$

Now $m/2.5 = \log I$ means that $I = 10^{m/2.5}$.

Solution to Part 4:   We can calculate the relative intensity of light arriving from Sirius by using the relationship we found in part 3:

$$I = 10^{m/2.5}$$

$$I = 10^{-1.45/2.5} = 0.26 .$$

Thus light from Vega is 0.26 times as intense as light from Sirius.

■ ■ ■

ANOTHER LOOK

Solving Exponential Equations

We noted that the solution for t of the equation $10^t = a$ is $t = \log a$. In fact we can use the logarithm (either common or natural) to solve more complicated exponential equations. We note first that the basic rules that apply to natural logarithms also apply to common logarithms. Their justification follows exactly the same course as the justification for the basic laws of natural logarithms, except that 10 replaces e in the arguments. The laws are

Product law: $\log xy = \log x + \log y$ if $x, y > 0$

Quotient law: $\log \left(\dfrac{x}{y} \right) = \log x - \log y$ if $x, y > 0$

Power law: $\log x^y = y \log x$ if $x > 0$

Let's see how to use these laws to solve some exponential equations. Let's look first at $5 \times 3^t = 8$. First we divide each side by 5.

$$3^t = \frac{8}{5} = 1.6$$

Next we take the logarithm of each side.

$$\log 3^t = \log 1.6$$

Now we use the power rule to bring the t outside the logarithm.

$$t \log 3 = \log 1.6$$

Finally, we divide each side by $\log 3$.

$$t = \frac{\log 1.6}{\log 3} = 0.43$$

The same thing can be accomplished by using the natural logarithm rather than the common logarithm, and the steps are virtually identical. The reader is urged to carry out this calculation and show that the resulting answer is

$$t = \frac{\ln 1.6}{\ln 3}$$

and that this gives the same numerical value for t that we found above.

Even more complex equations such as $5^t 3^t = 4 \times 2^{2t+1}$ are not difficult to handle. The first step is to combine $5^t 3^t$ to $(5 \times 3)^t = 15^t$. Next we take the logarithm of each side. You should note which law of logarithms is being used in each step below.

$$\log 15^t = \log\left(4 \times 2^{2t+1}\right)$$
$$t \log 15 = \log 4 + \log 2^{2t+1}$$
$$t \log 15 = \log 4 + (2t + 1)\log 2$$
$$t \log 15 = \log 4 + 2t \log 2 + \log 2$$
$$t \log 15 - 2t \log 2 = \log 4 + \log 2$$
$$t(\log 15 - 2\log 2) = \log 4 + \log 2$$
$$t = \frac{\log 4 + \log 2}{\log 15 - 2\log 2}$$
$$t = 1.57$$

These laws can also be used to produce alternative solutions to many of the examples in this section. Let's look back at Example 4.9, which discussed the decibel as a measure of loudness.

Alternative Solution to Part 1: Let one sound have relative intensity I. Then the second sound has relative intensity $100I$. If D_1 is the first decibel reading and D_2 is the second, then $D_1 = 10\log I$ and

$$D_2 = 10\log(100I) = 10(\log 100 + \log I)$$
$$= 10(2 + \log I) = 20 + 10\log I = 20 + D_1.$$

Thus the second sound has a decibel reading 20 units higher than the first.

Alternative Solution to Part 2: Let I denote the relative intensity of a 60-decibel speaker. Then $60 = 10\log I$. We want to know the decibel level if the relative intensity is $2I$. The calculation is

$$10\log 2I = 10(\log 2 + \log I) = 10\log 2 + 10\log I = 3 + 60 = 63 \text{ decibels}.$$

(Here we have approximated $10\log 2$ by 3.) Thus the reading is 63 decibels.

■ ■ ■

Historically, the common logarithm was invented to act as an early "calculator." A calculation such as

$$\frac{44.7^{1/3}}{744^{1/4}}$$

is very difficult to carry out by hand. You may consider how you would even begin without the aid of your calculator. Here is how you can use logarithms to make the calculation. First take the logarithm and apply the basic laws:

$$\log\left(\frac{44.7^{1/3}}{744^{1/4}}\right) = \log 44.7^{1/3} - \log 744^{1/4}$$

$$= \frac{1}{3}\log 44.7 - \frac{1}{4}\log 744.$$

The next step is to consult a table of logarithms to find that $\log 44.7 = 1.65$ and $\log 744 = 2.87$. Now the calculation is reduced to

$$\frac{1}{3} \times 1.65 - \frac{1}{4} \times 2.87.$$

We get -0.17. But this is the logarithm of the answer, not the answer itself. To finish, we look in the logarithm table once more to find which number has logarithm -0.17. This amounts to calculating $10^{-0.17}$, which is 0.68.

The magic of the logarithm is that it enables us to replace exponentiation by multiplication and to replace division by subtraction, both of which are much easier calculations. The most recent precursor to the calculator, the slide rule, was nothing but a quick way of calculating as above. Quick and accurate calculations via calculators or computers are very new on the mathematics scene and have had profound implications for science, engineering, mathematics, and virtually every other academic area, not to mention our own personal lives.

Enrichment Exercises

E-1. **Solving equations with logarithms:** Use the common logarithm to solve the following equations for t.

a. $5 \times 4^t = 7$

b. $2^{t+3} = 7^{t-1}$

c. $\dfrac{12^t}{9^t} = 7 \times 6^t$

d. $a^{2t} = 3a^{4t}$ where $a \neq 1$

E-2. **Solving equations using the natural logarithm:** Solve the equations in Exercise E-1 using the natural logarithm.

E-3. **Spectroscopic parallax again:** Solve Exercise 13 from Exercise Set 4.4 below by using the laws of logarithms.

E-4. **Calculating with logarithms:** Use logarithms to calculate the value of

$$22.7^{0.7} \times 17.6^{1.1},$$

as follows: Take the logarithm of the expression and use the laws of logarithms to simplify the calculation. You will need to use your calculator to find $\log 22.7$ and $\log 17.6$. Once you have the logarithm L calculated, you will need to use your calculator to find 10^L.

E-5. **Substitutions:** Solve the following equations by first making the indicated substitution. In each case, the substitution will lead to a quadratic that you can solve by factoring or by the quadratic formula.

a. $4^x - 5 \times 2^x + 6 = 0$. Substitute $y = 2^x$. *Note:* $4^x = (2^x)^2$.

b. $25^x - 7 \times 5^x + 12 = 0$. Substitute $y = 5^x$.

c. $9^x - 3^{x+1} + 2 = 0$. Substitute $y = 3^x$.

E-6. **Solving exponential equations:** Solve the following equations.

a. $3^{(2^x)} = 5$

b. $3^{(4^x)} = 7^{(2^x)}$

c. $2^{(5^{x+1})} = 4^{(3^x)}$

E-7. **Solving logarithmic equations:** Solve the following logarithmic equations. The identity $e^{\ln x} = x$ will be important here.

a. $\ln(x + 2) = 5$

b. $\ln(x + 5) = 2\ln(x - 1)$

c. $\ln(3x - 10) = \ln(x + 2) + \ln(x - 5)$

E-8. **Moore's law:**

a. Use the common logarithm to solve the equation $10^k = 2^{1/1.5}$ for k.

b. According to *Moore's law,* the speed of computer chips doubles about every 1.5 years.[27] Use the laws of exponents to show that an exponential function with yearly growth factor $2^{1/1.5}$ has this property.

c. Explain using part b why Moore's law says that every year the speed of computer chips increases by a factor of 10^k, where k is the solution to the equation in part a.

---

[27] See Exercise 3 at the end of Section 4.4.

4.4   *SKILL BUILDING EXERCISES*

S-1.   **The Richter scale:** One earthquake measures 4.2 on the Richter scale and another measures 7.2. How do the two quakes compare?

S-2.   **The Richter scale again:** One earthquake has a Richter scale reading of 6.5. A second is 100 times stronger. What is its Richter scale reading?

S-3.   **The decibel scale:** If one sound has a relative intensity 1000 times that of another, how do their decibel levels compare?

S-4.   **More decibels:** One sound has a decibel reading of 5.5. Another has a decibel reading of 7.5. How do the relative intensities of the sounds compare?

S-5.   **Calculating logarithms:** Calculate the following common logarithms by hand.

a. $\log 1000$

b. $\log 1$

c. $\log \dfrac{1}{10}$
  (*Suggestion:* Recall that $\frac{1}{10} = 10^{-1}$.)

S-6.   **Solving logarithmic equations:** Solve the following equations by hand.

a. $\log x = 2$

b. $\log x = 1$

c. $\log x = -2$

S-7.   **How the logarithm increases:** Suppose we know that $\log x = 6.6$. Find the following:

a. $\log(10x)$

b. $\log(1000x)$

c. $\log \dfrac{x}{10}$

S-8.   **How the logarithm increases:** If $\log x = 8.3$ and $\log y = 10.3$, how do x and y compare?

S-9.   **Solving exponential equations:** Solve the following exponential equations for t.

a. $10^t = 5$

b. $10^t = a$

c. $a^t = b$
  (*Suggestion:* Note first that $a = 10^{\log a}$.)

S-10.   **The graph of the logarithm:** The logarithm increases very slowly. But the graph eventually crosses the horizontal line $y = 20$ because $\log 10^{20} = 20$. Explain why the graph of the logarithm eventually crosses the line $y = K$ for every value of K.

4.4   *EXERCISES*

1.   **Earthquakes in Alaska and Chile:** In 1964 an earthquake measuring 8.4 on the Richter scale occurred in Alaska.

a. How did the power of the New Madrid earthquake described in Example 4.8 compare with that of the 1964 Alaska earthquake?

b. In 1960 an earthquake occurred in Chile which was 1.4 times more powerful than the Alaska quake. What was the Richter scale reading for the Chilean earthquake?

2.   **State quakes:** The largest recorded earthquake centered in Idaho measured 7.2 on the Richter scale.

a. The largest recorded earthquake centered in Montana was 3.16 times more powerful than the Idaho

earthquake. What was the Richter scale reading for the Montana earthquake?

b. The largest recorded earthquake centered in Arizona measured 5.6 on the Richter scale. How did the power of the Idaho quake compare with that of the Arizona quake?

3.   **Moore's law:** The speed of computer chips has increased with time in a remarkably consistent way. In fact, in the year 1965 Dr. Gordon E. Moore (now Chairman Emeritus of Intel Corporation) observed a trend and predicted that it would continue for a time. His observation, now known as *Moore's law*, is that every 18 months or so, a chip is introduced whose speed is double that of its fastest predecessor. This

law can be restated[28] in the following way: If time increases by 1 year, then the best speed available is multiplied by $10^{0.2}$. More generally, the rule is that if time increases by t years, then the speed is multiplied by $10^{0.2t}$. For example, after 3 years the speed is multiplied by $10^{0.2 \times 3}$, or about 4. The Pentium 4 processor was introduced by Intel Corporation in the year 2000.

a. If a chip were introduced in the year 2008, how much faster than the Pentium 4 would you expect it to be? Round your answer to the nearest whole number.

b. The speed of a chip is determined by the number of transistors on the chip. The limit of conventional computing will be reached when the size of a transistor is on the scale of an atom. At that point the speed of a chip will be 10,000 times the speed of the Pentium 4. When, according to Moore's law, will that limit be reached?

c. Even for unconventional computing, the laws of physics impose a limit on the speed of computation.[29] The fastest speed possible is about 10^{40} times that of the Pentium 4. Assume that Moore's law will continue to be valid even for unconventional computing, and determine when this limit will be reached.

4. **Brightness of stars:** Refer to Example 4.10 for the relationship between relative intensity of light and apparent magnitude of stars.

a. The light striking Earth from Vega is 2.9 times as bright as that from the star Fomalhaut. What is the apparent magnitude of Fomalhaut?

b. The star Antares has an apparent magnitude of 0.92. How does the intensity of light reaching Earth from Antares compare with that of light from Vega?

c. If the intensity of light striking Earth from one star is twice that of light from another, how do the stars' magnitudes compare?

5. **The pH scale:** Acidity of a solution is determined by the concentration H of hydrogen ions in the solution (measured in moles per liter of solution). Chemists use the negative of the logarithm of the concentration of hydrogen ions to define the pH scale:

$$\text{pH} = -\log H.$$

Lower pH values indicate a more acidic solution.

a. Normal rain has a pH value of 5.6. Rain in the eastern United States often has a pH level of 3.8. How much more acidic is this than normal rain?

b. If the pH of water in a lake falls below a value of 5, fish often fail to reproduce. How much more acidic is this than normal water with a pH of 5.6?

6. **Dispersion models:** Animal populations move about and disperse. A number of models for this dispersion have been proposed, and many of them involve the logarithm. For example, in 1965 O. H. Paris[30] released a large number of pill bugs and after 12 hours recorded the number n of individuals that could be found within r meters from the point of release. He reported that the most satisfactory model for this dispersion was

$$n = -0.772 + 0.297 \log r + \frac{6.991}{r}.$$

a. Make a graph of n against r for the circle around the release point with radius 15 meters.

b. How many pill bugs were to be found within 2 meters from the release point?

c. How far from the release point would you expect to find only a single individual?

7. **Weight gain:** Zoologists have studied the daily rate of gain in weight G as a function of daily milk-energy intake M during the first month of life in several ungulate (that is, hoofed mammal) species.[31] (Both M and G are measured per unit of mean body weight.) They developed the model

$$G = 0.067 + 0.052 \log M,$$

with appropriate units for M and G.   →

---

[28] See Enrichment Exercise E-8 for this restatement.

[29] S. Lloyd, "Ultimate physical limits to computation," *Nature* **406** (2000), 1047–1054.

[30] "Vagility of **P**[32]-labeled isopods in grassland," *Ecology* **46** (1965), 635–648.

[31] C. Robbins et al., "Growth and nutrient consumption of elk calves compared to other ungulate species," *J. Wildlife Management* **45** (1981), 172–186.

a. Draw a graph of G versus M. Include values of M up to 0.4 unit.

b. If the daily milk-energy intake M is 0.3 unit, what is the daily rate of gain in weight?

c. A zookeeper wants to bottle-feed an elk calf so as to maintain a daily rate of gain in weight G of 0.03 unit. What must the daily milk-energy intake be?

d. The study cited above noted that "the higher levels of milk ingested per unit of body weight are used with reduced efficiency." Explain how the shape of the graph supports this statement.

8. **Reaction time:** For certain decisions, the time it takes to respond is a logarithmic function of the number of choices faced.[32] One model is

$$R = 0.17 + 0.44 \log N ,$$

where R is the reaction time in seconds and N is the number of choices.

a. Draw a graph of R versus N. Include values of N from 1 to 10 choices.

b. Express using functional notation the reaction time if there are 7 choices, and then calculate that time.

c. If the reaction time is to be at most 0.5 second, how many choices can there be?

d. If the number of choices increases by a factor of 10, what happens to the reaction time?

e. Explain in practical terms what the concavity of the graph means.

9. **Age of haddock:** The age T, in years, of a haddock can be thought of as a function of its length L, in centimeters. One common model uses the natural logarithm:

$$T = 19 - 5 \ln (53 - L) .$$

a. Draw a graph of age versus length. Include lengths between 25 and 50 centimeters.

b. Express using functional notation the age of a haddock that is 35 centimeters long, and then calculate that value.

c. How long is a haddock that is 10 years old?

10. **Growth rate:** An animal grows according to the formula

$$L = 0.6 \log (2 + 5T) .$$

Here L is the length in feet and T is the age in years.

a. Draw a graph of length versus age. Include ages up to 20 years.

b. Explain in practical terms what $L(15)$ means, and then calculate that value.

c. How old is the animal when it is 1 foot long?

d. Explain in practical terms what the concavity of the graph means.

e. Use a formula to express the age as a function of the length.

11. **Stand density:** Forest managers are interested in measures of how crowded a given forest stand is.[33] One measurement used is the *stand-density index,* or SDI. It can be related to the number N of trees per acre and the diameter D, in inches, of a tree of average size (in terms of cross-sectional area at breast height) for the stand. The relation is

$$\log SDI = \log N + 1.605 \log D - 1.605 .$$

a. A stand has 500 trees per acre, and the diameter of a tree of average size is 7 inches. What is the stand-density index? (Round your answer to the nearest whole number.)

b. What is the effect on the stand-density index of increasing the number of trees per acre by a factor of 10, assuming that the average size of a tree remains the same?

c. What is the relationship between the stand-density index and the number of trees per acre if the diameter of a tree of average size is 10 inches?

12. **Fully stocked stands:** *This is a continuation of Exercise 11. In this exercise we study one of the ingredients used in formulating the relation given in the preceding exercise between the stand-density index SDI, the number N of trees per acre, and the*

---

diameter D, in inches, of a tree of average size.[34] This ingredient is an empirical relationship between N and D for *fully stocked* stands—that is, stands for which the tree density is in some sense optimal for the given size of the trees. This relationship, which was observed by L. H. Reineke in 1933, is

$$\log N = -1.605 \log D + k,$$

where k is a constant that depends on the species in question.

a. Assume that for loblolly pines in an area the constant k is 4.1. If in a fully stocked stand the diameter of a tree of average size is 8 inches, how many trees per acre are there? (Round your answer to the nearest whole number.)

b. For fully stocked stands, what effect does multiplying the average size of a tree by a factor of 2 have on the number of trees per acre?

c. What is the effect on N of increasing the constant k by 1 if D remains the same?

13. **Spectroscopic parallax:** Stars have an apparent magnitude m, which is the brightness of light reaching Earth. They also have an *absolute magnitude* M, which is the intrinsic brightness and does not depend on the distance from Earth. The difference $S = m - M$ is the *spectroscopic parallax*. Spectroscopic parallax is related to the distance D from Earth, in *parsecs*,[35] by

$$S = 5 \log D - 5.$$

a. The distance to the star Kaus Astralis is 38.04 parsecs. What is its spectroscopic parallax?

b. The spectroscopic parallax for the star Rasalhague is 1.27. How far away is Rasalhague?

c. How is spectroscopic parallax affected when distance is multiplied by 10?

d. The star Shaula is 3.78 times as far away as the star Atria. How does the spectroscopic parallax of Shaula compare to that of Atria?

14. **Rocket flight:** The velocity v attained by a launch vehicle during launch is a function of c, the exhaust velocity of the engine, and R, the *mass ratio* of the spacecraft.[36] The mass ratio is the vehicle's takeoff weight divided by the weight remaining after all the fuel has been burned, so the ratio is always greater than 1. It is close to 1 when there is room for only a little fuel relative to the size of the vehicle, and one goal in improving the design of spacecraft is to increase the mass ratio. The formula for v uses the natural logarithm:

$$v = c \ln R.$$

Here we measure the velocities in kilometers per second, and we assume that $c = 4.6$ (which can be attained with a propellant that is a mixture of liquid hydrogen and liquid oxygen).

a. Draw a graph of v versus R. Include mass ratios from 1 to 20.

b. Is the graph in part a increasing or decreasing? In light of your answer, explain why increasing the mass ratio is desirable.

c. To achieve a stable orbit, spacecraft must attain a velocity of 7.8 kilometers per second. With $c = 4.6$, what is the smallest mass ratio that allows this to happen? (*Note:* For such a propellant, the mass ratio needed for orbit is usually too high, and that is why the launch vehicle is divided into stages. The next exercise shows the advantage of this.)

15. **Rocket staging:** *This is a continuation of Exercise 14.* One way to raise the effective mass ratio is to divide the launch vehicle into different stages. See Figure 4.37. After one stage has burned its fuel, it drops away, and the next stage begins to fire. We assume that there are two stages and that each stage has the same exhaust velocity c. Then the total velocity v attained is the sum of the velocities attained by each stage:

$$v = c \ln R_1 + c \ln R_2, \qquad \longrightarrow$$

---

[34] The other main ingredient is implicit in part c of Exercise 11.

[35] One parsec is the distance from Earth that would show a 1-second parallax angle. That is the apparent angle produced by movement of Earth from one extreme of its orbit to the other. One parsec is about 3.26 light years.

[36] This exercise and the next are based on *Space Mathematics* by B. Kastner, published in 1985 by NASA.

where R_1 and R_2 are the mass ratios for stages 1 and 2, respectively. In this exercise we assume that $c = 3.7$ kilometers per second, and we assume that each stage has a mass ratio of 3.4.

a. What is the total velocity attained by this two-stage craft?

b. Can this craft achieve a stable orbit, as described in part c of Exercise 14?

FIGURE 4.37

16. **Stereo speakers:** See Example 4.9 for the relationship between the decibel scale and relative intensity of sound.

a. In this part assume that four stereo speakers, each producing 40 decibels, are placed side by side, multiplying by a factor of 4 the relative intensity of sound produced by only one of the speakers. What is the decibel level of the four speakers together?

b. Find an exponential formula giving relative intensity I as a function of decibel level D.

c. What is the growth factor for relative intensity as an exponential function of decibels? *Reminder:* $a^{b/c} = \left(a^{1/c}\right)^b$.

d. If the decibel level increases by 3 units, what happens to relative intensity?

17. **Relative abundance of species:** A collection of animals may contain a number of species. Some rare species may be represented by only 1 individual in the collection. Others may be represented by more. In 1943 Fisher, Corbert, and Williams[37] proposed a model for finding the number of species in certain types of collections represented by a fixed number of individuals. They associate with an animal collection two constants, α and x, with the following property: The number of species in the sample represented by a single individual is αx; the number in the sample represented by 2 individuals is $\alpha(x^2/2)$; and, in general, the number of species represented by n individuals is $\alpha(x^n/n)$. If we add all these numbers up, we get the total number S of species in the collection. It is an important fact from advanced mathematics that this sum also yields a natural logarithm:

$$S = \alpha x + \alpha \frac{x^2}{2} + \alpha \frac{x^3}{3} + \cdots = -\alpha \ln(1 - x).$$

If N is the total number of individuals in the sample, then it turns out that we can find x by solving the equation

$$\frac{x - 1}{x} \ln(1 - x) = \frac{S}{N}.$$

We can then find the value of α using

$$\alpha = \frac{N(1 - x)}{x}.$$

In 1935 at the Rothamsted Experimental Station in England, 6814 moths representing 197 species were collected and catalogued.

a. What is the value of S/N for this collection? (Keep four digits beyond the decimal point.)

---

b. Draw a graph of the function

$$\frac{x-1}{x} \ln(1-x)$$

using a horizontal span of 0 to 1.

c. Use your answer to part a and your graph in part b to determine the value of x for this collection. (Keep four digits beyond the decimal point.)

d. What is the value of α for this collection? (Keep two digits beyond the decimal point.)

e. How many species of moths in the collection were represented by 5 individuals?

f. For this collection, plot the graph of the number of species represented by n individuals against n. Include values of n up to 20.

Summary

Exponential functions are almost as pervasive as linear functions. They are commonly used to describe population growth, radioactive decay, free fall subject to air resistance, bank loans, inflation, and many other familiar phenomena. Their defining property is similar to that of linear functions. Linear functions are those with a constant rate of change, whereas exponential functions have a constant percentage or proportional rate of change. Any phenomenon that can be described in terms of yearly (monthly, daily, etc.) percentage growth (or decay) is properly modeled with an exponential function.

Exponential Growth and Decay

A linear function with slope m changes by constant sums of m. That is, if the variable is increased by 1 unit, the function is increased by adding m units. By contrast, an exponential function with base a changes by constant multiples of a. That is, when the variable is increased by 1, the function value is multiplied by a. Exponential functions are of the form

$$N = Pa^t,$$

where a is the *base* or *growth/decay factor*, and P is the initial value. When a is larger than 1, the exponential function grows rapidly, eventually becoming exceptionally large. When the base a is less than 1, the exponential function decays rapidly toward zero.

Many times, exponential functions are described in terms of constant percentage growth or decay. If r is a decimal showing a percentage increase, then the growth factor is $1 + r$; if r shows a percentage decrease, then the decay factor is $1 - r$.

The growth factor for an exponential function is tied to a time period. It is sometimes important to know the growth factor for a different time period. We do this with unit conversion.

Unit conversion for exponential growth factors: If the growth factor for one period of time is a, then the growth factor A for k periods of time is given by $A = a^k$.

As an example, U.S. population in the early 1800s grew at a rate of about 2.9% per year. Thus population is an exponential function with yearly growth factor 1.029. The appropriate exponential function is of the form $N = P \times 1.029^t$, where t is measured in years. If we want to express population growth in terms of decades, we need a new growth factor:

$$1.029^{10} = 1.331.$$

Thus we can say that U.S. population in the early 1800s grew at a rate of 33.1% per decade, and $N = P \times 1.331^d$, where d is measured in decades.

Modeling Exponential Data

Data that are evenly spaced for the variable are linear if the function values show constant successive differences. They are exponential if the function values show constant *successive quotients*. When the increment for the data is 1 unit, the common successive quotient is the growth (or decay) factor for the exponential function.

A simple example is provided by a hypothetical well that is being contaminated by seepage of water containing a pollutant at a level of 64 milligrams per liter. Each month the difference D between 64 and the contaminant level of the well is recorded.

t = months	0	1	2	3	4	5
D = contaminant level difference	64	45.44	32.26	22.91	16.26	11.55

The first quotient is calculated as $\frac{45.44}{64} = 0.71$. It can be verified that each successive quotient in the table gives this same value. We conclude that the data are exponential with decay factor 0.71. Since the initial value is 64, we obtain $D = 64 \times 0.71^t$.

Modeling Nearly Exponential Data

As with linear data, sampling error and other factors make it rare that experimentally gathered data can be modeled exactly by an exponential function. But as with linear functions, it may be appropriate to use the exponential model that most closely approximates the data. The key to making this happen is the *logarithm function*. The natural logarithm function $\ln x$ is the *inverse* of e^x. In other words, it is the function that "undoes" exponentiation: $\ln e^x = x$. This means that if the logarithm is applied to the function values of exponential data, the result will be linear, and linear regression can be applied. The required procedure is as follows:

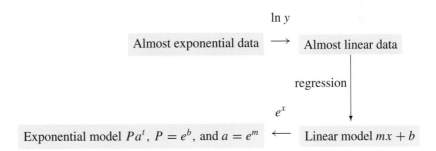

Logarithmic Functions

The common logarithm $\log x$ is the power of 10 that gives x. Thus, if we can express x as a power of 10, we get the logarithm immediately using

$$\log 10^t = t.$$

The formula above tells us that the common logarithm is the *inverse* of the exponential function 10^t, and we can think of the logarithm of x as the solution for t of the equation

$$x = 10^t.$$

This in turn provides us with a direct way of solving certain exponential equations. For example, the solution of the equation

$$10^t = 7.6$$

is $\log 7.6$, which is about 0.88.

Increasing x by a factor of 10 increases the logarithm by 1 unit, and in general if x is multiplied by 10^t, then t units are added to the logarithm. This can be expressed using the formula

$$\log(10^t x) = t + \log x .$$

Many common scales, such as the Richter scale and the decibel scale, are logarithmic in nature.

Chapter

5

A Survey of Other Common Functions

Although linear and exponential functions probably are the mathematical functions that are most commonly found in applications, other important functions occur as well. We will look closely at *power functions* in the first two sections of this chapter. The third section treats composition of functions and piecewise defined functions. The fourth section is devoted to *quadratic functions,* and the fifth considers *polynomials* and *rational functions.*

5.1 *Power Functions*

Recall that an exponential function has the form $f(x) = Pa^x$, where the base a is fixed and the exponent x varies. For a power function these properties are reversed—the base varies and the exponent remains constant—so a power function has the form $f(x) = cx^k$. The number k is called the *power*, the most significant part of a power function, and the coefficient c is equal to $f(1)$. For most applications, we are interested only in positive values of the variable x and of c, but we allow k to be any number.

We can use the graphing calculator to illustrate how power functions work and the role of k. In the exercises at the end of this section, you will have an opportunity to explore the role of c graphically. We look first at what happens when the power k is positive. In Figure 5.1 we show the graphs of x, x^2, x^3, and x^4. For the viewing window, we have used both a horizontal and a vertical span of 0 to 3. The first thing we note is that when the power is positive, the graph of the power function is increasing. We see also that, when $k = 1$, the graph is a straight line. This tells us that a power function with power 1 is a linear function. Finally, we observe that, for values of x larger than 1 (the common crossing point in Figure 5.1), larger powers cause the power function to grow faster.

In Figure 5.2, we look at what happens when the power k is negative. Here we show the graphs of x^{-1}, x^{-2}, x^{-3}, and x^{-4} using both a horizontal and a vertical span of 0 to 3. We see that for negative powers, power functions decrease toward 0. Furthermore, negative powers that are larger in size cause the graph to approach the horizontal axis more rapidly than do negative powers that are smaller in size.

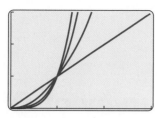

FIGURE 5.1 Power functions with positive powers are increasing functions

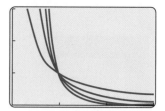

FIGURE 5.2 Power functions with negative powers decrease toward 0

KEY IDEA 5.1

Power Functions

For a power function $f(x) = cx^k$ with c and x positive:

1. If k is positive, then f is increasing. Larger positive values of k cause f to increase more rapidly.

2. If k is negative, then f decreases toward zero. Negative values of k that are larger in size cause f to decrease more rapidly.

EXAMPLE 5.1   Distance Fallen as a Function of Time

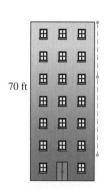

70 ft

FIGURE 5.3

When a rock is dropped from a tall structure, it will fall $D = 16t^2$ feet in t seconds. (See Figure 5.3.)

1. Make a graph that shows the distance the rock falls versus time if the building is 70 feet tall.

2. How long does it take the rock to strike the ground?

Solution to Part 1:   The first step is to enter the function 5.1 in the calculator function list and to record the variable associations:

$$\mathsf{Y_1} = D, \ \text{distance on vertical axis}$$

$$\mathsf{X} = t, \ \text{time on horizontal axis}.$$

Since the building is 70 feet tall, we allow a bit of extra room and set the vertical span from 0 to 100. Since our everyday experience tells us that it will take only a few seconds for the rock to reach the ground, we set the horizontal span from 0 to 5 5.2 . The graph appears in Figure 5.4.

Solution to Part 2:   We want to know the value of t when the rock strikes the ground—that is, when $D = 70$ feet. Thus we need to solve the equation

$$16t^2 = 70.$$

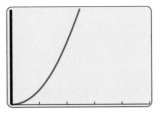

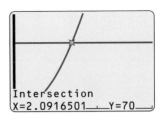

Intersection
X=2.0916501   Y=70

FIGURE 5.4 Graph of distance
fallen versus time

FIGURE 5.5 When the rock
strikes the ground

We proceed with the crossing-graphs method. We enter the target distance, 70, on
the function list and plot to see the picture in Figure 5.5. We use the calculator to
find the crossing point $\boxed{5.3}$ at $t = 2.09$ seconds as shown in Figure 5.5.

Homogeneity Property of Power Functions

Many times, the qualitative nature of a function is as important as the exact formula that
describes it. It is, for example, crucial to physicists to understand that distance fallen
is proportional to the square of the time (as opposed to an exponential, linear, or some
other type of function). This is an important qualitative observation about how gravity
acts on objects near the surface of the Earth. To make it clear what we mean, we look at
an important mathematical property of power functions that is known as *homogeneity* .

Suppose that in a power function $f(x) = cx^k$, the value of x is increased by a factor
of t. What happens to the value of f? To help answer this question, let's first look at
some specific examples.

How tripling the side of a square affects its area: The area A of a square of side s is
equal to the square of s. Thus $A = s^2$. Suppose a square initially has sides of length 4
feet. If the length of the sides of the square is tripled, how is the area affected? To answer
this, we calculate the original area and compare it to the area after the sides of the square
are tripled:

$$\text{Original area} = 4^2 = 16 \text{ square feet}.$$

To get the new area, we use $s = 12$:

$$\text{New area} = 12^2 = 144 \text{ square feet}.$$

Thus, if the side of the square is tripled, the area increases from 16 square feet to 144
square feet; that is, it increases by a factor of 9. The key thing to note here is that 9 is 3^2.
Increasing the side by a factor of 3 increased the area by a factor of 3^2—that is, 3 raised
to the same power as the function.

How doubling the radius of a sphere affects its volume: A slightly more complicated
calculation will show the same phenomenon. The volume V inside a sphere, such as
a tennis ball or basketball, depends on its radius r and is proportional to the cube of
the radius. Specifically, from elementary geometry we know that $V = \frac{4\pi}{3}r^3$. Suppose
a balloon initially has a radius of 5 inches. If air is pumped into the balloon until the
radius doubles, what is the effect on the volume? We proceed as before, calculating the
original volume and comparing it with the volume after the radius is doubled. Initially

the radius is $r = 5$. Accuracy is important in this calculation, so in this case we report all the digits given by the calculator:

$$\text{Original volume} = \frac{4\pi}{3}5^3 = 523.5987756.$$

To get the new volume, we use $r = 10$:

$$\text{New volume} = \frac{4\pi}{3}10^3 = 4188.790205.$$

To understand how the volume has changed, we divide:

$$\frac{\text{New volume}}{\text{Old volume}} = \frac{4188.790205}{523.5987756} = 8.$$

Increasing the radius by a factor of 2 results in an increase in volume by a factor of 8. Once again, the key thing to observe is that 8 is 2^3. Summarizing, we see that if the radius is increased by a factor of 2, then the volume is increased by a factor of 2^3—that is, 2 raised to the power of the function.

The phenomenon we observed in these two examples is characteristic of power functions, and we can show this using some elementary properties of exponents. Let's return to our original question. Suppose $f = cx^k$. If x is increased by a factor of t, what is the effect on f? We have

$$\text{Old value} = cx^k.$$

To get the new value, we replace x by tx:

$$\text{New value} = c(tx)^k = ct^k x^k = t^k(cx^k) = t^k \times \text{Old value}.$$

Thus, just as we observed in our examples, increasing x by a factor of t increases f by a factor of t^k.

KEY IDEA 5.2

Homogeneity Property of Power Functions

For a power function $f = cx^k$, if x is increased by a factor of t, then f is increased by a factor of t^k.

EXAMPLE 5.2   Expansion of a Shock Wave

The shock wave produced by a large explosion expands rapidly. (See Figure 5.6.) The radius R of the wave 1 second after the explosion is a power function of the energy E released by the explosion, and the formula is

$$R = 4.16E^{0.2}.$$

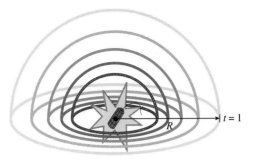

FIGURE 5.6

Here R is measured in centimeters and E in ergs.

1. If one explosion releases 1000 times as much energy as another, how much larger will the radius of the shock wave be after 1 second?

2. If one explosion releases half as much energy as another, how much smaller will the radius of the shock wave be after 1 second?

Solution to Part 1:   We use the homogeneity property of power functions. Here R is a power function of E with power $k = 0.2$. Thus when E is increased by a factor of 1000, the function R is increased by a factor of $1000^{0.2}$. Since $1000^{0.2} = 3.98$, after 1 second the radius of the shock wave for the larger explosion will be 3.98 times as large as that for the smaller explosion.

Solution to Part 2:   When E is multiplied by 0.5, the function R is multiplied by $0.5^{0.2}$, or about 0.87. Thus after 1 second the radius of the shock wave for the smaller explosion will be 0.87 times as large as that for the larger explosion.   ■ ■ ■

Note that in this example the coefficient $c = 4.16$ did not enter into the computations. The next example further illustrates that the homogeneity property for power functions can be useful in situations where only the power is known.

EXAMPLE 5.3   **Number of Species Versus Available Area**

Ecologists have studied how the number S of species that make up a given group existing in a closed environment (often an island) varies with the area A that is available.[1] They use the approximate *species-area relation*

$$S = cA^k$$

to estimate, among similar habitats, the number of species as a function of available area. For birds on islands in the Bismarck Archipelago near New Guinea, the value of k is estimated to be about $k = 0.18$.

---

[1] See J. Diamond and R. M. May, "Island biogeography and the design of natural reserves," in R. M. May (ed.), *Theoretical Ecology*, 2nd ed., 1981 (Sunderland, MA: Sinauer Associates, 1981). See also the references therein.

1. If one island in the Bismarck Archipelago were twice as large as another, how many more species of birds would it have? Interpret your answer in terms of percentages.

2. If there are 50 species of birds on an island in the Bismarck Archipelago of area 100 square kilometers, find the value of c, and then make a graph of the number of species as a function of available area for islands in the range of 50 to 200 square kilometers. (Here we measure A in square kilometers.)

3. It is thought that a power relation $S = cA^k$ also applies to the Amazon rain forest, whose size is being reduced by (among other things) burning in preparation for farming. Ecologists use the estimate $k = 0.30$ for the power. If the rain forest were reduced in area by 20%, by what percentage would the number of surviving species be expected to decrease?

4. Experience has shown that, in a certain chain of islands, a 10% reduction in area leads to a 4% reduction in the number of species. Find the value of k in the species-area relation. What would be the result of a 25% reduction in the usable area of one of these islands?

Solution to Part 1:   The number of species S is a power function of the area A, with power $k = 0.18$. If the variable A is doubled, then, by the homogeneity property of power functions, the function S increases by a factor of $2^{0.18} = 1.13$. Thus the number of species of birds increases by a factor of 1.13. In terms of percentages, doubling A is a 100% increase, and changing S by a factor of 1.13 represents a 13% increase. Thus increasing the area by 100% has the effect of increasing the number of species by 13%.

Solution to Part 2:   We know that $S = cA^{0.18}$ and that an area of 100 square kilometers supports 50 species. That is, $S = 50$ when $A = 100$:

$$50 = c \times 100^{0.18}$$

$$50 = c \times 2.29 .$$

This is a linear equation, and we can solve for c by hand calculation:

$$c = \frac{50}{2.29} = 21.83 .$$

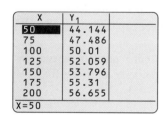

X	Y₁
50	44.144
75	47.486
100	50.01
125	52.059
150	53.796
175	55.31
200	56.655

X=50

FIGURE 5.7 A table of values for number of species

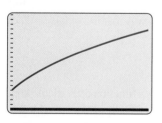

FIGURE 5.8 Number of species versus area in the Bismarck Archipelago

We conclude that, in the Bismarck Archipelago, the number of species is related to area by $S = 21.83A^{0.18}$. We first enter $\boxed{5.4}$ this on our calculator function list and record variable correspondences:

$$Y_1 = S, \text{ number of species on vertical axis}$$

$$X = A, \text{ area on horizontal axis} .$$

We are asked to make the graph for islands from 50 to 200 square kilometers in area. Thus we use a horizontal span of 50 to 200. For the vertical span, we look at the table of values in Figure 5.7. Allowing a little extra room, we use a vertical span of 40 to 60. The completed graph is shown in Figure 5.8.

Solution to Part 3:   To say that there is a 20% decrease in area means that the new area is 80% of its original value. Thus A has changed by a factor of 0.8. Using the homogeneity property of power functions, we conclude that S, the number of species, changes by a factor of $0.8^{0.30} = 0.94$. Thus 94% of the original species can be expected to survive, so the number of species has been reduced by 6%.

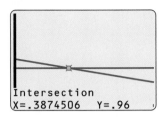

Intersection
X=.3874506   Y=.96

FIGURE 5.9 Finding the power in the species-area relation

Solution to Part 4:   We know that a 10% reduction in the area of an island will result in a 4% reduction in the number of species. In other words, if A is changed by a factor of 0.9, then S is changed by a factor of 0.96. The homogeneity property of power functions tells us that $0.96 = 0.9^k$. We can use the crossing-graphs method to solve this equation as is shown in Figure 5.9, where we have used a horizontal span of 0 to 1 and a vertical span of 0.8 to 1.2. We read from the prompt at the bottom of the figure that $k = 0.39$, rounded to two decimal places.

If the area is reduced by 25%, we are changing A by a factor of 0.75. Thus S changes by a factor of $0.75^{0.39} = 0.89$. That is an 11% reduction in the number of species. ■ ■ ■

Comparing Exponential and Power Functions

Over limited ranges, the graphs of exponential functions and power functions may appear to be similar. We see this in Figure 5.10, where we have graphed the exponential function 2^x and the power function x^2 using a horizontal span of 1 to 4 and a vertical span of 0 to 20. As the figure shows, the two graphs are nearly identical on this span. This sometimes makes it difficult to determine whether observed data should be modeled with an exponential function or with a power function. We will return in the next section to the problem of how to make an appropriate choice of model. For now, we want to illustrate the consequences of an inappropriate choice. Figure 5.10 shows the similarity of the functions over the displayed range, but Figure 5.11 shows a dramatic difference if we view the graphs with a horizontal span of 1 to 7 and a vertical span of 0 to 70. The exponential function 2^x is the graph in Figure 5.11 that rises rapidly above the graph of the power function.

If you view these graphs on an even larger horizontal span, the differences will be even more dramatic. This behavior is typical of the comparison of any exponential function with base larger than 1 with any power function, no matter how large the power. Power functions may look similar to exponential functions over a limited range, and they may even grow more rapidly for brief periods. But eventually, exponential functions always grow many times faster than power functions. This makes the consequences of choosing the wrong model quite serious. For example, if we were to choose an exponential model for federal spending when a power model was in fact appropriate, our mistake would lead us to predict federal spending at levels many orders of magnitude too large.

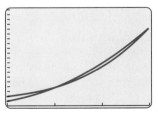

FIGURE 5.10 A limited span where a power function and an exponential function appear similar

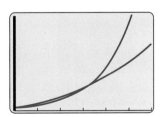

FIGURE 5.11 The characteristic dominance of exponential functions over power functions in the long term

Or, if we chose a power model when in fact an exponential model was appropriate, we would be led to predict future federal spending at levels far too low.

The fact that exponential functions eventually grow many times faster than power functions is one of the most important qualitative distinctions between these two types of functions.

KEY IDEA 5.3

Exponential Functions Grow Faster Than Power Functions

Over a sufficiently large horizontal span, an exponential function (with base larger than 1) will increase much more rapidly than a power function.

EXAMPLE 5.4 Exponential versus Power Models

In Chapter 4 we modeled early U.S. population using the exponential model $N = 5.34 \times 1.03^t$, where t is the number of years since 1800 and N is population in millions. Recall that an exponential model was chosen not only because it fit the data, but also because populations are expected (at least for brief periods) to grow at a constant percentage rate. Suppose we had used the power model $P = 0.04t^{1.5} + 5.34$, where t is years since 1800 and P is U.S. population in millions.[2]

X	Y₁	Y₂
0	5.34	5.34
5	6.1905	5.7872
10	7.1765	6.6049
15	8.3195	7.6638
20	9.6446	8.9177
25	11.181	10.34
30	12.962	11.913

X=0

FIGURE 5.12 A table of values for two population models from 1800 to 1830

1. Graph the exponential model and the power model over the years from 1800 to 1830. Do the models yield similar predictions over this time period?

2. Graph the exponential model and the power model over the years from 1800 to 1870. Do the models yield similar predictions over this time period? The actual population in 1860 was 31.44 million. Which model gives the more accurate estimate for 1860?

Solution to Part 1: First we enter 5.5 the two formulas in the calculator function list and record variable correspondences:

$$Y_1 = N, \text{ exponential model population on vertical axis}$$

$$Y_2 = P, \text{ power model population on vertical axis}$$

$$X = t, \text{ years since 1800 on horizontal axis}.$$

We want the horizontal span to go from 0 to 30. We consult the table 5.6 of values in Figure 5.12 to get a vertical span. Allowing a bit of extra room, we set the vertical span from 0 to 15. The graphs appear in Figure 5.13, and we note that they lie very

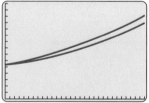

FIGURE 5.13 Comparing population models from 1800 to 1830

---

[2]Strictly speaking, this is not a power model since the extra constant term 5.34 is added. This addition is often made to satisfy an initial condition.

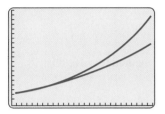

X	Y_1	Y_2
10	7.1765	6.6049
20	9.6446	8.9177
30	12.962	11.913
40	17.419	15.459
50	23.41	19.482
60	31.461	23.93
70	42.281	28.766

X=70

FIGURE 5.14 A table of values for two population models from 1800 to 1870

FIGURE 5.15 Comparing population models from 1800 to 1870

close together. This shows that the two models give very similar predictions from 1800 to 1830, a fact that is borne out by the table of values in Figure 5.12.

Solution to Part 2: We want to change the horizontal span so that it goes from 0 to 70. The table of values in Figure 5.14 led us to choose a vertical span of 0 to 45, and we used these settings to make the graphs in Figure 5.15. The exponential model is the graph that is on top in Figure 5.15. The graph shows that the exponential model predicts a larger population in the mid-1800s than does the power model. If you look on a larger horizontal span, you will observe greater separation in the curves, as is characteristic of the more rapid growth of exponential functions. Consulting the table of values in Figure 5.14, we see that the exponential model predicts an 1860 population of 31.46 million, whereas the power model predicts an 1860 population of only 23.93 million. Clearly, the prediction that the exponential model yields is much closer to the true population of 31.44 million. ■ ■ ■

ANOTHER LOOK

Homogeneity, More on Power and Exponential Functions

In this section we first deal with power functions in a more traditional algebraic way. The procedure amounts to a new derivation of the homogeneity property each time a problem is solved. Let's look, for instance, at an alternative solution of Example 5.3.

Alternative Solution to Part 1: Let the first island have area A_1. Then the number of species is given by

$$S_1 = cA_1^{1.8}.$$

If a second island has twice the area, then its area is $2A_1$. The number of species is given by

$$S_2 = c(2A_1)^{0.18} = 2^{0.18} \left(cA_1^{0.18} \right) = 1.13S_1 \,.$$

Thus the second island has 1.13 times as many species as the second, or 13% more.

The solution to part 2 is unchanged.

Alternative Solution to Part 3: Let the original area be A_0. Then the original number of species is

$$S_0 = cA_0^{0.3}.$$

If the area is reduced by 20%, then the new area is $0.8A_0$. Thus the number of species is now

$$S_1 = c (0.8A_0)^{0.3} = 0.8^{0.3} \left(cA_0^{0.3} \right) = 0.94S_0 \,.$$

The new number of species is 94% of its original value, a 6% decrease.

Alternative Solution to Part 4: Let A_0 denote the original area. We know that if A_0 is replaced by $0.9A_0$ then the number of species is 96% of its original value. The original value is cA_0^k, and the new value is $c(0.9A_0)^k$. From this we get the equation

$$0.96cA_0^k = c(0.9A_0)^k.$$

Using the laws of exponents, we fnd that this simplifies to

$$0.96cA_0^k = c0.9^kA_0^k.$$

We divide each side by cA_0^k to obtain

$$0.96 = 0.9^k.$$

We can use logarithms rather than the crossing-graphs method to solve this:

$$\ln 0.96 = \ln 0.9^k$$

$$\ln 0.96 = k\ln 0.9$$

$$\frac{\ln 0.96}{\ln 0.9} = k$$

$$0.39 = k.$$

We now have that $S = cA^{0.39}$. Let A_0 denote the original area. If the area is reduced by 25%, then the new area is $0.75A_0$. Thus the new number of species is

$$S_1 = c(0.75A)^{0.39} = 0.75^{0.39}cA^{0.39} = 0.89S_0,$$

if S_0 is the original number of species. The number of species is reduced by 11%. ∎

It is good to be able to perform the algebraic manipulations required above, but remember that the concept of homogeneity is itself important. It is central to many scientific studies involving power functions.

We can be more precise about how exponential functions grow in relation to power functions. Recall that $\lim_{x\to\infty} f(x)$ is the value that $f(x)$ nears as x increases without bound. Suppose we look at a power function divided by an exponential function, namely

$$\lim_{x\to\infty}\frac{x^p}{a^x},$$

where $a > 1$. We know that increasing exponential functions grow much faster than power functions, so for large values of x we have that x^p is smaller than a^x. Since the numerator is smaller than the denominator, the fraction is less than 1. How much less? Let's look at an example with the calculator to get an idea. In Figure 5.16 we have made the graph of $x^2/2^x$. We see that the graph grows initially but then decreases to zero.

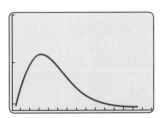

FIGURE 5.16 The graph of a power function divided by an exponential function

In general, the exponential function grows so much faster than the power function that the fraction will always approach 0. Thus when $a > 1$, we have

$$\lim_{x \to \infty} \frac{x^p}{a^x} = 0.$$

Enrichment Exercises

E-1. **More skid marks:** Do Exercise 5 at the end of this section using an algebraic method rather than using homogeneity.

E-2. **Kepler's third law again:** Do Exercise 16 at the end of this section using an algebraic method rather than using homogeneity.

E-3. **Calculating a limit:** Calculate $\lim_{x \to \infty} (2 \times 3^x + x^{1000})/3^x$. *Suggestion:* Your calculator will not be able to show an accurate picture of the graph, so hand calculation is necessary. Break the fraction up into two pieces as

$$\frac{2 \times 3^x}{3^x} + \frac{x^{1000}}{3^x}.$$

E-4. **Algebraic justification of homogeneity:** Show algebraically that if $x, t > 0$ and x is multiplied by t, then $y = Ax^k$ is multiplied by t^k.

E-5. **Uniqueness of homogeneity:** Show that if $f(tx) = t^k f(x)$ for all $x > 0$ and all $t > 0$, then f is a power function. Thus power functions are the only ones that have the homogeneity property. *Suggestion:* The equation is true when $x = 1$.

5.1   SKILL BUILDING EXERCISES

S-1. Graph of a power function: If k is negative, is the graph of x^k increasing or decreasing for $x > 0$?

S-2. Graph of a power function: If k is positive, is the graph of x^k increasing or decreasing for $x > 0$?

S-3. Homogeneity: Let $f(x) = cx^{1.47}$. If x is tripled, by what factor is f increased?

S-4. Homogeneity: Let $f(x) = cx^{2.53}$. By what factor must x be increased in order to triple the value of f?

S-5. Homogeneity: Let $f(x) = cx^{3.11}$. How do the function values $f(3.6)$ and $f(5.5)$ compare?

S-6. Homogeneity: Let $f(x) = cx^{3.11}$. Suppose that $f(y)$ is 9 times as large as $f(z)$. How do y and z compare?

S-7. Homogeneity: Let $f(x) = cx^k$. Suppose that $f(6.6)$ is 6.2 times as large as $f(1.76)$. What is the value of k?

S-8. Constant term: Let $f(x) = cx^{4.2}$ and suppose that $f(4) = 8$. Find the value of c.

S-9. Constant term: Let $f(x) = cx^{-1.32}$ and suppose that $f(5) = 11$. Find the value of c.

S-10. Exponential versus power functions: Which function is eventually larger: x^{1023} or 1.0002^x?

5.1   EXERCISES

1. **The role of the coefficient in power functions with positive power:** Consider the power functions $f(x) = cx^2$. On the same screen, make graphs of f versus x for $c = 1$, $c = 2$, $c = 3$, and $c = 4$. We suggest a horizontal span of 0 to 5. A table of values will be helpful in choosing a vertical span. On the basis of the plots you make, discuss the effect of the coefficient c on a power function when the power is positive.

2. **The role of the coefficient in power functions with a negative power:** Consider the power functions $f(x) = cx^{-2}$. On the same screen, make graphs of f versus x for $c = 1$, $c = 2$, $c = 3$, and $c = 4$. We suggest a horizontal span of 0 to 5. A table of values will be helpful in choosing a vertical span. On the basis of the plots you make, discuss the effect of the coefficient c on a power function when the power is negative.

3. **Speed and stride length:** The speed at which certain animals run is a power function of their stride length, and the power is $k = 1.7$. (See Figure 5.17.) If one animal has a stride length three times as long as another, how much faster does it run?

Stride length

FIGURE 5.17

4. **Weight and length:** A biologist has discovered that the weight of a certain fish is a power function of its length. He also knows that when the length of the fish is doubled, its weight increases by a factor of 8. What is the power k?

5. **Length of skid marks versus speed:** When a car skids to a stop, the length L, in feet, of the skid marks is related to the speed S, in miles per hour, of the car by the power function $L = \frac{1}{30h}S^2$. Here the constant h is the *friction coefficient*, which depends on the road surface.[3] For dry concrete pavement, the value of h is about 0.85.

   a. If a driver going 55 miles per hour on dry concrete jams on the brakes and skids to a stop, how long will the skid marks be?

---

[3]See J. C. Collins, *Accident Reconstruction* (Springfield, IL: Charles C Thomas, 1979). Collins notes that the friction coefficient is actually reduced at higher speeds.

b. A policeman investigating an accident on dry concrete pavement finds skid marks 230 feet long. The speed limit in the area is 60 miles per hour. Is the driver in danger of getting a speeding ticket?

c. This part of the problem applies to any road surface, so the value of h is not known. Suppose you are driving at 60 miles per hour but, because of approaching darkness, you wish to slow to a speed that will cut your emergency stopping distance in half. What should your new speed be? (*Hint:* You should use the homogeneity property of power functions here. By what factor should you change your speed to ensure that L changes by a factor of 0.5?)

6. **Binary stars:** *Binary stars* are pairs of stars that orbit each other. The *period p* of such a pair is the time, in years, required for a single orbit. The separation *s* between such a pair is measured in seconds of arc. The *parallax* angle *a* (also in seconds of arc) for any stellar object is the angle of its apparent movement as the Earth moves through one half of its orbit around the sun. Astronomers can calculate the total mass M of a binary system using

$$M = s^3 a^{-3} p^{-2}.$$

Here M is the number of *solar masses*.

a. Alpha Centauri, the nearest star to the sun, is in fact a binary star. The separation of the pair is $s = 17.6$ seconds of arc, its parallax angle is $a = 0.76$ second of arc, and the period of the pair is 80.1 years. What is the mass of the Alpha Centauri pair?

b. How would the mass change if the separation angle were doubled but parallax and period remained the same as for the Alpha Centauri system?

c. How would the mass change if the parallax angle were doubled but separation and period remained the same?

d. How would the mass change if the period doubled but parallax angle and separation remained the same?

7. **Life expectancy of stars:** The life expectancy E of a main-sequence star[4] depends on its mass M. The relation is given by

$$E = M^{-2.5},$$

where M is solar masses and E is solar lifetimes. The sun is thought to be at the middle of its life, with a total life expectancy of about 10 billion years. Thus the value $E = 1$ corresponds to a life expectancy of 10 billion years.

a. Does a more massive star have a longer or a shorter life expectancy than a less massive star?

b. Spica is a main-sequence star that is about 7.3 solar masses. What is the life expectancy of Spica?

c. Express using functional notation the life expectancy of a main-sequence star with mass equal to 0.5 solar mass, and then calculate that value.

d. Vega is a main-sequence star that is expected to live about 6.36 billion years. What is the mass of Vega?

e. If one main-sequence star is twice as massive as another, how do their life expectancies compare?

8. **Height of tsunami waves:** When waves generated by tsunamis approach shore, the height of the waves generally increases. Understanding the factors that contribute to this increase can aid in controlling potential damage to areas at risk.

   Green's law tells how water depth affects the height of a tsunami wave. If a tsunami wave has height H at an ocean depth D, and the wave travels to a location of water depth d, then the new height h of the wave is given by $h = HR^{0.25}$, where R is the water depth ratio given by $R = \frac{D}{d}$.

a. Calculate the height of a tsunami wave in water 25 feet deep if its height is 3 feet at its point of origin in water 15,000 feet deep.

b. If water depth decreases by half, the depth ratio R is doubled. How is the height of the tsunami wave affected?

9. **Tsunami waves and breakwaters:** *This is a continuation of Exercise 8.* Breakwaters affect wave height by reducing energy. (See Figure 5.18.) If a tsunami

___

[4]About 90% of all stars are main-sequence stars.

wave of height H in a channel of width W encounters a breakwater that narrows the channel to a width w, then the height h of the wave beyond the breakwater is given by $h = HR^{0.5}$, where R is the width ratio $R = \frac{w}{W}$.

a. Suppose a wave of height 8 feet in a channel of width 5000 feet encounters a breakwater that narrows the channel to 3000 feet. What is the height of the wave beyond the breakwater?

b. If a channel width is cut in half by a breakwater, what is the effect on wave height?

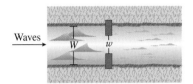

FIGURE 5.18

10. **Bores:** Under certain conditions, tsunami waves encountering land will develop into *bores*. A bore is a surge of water much like what would be expected if a dam failed suddenly and emptied a reservoir into a river bed. In the case of a bore traveling from the ocean into a dry river bed, one study[5] shows that the velocity V of the tip of the bore is proportional to the square root of its height h. Expressed in a formula, this is

$$V = kh^{0.5},$$

where k is a constant.

a. A bore travels up a dry river bed. How does the velocity of the tip compare with its initial velocity when its height is reduced to half of its initial height?

b. How does the height of the bore compare with its initial height when the velocity of the tip is reduced to half its initial velocity?

c. If the tip of one bore surging up a dry river bed is three times the height of another, how do their velocities compare?

11. **Terminal velocity:** By comparing the surface area of a sphere with its volume and assuming that air resistance is proportional to the square of velocity, it is possible to make a heuristic argument to support the following premise: *For similarly shaped objects, terminal velocity varies in proportion to the square root of length.* Expressed in a formula, this is

$$T = kL^{0.5},$$

where L is length, T is terminal velocity, and k is a constant that depends on shape, among other things. This relation can be used to help explain why small mammals easily survive falls that would seriously injure or kill a human.

a. A 6-foot man is 36 times as long as a 2-inch mouse (neglecting the tail). How does the terminal velocity of a man compare with that of a mouse?

b. If the 6-foot man has a terminal velocity of 120 miles per hour, what is the terminal velocity of the 2-inch mouse?

c. Neglecting the tail, a squirrel is about 7 inches long. Again assuming that a 6-foot man has a terminal velocity of 120 miles per hour, what is the terminal velocity of a squirrel?

12. **Dropping rocks on other planets:** It is a consequence of Newton's law of gravitation that near the surface of any planet, the distance D fallen by a rock in time t is given by $D = ct^2$. That is, distance fallen is proportional to the square of the time, no matter what planet one may be on. But the value of c depends on the mass of the planet. For Earth, if time is measured in seconds and distance in feet, the value of c is 16.

a. Suppose a rock is falling near the surface of a planet. What is the comparison in distance fallen from 2 seconds to 6 seconds into the drop? (*Hint:* This question may be rephrased as follows: "If time increases by a factor of 3, by what factor will distance increase?")

b. For objects falling near the surface of Mars, if time is measured in seconds and distance in feet, the value of c is 6.4. If a rock is dropped from 70

[5]Y. M. Fukui, H. S. Nakamura, and Y. Sasaki, "Hydraulic study on tsunami," *Coastal Engr. Japan* **6** (1963), 67–82.

feet above the surface of Mars, how long will it take for the rock to strike the ground?

c. On Venus, a rock dropped from 70 feet above the surface takes 2.2 seconds to strike the ground. What is the value of c for Venus?

13. **Newton's law of gravitation:** According to Newton's law of gravity, the gravitational attraction between two massive objects such as planets or asteroids is proportional to d^{-2}, where d is the distance between the centers of the objects. Specifically, the gravitational force F between such objects is given by $F = cd^{-2}$, where d is the distance between their centers. The value of the constant c depends on the masses of the two objects and on the *universal gravitational constant*.

a. Suppose the force of gravity is causing two large asteroids to move toward each other. What is the effect on the gravitational force if the distance between their centers is halved? What is the effect on the gravitational force if the distance between their centers is reduced to one-quarter of its original value?

b. Suppose that for a certain pair of asteroids whose centers are 300 kilometers apart, the gravitational force is 2,000,000 newtons. (One newton is about one-quarter of a pound.) What is the value of c? Find the gravitational force if the distance between the centers of these asteroids is 800 kilometers.

c. Using the value of c you found in part b, make a graph of gravitational force versus distance between the centers of the asteroids for distances from 0 to 1000 kilometers. What happens to the gravitational force when the asteroids are close together? What happens to the gravitational force when the asteroids are far apart?

14. **Geostationary orbits:** For communications satellites to work properly, they should appear from the surface of the Earth to remain stationary. That is, they should orbit the Earth exactly once each day. For *any* satellite, the *period* P (the length of time required to complete an orbit) is determined by its mean distance A from the center of the Earth. For a satellite of negligible mass, P and A are related by a power function $A = cP^{2/3}$.

a. The moon is 239,000 miles from the center of the Earth and has a period of about 28 days. How high above the center of the Earth should a geostationary satellite be? (*Hint:* You want the distance A for a satellite with period $\frac{1}{28}$ that of the moon. The homogeneity property of power functions is applicable.)

b. The radius of the Earth is about 3963 miles. How high above the surface of the Earth should a geostationary satellite be?[6]

15. **Giant ants and spiders:** Many science fiction movies feature animals such as ants, spiders, or apes growing to monstrous sizes and threatening defenseless Earthlings. (Of course, they are in the end defeated by the hero and heroine.) Biologists use power functions as a rough guide to relate body weight and cross-sectional area of limbs to length or height. Generally, weight is thought to be proportional to the cube of length, whereas cross-sectional area of limbs is proportional to the square of length. Suppose an ant, having been exposed to "radiation," is enlarged to 500 times its normal length. (Such an event can occur only in Hollywood fantasy. Radiation is utterly incapable of causing such a reaction.)

a. By how much will its weight be increased?

b. By how much will the cross-sectional area of its legs be increased?

c. Pressure on a limb is weight divided by cross-sectional area. By how much has the pressure on a leg of the giant ant increased? What do you think is likely to happen to the unfortunate ant?[7] *Note:* The factor by which pressure increases is given by

$$\frac{\text{Factor of increase in weight}}{\text{Factor of increase in area}}.$$

16. **Kepler's third law:** By 1619 Johannes Kepler had completed the first accurate mathematical model

---

[6]The actual height is 22,300 miles above the surface of the Earth. You will get a slightly different answer because we have neglected the mass of the moon.

[7]Similar arguments about creatures of extraordinary size go back at least to Galileo. See his famous book *Two New Sciences*, published in 1638.

describing the motion of planets around the sun. His model consisted of three laws that, for the first time in history, made possible the accurate prediction of future locations of planets. Kepler's third law related the period (the length of time required for a planet to complete a single trip around the sun) to the mean distance D from the planet to the sun. In particular, he stated that the period P is proportional to $D^{1.5}$.

a. Neptune is about 30 times as far from the sun as is the Earth. How long does it take Neptune to complete an orbit around the sun? (*Hint:* The period for Earth is 1 year. If the distance is increased by a factor of 30, by what factor will the period be increased?)

b. The period of Mercury is about 88 days. The Earth is about 93 million miles from the sun. How far is Mercury from the sun? (*Hint:* The period of Mercury is different from that of Earth by a factor of $\frac{88}{365}$.)

17. **Time to failure:** A building that is subjected to shaking (caused, for example, by an earthquake) may collapse. Failure depends both on intensity and on duration of the shaking. If an intensity I_1 causes a building to collapse in t_1 seconds, then an intensity I_2 will cause the collapse in t_2 seconds, where

$$\frac{t_1}{t_2} = \left(\frac{I_2}{I_1}\right)^2.$$

If a certain building collapses in 30 seconds at one intensity, how long would it take the building to collapse at triple that intensity?

5.2 Modeling Data with Power Functions

In this section, we will learn how to construct power function models much as we did exponential models in the previous chapter.

The Connection Between Power Data and Linear Data

L2	L3	L4 4
------	1	3
	2	12
	3	27
	4	48
	5	75
	6	108
	7	147

L4(1) = 3

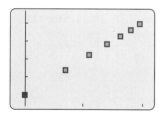

FIGURE 5.19 Data for x and $3x^2$

L1	L2	L3 2
0	1.0986	1
.69315	2.4849	2
1.0986	3.2958	3
1.3863	3.8712	4
1.6094	4.3175	5
1.7918	4.6821	6
1.9459	4.9904	7

L2(1)=1.098612288...

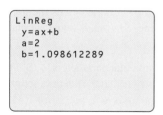

FIGURE 5.20 Data for $\ln x$ and $\ln(3x^2)$

Linear regression makes it relatively easy to construct linear models, and the procedure is so reliable that in constructing other types of models, scientists and mathematicians often seek links with linear data so that regression can be used. For example, to construct an exponential model for observed phenomena as we did in the previous chapter, the needed link was the logarithm. We began with exponential data, used the logarithm of function values to convert them to linear data, applied regression to get a linear model for the transformed data, and finally converted the linear model to an exponential model for the original data. In order to make power function models, a similar link between power data and linear data is needed. As we shall see, we can get it by using the logarithm in a slightly different way.

To see the connection, we look at the familiar power function $f(x) = 3x^2$. In Figure 5.19 we have entered $x = 1, 2, \ldots, 7$ in the third column and $f(x) = 3 \times 1^2, 3 \times 2^2, \ldots, 3 \times 7^2$ in the fourth column [5.7]. Thus in columns 3 and 4 of Figure 5.19, we see data from the power function $f(x) = 3x^2$. In Figure 5.20 we have entered [5.8] the natural logarithm of x in the first column and the natural logarithm of $f(x)$ in the second [5.9] column.

If we plot [5.10] the data for $\ln(3x^2)$ versus the data for $\ln x$, we see in Figure 5.21 that the points fall on a straight line, which indicates that a linear relation holds. This is the link we had hoped to find. If f is a power function of x, then $\ln f$ is a linear function of $\ln x$.

FIGURE 5.21 $\ln f$ versus $\ln x$

Getting a Power Model from Data

In practice, what is often wanted is to begin with observed data, check to see whether they are appropriately modeled with a power function, and then actually build the model. Looking more closely at what we have already done, we can see how to do that. You may wish to refer to the *Technology Guide* for the exact keystrokes used to produce the following work.

We want to take the data we have and see how to recover the power function $f = 3x^2$ that generated them. We have arranged things so that data for $\ln x$ are already in column 1 and data for $\ln f$ are in column 2. Thus we can get the regression line parameters [5.11] shown in Figure 5.22 in the usual way. We see that $\ln f$ is related to $\ln x$ by the linear function $\ln f = 2 \ln x + 1.1$.

```
LinReg
 y=ax+b
 a=2
 b=1.098612289
```

FIGURE 5.22 Regression line parameters

Observe that the slope 2 for the regression line turns out to be the same as the power for $f(x) = 3x^2$. Also, 1.1 is the vertical intercept of the line, and $3 = e^{1.1}$ is the c value for the power function. The connection we see here is in fact characteristic of the link between linear and power functions.

Power Functions on a Logarithmic Scale

If $\ln f(x)$ is a linear function of $\ln x$ with slope k, then f is a power function with power k.

The number c in the formula $f(x) = cx^k$ is $c = e^b$, where b is the vertical intercept of the line.

EXAMPLE 5.5   The Volume Inside a Sphere

FIGURE 5.23

The following table of values gives the volume V in cubic inches inside a sphere of radius r inches. (See Figure 5.23.)

Radius r	1	2	3	4	5
Volume V	4.19	33.51	113.10	268.08	523.60

1. Plot the graph of $\ln V$ versus $\ln r$ to determine whether it is reasonable to think that the volume is related to the radius by a power function.[8]

2. Use regression to find a formula for the volume of a sphere as a function of the radius.

3. To check your work, plot the graph of the original data points together with the function you found in part 2.

```
L2      L3      L4    4
------  1       4.19
        2       33.51
        3       113.1
        4       268.08
        5       523.6
        ------  ████
L4(6) =
```

FIGURE 5.24 Radius and volume of spheres

```
L1      L2      L3    2
0       1.4327  1
.69315  3.5118  2
1.0986  4.7283  3
1.3863  5.5913  4
1.6094  6.2607  5
------  ------  ------
L2(1)=1.432700733...
```

FIGURE 5.25 Logarithm of radius and logarithm of volume

Solution to Part 1:   The first step is to enter 5.12 the data into the calculator. Figure 5.24 shows the radius in column 3 and the corresponding volume in column 4, and Figure 5.25 shows 5.13 the logarithm of the radius in column 1 and the logarithm of the volume in column 2.

In Figure 5.26 we have plotted 5.14 the logarithm of volume versus the logarithm of radius. The points clearly fall on a straight line, and this is evidence that the volume V is indeed a power function of the radius r.

Solution to Part 2:   We now proceed to calculate the regression line parameters 5.15 as shown in Figure 5.27. Rounding to two decimal places, we see that the linear function relating the logarithm of volume to the logarithm of the radius is $\ln V = 3 \ln r + 1.43$. The power we need is the slope 3 of this line, and the value of c is $e^{1.43} = 4.18$. We conclude that $V = 4.18r^3$.

---

[8]As we noted in the preceding section, we know from elementary geometry that such a relationship is indeed valid and that $V = \frac{4\pi}{3}r^3$.

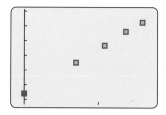

FIGURE 5.26 Logarithm of volume versus logarithm of radius

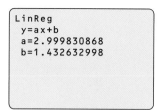

FIGURE 5.27 Regression line parameters

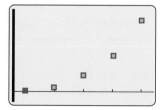

FIGURE 5.28 Plot of data for volume versus radius

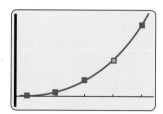

FIGURE 5.29 Adding the power function model

Solution to Part 3: To check our work, we plotted 5.16 the original data for volume versus radius in Figure 5.28 and added the graph of $4.18r^3$ in Figure 5.29. This shows excellent agreement between the given data and the power model we constructed.

EXAMPLE 5.6 Kepler's Third Law

Johannes Kepler was the first to give an accurate description of the motion of the planets about the sun. He presented his model in the form of three laws. His third law states, "It is absolutely certain and exact that the ratio which exists between the periodic times of any two planets is precisely the ratio of the $\frac{3}{2}$th power of the mean distances [of the planet to the sun]."[9] Kepler formulated his third law by examining carefully the recorded measurements of the astronomer Tycho Brahe, but he never gave any derivation of it from other principles. That is to say, Kepler believed there was a power relation between the period P of a planet (the time required to complete a revolution about the sun) and its mean distance D from the sun, and he found the correct power by looking at data. We will use more accurate data and much more powerful calculation techniques than were available to Kepler to arrive at similar conclusions. The following table gives distances measured in millions of miles and periods measured in years for the nine planets.

---

[9]From *The Harmonies of the World*, translated by Charles Glenn Wallis in the *Great Books*. Cited by Victor J. Katz in *A History of Mathematics* (New York: HarperCollins, 1993).

Planet	Distance D	Period P
Mercury	36.0	0.24
Venus	67.1	0.62
Earth	92.9	1
Mars	141.7	1.88
Jupiter	483.4	11.87
Saturn	886.1	29.48
Uranus	1782.7	84.07
Neptune	2793.1	164.90
Pluto	3666.1	249

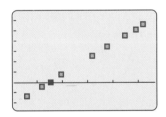

FIGURE 5.30 Data for orbital period versus distance from the sun

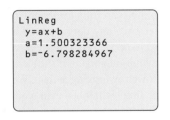

FIGURE 5.31 Logarithms of planetary data

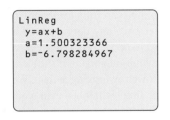

FIGURE 5.32 Plot of logarithms of data

```
LinReg
  y=ax+b
  a=1.500323366
  b=-6.798284967
```

FIGURE 5.33 Regression line parameters

1. Plot the logarithms of the data points and determine whether it is reasonable to model this data using a power function.

2. Use a formula to express P as a power function of D.

3. If one planet were twice as far from the sun as another, how would their periods compare?

Solution to Part 1:   The first step is to enter [5.17] the data into the graphing calculator as we have done in Figure 5.30. Distance appears in the third column, and period appears in the fourth. In Figure 5.31 the logarithm of distance [5.18] is in the first column and the logarithm of period is in the second column. When we plot the logarithms as we have done in Figure 5.32, we see that the points line up nicely, supporting the idea that distance and period are related by a power function.

Solution to Part 2:   In Figure 5.33 we have calculated the regression line for $\ln P$ versus $\ln D$. We see that $\ln P = 1.5 \ln D - 6.8$. Thus the power we need is $1.5 = \frac{3}{2}$, precisely the power proposed by Kepler. The c value for the power function is $e^{-6.8} = 0.0011$. We conclude that if we measure periods in years and distances in millions of miles, then $P = 0.0011 D^{1.5}$.

Solution to Part 3:   Here we use the homogeneity property of power functions. If the distance is increased by a factor of 2, then period will be increased by a factor of $2^{1.5} = 2.83$. Thus, if one planet is twice as far away from the sun as another, its period will be 2.83 times as long.   ■ ■ ■

Almost Power Data

As with linear and exponential models, many times observed data cannot be modeled exactly by a power function, but linear regression enables us to make power models that approximately fit the data.

EXAMPLE 5.7   Generation Time as a Power Function of Length

The *generation time* for an organism is the time it takes to reach reproductive maturity. Biologists have observed that generation time depends on size, and in particular on the

Organism	Length L	Generation time T
House fly	0.023	0.055 (20 days)
Cotton deermouse	0.295	0.192 (70 days)
Tiger salamander	0.673	1
Beaver	2.23	2.8
Grizzly bear	5.91	4
African elephant	11.5	12.3
Yellow birch	72.2	40
Giant sequoia	262	60

TABLE 5.1 Length and generation time

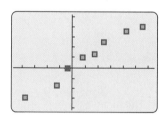

FIGURE 5.34 Data for L and T

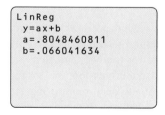

FIGURE 5.35 Data for ln L and ln T

length of an organism. Table 5.1 gives the length L measured in feet and generation time T measured in years for various organisms.[10]

1. By plotting $\ln T$ against $\ln L$, determine whether it is reasonable to model T as a power function of L.

2. Find a formula for the regression line of $\ln T$ against $\ln L$.

3. Find a formula that models T as a power function of L and plot the function along with the given data.

4. If one organism is 5 times as long as another, what would be the expected comparison of generation times? (*Suggestion:* Use the homogeneity property of power functions.)

Solution to Part 1:   In the third column we enter 5.19 the L values, and in the fourth column we enter 5.20 the T values. The result is shown in Figure 5.34. Next we put the values of $\ln L$ in the first column 5.21 and the values of $\ln T$ in the second column 5.22 , as shown in Figure 5.35.

Now we plot 5.23 the data for $\ln T$ against $\ln L$. The result is shown in Figure 5.36. The data do not lie exactly in a straight line but are approximately linear, so it is reasonable to model T as a power function[11] of L.

Solution to Part 2:   The regression line parameters 5.24 for the data in Figure 5.36 are shown in Figure 5.37. Rounding, we get

$$\ln T = 0.8 \ln L + 0.066 .$$

Solution to Part 3:   Since the slope of the regression line for $\ln T$ as a function of $\ln L$ is 0.8, that is also the power we use for our power model. Thus $T = cL^{0.8}$, and

FIGURE 5.36 Plotting the data for ln L against ln T

```
LinReg
  y=ax+b
  a=.8048460811
  b=.066041634
```

FIGURE 5.37 Finding the regression line

---

[10]These data are adapted from J. T. Bonner, *Size and Cycle* (Princeton, NJ: Princeton University Press, 1965). We have presented only a sample of the 46 organisms for which Bonner gives data.

[11]It is important to note that in a setting such as this, exact modeling is unreasonable to expect, and it is striking that any kind of consistent relationship can be found among such a diverse collection of species.

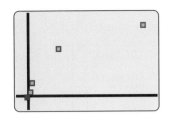

FIGURE 5.38 Plot of data for generation time versus length

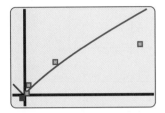

FIGURE 5.39 Adding the power model

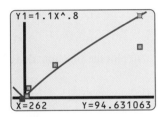

FIGURE 5.40 A closer look at the sequoia

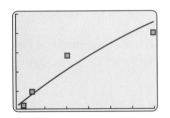

FIGURE 5.41 Changing the window to a smaller range

we need to find c. We get this from the vertical intercept 0.066 found in part 2: $c = e^{0.066}$, which is about 1.1, so the power model is $T = 1.1L^{0.8}$. In Figure 5.38 we have plotted the original data, and we added our power model in Figure 5.39.

We note that the generation time for the giant sequoia, the last data point in Figure 5.39, seems to lie well below the power function model. In Figure 5.40 we have traced the graph and set the cursor to **X=262**. We see that the power model shows a generation time of 94 years, in contrast to the actual time of 60 years. In some settings, this might cause us to question the validity of the model. Such questions are always appropriate, but in this case we are happy with a model that can be a starting point for further analysis rather than an exact relationship, such as that which we had for Kepler's third law. We would be more concerned if the value given by the model were off by a factor of 10. In fact, one reason for the difference is the extreme variation in both length and generation time that is to some degree characteristic of data to which power function models are applied.[12] The 262-foot sequoia is over 10,000 times longer than the house fly, and its 60-year generation time is over 1000 times longer than that of the house fly. This is the cause of the bunching of data in Figure 5.39. In Figure 5.41 the plot is shown with a horizontal span of 0 to 6 and a vertical span of 0 to 5. It shows more of the smaller data points but leaves out the elephant, birch, and sequoia.

Solution to Part 4:   The homogeneity property of a power function such as $T = 1.1L^{0.8}$ tells us that if length L is increased by a factor of 5, then generation time T will be increased by a factor of $5^{0.8} = 3.63$. Thus we would expect the longer organism to have a generation time 3.63 times as long as the shorter organism. ▪ ▪ ▪

Graphing on a Logarithmic Scale: Common versus Natural Logarithms

Throughout this text we have used the natural logarithm $\ln x$. But many scientists prefer the common logarithm $\log x$, which is associated with 10 rather than e, and this may be the logarithm that you encounter in applications. Traditionally, scientists have tested data to see whether a power model is appropriate by plotting points on *log log* graphing paper, which has both the horizontal and the vertical axis marked in powers of 10. Figure 5.42 shows such a plot that might have been used in Example 5.7.

Graphing with such paper is really the same as the procedure we have used here except that the common logarithm is used instead of the natural logarithm.[13] The only substantive difference is that when calculating the c value for a power function, we use 10^b rather than e^b, where b is the vertical intercept of the regression line for the logarithm. To show this, we work through the Kepler example once more using the common logarithm rather than the natural logarithm. As we shall see, the final answers are the same.

_____

[12]In contrast, data that should be modeled exponentially may show relatively small variation in the horizontal direction but much larger variation in the vertical direction.

[13]If the graphing paper were marked in powers of e rather than in powers of 10, the procedure would be exactly the same.

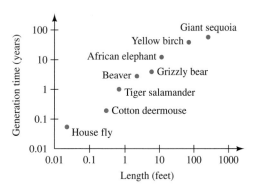

FIGURE 5.42 Generation time against length, on a logarithmic scale

Alternative solution of Example 5.6 using the common logarithm: We enter the data for D and P in the third and fourth columns, just as we did in Example 5.6, but this time we put the common logarithm of D in the first column and the common logarithm of P in the second column 5.25 . We plot the common logarithm of the data as shown in Figure 5.43 and note that this is similar to Figure 5.32.

Next we get the regression line parameters shown in Figure 5.44 exactly as we did in Example 5.6. We see that $\log P = 1.5 \log D - 2.95$. The power we use is the slope 1.5 of this line, the same as with the natural logarithm. But to get the c value, we use 10 rather than e. Since $10^{-2.95} = 0.0011$, we arrive at the same answer ($P = 0.0011 D^{1.5}$) as we did in Example 5.6.

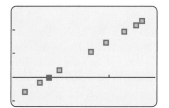

FIGURE 5.43 Plot of common logarithms of planetary data

FIGURE 5.44 Regression line parameters for common logarithms

ANOTHER LOOK

More on Power Regression

We have relied heavily in this section on the statement that if f is a power function of x, then $\ln f$ is a linear function of $\ln x$. We want to use the laws of logarithms to verify this and to show the details of the connection.

Suppose $f = cx^k$ is a power function. If we take the natural logarithm of each side, we get

$$\ln f = \ln \left(cx^k\right) = \ln c + \ln x^k = \ln c + k \ln x \, .$$

Let $y = \ln f$, $A = \ln c$, and $z = \ln x$. Then the above equation becomes $y = A + kz$, and clearly $y = \ln f$ is a linear function of $z = \ln x$ with slope k and initial value $A = \ln c$.

Suppose now that we start with data generated by some (unknown) power function $f = cx^k$. If we take the logarithm of both f and x, we can fit the new data with a linear relation (in practice the regression line) $y = A + kz$. The derivation above shows us how to use this linear equation to get the power function $f = cx^k$ that models the original data. Since $A = \ln c$, we have $c = e^A$. Thus we get the coefficient for the power function by exponentiating the initial value of the regression line. Further, the power k is the slope of the regression line. This verifies the relationship we pointed out earlier.

We should note that given two points with all four coordinates positive,[14] there is a unique power function that passes through the two points. For example, let's find the unique power function $y = cx^k$ that passes through $(2, 3)$ and $(8, 9)$. If the graph of this function passes through the two given points, then

$$3 = c \times 2^k$$

$$9 = c \times 8^k.$$

Dividing the second equation by the first gives

$$\frac{9}{3} = \frac{c \times 8^k}{c \times 2^k} = \left(\frac{8}{2}\right)^k,$$

or

$$3 = 4^k.$$

We can solve this equation using logarithms:

$$\ln 3 = \ln 4^k$$

$$\ln 3 = k \ln 4$$

$$\frac{\ln 3}{\ln 4} = k$$

$$0.79 = k.$$

We put this value back into the first equation and solve for c:

$$3 = c \times 2^{0.79}$$

$$\frac{3}{2^{0.79}} = c$$

$$1.74 = c.$$

Thus the power function we seek is $y = 1.74x^{0.79}$.

Enrichment Exercises

E-1. **Finding power functions:**

    a. Find the power function that passes through $(2, 7)$ and $(6, 56)$.

    b. Find the power function that passes through $(a, 12)$ and $(10a, 1200)$ for a given $a \neq 0$. *Suggestion:* This exercise will work out better if you use the common logarithm.

---

[14]Many power functions are not defined for negative values of x.

E-2. **Finding more power functions:** Find power functions passing through the given points.

a. $(3, 7)$ and $(4, 9)$
b. $(2, 5)$ and $(6, 1)$
c. $(3, 2)$ and $(5, 3)$

E-3. **Linear to power functions:** Show that if $\ln y = \ln c + k \ln x$, then $y = cx^k$.

E-4. **Finding a power function through general points:** Find a power function through the points (x_1, y_1) and (x_2, y_2). Assume that each x_j, y_j is positive.

5.2   *SKILL BUILDING EXERCISES*

S-1. **Logarithmic conversion of both x and y:** If the values in the table

Input value	a	b	$\cdots$
Function value	z	w	$\cdots$

are power data, then what kind of data would appear in the following table?

Input value	$\ln a$	$\ln b$	$\cdots$
Function value	$\ln z$	$\ln w$	$\cdots$

S-2. **Formula conversion:** Suppose $\ln f = 3 \ln x + 2$. What kind of function is $f = f(x)$? Find a formula for f.

S-3. **Getting the power:** If the regression line for $\ln y$ as a linear function of $\ln x$ has slope 3, what is the power used to express y as a power function of x?

S-4. **Getting c:** If the regression line for $\ln y$ as a linear function of $\ln x$ has initial value 4, what is the value of c in the formula $y = cx^k$?

S-5. **Modeling power data:** The following data table was generated by a power function f. Find a formula for f and plot the data points along with the graph of the formula.

x	f
1	3.6
2	8.86
3	15.02
4	21.83
5	29.17

S-6. **Modeling almost power data:** Model the following data with a power function. Give the formula and plot the data points along with the model.

x	1	2	3	4	5
f	6.3	1.9	0.6	0.2	0.07

S-7. **Modeling almost power data:** Model the following data with a power function. Give the formula and plot the data points along with the model.

x	f
1	2.2
2	51.7
3	338.9
4	1236.4
5	3177.8

S-8. **Modeling almost power data:** Model the following data with a power function. Give the formula and plot the data points along with the model.

x	1	2	3	4	5
f	5.5	15.1	28.8	63.1	84.2

S-9. **Modeling almost power data:** Model the following data with a power function. Give the formula and plot the data points along with the model.

x	0.3	1.3	2.2	3.3	4.1
f	5.6	2	0.92	0.77	0.51

S-10. **The common logarithm:** Repeat Exercise S-6 using the common logarithm.

5.2 EXERCISES

1. **Hydroplaning:** On wet roads, under certain conditions the front tires of a car will *hydroplane*, or run along the surface of the water. The critical speed V at which hydroplaning occurs is a function of p, the tire inflation pressure.[15] The following table shows hypothetical data for p, in pounds per square inch, and V, in miles per hour.

Tire inflation pressure p	Critical speed V for hydroplaning
20	46.3
25	51.8
30	56.7
35	61.2

a. Find a formula for the regression line of $\ln V$ against $\ln p$.

b. Find a formula that models V as a power function of p.

c. In the rain a car (with tires inflated to 35 pounds per square inch) is traveling behind a bus (with tires inflated to 60 pounds per square inch), and both are moving at 65 miles per hour. If they both hit their brakes, what might happen?

2. **Urban travel times:** The accompanying table shows the 1960 population N, in thousands, for several cities, together with the average time T, in minutes, spent by residents driving to work.

City	Population N	Driving time T
Los Angeles	6489	16.8
Pittsburgh	1804	12.6
Washington	1808	14.3
Hutchinson	38	6.1
Nashville	347	10.8
Tallahassee	48	7.3

An analysis[16] of these data, along with data from 17 other cities in the United States and Canada, led to a power model of average driving time as a function of population.

a. Construct a power model of driving time in minutes as a function of population measured in thousands.

b. Is average driving time in Pittsburgh more or less than would be expected from its population?

c. If you wish to move to a smaller city to reduce your average driving time to work by 25%, how much smaller should the city be?

3. **Mass-luminosity relation:** Roughly 90% of all stars are *main-sequence stars*. Exceptions include supergiants, giants, and dwarfs. For main-sequence stars (including the sun) there is an important relationship called the *mass-luminosity relation* between the relative luminosity[17] L and the mass M in terms of solar masses. Relative masses and luminosities of several main-sequence stars are reported in the accompanying table. →

---

[15] See J. C. Collins, *Accident Reconstruction* (Springfield, IL: Charles C Thomas, 1979). The critical speed for hydroplaning also increases with the amount of tire tread available.

[16] A. M. Voorhees, S. J. Bellomo, J. L. Schofer, and D. E. Cleveland, "Factors in work trip lengths," Highway Research Record No. 141 (Washington, DC: Highway Research Board, 1966), 24–46. The analysis is based on data reported by Alan M. Voorhees & Associates.

[17] Relative luminosity is the ratio of the luminosity of a star to that of the sun.

Star	Solar mass M	Luminosity L
Spica	7.3	1050
Vega	3.1	55
Altair	1	1.1
The Sun	1	1
61 Cygni A	0.17	0.002

a. Find a power model for the data in this table. (Round the power and the coefficient to one decimal place.) The function you find is known to astronomers as the mass-luminosity relation.

b. Kruger 60 is a main-sequence star that is about 0.11 solar mass. Use functional notation to express the relative luminosity of Kruger 60, and then calculate that value.

c. Wolf 359 has a relative luminosity of about 0.0001. How massive is Wolf 359?

d. If one star is 3 times as massive as another, how do their luminosities compare?

4. **Growth rate versus weight:** Ecologists have studied how a population's intrinsic exponential growth rate r is related to the body weight W for herbivorous mammals.[18] In Table 5.2, W is the adult weight measured in pounds, and r is growth rate per year.

Animal	Weight W	r
Short-tailed vole	0.07	4.56
Norway rat	0.7	3.91
Roe deer	55	0.23
White-tailed deer	165	0.55
American elk	595	0.27
African elephant	8160	0.06

TABLE 5.2 Weight and exponential growth rate

a. Make a plot of $\ln r$ against $\ln W$. Is it reasonable to model r as a power function of W?

b. Find a formula that models r as a power function of W, and draw a graph of this function.

5. **Speed in flight versus length:** Table 5.3 gives the length L, in inches, of a flying animal and its maximum speed F, in feet per second, when it flies.[19] (For comparison, 10 feet per second is about 6.8 miles per hour.)

Animal	Length L	Flying speed F
Fruit fly	0.08	6.2
Horse fly	0.51	21.7
Ruby-throated hummingbird	3.2	36.7
Willow warbler	4.3	39.4
Flying fish	13	51.2
Bewick's swan	47	61.7
White pelican	62	74.8

TABLE 5.3 Length and flying speed

a. Judging on the basis of this table, is it generally true that larger animals fly faster?

b. Find a formula that models F as a power function of L.

c. Make the graph of the function in part b.

d. Is the graph you found in part c concave up or concave down? Explain in practical terms what your answer means.

e. If one bird is 10 times longer than another, how much faster would you expect it to fly? (Use the homogeneity property of power functions.)

 6. **Speed swimming versus length:** *This is a continuation of Exercise 5.* Table 5.4 gives the length L,

[18] G. Caughley and C. J. Krebs, "Are big mammals simply little mammals writ large?" *Oecologia* **59** (1983), 7–17.

[19] The tables in this exercise and the next are adapted from J. T. Bonner, op. cit.

in inches, of a swimming animal and its maximum speed S, in feet per second, when it swims.

Animal	Length L	Swimming speed S
Bacillus	9.8×10^{-5}	4.9×10^{-5}
Paramecium	0.0087	0.0033
Water mite	0.051	0.013
Flatfish larva	0.37	0.38
Goldfish	2.8	2.5
Dace	5.9	5.7
Adélie penguin	30	12.5
Dolphin	87	33.8

TABLE 5.4 Length and swimming speed

a. Find a formula for the regression line of $\ln S$ against $\ln L$. (Round the slope to one decimal place.)

b. Find a formula that models S as a power function of L. On the basis of the power you found, what special type of power function is this?

c. Add the graph of S against L to the graph of F that you drew in Exercise 5.

d. Is flying a significant improvement over swimming if an animal is 1 foot long (the approximate length of a flying fish)?

e. Would flying be a significant improvement over swimming for an animal 20 feet long?

f. A blue whale is about 85 feet long, and its maximum speed swimming is about 34 feet per second. Judging on the basis of these facts, do you think the trend you found in part b continues indefinitely as the length increases?

7. **Metabolism:** Physiologists who study warm-blooded animals are interested in the *basal metabolic rate*, which is one measure of the energy

needed for survival. It can be measured from the volume and composition of expired air. Table 5.5 gives the weight W, in pounds, of an animal and its basal metabolic rate B, in kilocalories per day.[20]

Animal	Weight W	Basal metabolic rate B
Rat	0.38	20.2
Pigeon	0.66	30.8
Hen	4.3	106
Dog	25.6	443
Sheep	101	1220
Cow	855	6421

TABLE 5.5 Weight and basal metabolic rate

a. Find a formula that models B as a power function of W.

b. Define the *metabolic weight* of a warm-blooded animal to be $W^{0.75}$ if W is its weight in pounds. Direct comparison of food intake among animals of different sizes is not easy. Clearly, large animals will consume more food than small ones, and to compensate for this, we should take account of the energy needed for survival—that is, the basal metabolic rate B. To compare maximum daily food intake among animals of different sizes, we divide this food intake by the metabolic weight.

i. Explain why dividing by the metabolic weight is more meaningful than dividing by the weight in comparing maximum daily food intake.

ii. Find the metabolic weight of a 2.76-pound animal and that of a 126.8-pound animal.

iii. An ecologist found the maximum daily food intake of a 2.76-pound rabbit to be about 0.18 pound and that of a 126.8-pound merino

[20]The table is adapted from M. Kleiber, "Body size and metabolism," *Hilgardia* **6** (1932), 315–353. See also his book *The Fire of Life*, rev. ed. (Huntington, New York: Robert E. Krieger, 1975).

sheep to be about 2.8 pounds.[21] Divide each intake by the corresponding metabolic weight. How do the daily consumption levels compare on this basis?

8. **Metabolism and surface area:** *This is a continuation of Exercise 7.* Table 5.6 gives the weight W, in pounds, the basal metabolic rate B, in kilocalories per day, and the surface area A, in square inches, of a variety of marsupials.[22]

Animal	Weight W	Basal metabolic rate B	Surface area A
Fat-tailed marsupial mouse	0.03	2.16	12.4
Brown marsupial mouse	0.08	4.18	20.5
Long-nosed bandicoot	1.5	37.0	122
Brush-tailed possum	4.4	71.8	260
Tammar wallaby	10.6	159	465
Red kangaroo	71.6	643	1796

TABLE 5.6 Weight, basal metabolic rate, and surface area

a. Find a formula that models B as a power function of W for this group of marsupials.

b. How does the formula you found in part a compare with the one you found in part a of the preceding exercise? What does this tell you about the metabolic rates of marsupials?

c. Make a plot of $\ln A$ against $\ln W$.

d. Find a formula for the regression line of $\ln A$ against $\ln W$, and add this line to the plot you found in part c.

e. The surface area of a 0.26-pound sugar glider (a marsupial similar to a flying squirrel) is about 95 square inches. Use your plot in part d to compare this with the trend for surface area versus weight in the table. Can you explain the deviation?

f. Find a formula that models A as a power function of W.

g. Often the power $k = \frac{2}{3}$ is used to model A as a power function of W. Compare this with the power you found in part f.

h. In early studies of metabolic rates, scientists often assumed that the basal metabolic rate of an animal was proportional to its surface area. Is this assumption supported by the data in the table?

9. **Proportions of trees:** Table 5.7 gives the diameter d and height h, both in feet, of some "champion" trees (largest American specimens) of a variety of shapes.[23] (See Figure 5.45.)

Tree	Diameter d	Height h
Plains cottonwood	2.9	80
Hackberry	5.7	113
Weeping willow	6.2	95
Ponderosa pine	8.6	162
Douglas fir	14.4	221

TABLE 5.7 Diameter and height of champion trees

[21]J. Short, "Factors affecting food intake of rangelands herbivores." In G. Caughley, N. Shepherd, and J. Short, eds., *Kangaroos* (Cambridge, England: Cambridge University Press, 1987).

[22]The data in this table are adapted from T. J. Dawson and A. J. Hulbert, "Standard metabolism, body temperature, and surface areas of Australian marsupials," *Am. J. Physiol.* **218** (1970), 1233–1238.

[23]This exercise is based on the work of T. McMahon, "Size and shape in biology," *Science* **179** (1973), 1201–1204. He considers data for 576 trees, primarily from the American Forestry Association lists of champions.

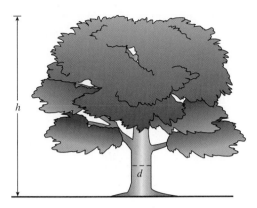

FIGURE 5.45

L	W
28.5	213
30.5	259
32.5	308
34.5	363
36.5	419
38.5	500
40.5	574
42.5	674
44.5	808
46.5	909
48.5	1124

a. Make a plot of $\ln h$ against $\ln d$.

b. Find a formula for the regression line of $\ln h$ against $\ln d$, and add this line to the plot you found in part a.

c. Which is taller *for its diameter*: the plains cottonwood or the weeping willow?

d. Find a formula that models h as a power function of d.

e. It has been determined that the critical height at which a column made from green wood of diameter d, in feet, would buckle under its own weight is $140d^{2/3}$ feet.

  i. How does your answer to part d compare with this formula?

  ii. Are any of the trees in the table taller than their critical buckling height?

10. **Weight versus length:** The accompanying table shows the relationship between the length L, in centimeters, and the weight W, in grams, of the North Sea plaice (a type of flatfish).[24]

  a. Find a formula that models W as a power function of L. (Round the power to one decimal place.)

  b. Explain in practical terms what $W(50)$ means, and then calculate that value.

  c. If one plaice were twice as long as another, how much heavier than the other should it be?

11. **Self-thinning:** When seeds of a plant are sown at high density in a plot, the seedlings must compete with each other. As time passes, individual plants grow in size, but the density of the plants that survive decreases.[25] This is the process of *self-thinning*. In one experiment, horseweed seeds were sown on October 21, and the plot was sampled on successive dates. The results are summarized in Table 5.8, which gives for each date the density p, in number per square meter, of surviving plants and the average dry weight w, in grams, per plant.

  a. Explain how the table illustrates the phenomenon of self-thinning.

  b. Find a formula that models w as a power function of p.   →

---

[24]The table is from Lowestoft market samples in 1946, as described by R. J. H. Beverton and S. J. Holt, *On the Dynamics of Exploited Fish Populations*, Fishery Investigations, Series 2, Volume 19 (London: Ministry of Agriculture, Fisheries and Food, 1957).

[25]This exercise is based on the work of K. Yoda, T. Kira, H. Ogawa, and K. Hozumi, "Self-thinning in overcrowded pure stands under cultivated and natural conditions," *J. Biol. Osaka City Univ.* **14** (1963), 107–129. See also Chapter 6 of John L. Harper, *Population Biology of Plants* (London: Academic Press, 1977).

c. If the density decreases by a factor of $\frac{1}{2}$, what happens to the weight?

d. The *total plant yield y* per unit area is defined to be the product of the average weight per plant and the density of the plants: $y = w \times p$. As time goes on, the average weight per plant increases while the density decreases, so it's unclear whether the total yield will increase or decrease. Use the power function you found in part b to determine whether the total yield increases or decreases *with time*. Check your answer using the table.

Date	Density p	Weight w
November 7	140,400	1.6×10^{-4}
December 16	36,250	7.7×10^{-4}
January 30	22,500	0.0012
April 2	9100	0.0049
May 13	4510	0.018
June 25	2060	0.085

TABLE 5.8 Density and weight of surviving plants

12. **Species-area relation:** Ecologists have studied the relationship between the number S of species of a given taxonomic group within a given habitat (often an island) and the area A of the habitat.[26] They have discovered a consistent relationship: Over similar habitats, S is approximately a power function of A, and for islands the powers fall within the range 0.2 to 0.4. Table 5.9 gives, for some islands in the West Indies, the area in square miles and the number of species of amphibians and reptiles.

a. Find a formula that models S as a power function of A.

b. Is the graph of S against A concave up or concave down? Explain in practical terms what your answer means.

Island	Area A	Number S of species
Cuba	44,000	76
Hispaniola	29,000	84
Jamaica	4200	39
Puerto Rico	3500	40
Montserrat	40	9
Saba	5	5

TABLE 5.9 Area and number of species of amphibians and reptiles

c. The species-area relation for the West Indies islands can be expressed as a rule of thumb: If one island is 10 times larger than another, then it will have ____ times as many species. Use the homogeneity property of the power function you found in part a to fill in the blank in this rule of thumb.

d. In general, if the species-area relation for a group of islands is given by a power function, the relation can be expressed as a rule of thumb: If one island is 10 times larger than another, then it will have ____ times as many species. How would you fill in the blank? (*Hint:* The answer depends only on the power.)

13. **Cost of transport:** Physiologists have discovered that steady-state oxygen consumption (measured per unit of mass) in a running animal increases linearly with increasing velocity. The slope of this line is called the *cost of transport* of the animal, since it measures the energy required to move a unit mass 1 unit of distance. Table 5.10 gives the weight W, in grams, and the cost of transport C, in milliliters of oxygen per gram per kilometer, of seven animals.[27]

a. Judging on the basis of the table, does the cost of transport generally increase or decrease with increasing weight? Are there any exceptions to this trend?

---

[26] See the discussion in Example 5.3 of Section 5.1. The data in this exercise are adapted from Philip J. Darlington, Jr., *Zoogeography* (Huntington, New York: Robert E. Krieger, 1980 reprint).

[27] The data in this table are taken from C. R. Taylor, K. Schmidt-Nielsen, and J. L. Raab, "Scaling of energetic cost of running to body size in mammals," *Am. J. Physiol.* **219** (1970), 1104–1107. For an extensive collection of data with references, see M. A. Fedak and H. J. Seeherman, "Reappraisal of energetics of locomotion shows identical cost in bipeds and quadrupeds including ostrich and horse," *Nature* **282** (1979), 713–716.

Animal	Weight W	Cost of transport C
White mouse	21	2.83
Kangaroo rat	41	2.01
Kangaroo rat	100	1.13
Ground squirrel	236	0.66
White rat	384	1.09
Dog	2600	0.34
Dog	18,000	0.17

TABLE 5.10 Weight and cost of transport

Star	Parallax angle	Distance
Markab	0.030	109
Al Na'ir	0.051	64
Alderamin	0.063	52
Altair	0.198	16.5
Vega	0.123	26.5
Rasalhague	0.056	58

b. Make a plot of $\ln C$ against $\ln W$.

c. Find a formula for the regression line of $\ln C$ against $\ln W$, and add this line to the plot you found in part b.

d. The cost of transport for a 20,790-gram emperor penguin is about 0.43 milliliter of oxygen per gram per kilometer. Use your plot in part c to compare this with the trend for cost of transport versus weight in the table. Does this confirm the stereotype of penguins as awkward waddlers?

e. Find a formula that models C as a power function of W.

14. **Parallax angle:** If we view a star now, and then view it again 6 months later, our position will have changed by the diameter of the Earth's orbit around the sun. (See Figure 5.46.) For stars within about 100 light-years of Earth, the change in viewing location is sufficient to make the star appear to be in a different location in the sky. Half of the angle from one location to the next is known as the *parallax angle*. Even for nearby stars, the parallax angle is very small[28] and is normally measured in seconds of arc. The distance to a star can be determined from the parallax angle. The table below gives parallax angle p measured in seconds of arc and the distance d from the sun measured in light-years.

a. Make a plot of $\ln d$ against $\ln p$ and determine whether it is reasonable to model the data with a power function.

b. Make a power function model of the data for d in terms of p.

c. If one star has a parallax angle twice that of a second, how do their distances compare?

d. The star Mergez has a parallax angle of 0.052 second of arc. Use functional notation to express how far away Mergez is, and then calculate that value.

e. The star Sabik is 69 light-years from the sun. What is its parallax angle?

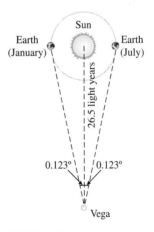

FIGURE 5.46

---

[28] Early astronomers were caught up in the problem of determining whether the Earth or the sun should be thought of as the center of motion of the solar system. They were well aware that if the Earth orbited the sun, then stars would show a parallax angle. But the angle is so small that they were unable to detect it. This was presented as justification for considering Earth a stationary object.

15. **Exponential growth rate and generation time:** In this exercise, we will examine the relationship between a population's intrinsic exponential growth rate r and the generation time T—that is, the time it takes an organism to reach reproductive maturity. The following table gives values of the generation time and the exponential growth rate for a selection of lower organisms.[29] The basic unit of time is a day.

Organism	Generation time T	r
Bacterium *E. coli*	0.014	60
Protozoan *Paramecium aurelia*	0.5	1.24
Spider beetle *Eurostus hilleri*	110	0.01
Golden spider beetle	154	0.006
Spider beetle *Ptinus sexpunctatus*	215	0.006

a. By plotting $\ln r$ against $\ln T$, determine whether it is reasonable to model r as a power function of T.

b. Find a formula for the regression line of $\ln r$ against $\ln T$. (Round the slope to one decimal place.)

c. Find a formula that models r as a power function of T, and make a graph of this function. (Use a horizontal span from 0 to 10.)

d. In Example 5.7 we found a power relation between generation time T, in years, and length L. Use this relation and the results of part c to find a formula for r as a function of length L for this group of lower organisms. (Convert the power relation from Example 5.7 to measure time in days, and remember that $T^{-1} = 1/T$.)

16. **Using the common logarithm:** Solve Exercise 10 using the common logarithm.

 17. **Laboratory experiment:** This lab uses a motion detector and a calculator-based laboratory (CBL™) unit. The motion detector measures the distance from the detector to an object in front of it, and the CBL records the data. After collecting the data, the CBL unit sends the data to the calculator, which displays a graph of the distance recorded with respect to time. In this lab, we will establish the relationship between the period and the length of a pendulum. For a detailed description, go to

http://math.college.hmco.com/students

and then, from the book's website, go to the Pendulum experiment.

 18. **Laboratory experiment:** This lab uses a pressure sensor and a calculator-based laboratory (CBL™) unit. In this lab we investigate Boyle's law, which describes the relationship between the pressure and the volume of a fixed mass of gas at a constant temperature. We test this relationship by using a large syringe with a plunger that is attached to the pressure sensor. For a detailed description, go to

http://math.college.hmco.com/students

and then, from the book's website, go to the Pressure and Volume experiment.

---

[29]The table is adapted from E. R. Pianka, *Evolutionary Ecology*, 4th ed. (New York: Harper & Row, 1988).

5.3   *Combining and Decomposing Functions*

We have up to now looked most closely at linear, exponential, and power functions. But we have also encountered various combinations of these and other functions. In this section we look more carefully at how functions are combined to make new ones and at the meanings of the individual pieces.

Sums, Products, and Limiting Values

In looking at linear and exponential functions, we have already emphasized the meanings of various parts of functions. Suppose, for example, that the cost c in dollars of joining a music club and purchasing n CDs is given by the linear function

$$c(n) = 20 + 12n \, .$$

The pieces of the formula have important meanings with which we are familiar. The first piece, 20, is the initial value, and it tells us that it costs $20 to join the club. The second piece, $12n$, shows the slope, 12, and it tells us that once we join the music club, we can buy CDs for $12 each. Thus the total cost function c is made up of two pieces, the price of joining the club and the cost of buying n CDs. Furthermore, we can glean this information by looking at pieces of the formula.

In the previous chapter we used the exponential function

$$N(t) = 5.34 \times 1.03^t$$

to model the U.S. population from 1800 through 1860. Here t is the number of years since 1800, and N is the population in millions. Once again, the pieces of the formula have familiar meanings. The number 5.34 is the initial value of the function, and it tells us that the U.S. population in 1800 was about 5.34 million. The second part of the exponential function, 1.03^t, gives the growth factor 1.03, and it tells us that population grew by about 3% per year. The total population N is thus made up of two pieces, the initial population and the effect of t years of growth.

Many models are obtained by putting together linear, exponential, or power functions in various ways, and often we can get desired information by looking at these pieces.

If a yam, initially at room temperature, is placed in a preheated oven, its temperature P after t minutes may be given by $P = 400 - D$, where $D = 325 \times 0.98^t$. Note that P is made up of two pieces: the constant 400 and the exponential function D. Examining the role that each of these plays will help us better understand how the temperature of the yam changes. You should check by looking at a table of values that D has a limiting value of 0. Thus $P = 400 - D$ has a limiting value of 400. Since we expect the temperature of the yam eventually to match that of the oven, we conclude that the oven temperature is 400 degrees.

To understand the function D better, we rearrange the formula for P to obtain

$$D = 400 - P \, .$$

Thus D represents the difference between the temperature of the oven, which is 400 degrees, and that of the yam. Hence the pieces of the temperature function represent the limiting value (oven temperature) and the difference between oven temperature and that of the yam:

$$P = \text{Limiting value} - \text{Temperature difference}$$
$$P = \text{Oven temperature} - \text{Temperature difference}.$$

We noted that the limiting value of the exponential function $D = 325 \times 0.98^t$ is 0. Indeed, as we noted in Section 4.1, the limiting value of any function that represents exponential decay is 0.

EXAMPLE 5.8   Baking a Cake

A pan of cake batter is initially at a room temperature of 75 degrees. The pan is placed in a 350-degree oven to bake. Let $C = C(t)$ denote the temperature of the cake batter t minutes after it is placed in the oven. In this example, all temperatures are measured in degrees Fahrenheit.

1. The temperature of the cake batter is given by

$$C = \text{Limiting value} - \text{Temperature difference}.$$

   What is the limiting value of C?

2. Let D denote the difference between oven temperature and that of the batter. Then *Newton's law of cooling* tells us that D is an exponential function. What is the initial value of D?

3. After 10 minutes, we find that the cake batter has heated to 165 degrees. Find a formula for D.

4. Find a model for the temperature of the cake batter t minutes after it is placed in the oven.

Solution to Part 1:   If the cake batter is left in the oven for a long time, its temperature will match that of the oven. Thus the limiting value of C is 350 degrees.

Solution to Part 2:   Since the initial temperature of the cake batter is room temperature, or 75 degrees, the initial value of D is $350 - 75 = 275$ degrees.

Solution to Part 3:   We know from part 2 that D can be written as

$$D = 275a^t,$$

where a is the decay factor for D. After 10 minutes the temperature of the batter is 165 degrees, so the temperature difference after 10 minutes is $350 - 165 = 185$ degrees. That is, $D = 185$ when $t = 10$. We can find the decay factor a by solving the equation

$$275a^{10} = 185.$$

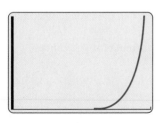

FIGURE 5.47 Temperature difference versus decay factor

We use the crossing-graphs method to get the solution. In Figure 5.47 we have graphed the temperature difference $D = 275a^{10}$ using a horizontal span of 0 to 1 and a vertical span of 0 to 225. Note that in this figure the horizontal axis represents

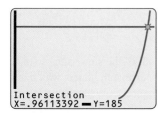

FIGURE 5.48 Finding the decay factor

the decay factor, not time. In Figure 5.48 we have added the horizontal line 185, and we see that the graphs cross at $a = 0.96$. We conclude that

$$D = 275 \times 0.96^t.$$

Solution to Part 4: To get a model for C, we use the information found in parts 1 and 3:

$$C = \text{Limiting value} - \text{Temperature difference}$$

$$C = 350 - 275 \times 0.96^t.$$ ∎∎∎

Composition of Functions

Suppose we know that the weight w of a certain young boy from ages 6 through 16 depends on his height h in the following way:

$$w = 4.4h - 150.$$

Here weight is in pounds and height in inches. Suppose we also know that the boy's height in inches depends on his age t in years according to

$$h = 1.9t + 40.$$

Thus we have weight $w = w(h)$ expressed as a function of height, and in turn height $h = h(t)$ is expressed as a function of age. Suppose we wish to know the boy's weight at age 11. We can do this by first calculating the height at age 11:

$$h = h(11) = 60.9 \text{ inches}.$$

Next we use this value to calculate the weight:

$$w = w(60.9) = 117.96 \text{ pounds}.$$

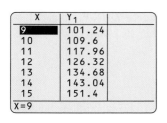

FIGURE 5.49 Table of values for w composed with h

An alternative method for making the same calculation is to *compose* the two functions that we have to make a new function that enables us to calculate the weight directly as a function of the age. We get the formula for this new function by simply replacing h in the formula for w by its expression in terms of t as shown below:

$$w = w(h) = 4.4h - 1.50$$

$$w = w(h(t)) = 4.4(1.9t + 40) - 150.$$

This new function is known as w *composed with* h, and it enables us to calculate the weight at age 11 directly in a single calculation:

$$w = w(h(11)) = 4.4(1.9 \times 11 + 40) - 150 = 117.96.$$

Putting together, or composing, functions in this way is important not only for making direct calculations but also for clearer analysis. After composition, we can, for example, view the graph of weight versus time or make a table of values. A partial table of values is shown in Figure 5.49, and the graph, where we have used a horizontal span of 10 to 15 years and a vertical span of 100 to 200 pounds, is shown in Figure 5.50.

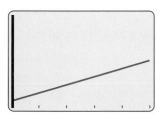

FIGURE 5.50 Graph of $w(h(t))$

EXAMPLE 5.9   North Sea Plaice

In biological studies a *cohort* is a collection of individuals all of the same age. A study by Beverton and Holt[30] contains information on a cohort of North Sea plaice, which is a type of flatfish. Some of that information is adapted for this exercise. We measure length in inches and weight in pounds. Also, the variable t measures the so-called recruitment age in years, which we refer to simply as the age.

1. It was found that plaice grow to a maximum length of about 27 inches. Furthermore, the difference $D = D(t)$ between maximum length and length $L = L(t)$ at age t is an exponential function. If initially a plaice is 1.9 inches long, find the initial value of D.

2. It was found that a 5-year-old plaice is about 11.3 inches long. Find a formula that gives L in terms of t.

3. The weight W of a plaice can be calculated from its length L using

$$W = 0.000322L^3.$$

Use function composition to find a formula for weight W as a function of age t.

4. After t years, the total number of fish in the cohort is given by the exponential function

$$N = 1000e^{-0.1t}.$$

The total *biomass* of the cohort is the number of fish times the weight per fish. Find a formula for the total biomass B of the cohort as a function of the age of the cohort.

5. How old is the cohort when biomass is at a maximum?

Solution to Part 1:   Since the initial length is 1.9 inches, the initial difference is $27 - 1.9 = 25.1$ inches. This is the initial value of D.

Solution to Part 2:   We first find a formula for length difference $D(t)$. From part 1 we know that D is an exponential function with initial value 25.1, so it can be written as

$$D = 25.1a^t,$$

where a is the decay factor. A 5-year-old plaice is about 11.3 inches long, and this means the length difference is $27 - 11.3 = 15.7$ inches. That is, $D = 15.7$ when $t = 5$. Hence we can find the decay factor a by solving the equation

$$15.7 = 25.1a^5.$$

In Figure 5.51 we have used the crossing-graphs method with a horizontal span of 0 to 1 and a vertical span of 0 to 25. We find that $a = 0.91$.

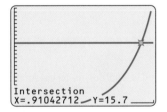

FIGURE 5.51 Finding the decay factor for length difference

---

[30]R. J. H. Beverton and S. J. Holt, *On the Dynamics of Exploited Fish Populations*, Fishery Investigations, Series 2, Volume 19 (London: Ministry of Agriculture, Fisheries and Food, 1957).

We conclude that $D = 25.1 \times 0.91^t$. Finally, from rearranging $D = 27 - L$ we obtain

$$L = 27 - D = 27 - 25.1 \times 0.91^t.$$

Solution to Part 3:   We simply need to replace L in the formula $W = 0.000322L^3$ by its expression in terms of t, which is $L = 27 - 25.1 \times 0.91^t$. We obtain

$$W = 0.000322(27 - 25.1 \times 0.91^t)^3.$$

Solution to Part 4:   The total biomass is the number of fish times the weight of each. That is, $B = NW$. Replacing N and W by their expressions in terms of t gives

$$B = 1000e^{-0.1t}[0.000322(27 - 25.1 \times 0.91^t)^3].$$

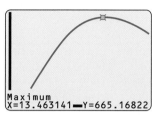

Solution to Part 5:   In Figure 5.52 we have graphed B versus t using a horizontal span of 0 to 20 years and a vertical span of 0 to 700 pounds. We read from the prompt at the bottom of the screen that a maximum biomass of about 665 pounds occurs at an age of about 13.5 years. ■ ▧ ▨

FIGURE 5.52 Finding maximum biomass

Piecewise-defined Functions

Piecewise-defined functions are functions defined by different formulas for different values of the variable. A simple example is the cost of first-class U.S. postage for a letter as a function of the weight of the letter. For a letter weighing 1 ounce or less, the postage at the time of the printing of this text is 34 cents; for a letter weighing more than 1 ounce but no more than 2 ounces, the postage is 55 cents; for a letter weighing more than 2 ounces but no more than 3 ounces, the postage is 76 cents, and so on (an additional 21 cents for each ounce). Thus the expression for postage P in cents as a function of weight w in ounces involves several different formulas: $P(w) = 34$ for $w \leq 1$, $P(w) = 55$ for $1 < w \leq 2$, $P(w) = 76$ for $2 < w \leq 3$, and so on. This can be written somewhat less awkwardly by using a brace:

$$P = \begin{cases} 34 & \text{for} \quad w \leq 1 \\ 55 & \text{for} \quad 1 < w \leq 2 \\ 76 & \text{for} \quad 2 < w \leq 3. \end{cases}$$

Piecewise-defined functions are used to describe functions that are best described, or most naturally described, by different formulas for different intervals of variable values. Here are a couple of examples showing how this idea can be used.

EXAMPLE 5.10   Roof Lines

The outside wall of a house is 10 feet high. From the western wall, the roof rises 4 feet over a horizontal span of 16 feet and then descends at the same rate over another horizontal span of 16 feet to the eastern wall (see Figure 5.53).

1. Write a formula for the line that describes the ascending part of the roof.

2. Write a formula for the line that describes the descending part of the roof.

FIGURE 5.53 Sketch of a roof line

3. Write a piecewise-defined function describing the entire roof line.

4. Graph the piecewise-defined function from part 3.

Solution to Part 1:   If we measure the horizontal distance from the western wall, then the line has an initial value of 10 feet. The slope is the rise divided by the run, which is $\frac{4}{16} = 0.25$, so a formula for the line describing the ascending part is $R = 0.25h + 10$. Here h is the horizontal distance from the western wall, R is the height of the roof (both in feet), and the formula is valid for h up to 16 feet.

Solution to Part 2:   This line descends at the same rate at the ascending line from part 1, so this time the slope is -0.25. To find the initial value, we imagine this roof line extending back west beyond the peak. We can see that it would lie 8 feet over the top of the western wall, or 18 feet above ground. Thus a formula for this part of the roof line is $R = -0.25h + 18$, with h and R as in part 1. This formula is valid for h between 16 and 32 feet.

FIGURE 5.54 Roof line graph

Solution to Part 3:   The roof is ascending for $h \leq 16$ and it is descending for $16 \leq h \leq 32$, so the roof line itself can be described as

$$R = \begin{cases} 0.25h + 10 & \text{for} \quad h \leq 16 \\ -0.25h + 18 & \text{for} \quad 16 \leq h \leq 32. \end{cases}$$

Solution to Part 4:   Graphing the parts of the two lines gives Figure 5.54.[31] The horizontal span is from 0 to 32, and the vertical span is from 5 to 20.   ▪ ▪ ▪

In the next example, we will need to weave together two formulas to make a single piecewise-defined function.

EXAMPLE 5.11   **Public High School Enrollments 1965–1996**

In Exercises 2 and 3 of Section 2.1, two different models are given for U.S. public high school enrollments. In Exercise 3 the model is $N = -0.02t^2 + 0.44t + 11.65$, where N is enrollment in millions of students, t is time in years since 1965, and the formula is valid from 1965 through 1985. On the other hand, in Exercise 2 the model is $N = 0.05t^2 - 0.42t + 12.33$, where N is enrollment in millions of students, t is time in years since 1986, and the formula is valid from 1986 to 1996.

1. Determine whether these two different formulas give contradictory information.

2. Find a way to write these formulas in terms of the same variable.

---

[31] See the *Technology Guide* for the keystrokes needed to graph functions defined piecewise.

3. Write N as a single piecewise-defined function.

4. Graph N for the years 1965 through 1996.

Solution to Part 1:   Each formula is valid only for a specified interval of years: 1965 through 1985 for one and 1986 through 1996 for the other. Since each is not valid in the other's years, there can be no contradiction, despite the different formulas that could give different values for the same year.

Solution to Part 2:   As written, each formula uses t as the variable, but t means something different for each formula: For the model from Exercise 3, t is years since 1965, whereas for the model from Exercise 2, t is years since 1986. We need a new variable. Since the final model will cover the years from 1965 through 1996, it makes sense to use years since 1965 as the variable, but we need a new name. Let's denote the years since 1965 by T. Then $T = t$ for the 1965–1985 model, but we need to relate T to t for the 1986–1996 model. A table comparing the two for a few years is helpful:

Year	1986	1987	1988	1989
t years since 1986	0	1	2	3
T years since 1965	21	22	23	24

Clearly $T = t + 21$; equivalently, $t = T - 21$. Substituting $t = T - 21$ into the 1986–1996 model—that is, writing N as a composition of functions—enables us to put both models in terms of the same variable T, years since 1965.

Solution to Part 3:   Using the variable T and substituting, we find that

$$N = \begin{cases} -0.02T^2 + 0.44T + 11.65 & \text{for } 0 \le T \le 20 \\ 0.05(T - 21)^2 - 0.42(T - 21) + 12.33 & \text{for } 21 \le T \le 31. \end{cases}$$

Solution to Part 4:   Graphing the two functions gives Figure 5.55. Here the horizontal span is from 0 to 31, and the vertical span is from 10 to 15.[32]   ■ ■ ■

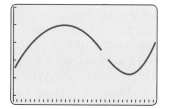

FIGURE 5.55 High school enrollments 1965–1996

ANOTHER LOOK

Chaotic Behavior

Studying compositions of certain functions is one of the easiest ways to observe *chaotic behavior*. We begin with a very simple function $f(x) = x^2 + c$. For various values of c we will look at $f(0)$, $f(f(0))$, $f(f(f(0)))$, and so on. For convenience, let's use the notation $f^n(0)$ to denote f composed with itself n times and evaluated at 0. Thus, for example, $f^3(0)$ means $f(f(f(0)))$.

---

[32]For functions defined piecewise, it is often wise to set the graphing mode so that the graph is not connected. On your calculator this mode may be called "dot" or something similar.

Let's look at what happens when $c = 1$. Thus we are using $f(x) = x^2 + 1$. Now $f(0) = 1$, and $f^2(0) = f(f(0)) = f(1) = 2$. Continuing, we get the following values.

$f(0)$	$f^2(0)$	$f^3(0)$	$f^4(0)$	$f^5(0)$
1	2	5	26	677

Thus for $c = 1$, it is clear that the values just keep increasing without bound.

Something more interesting happens when $c = 0.2$. In that case, we are looking at $f(x) = x^2 + 0.2$. We record the results to five decimal places.

$f(0)$	$f^2(0)$	$f^3(0)$	$f^4(0)$	$f^5(0)$	$f^6(0)$
0.2	0.24	0.2576	0.26636	0.27095	0.27341

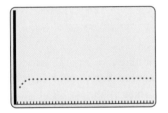

FIGURE 5.56 $c = 0.2$, a single convergent branch

In this case, it appears that the points are increasing but may have a limiting value. For further verification we look at a plot of the first 50 iterates in Figure 5.56. Your calculator must be set to a special mode called *sequential mode* in order to make such graphs. Consult your *Technology Guide* for instructions. Figure 5.56 shows clearly that the sequence of values levels off. In this figure the vertical span is 0 to 1. In all but the last of the remaining figures, the vertical span is -2 to 2.

In Figure 5.57 we have plotted $f^n(0)$ for $c = -0.7$. Here the sequence seems to have split into two branches, but the branches appear to come together. If we look at $c = -0.75$ in Figure 5.58, we again see two branches to the sequence, but it is unclear whether the two branches might later come together. Next (see Figure 5.59) we look at the case of $c = -0.9$, where the two branches definitely appear to remain apart. Indeed they do, but it is difficult to provide more than graphical evidence of this at this level. This phenomenon is called *bifurcation*.

If we look still further, something dramatic happens. In Figure 5.60 we use $c = -1.7$, and it appears that the points of the sequence are scattered in a totally random, or *chaotic*, pattern. In Figure 5.61 we look at the first 500 points in the sequence, and the chaotic behavior appears to persist.

Interestingly enough, when we continue down to $c = -2$, convergence is restored, as shown in Figure 5.62 (where the vertical span is -3 to 3).

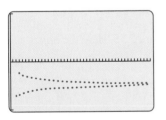

FIGURE 5.57 $c = -0.7$, two branches that appear to come together

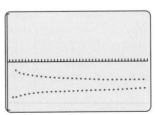

FIGURE 5.58 $c = -0.75$, two branches that may or may not come together

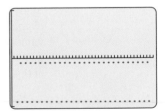

FIGURE 5.59 $c = -0.9$, two branches that remain separate

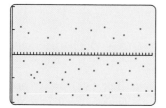

FIGURE 5.60 $c = -1.7$, a
chaotic pattern

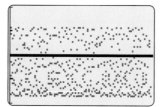

FIGURE 5.61 Expanding the
view to include 500 points

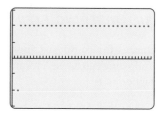

FIGURE 5.62 $c = -2$,
convergence restored

The chaotic behavior seen above is not just a mathematical curiosity. In recent years the study of chaotic behavior has been applied to many different fields. For example, the growth of some insect populations has been modeled by sequences similar to those studied here.

Enrichment Exercises

E-1. **Further investigation of the case $c = -0.75$:** Look at a plot of 500 points of $f^n(0)$ in the case $f(x) = x^2 - 0.75$. Try to determine whether the two branches of the sequence come together.

E-2. **Other values of c:** Do a graphical exploration to determine what happens to $f^n(0)$ in the case $f(x) = x^2 + c$ when $c < -2$.

E-3. **Graphical analysis of another sequence:** Let $g(x) = cx(1 - x)$ with $c > 0$, and consider the sequence of points $g^n(0.1)$. Look at a plot of the sequence for various values of c. Find a value of c where the sequence appears to converge. Find a value of c where the sequence seems to split into two separate branches. Find a value of c where the sequence appears to be chaotic.

5.3   SKILL BUILDING EXERCISES

S-1. Formulas for composed functions: In each of the following cases, use a formula to express w as a function of t.

   a. $w = s^2 + 1$ and $s = t - 3$

   b. $w = \dfrac{s}{s+1}$ and $s = t^2 + 2$

   c. $w = \sqrt{2s + 3}$ and $s = e^t - 1$

S-2. Formulas for composed functions: For each of the following functions, find $f(g(x))$ and $g(f(x))$.

   a. $f(x) = 3x + 1$ and $g(x) = \dfrac{2}{x}$

   b. $f(x) = x^2 + x$ and $g(x) = x - 1$

   c. $f(x) = \dfrac{1}{x}$ and $g(x) = \dfrac{1}{x}$

S-3. Limiting values: Find the limiting value of $7 + a \times 0.6^t$.

S-4. Multiplying functions: A certain function f is the product of weight, given by $t + (1/t)$, and height, given by t^2. Find a formula for f in terms of t.

S-5. Adding functions: A certain function f is the sum of two temperatures, one given by $t^2 + 3$, and the other, given by $t/(t^2 + 1)$. Find a formula for f in terms of t.

S-6. Decomposing functions: Let $f(x) = x^2$ and $g(x) = x + 1$. Express $(x + 1)^2 + 1$ in terms of compositions of f and g.

S-7. Composing a function with itself: Let $f(x) = x^2 + 1$. Find formulas for $f(f(x))$ and $f(f(f(x)))$.

S-8. Decomposing functions: To join a book club, you pay an initial fee and then a fixed price each month for a book. The total cost in dollars of joining the club and buying n books is given by $C = 30 + 17n$. What is the initial fee? What is the cost of each book after you are a club member?

S-9. Decomposing functions: The population of a certain species is given by $N = 128 \times 1.07^t$, where t is measured in years. What is the initial population? By what percentage does the population grow each year?

S-10. Combining functions: Let $f(x) = x^2 - 1$ and $g(x) = 1 - x$. Find a formula for $f(g(x)) + g(f(x))$ in terms of x.

5.3   EXERCISES

1. **A skydiver:** If a skydiver jumps from an airplane, his velocity v, in feet per second, starts at 0 and increases toward terminal velocity. An average-size man has a terminal velocity T of about 176 feet per second. The difference $D = T - v$ is an exponential function of time.

   a. What is the initial value of D?

   b. Two seconds into the fall, the velocity is 54.75 feet per second. Find an exponential formula for D.

   c. Find a formula for v.

   d. Express using functional notation the velocity 4 seconds into the fall, and then calculate that value.

2. **Present value:** If you invest P dollars (the *present value* of your investment) in a fund that pays an interest rate of r, as a decimal, compounded yearly, then after t years your investment will have a value F dollars, which is known as the *future value*. The *discount rate* D for such an investment is given by

$$D = \frac{1}{(1 + r)^t},$$

where t is the life in years of the investment. The present value of an investment is the product of the future value and the discount rate. Find a formula that gives the present value in terms of the future value, the interest rate, and the life of the investment.

3. **Immigration:** Suppose a certain population is initially absent from a certain area but begins migrating there at a rate of v individuals per day. Suppose further that this is an animal group that would normally

grow at an exponential rate. Then the population after t days in the new area is given by

$$N = \frac{v}{r}(e^{rt} - 1),$$

where r is a constant that depends on the species and the environment. If the new location proves unfavorable, then the value of r may be negative. In such a case, we can rewrite the population function as

$$N = \frac{v}{r}(a^t - 1),$$

where a is less than 1. Under these conditions, what is the limiting value of the population?

4. **Correcting respiration rate for temperature:** In a study by R. I. Van Hook[33] of a grassland ecosystem in Tennessee, the rate O of energy loss to respiration of consumers and predators was initially modeled using

$$O = aW^B,$$

where W is weight and a and B are constants. The model was then corrected for temperature by multiplying O by $C = 1.07^{T-20}$, where T is temperature in degrees Celsius.

a. What effect does the correction factor have on energy loss due to respiration if the temperature is larger than 20 degrees Celsius?

b. What effect does the correction factor have on energy loss due to respiration if the temperature is exactly 20 degrees Celsius?

c. What effect does the correction factor have on energy loss due to respiration if the temperature is less than 20 degrees Celsius?

5. **Biomass of haddock:** In the study by Beverton and Holt[34] described in Example 5.9, information is also provided on haddock. We have adapted it for this exercise. We measure length in inches, weight in pounds, and age t in years.

a. It was found that haddock grow to a maximum length of about 21 inches. Furthermore, the dif-

ference $D = D(t)$ between maximum length and length $L = L(t)$ at age t is an exponential function. If initially a haddock is 4 inches long, find the initial value of D.

b. It was found that a 6-year-old haddock is about 15.8 inches long. Find a formula that gives L in terms of t.

c. The weight W in pounds of a haddock can be calculated from its length L using

$$W = 0.000293L^3.$$

Use function composition to find a formula for weight W as a function of age t.

d. After t years, the total number of fish in the cohort is given by the exponential function

$$N = 1000e^{-0.2t}.$$

The total biomass of the cohort is the number of fish times the weight per fish. Find a formula for the total biomass B of the cohort as a function of the age of the cohort.

e. How old is the cohort when biomass is at a maximum? (Consider ages up to 10 years.)

6. **Traffic flow:** For traffic moving along a highway, we use q to denote the *mean flow rate*. That is the average number of vehicles per hour passing a certain point. We let q_m denote the maximum flow rate, k the *mean traffic density* (that is, the average number of vehicles per mile), and k_m the density at which flow rate is a maximum (that is, the value of k when $q = q_m$).

a. An important measurement of traffic on a highway is the *relative density* R, which is defined as

$$R = \frac{k}{k_m}.$$

i. What does a value of $R < 1$ indicate about traffic on a highway? →

---

[33]"Energy and nutrient dynamics of spider and orthopteran populations in a grassland ecosystem," *Ecol. Monogr.* **41** (1971), 1–26.

[34]R. J. H. Beverton and S. J. Holt, *On the Dynamics of Exploited Fish Populations*, Fishery Investigations, Series 2, Volume 19 (London: Ministry of Agriculture, Fisheries and Food, 1957).

ii. What does a value of $R > 1$ indicate about traffic on a highway?

b. Let u denote the mean speed of vehicles on the road and u_f the *free speed*—that is, the speed when there is no traffic congestion at all. One study[35] proposes the following relation between density and speed:

$$u = u_f e^{-0.5R^2}.$$

Use function composition to find a formula that directly relates mean speed to mean traffic density.

c. Make a graph of mean speed versus mean traffic density, assuming that k_m is 122 cars per mile and u_f is 75 miles per hour. (Include values of mean traffic density up to 250 vehicles per mile.) Paying particular attention to concavity, explain the significance of the point $k = 122$ on the graph.

d. Traffic is considered to be seriously congested if the mean speed drops to 35 miles per hour. Use the graph from part c to determine what density will result in serious congestion.

7. **Waiting at a stop sign:** Consider a side road connecting to a major highway at a stop sign. According to a study by D. R. Drew[36] the average delay D, in seconds, for a car waiting at the stop sign to enter the highway is given by

$$D = \frac{e^{qT} - 1 - qT}{q},$$

where q is the *flow rate*, or the number of cars per second passing the stop sign on the highway, and T is the *critical headway*, or the minimum length of time in seconds between cars on the highway that will allow for safe entry. We assume that the critical headway is $T = 5$ seconds.

a. What is the average delay time if the flow rate is 500 cars per hour (0.14 car per second)?

b. The *service rate s* for a stop sign is the number of cars per second that can leave the stop sign. It is related to the delay by

$$s = D^{-1}.$$

Use function composition to represent the service rate as a function of flow rate. *Reminder:* $(a/b)^{-1} = b/a$.

c. What flow rate will permit a stop sign service rate of 5 cars per minute (0.083 car per second)?

8. **Average traffic spacing:** The *headway h* is the average time between vehicles. On a highway carrying an average of 500 vehicles per hour, the probability P that the headway is at least t seconds is given[37] by

$$P = 0.87^t.$$

a. What is the limiting value of P? Explain what this means in practical terms.

b. The headway h can be calculated as the quotient of the *spacing f*, in feet, which is the average distance between vehicles, and the average speed v, in feet per second, of traffic. Thus the probability that spacing is at least f feet is the same as the probability that the headway is at least f/v seconds. Use function composition to find a formula for the probability Q that the spacing is at least f feet. *Note:* Your formula will involve both f and v.

c. If the average speed is 88 feet per second (60 miles per hour), what is the probability that the spacing between two vehicles is at least 40 feet?

9. **Probability of extinction:** The assumption that the yearly rate of change in a population is $(b - d)N$, where b is births per year, d is deaths per year, and N is current population, leads to an exponential model of population growth. The increase is in fact to some degree probabilistic in nature. If we assume that pop-

[35] J. Drake, A. D. May, and J. L. Schofer, "A statistical analysis of speed density hypotheses," *Traffic Flow Characteristics*, Highway Research Record No. 154 (Washington, DC: Highway Research Board, 1967), 53–87.

[36] *Traffic Flow Theory and Control* (New York: McGraw-Hill, 1968).

[37] See Institute of Traffic Engineers, *Transportation and Traffic Engineering Handbook*, ed. by John E. Baerwald (Englewood Cliffs, NJ: Prentice-Hall, 1976), 102.

ulation increase is *normally distributed* around rN, where $r = b - d$, then we can discuss the probability of extinction of a population.

a. If the population begins with a single individual, then the probability of extinction by time t is given by

$$P(t) = \frac{d(e^{rt} - 1)}{be^{rt} - d}.$$

If $d = 0.24$ and $b = 0.72$, what is the probability that this population will eventually become extinct? *Hint:* The probability that the population will eventually become extinct is the limiting value for P.

b. If the population starts with k individuals, then the probability of extinction by time t is

$$Q = P^k,$$

where P is the function in part a. Use function composition to obtain a formula for Q in terms of t, b, d, r, and k.

c. If $b > d$ (births greater than deaths), so that $r > 0$, then the formula obtained in part b can be rewritten as

$$Q = \left(\frac{d(1 - a^t)}{b - da^t} \right)^k,$$

where $a < 1$. What is the probability that a population starting with k individuals will eventually become extinct?

d. If b is twice as large as d, what is the probability of eventual extinction if the population starts with k individuals?

e. What is the limiting value of the expression you found in part d as a function of k? Explain what this means in practical terms.

10. **Decay of litter:** Litter such as leaves falls to the forest floor, where the action of insects and bacteria initiates the decay process. Let A be the amount of litter present, in grams per square meter, as a function of time t in years. If the litter falls at a constant rate of L grams per square meter per year, and if it decays at a constant proportional rate of k per year, then the limiting value of A is $R = L/k$. For this exercise and the next, we suppose that at time $t = 0$, the forest floor is clear of litter.

a. If D is the difference between the limiting value and A, so that $D = R - A$, then D is an exponential function of time. Find the initial value of D in terms of R.

b. The yearly decay factor for D is e^{-k}. Find a formula for D in terms of R and k. *Reminder:* $\left(a^b \right)^c = a^{bc}$.

c. Explain why $A = R - Re^{-kt}$.

11. **Applications of the litter formula:** *This is a continuation of Exercise 10. Here we make use of the formula* $A = R - Re^{-kt}$ *obtained in the previous exercise.*

a. Show that 95% of the limiting value is reached when $t = 3/k$.

b. A study[38] of sand dune soils near Lake Michigan found the decay rate to be $k = 0.003$ per year. How long is required to reach 95% of the limiting value?

c. The decay rate for forest litter in the Sierra Nevada Mountains in California is about 0.063 per year, whereas the decay rate in the Congo[39] is about 4 per year. How does the time required to reach 95% of the limiting value compare for the Congo and the Sierra Nevada Mountains?

12. **High school graduates:** The following table shows the number, in millions, graduating from high school in the United States for selected years.[40] →

---

[38]J. S. Olson, "Rates of succession and soil changes on southern Lake Michigan sand dunes," *Bot. Gaz.* **119** (1958), 125–170.

[39]J. S. Olson, "Energy storage and the balance of producers and decomposers in ecological systems," *Ecology* **44** (1963), 322–331.

[40]The data in this exercise are adapted from the *1999 Statistical Abstract of the United States.*

Year	Number
1986	2.79
1987	2.65
1989	2.45
1990	2.36
1991	2.28
1992	2.40
1994	2.52
1996	2.66
1998	2.80

Year	Wolves
1990	6
1991	17
1992	21
1993	30
1994	57
1995	80
1996	116
1997	112
1998	140
1999	174
2000	216

a. Make a plot of the data and explain why a linear model is not appropriate.

b. Use regression to find a linear model for the years 1986 through 1991. (In this part and the next, round regression line parameters to three decimal places.)

c. Use regression to find a linear model for the years 1992 through 1998.

d. Write a formula for a model of the number, in millions, graduating as a piecewise-defined function using the linear models from part b and part c.

e. Make a graph of the formula you found in part d.

f. The number graduating in 1993 was 2.34 million. On the basis of your graph in part e, determine whether this is more or less than would be expected from your formula.

13. **Gray wolves in Michigan:** Gray wolves recolonized in the Upper Peninsula of Michigan beginning in 1990. Their population has been documented as shown in the accompanying table.[41]

a. Explain why one would expect an exponential model to be appropriate for these data.

b. Find an exponential model for the data given.

c. Graph the data and the exponential model. For what year is this model suspect?

d. Find an exponential model for 1990 through 1996 and another for 1997 through 2000.

e. Write a formula for the number of wolves as a piecewise-defined function using the two exponential models. Is the combined model a better fit?

14. **Mother-to-child transmission of HIV:** The retrovirus HIV can be transmitted from mother to child during breastfeeding. But under conditions of poverty or poor hygiene, the alternative to breastfeeding carries its own risk for infant mortality. These risks differ in that the risk of HIV transmission is the same regardless of the age of the child, whereas the risk of mortality due to artificial feeding is high for newborns but decreases with the age of the child.[42] Figure 5.63 illustrates the additional risk of death due to artificial feeding and due to HIV transmission from breastfeeding.

---

[41] Data courtesy of the International Wolf Center, Ely, Minnesota. For more information about wolves and their recovery, go to **http://www.wolf.org**.

[42] Adapted from J. Ross, "A spreadsheet model to estimate the effects of different infant feeding strategies on mother-to-child transmission of HIV and on overall infant mortality," preprint, 1999, Academy for Educational Development, Washington, DC. For simplicity, we have suppressed the units for additional risk.

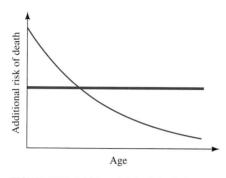

FIGURE 5.63 Additional risk of death from breastfeeding and from artificial feeding as a function of the age of the child

Here the lighter curve is additional risk of death from artificial feeding, and the darker curve is additional risk of death from breastfeeding.

For the purposes of this exercise, assume that the additional risk of death from artificial feeding is $A = 5 \times 0.785^t$ and that the additional risk from breastfeeding is $B = 2$, where t is the child's age in months.

a. At what age is the additional risk of HIV transmission from breastfeeding the same as the additional risk from artificial feeding?

b. For which ages of the child does breastfeeding carry the smaller additional risk?

c. What would be the optimal plan for feeding a child to minimize additional risk of death?

d. Write a formula for a piecewise-defined function R of t giving the additional risk of death under the optimal plan from part c.

15. **Quarterly pine pulpwood prices:** In southwest Georgia, the average pine pulpwood prices vary predictably over the course of the year, primarily because of weather. From 1993 through 1997 prices followed a similar pattern. In the first quarter of each year, the average price P was $18.50 per ton . It decreased at a steady rate to $14 in the second quarter

and then increased at a steady rate up to $18 by the fourth quarter.[43]

a. Sketch a graph of pulpwood prices as a function of the quarter in the year.

b. What type of function is P from the first to the second quarter?

c. What formula for price P as a function of t, the quarter, describes the price from the first to the second quarter?

d. What type of function is P from the second to the fourth quarter?

e. What formula for price P as a function of t, the quarter, describes the price from the second to the fourth quarter?

f. Write a formula for price P throughout the year as a piecewise-defined function of t, the quarter.

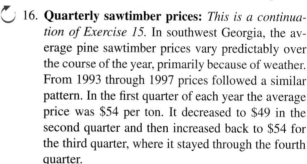

16. **Quarterly sawtimber prices:** *This is a continuation of Exercise 15.* In southwest Georgia, the average pine sawtimber prices vary predictably over the course of the year, primarily because of weather. From 1993 through 1997 prices followed a similar pattern. In the first quarter of each year the average price was $54 per ton. It decreased to $49 in the second quarter and then increased back to $54 for the third quarter, where it stayed through the fourth quarter.

a. Sketch a graph of sawtimber prices as a function of the quarter in the year.

b. Assume that $5t^2$ accurately describes the shape of the curve during the first three quarters. Compose the function with an appropriate translation $t - a$ for some a, and add a constant to the function to obtain a function P that describes the price throughout the three quarters.

c. What formula for price P describes the price during the third to fourth quarters?

d. Write a formula for price P throughout the year as a piecewise-defined function of t, the quarter.

---

[43]The data in this exercise and the next are adapted from E. Frazer and S. Moss, "Timing timber sales," *Forest Landowner* **57** no. 3 (1998), 12–14.

5.4 *Quadratic Functions and Parabolas*

In this section, we will look at quadratic functions, their relation to linear functions, quadratic models, quadratic regression, and the traditional use of the quadratic formula in solving second-degree polynomial equations.

A *quadratic* function is a function whose formula can be written as a sum of power functions where each of the powers is 0, 1, or 2. For example, $x^2 + 3x + 5$ is a quadratic function because it is a sum of the power functions x^2, $3x$, and $5 = 5x^0$. Another example of a quadratic function is $(2x - 1)^2$ since it can be written as $4x^2 - 4x + 1$.

Quadratic functions are useful in many applications of mathematics when a single power function is not sufficient. For example, when a rock is dropped, the distance D, in feet, that it falls in t seconds is given by the power function $D = 16t^2$. But if the rock is thrown downward with an initial velocity of 5 feet per second, then the distance it falls is no longer governed by a power function. Instead, the quadratic function $D = 16t^2 + 5t$ is needed to describe the distance traveled by the rock.

The graph of a quadratic function is known as a *parabola* and has a distinctive shape that occurs often in nature. In Figure 5.64 we show the parabola that is the graph of $x^2 - 3x + 7$, and in Figure 5.65 we show the parabola that is the graph of $-x^2 + 3x + 7$. A positive sign on x^2 makes the parabola open upward, as in Figure 5.64, and a negative sign on x^2 makes the parabola open downward, as in Figure 5.65.

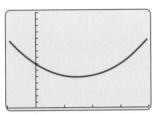

FIGURE 5.64 The graph of $x^2 - 3x + 7$, an upward-opening parabola

EXAMPLE 5.12 **Flight of a Cannonball**

A cannonball fired from a cannon will follow the path of the parabola[44] that is the graph of $y = y(x)$, where $y = -16(1 + s^2)(x/v_0)^2 + sx$ feet. Here s is the slope of inclination of the cannon barrel, v_0 is the initial velocity in feet per second, and the variable x is the distance downrange in feet. Suppose that a cannon is elevated with a slope of 0.5 (corresponding to an angle of inclination of about 27 degrees) and the cannonball is given an initial velocity of 250 feet per second. (See Figure 5.66.)

1. How far will the cannonball travel?

2. What is the maximum height that the cannonball will reach?

3. Suppose the cannon is firing at a fort whose 20-foot-high wall is 1540 feet downrange. Show that the cannonball will not clear the fort wall. There are two possible adjustments that might be made so that the cannonball will clear the wall: Increase the initial velocity (add powder) or change the slope of inclination.

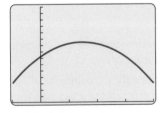

FIGURE 5.65 The graph of $-x^2 + 3x + 7$, a downward-opening parabola

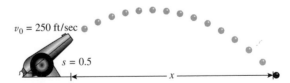

$v_0 = 250$ ft/sec

$s = 0.5$

x

FIGURE 5.66 Cannonball's inclination and initial velocity

---

[44]There is nothing special about a cannonball here. If air resistance is neglected, then any object, such as a rifle bullet, golf ball, baseball, or basketball, ejected near the surface of the Earth (or any other planet, for that matter) will follow a parabolic path.

a. How would you adjust the initial velocity so that the wall will be cleared?

b. How would you adjust the slope of inclination so that the wall will be cleared?

Solution to Part 1: Since the slope of inclination is 0.5 and the initial velocity is 250 feet per second, the function we wish to graph is

$$-16\left(1 + 0.5^2\right)\left(\frac{x}{250}\right)^2 + 0.5x = -20\left(\frac{x}{250}\right)^2 + 0.5x\,.$$

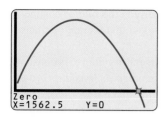

FIGURE 5.67 Where the cannonball will land

After consulting a table of values, we chose a horizontal span of 0 to 1800 and a vertical span of -50 to 220 to make the graph shown in Figure 5.67. We want to know how far downrange the cannonball will land, so we use the calculator to get the crossing point $x = 1562.5$ feet shown in Figure 5.67.

Solution to Part 2: To get the maximum height of the cannonball, we use the calculator to locate the peak of the parabola as shown in Figure 5.68. We see that the cannonball will reach a maximum height of 195.31 feet at 781.25 feet downrange.

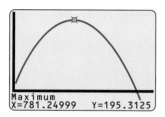

FIGURE 5.68 The maximum height of the cannonball

Solution to Part 3: In Figure 5.69 we have added the horizontal line $y = 20$ to the picture and calculated the intersection point. We see that the cannonball will be 20 feet high at 1521.42 feet downrange. Since the 20-foot-high wall is several feet farther downrange, the cannonball will not clear it.

For part a, we want to choose an initial velocity v_0 so that the height

$$-16\left(1 + 0.5^2\right)\left(\frac{x}{v_0}\right)^2 + 0.5x$$

will be 20 when the distance x is 1540. That is, we want to solve the equation

$$-20\left(\frac{1540}{v_0}\right)^2 + 0.5 \times 1540 = 20$$

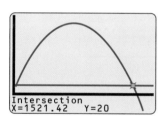

FIGURE 5.69 The cannonball won't clear the 20-foot wall

for v_0. We can do this by using the crossing-graphs method as shown in Figure 5.70. Since the cannonball doesn't need much increase in initial velocity to clear the wall, we use a horizontal span of 240 to 260 and a vertical span of 0 to 40. We see that an initial velocity of 251.48 feet per second is needed to clear the wall.

For part b, we want to leave the initial velocity at 250 feet per second and change the slope of inclination, so we need to solve the equation

$$-16(1 + s^2)\left(\frac{1540}{250}\right)^2 + s \times 1540 = 20$$

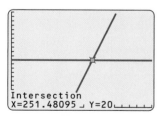

FIGURE 5.70 Adjusting initial velocity

for s. Once again, we don't need to change the slope much, so we use a horizontal span of 0.4 to 0.6 and a vertical span of 0 to 40. When we calculate the intersection, we find that a slope of 0.51 is required. ▪ ▪ ▪

Linear Rates of Change and Second-order Differences

We know that a linear function is one with a constant rate of change and that an exponential function is one with a constant proportional rate of change. Quadratic functions can also be characterized in terms of rates of change; they are functions with a linear rate of change. To illustrate this, we consider a falling rock. Its velocity t seconds after it is released is given by the linear function $32t$ feet per second. The distance the rock falls will be $16t^2$ feet at the end of t seconds.[45] Since velocity is the rate of change in distance, we see that in this case, the distance is a quadratic and that its rate of change, the velocity, is linear.

Carrying this one step further, since velocity is linear, its rate of change is constant. That is, for a quadratic function *the second-order rate of change is constant*. In fact, a table of data (with evenly spaced values for the variable) is quadratic whenever the *second-order differences* in function values are constant. The following more complex example will illustrate this.

If we stand atop a 135-foot-high building and toss a rock upward with an initial velocity of 38 feet per second, then the rock will travel upward for a while and then eventually be pulled by gravity down to the ground. Elementary physics can be used to show that the distance up from the ground, in feet, is given by

$$D = -16t^2 + 38t + 135,$$

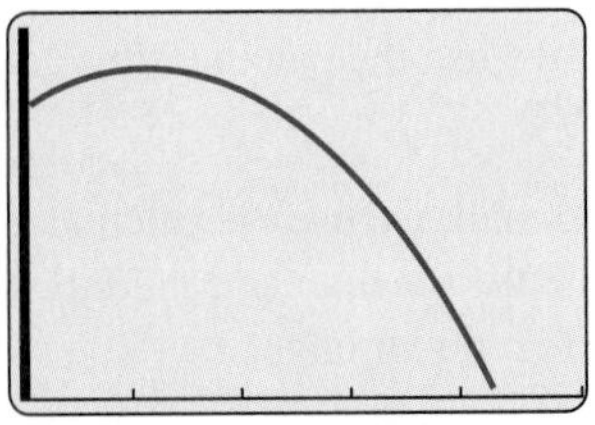

where t is the time in seconds after the rock is thrown. The graph of D versus t will be part of a parabola opening downward. The graph is shown in Figure 5.71, where we have used a horizontal span of $t = 0$ to $t = 5$ and a vertical span of $D = 0$ to $D = 180$.

We'll use this function to illustrate the properties of rates of change of quadratic functions. To begin, consider a table of values of D:

FIGURE 5.71 A rock tossed upward from the top of a building

Seconds t	0	1	2	3	4
Distance D	135	157	147	105	31

and then a table of first-order differences in D:

Change in t	First-order change in D
0 to 1	$157 - 135 = 22$
1 to 2	$147 - 157 = -10$
2 to 3	$105 - 147 = -42$
3 to 4	$31 - 105 = -74$

To see that this rate of change is linear, compute the second-order differences in D, which are the same as the first-order differences in the rate of change of D. The first-order differences are 22, -10, -42, and -74, so each term differs from its predecessor by -32. Thus the second-order differences are all -32. We see that the quadratic function $D = -16t^2 + 38t + 135$ has constant second-order differences or, equivalently, a linear rate of change; moreover, here are important values we computed:

---

[45] Since the velocity changes linearly from 0 to $32t$, the rock's distance after t seconds is the same as it would be if its velocity were the average $16t$ throughout the fall. Thus the distance is $16t \times t = 16t^2$.

Initial value of D	135
Initial first-order difference	22
Second-order difference	−32

We can reconstruct the formula for D from this information. Since all second-order differences are the same, D is quadratic and so may be written in the form $at^2 + bt + c$. The relation turns out to be

$$a = \tfrac{1}{2} \times \text{second-order difference}$$
$$b = \text{initial first-order difference} - a$$
$$c = \text{initial value}.$$

In our case, $a = \tfrac{1}{2} \times -32 = -16$, $b = 22 - a = 22 - (-16) = 38$, and $c = 135$, as we expected. In fact, the relationship above gives the formula for quadratic functions in general if we have data for $t = 0$, $t = 1$, and $t = 2$.

EXAMPLE 5.13 **A Five-Gallon Water Jug**

In general, when a large cylinder is filled with a fluid and then drained through a small hole, the depth of the fluid decreases as a quadratic function of time. For example, consider a standard 5-gallon water jug with a spigot in it. (See Figure 5.72.) Certainly, it fills a cup faster when it is full than when it is nearly empty. A completely full 5-gallon water jug has a water depth of 15.5 inches (above the spigot level). Here is the depth as a function of time:

Minutes t that the spigot is open	0	1	2	3	4
Depth D above the spigot, in inches	15.50	11.71	8.45	5.72	3.52

1. Show that D can be modeled as a quadratic function of t.

2. Write the formula for D as a quadratic function of t.

3. How long will it take for the jug to be completely drained?

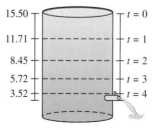

FIGURE 5.72

4. How long does it take for the depth to drop the first $\frac{1}{2}$ inch?

5. How long does it take for the depth to drop the final $\frac{1}{2}$ inch?

Solution to Part 1:   To show that D can be modeled by a quadratic function, it suffices to show that the second-order differences are constant:

t	0	1	2	3	4
D	15.50	11.71	8.45	5.72	3.52
First-order difference	−3.79	−3.26	−2.73	−2.20	
Second-order difference	0.53	0.53	0.53		

Since the second-order differences are all 0.53, we see that D can be modeled by a quadratic function.

Solution to Part 2:   To get the formula for the quadratic function, we use the standard form $D = at^2 + bt + c$ and the relations

$$a = \tfrac{1}{2} \times \text{second-order difference}$$

$$b = \text{initial first-order difference} - a$$

$$c = \text{initial value}.$$

Thus

$$a = \tfrac{1}{2} \times 0.53 = 0.265$$

$$b = -3.79 - a = -3.79 - 0.265 = -4.055$$

$$c = 15.50.$$

Hence $D = 0.265t^2 - 4.055t + 15.50$.

Solution to Part 3:   The water jug is completely drained when the water depth above the spigot is 0, so we want to solve $D = 0$, or

$$0.265t^2 - 4.055t + 15.50 = 0,$$

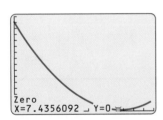

FIGURE 5.73 Finding the zero of the water depth function

for t. We use the crossing-graphs method to do this. In Figure 5.73 we have graphed D with a horizontal span from $t = 0$ to $t = 10$ and a vertical span from $D = 0$ to $D = 16$ and then solved by finding the zero at $t = 7.44$ minutes.

Solution to Part 4:   To find how long it takes to drop the first $\frac{1}{2}$ inch, which is from 15.5 inches down to 15 inches, we solve

$$0.265t^2 - 4.055t + 15.50 = 15$$

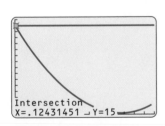

FIGURE 5.74 Finding when the water depth is 15 inches

for t. The crossing-graphs method for solving this is shown in Figure 5.74. It takes 0.124 minute, or about 7.44 seconds, for the water level to drop the first $\frac{1}{2}$ inch.

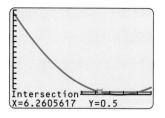

FIGURE 5.75 Finding when the water depth is $\frac{1}{2}$ inch

Solution to Part 5:   To find how long it takes to drop the last $\frac{1}{2}$ inch, we solve $D = 0.5$, or

$$0.265t^2 - 4.055t + 15.50 = 0.5 ,$$

for t. Adding a line at $D = 0.5$ to the graph in Figure 5.73 and finding the intersection in Figure 5.75, we find that $D = 0.5$ when $t = 6.26$ minutes. Since the jug is drained when $t = 7.44$ minutes, it takes $7.44 - 6.26 = 1.18$ minutes, or about 70.8 seconds, for the water level to drop the last $\frac{1}{2}$ inch.   ■ ■ ■

Quadratic Regression

The method used in Example 5.13, where we recover a quadratic formula from a table of data that shows constant second-order differences does not always apply. It requires that the input data include values of 0, 1, and 2 for the variable. Often such data is not available, and an alternative method is needed. One method available on most calculators is *quadratic regression*, which is illustrated in the following example.

EXAMPLE 5.14   **Cost per Student for Education**

In the study "Economies of scale in high school operation" by J. Riew,[46] the author studied data from the early 1960s on expenditures for high schools ranging from 150 to 2400 enrollment. The data he observed was similar to that in the table below.

Enrollment	200	600	1000	1400	1800
Cost per student, in dollars	667.4	545.0	461.0	415.4	408.2

1. Show that the cost per student C, in dollars, can be modeled as a quadratic function of the number n of students.

2. J. Riew used a quadratic function to model these data. Find a quadratic model that fits the data. Plot the graph of the data along with the quadratic model.

3. According to this model, what size enrollment is most efficient in terms of cost per student?

Solution to Part 1:   In order to show that the data are quadratic in nature, we need to show that the second-order differences are constant. We do this in two steps. To get

_____

[46] Published in *Review of Economics and Statistics* **48** (August 1966), 280–287. In our presentation of Riew's results, we have suppressed variables such as teacher salaries.

the first-order differences, we simply subtract each current entry from its preceding entry:

Enrollment	200	600	1000	1400
First-order difference	−122.4	−84.0	−45.6	−7.2

To get the second-order differences, we simply repeat the process, taking differences of the first-order differences:

First-order difference	−122.4	−84.0	−45.6
Second-order difference	38.4	38.4	38.4

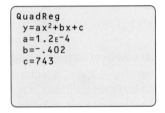

FIGURE 5.76 Quadratic regression coefficients

We see that the second-order differences are constant, and we conclude that the original data can be modeled with a quadratic function.

Solution to Part 2: We use the calculator to get the quadratic regression equation. The exact keystrokes required to do this are presented in the *Technology Guide*. We see from Figure 5.76 that the regression equation is

$$C = 0.00012n^2 - 0.402n + 743 .$$

The data and the quadratic model are shown in Figure 5.77.

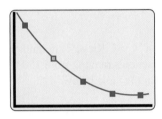

FIGURE 5.77 Data and quadratic model

Solution to Part 3: We want to minimize the function C. To do this, we use the calculator to graph and find the minimum. We choose a horizontal span of 0 to 2400 and a vertical span of 0 to 750. From Figure 5.78 we see that the most efficient enrollment is about 1675 students. ■ ■ ■

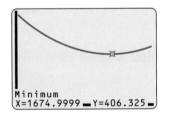

FIGURE 5.78 Minimum cost per student

Almost Quadratic Data

In Example 5.14 we used quadratic regression to get an exact fit for data. We can also find the quadratic function that best fits a data set that is almost quadratic in nature, just as we used linear regression and exponential regression. It's important to consider whether such a quadratic model is appropriate before we perform regression. Good reasons might include an expectation that the data ought to be quadratic—for example, in a case of acceleration due to gravity.

EXAMPLE 5.15 **Ospreys in the Twin Cities**

According to the Raptor Center 1995 Osprey Report[47] the numbers of introduced ospreys fledged in the Twin Cities, Minnesota, from 1992 through 1995 are as given in the accompanying table.

---

[47] See **http://www.raptor.cvm.umn.edu/raptor/meeen/osreport.html**.

Year	1992	1993	1994	1995
Ospreys fledged	8	18	18	7

1. Calculate the second-order differences of the data and determine whether the data are quadratic. Plot the number of ospreys fledged versus years since 1992.

2. Use quadratic regression to find a model for the data.

3. Add the graph of the model to the plot of data points from part a. Discuss the fit of the model.

4. What does this model predict for the number of ospreys fledged in 1996?

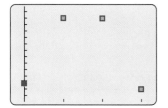

FIGURE 5.79 Plot of data

Solution to Part 1: The first-order differences are 10, 0, and −11, so the second-order differences are −10 and −11. Since the second-order differences are different, the data are not exactly quadratic. The plot of the data appears in Figure 5.79. Here the variable is t, the number of years since 1992.

Solution to Part 2: Quadratic regression yields the formula

$$I = -5.25t^2 + 15.45t + 7.95,$$

where I is the number of introduced ospreys fledged.

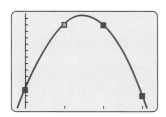

FIGURE 5.80 Quadratic fit

Solution to Part 3: The plot of the data points and the quadratic model is shown in Figure 5.80. The fit appears to be excellent.

Solution to Part 4: Now 1996 corresponds to $t = 4$, and the formula for the quadratic regression model gives $I(4) = -14.25$. The number, being negative, certainly could not be a reasonable prediction for the number of introduced ospreys. This is an example of the dangers of using models to extrapolate beyond the range of the data collected. ■ ▨ ▨

The Quadratic Formula

We have taken the point of view that linear functions should be solved by hand calculation and that others should be solved using the calculator. However, there is a well-known formula, the *quadratic formula*, which makes possible the solution of quadratic equations by hand calculation, and some instructors may prefer that their students use it. In general, a quadratic equation $ax^2 + bx + c = 0$ has two solutions, which differ only by a sign in one place:

$$x = \frac{-b + \sqrt{b^2 - 4ac}}{2a}$$

and

$$x = \frac{-b - \sqrt{b^2 - 4ac}}{2a}.$$

To illustrate the use of the quadratic formula, we look once more at the function C from Example 5.14. Suppose that we want to know what enrollment will produce a cost per student of $500. That is, we wish to solve the equation

$$0.00012n^2 - 0.402n + 743 = 500 \,.$$

To solve this using the quadratic formula, we first subtract 500 from each side:

$$0.00012n^2 - 0.402n + 243 = 0 \,.$$

Next we identify the values of a, b, and c:

$$a = 0.00012$$
$$b = -0.402$$
$$c = 243 \,.$$

Plugging these values into the quadratic formula yields

$$n = \frac{-b + \sqrt{b^2 - 4ac}}{2a} = \frac{0.402 + \sqrt{(-0.402)^2 - 4 \times 0.00012 \times 243}}{2 \times 0.00012} = 2558.53$$

$$n = \frac{-b - \sqrt{b^2 - 4ac}}{2a} = \frac{0.402 - \sqrt{(-0.402)^2 - 4 \times 0.00012 \times 243}}{2 \times 0.00012} = 791.47 \,.$$

Thus we see that an enrollment of either 791 or 2559 will produce a cost of about $500 per student. You may wish to check to see that you get the same answers using the crossing-graphs method with the calculator.

ANOTHER LOOK

Completing the Square, Complex Numbers

The quadratic formula can be derived from a method for dealing with quadratic functions known as *completing the square*. The key step is simply to note that if we add $b^2/4$ to $x^2 + bx$, then it becomes a perfect square. Indeed,

$$x^2 + bx + \frac{b^2}{4} = \left(x + \frac{b}{2}\right)^2 \,.$$

Let's use this method to solve the quadratic equation $x^2 + 4x = 9$. We want to make the left side into a perfect square. To do that, we need to add $4^2/4 = 4$. We add 4 to each side of the equation to get

$$x^2 + 4x + 4 = 13$$
$$(x + 2)^2 = 13 \,.$$

Now we can take the square root of each side of the equation to get the solution:

$$x + 2 = \pm\sqrt{13}$$
$$x = -2 \pm \sqrt{13} \,.$$

Thus we get the two solutions $x = -2 + \sqrt{13}$ and $x = -2 - \sqrt{13}$.

The quadratic formula is obtained by completing the square for a general quadratic as follows: Consider the quadratic equation $ax^2 + bx + c = 0$. We first rearrange the equation:

$$ax^2 + bx = -c$$

$$a\left(x^2 + \frac{b}{a}x\right) = -c$$

$$x^2 + \frac{b}{a}x = -\frac{c}{a}.$$

To complete the square for $x^2 + \frac{b}{a}x$ we need to add

$$\frac{\left(\frac{b}{a}\right)^2}{4} = \frac{b^2}{4a^2}.$$

When we do this, we find

$$x^2 + \frac{b}{a}x + \frac{b^2}{4a^2} = \frac{b^2}{4a^2} - \frac{c}{a}$$

$$\left(x + \frac{b}{2a}\right)^2 = \frac{b^2 - 4ac}{4a^2}$$

$$x + \frac{b}{2a} = \frac{\pm\sqrt{b^2 - 4ac}}{2a}$$

$$x = \frac{-b \pm \sqrt{b^2 - 4ac}}{2a}.$$

This is the classical quadratic formula.

For some quadratic equations, application of the quadratic formula results in taking the square root of a negative number. To accommodate this, we introduce a number i whose square is -1:

$$i^2 = -1.$$

The introduction of this new number enables us to find the square roots of any negative number: If a is positive, then the square roots of $-a$ are $\pm\sqrt{a}\,i$, since

$$(\pm\sqrt{a}\,i)^2 = ai^2 = -a.$$

Thus, for example, $\sqrt{-4} = \pm 2i$. Now we know how to interpret the result of applying the quadratic formula when it involves the square root of a negative number. For example, when we solve $x^2 + 3x + 5 = 0$ using the quadratic formula, we get

$$x = \frac{-3 \pm \sqrt{9 - 20}}{2} = \frac{-3 \pm \sqrt{11}\,i}{2}.$$

In general, *complex* numbers are numbers of the form $a + bi$, with a and b real numbers. We add complex numbers in the obvious way. For example, $(6 + 4i) + (2 + 5i) = 8 + 9i$. We perform multiplication of complex numbers such as

$(6 + 4i)(2 + 5i)$ in the same way we would compute $(6 + 4x)(2 + 5x)$, except that in the end we replace each occurrence of i^2 by -1:

$$(6 + 4i)(2 + 5i) = 12 + 30i + 8i + 20i^2 = 12 + 38i - 20 = -8 + 38i .$$

Enrichment Exercises

E-1. **Solving equations by completing the square:** Solve each of the following equations by completing the square.

a. $x^2 + 6x = 8$ b. $x^2 - 2x - 5 = 0$

c. $2x^2 = x + 4$ d. $(x + 1)^2 + (x + 2)^2 = (x + 4)^2$

E-2. **Special form for quadratics:** Show that every quadratic function can be written as $a(x + b)^2 + c$.

E-3. **Complex solutions:** Use the quadratic formula to solve the following equations.

a. $x^2 + 2x + 3 = 0$ b. $x^2 = x - 8$

c. $2x^2 + 3x + 4 = 0$

E-4. **Rearranging complex numbers:** Write $(1 + i)/(1 - i)$ in the form $a + bi$. *Suggestion:* Multiply top and bottom of the fraction by $1 + i$.

E-5. **High powers of i:** Calculate i^{233}. *Suggestion:* Calculate i, i^2, i^3, etc., and watch for a pattern.

E-6. **Standard form for complex numbers:** Show how to write a complex number of the form $(a + bi)/(c + di)$ in the form $p + qi$. *Suggestion:* Multiply top and bottom of the fraction by $c - di$.

E-7. **The square roots of i:** Show that the square roots of i are $\pm \left(\frac{1}{\sqrt{2}} + \frac{1}{\sqrt{2}}i \right)$.

E-8. **Complex conjugates:** If $z = a + bi$, with a and b real, then the *complex conjugate* of z is $\bar{z} = a - bi$. Thus z is a real number if and only if $z = \bar{z}$.

a. Show that $\overline{z + w} = \bar{z} + \bar{w}$. *Suggestion:* Let $z = a + bi$ and $w = c + di$, and calculate both sides of the desired equation in terms of a, b, c, and d.

b. Show that $\overline{zw} = \bar{z}\,\bar{w}$.

c. Show that if a, b, c, and d are real numbers and z is a zero of $P(x) = ax^3 + bx^2 + cx + d$, then $\bar{z}$ is also a zero of $P(x)$.

5.4 SKILL BUILDING EXERCISES

S-1. The rate of change: What can be said about the rate of change of a quadratic function?

S-2. Testing for quadratic data: Test the following data table to see whether the data are quadratic.

x	1	2	3	4	5
y	0	3	10	21	36

S-3. Testing for quadratic data: Test the following data table to see whether the data are quadratic.

x	1	2	3	4	5
y	1	5	9	24	37

S-4. Quadratic formula: Use the quadratic formula to solve $3x^2 - 5x + 1 = 0$.

S-5. Quadratic formula: Use the quadratic formula to solve $-2x^2 + 2x + 5 = 0$.

S-6. Quadratic formula: Use the quadratic formula to solve $5x^2 - 8 = 0$.

S-7. Single-graph method: Use the single-graph method to solve the quadratic equation in Exercise S-4.

S-8. Quadratic regression: Use quadratic regression to find a model for the following data set.

x	1	3	5	6	8
f(x)	2.2	9.7	27.7	35.2	62.1

S-9. Quadratic regression: Use quadratic regression to find a model for the following data set.

x	1	3	5	6	8
f(x)	1	3	7	5	2

S-10. Quadratic regression: Use quadratic regression to find a model for the following data set.

x	1	3	5	6	8
f(x)	5.5	11.3	8.7	6.1	1.2

5.4 EXERCISES

1. **Sales growth:** The rate of growth G, in thousands of dollars per year, in sales of a certain product is a function of the current sales level s, in thousands of dollars, and the model uses a quadratic function:

$$G = 1.2s - 0.3s^2.$$

The model is valid up to a sales level of 4 thousand dollars.

a. Draw a graph of G versus s.

b. Express using functional notation the rate of growth in sales at a sales level of $2260, and then estimate that value.

c. At what sales level is the rate of growth in sales maximized?

2. **Marine fishery:** A class of models for population growth rates in marine fisheries assumes that the harvest from fishing is proportional to the population size. One such model uses a quadratic function:

$$G = 0.3n - 0.2n^2.$$

Here G is the growth rate of the population, in millions of tons of fish per year, and n is the population size, in millions of tons of fish. →

a. Make a graph of G versus n. Include values of n up to 1.7 million tons.

b. Calculate $G(1.62)$ and explain what your answer means in practical terms.

c. At what population size is the growth rate the largest?

3. **Cox's formula:** Assume that a long horizontal pipe connects the bottom of a reservoir with a drainage area. Cox's formula provides a way of determining the velocity v of the water flowing through the pipe:

$$\frac{Hd}{L} = \frac{4v^2 + 5v - 2}{1200}.$$

Here H is the depth of the reservoir in feet, d is the pipe diameter in inches, L is the length of the pipe in feet, and the velocity v of the water is in feet per second. (See Figure 5.81.)

a. Graph the quadratic function $4v^2 + 5v - 2$ using a horizontal span from 0 to 10.

b. Judging on the basis of Cox's formula, is it possible to have a velocity of 0.25 foot per second?

c. Find the velocity of the water in the pipe if its diameter is 4 inches, its length is 1000 feet, and the reservoir is 50 feet deep.

d. If the water velocity is too high, there will be erosion problems. Assuming that the pipe length is 1000 feet and the reservoir is 50 feet deep, determine the largest pipe diameter that will ensure that the water velocity does not exceed 10 feet per second.

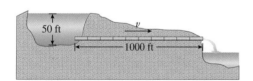

FIGURE 5.81

4. **Poiseuille's law for fluid velocities:** Poiseuille's law describes the velocities of fluids flowing in a tube—for example, the flow of blood in a vein. (See Figure 5.82.) This law applies when the velocities

are not too large, more specifically when the flow has no turbulence. In this case the flow is *laminar*, which means that the paths of the flow are all parallel to the tube walls. The law states that

$$v = k(R^2 - r^2),$$

where v is the velocity, k is a constant (which depends on the fluid, the tube, and the units used for measurement), R is the radius of the tube, and r is the distance from the centerline of the tube. Since k and R are fixed for any application, v is a function of r alone, and the formula gives the velocity at a point of distance r from the centerline of the tube.

a. What is r for a point along the walls of the tube? What is the velocity of the fluid along the walls of the tube?

b. Where in the tube does the fluid flow most rapidly?

c. Choose numbers for k and R and make a graph of v as a function of r. Be sure that the horizontal span for r goes from 0 to R.

d. Describe your graph from part c.

e. Explain why the horizontal span needed to go from 0 to R in order to describe the flow throughout the tube.

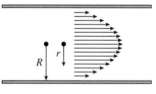

FIGURE 5.82

5. **Surveying vertical curves:** When a road is being built, it usually has straight sections, all with the same grade, that must be linked to each other by curves. (By this we mean curves up and down rather than side to side, which would be another matter.) It's important that as the road changes from one grade to another, the rate of change of grade between the

two be constant.[48] The curve linking one grade to another grade is called a *vertical curve*.

Surveyors mark distances by means of *stations* that are 100 feet apart. To link a straight grade of g_1 to a straight grade of g_2, the elevations of the stations are given by

$$y = \frac{g_2 - g_1}{2L}x^2 + g_1 x + E - \frac{g_1 L}{2}.$$

Here y is the elevation of the vertical curve in feet, g_1 and g_2 are percents, L is the length of the vertical curve in hundreds of feet, x is the number of the station, and E is the elevation in feet of the intersection where the two grades would meet. (See Figure 5.83.) The station $x = 0$ is the very beginning of the vertical curve, so the station $x = 0$ lies where the straight section with grade g_1 meets the vertical curve. The last station of the vertical curve is $x = L$, which lies where the vertical curve meets the straight section with grade g_2.

Assume that the vertical curve you want to design goes over a slight rise, joining a straight section of grade 1.35% to a straight section of grade -1.75%. Assume that the length of the curve is to be 500 feet (so $L = 5$) and that the elevation of the intersection is 1040.63 feet.

a. What phrase in the first paragraph of this exercise assures you that a quadratic model is appropriate?

b. What is the equation for the vertical curve described above? Don't round the coefficients.

c. What are the elevations of the stations for the vertical curve?

d. Where is the highest point of the road on the vertical curve? (Give the distance along the vertical curve and the elevation.)

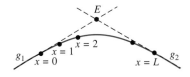

FIGURE 5.83

6. **Quadratic data:** Show that the following data can be modeled by a quadratic function.

x	0	1	2	3	4
$P(x)$	6	5	8	15	26

7. **Data that are not quadratic:** Show that the following data cannot be modeled by a quadratic function.

x	0	1	2	3	4
$P(x)$	5	8	17	38	77

8. **A quadratic model:** Show that the following data can be modeled by a quadratic function, and find a formula for a quadratic model.

x	0	1	2	3	4
$Q(x)$	5	6	13	26	45

9. **Linear and quadratic data:** One of the two tables below shows data that can be modeled by a linear function, and the other shows data that can be modeled by a quadratic function. Identify which is which, and find a formula for each model.

x	0	1	2	3	4
$f(x)$	10	17	26	37	50

Table A

x	0	1	2	3	4
$g(x)$	10	17	24	31	38

Table B

---

[48]Adapted from F. H. Moffitt and H. Bouchard, *Surveying*, 9th ed. (New York: HarperCollins, 1992).

10. **A leaking can:** The side of a cylindrical can full of water springs a leak, and the water begins to stream out. (See Figure 5.84.) The depth H, in inches, of water remaining is a function of the distance D in inches (measured from the base of the can) at which the stream of water strikes the ground. Here is a table of values of D and H:

Distance D in inches	Depth H in inches
0	1.00
1	1.25
2	2.00
3	3.25
4	5.00

a. Show that H can be modeled as a quadratic function of D.

b. Find the formula for H as a quadratic function of D.

c. When the depth is 4 inches, how far from the base of the can will the water stream strike the ground?

d. When the water stream strikes the ground 5 inches from the base of the can, what is the depth of water in the can?

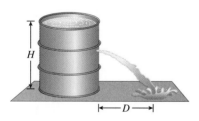

FIGURE 5.84

11. **Traffic accidents:** The following table shows the rate R of vehicular involvement in traffic accidents (per 100,000,000 vehicle-miles) as a function of vehicular speed s, in miles per hour, for commercial vehicles driving at night on urban streets.[49]

Speed s	Accident rate R
20	1600
25	700
30	250
35	300
40	700
45	1300

a. Use regression to find a quadratic model for the data.

b. Calculate $R(50)$ and explain what your answer means in practical terms.

c. At what speed is vehicular involvement in traffic accidents (for commercial vehicles driving at night on urban streets) at a minimum?

12. **Vehicles parked:** The following table shows the number, in thousands, of vehicles parked in the central business district of a certain city on a typical Friday as a function of the hour of the day.[50]

Hour of the day	Vehicles parked (thousands)
9 A.M.	6.2
11 A.M.	7.5
1 P.M.	7.6
3 P.M.	6.6
5 P.M.	3.9

---

[49]The data are adapted from J. C. Marcellis, "An economic evaluation of traffic movement at various speeds," *Travel Time and Vehicle Speed*, Record 35 (Washington, DC: Highway Research Board, 1963).

[50]The data are adapted from *Nashville Metropolitan Area Transportation Study—Downtown Parking* (Nashville, TN: Nashville Parking Board, 1960), 57.

a. Use regression to find a quadratic model for the data. (Round the regression parameters to three decimal places.)

b. Express using functional notation the number of vehicles parked on a typical Friday at 2 P.M., and then estimate that value.

c. At what time of day is the number of vehicles parked at its greatest?

13. **Women employed outside the home:** The following table shows the number, in millions, of women employed outside the home in the given year.[51]

Year	Number in millions
1942	16.11
1943	18.70
1944	19.17
1945	19.03
1946	16.78

a. Use regression to find a quadratic model for the data. (Round the regression parameters to three decimal places.)

b. Express using functional notation the number of women working outside the home in 1947, and then estimate that value.

c. The actual number of women working outside the home in 1947 was 16.90 million, whereas in 1948 the number was 17.58 million. In light of this, is a quadratic model appropriate for the period from 1942 through 1948?

14. **Resistance in copper wire:** Electric resistance in copper wire changes with the temperature of the wire. If $C(t)$ is the electric resistance at temperature t, in degrees Fahrenheit, then the resistance ratio $C(t)/C(0)$ can be measured.

Temperature t in degrees	$\dfrac{C(t)}{C(0)}$ ratio
0	1
10	1.0393
20	1.0798
30	1.1215
40	1.1644

a. On the basis of the data in the table, explain why the ratio $C(t)/C(0)$ can be reasonably modeled by a quadratic function.

b. Find a quadratic formula for the ratio $C(t)/C(0)$ as a function of temperature t.

c. At what temperature is the electric resistance double that at 0 degrees?

d. Suppose that you have designed a household appliance to be used at room temperature (72 degrees) and you need to have the wire resistance inside the appliance accurate to plus or minus 10% of the predicted resistance at 72 degrees.

 i. What resistance ratio do you predict at 72 degrees? (Use four decimal places.)

 ii. What range of resistance ratios represents plus or minus 10% of the resistance ratio for 72 degrees?

 iii. What temperature range for the appliance will ensure that your appliance operates within the 10% tolerance? Is this range reasonable for use inside a home?

15. **A falling rock:** A rock is thrown downward, and the distance D, in feet, that it falls in t seconds is given by $D = 16t^2 + 3t$. Find how long it takes for the rock to fall 400 feet by using

a. the quadratic formula.

b. the crossing-graphs method.

16. **The water jug revisited:** Solve part 3 of Example 5.13 using the quadratic formula.

---

[51] The data are taken from the *1950 Statistical Abstract of the United States*.

17. **Profit:** The *total cost C* for a manufacturer during a given time period is a function of the number N of items produced during that period. (In this exercise, we measure all monetary values in dollars.) To determine a formula for the total cost, we need to know the manufacturer's *fixed costs* (covering such things as plant maintenance and insurance) as well as the cost for each unit produced, which is called the *variable cost*. To find the total cost, we multiply the variable cost by the number of items produced during that period and then add the fixed costs. The *total revenue R* for a manufacturer during a given time period is also a function of the number N of items produced during that period. To determine a formula for the total revenue, we need to know the selling price p per unit of the item, which in general is also a function of N. To find the total revenue, we multiply this selling price by the number of items produced. The *profit P* for a manufacturer is the total revenue minus the total cost.

Suppose that a manufacturer of widgets has fixed costs of $2000 per month and that the variable cost is $30 per widget. Further, the manufacturer has developed the following table showing the highest price p, in dollars, of a widget at which N widgets can be sold.

Number *N*	Price *p*
200	41.00
250	40.50
300	40.00
350	39.50

a. Use a formula to express the total cost C of this manufacturer in a month as a function of N.

b. Judging on the basis of the table, should the price p be modeled by a linear function or by a quadratic function of N? Find the appropriate model for p.

c. Use your answer to part b to find a formula expressing the total revenue R in a month as a function of N.

d. Use your answers to part a and part c to find a formula expressing the profit P in a month as a function of N. What kind of function is the profit: linear or quadratic?

e. Find the two monthly production levels at which the manufacturer just breaks even (that is, where the profit is zero).

18. **More on profit:** *This is a continuation of Exercise 17.* A manufacturer of widgets has developed the following table showing the monthly profit P, in dollars, at a monthly production level of N widgets.

Number *N*	Profit *P*
200	900
250	1375
300	1800
350	2175
400	2500

a. Judging on the basis of the table, should the profit P be modeled by a linear function or by a quadratic function of N? Find the appropriate model for P.

b. Find the manufacturer's fixed costs.

c. Determine the maximum profit in a month if the manufacturer can produce at most 1000 widgets in a month.

5.5   *Higher-degree Polynomials and Rational Functions*

In addition to linear, exponential, logarithmic, power, and quadratic functions, many other types of functions occur in mathematics and its applications. In this section, we will look at higher-degree polynomials and rational functions. In particular, we will consider cubic functions, the cubic formula, cubic and quartic regression, graphing rational functions, and the use of rational functions as models.

A *polynomial* function is a function whose formula can be written as a sum of power functions where each of the powers is a non-negative whole number. For example, $5x^3 + 6x + 4$ is a polynomial because it is a sum of the power functions $5x^3$, $6x$, and $4 = 4x^0$. Another example of a polynomial is $(x + 3)^2$ since it can be written as $x^2 + 6x + 9$. This polynomial happens to be a quadratic function also. On the other hand, $\sqrt{x}$ and $1/x$ are not polynomials since the powers of x are not non-negative whole numbers. Also 2^x, e^x, and $\ln x$ cannot be written as sums of power functions and so are not polynomials either.

The graphs of polynomial functions usually have important geometric properties needed in models. For example, polynomials are the simplest functions having a minimum, or having a point of inflection and a maximum. The simplicity of polynomials makes them attractive models for data that have a minimum or a maximum or other geometric features.

Higher-degree Polynomials and Their Roots

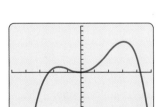

FIGURE 5.85 The cubic function $Y = X^3 - 4X^2 + X - 2$

Higher-degree polynomials are named for the highest power that occurs in the polynomials. For example, *cubic* polynomials have a highest term of degree 3, *quartic* polynomials have a highest term of degree 4, and so on. These polynomials often have interesting graphs. The cubic graph in Figure 5.85 has a horizontal span of -1 to 5 and a vertical span of -10 to 10. It displays both a maximum and a minimum, and this is typical of a cubic function. The quartic graph in Figure 5.86 has a horizontal span of -5 to 5 and a vertical span of -10 to 10. It shows two maxima and a minimum, and this is typical of the graph of a quartic function.

Just as for quadratic functions, the rates of change of cubic functions have a special property: The rate of change of a cubic function is a quadratic function. In particular, data may be modeled exactly by a cubic function precisely if, with evenly spaced values for the variable, the third-order differences in the function values are all the same. In general, data with evenly spaced values for the variable may be modeled exactly by a polynomial of degree n if the nth-order differences are constant—that is, all the same. Just as in the quadratic case, if you know the initial value and all the initial differences (first-order, second-order, and so on), then you can write out the formula for the polynomial; however, this gets progressively more complicated as the degree increases.

For cubic equations, just as for quadratic equations, there is a *cubic formula*, but it is considerably more complicated than the quadratic formula. For example, in the special case $x^3 + cx = d$, the formula gives one root as

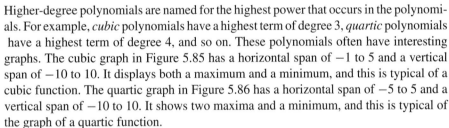

FIGURE 5.86 The quartic function $Y = -0.1X^4 + 0.2X^3 + X^2$

$$x = \sqrt[3]{\sqrt{\left(\frac{d}{2}\right)^2 + \left(\frac{c}{3}\right)^3} + \frac{d}{2}} - \sqrt[3]{\sqrt{\left(\frac{d}{2}\right)^2 + \left(\frac{c}{3}\right)^3} - \frac{d}{2}}.$$

The general cubic formula is more complicated. There is also a quartic formula, but it is quite difficult to use. This means that there are formulas to solve any polynomial equation of degree 4 or less. Amazingly, not only is there no known formula of this type for solving polynomial equations of degree 5, but it can be shown that there *cannot* be a formula of this type for solving polynomial equations of degree 5. This remarkable fact was proved in the mid-1820s by Niels Henrik Abel when he was in his early twenties.

Cubic and Other Polynomial Models

Cubic, quartic, and higher-degree polynomials have some desirable features that make them attractive as models. As we saw in Figure 5.85 and Figure 5.86, the graphs of cubic and quartic polynomials can exhibit interesting behavior, such as one or more minima or maxima as well as points of inflection. Cubic polynomials are the simplest functions that have a point of inflection and so may be an appropriate model of data that exhibit a change of concavity. Since we rarely see cubic or higher-order polynomials exactly fitting data, we can use cubic or quartic regression to find such models rather than checking third- or fourth-order differences.

EXAMPLE 5.16 **Natural Gas Prices**

Here is a table[52] of the average price P of natural gas for home use at certain times t.

Years t since 1982	Price P
0	5.17
1	6.06
2	6.12
3	6.12
4	5.83
5	5.54
6	5.47
7	5.64
8	5.77

Here prices are in dollars per thousand cubic feet and time is in years since 1982.

1. Plot the data points and determine what model might be most appropriate.

2. Find the equation of a cubic model and comment on its fit with the data.

3. What does the cubic model predict for future gas prices?

---

[52]Adapted from D. R. LaTorre, J. W. Kenelly, et al., *Calculus Concepts* (Boston: Houghton Mifflin, 2002).

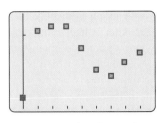

FIGURE 5.87 Plot of natural gas data

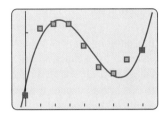

FIGURE 5.88 Cubic regression curve added to gas data

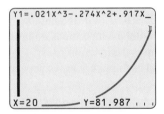

FIGURE 5.89 Using a cubic model to predict future prices

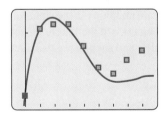

FIGURE 5.90 Quartic regression curve added to gas data

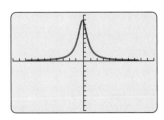

FIGURE 5.91 A graph of the rational function $y = 8/(x^2 + 1)$

4. Find the equation of a quartic model and comment on its fit with the data.

5. What does the quartic model predict for future gas prices?

Solution to Part 1:   Plotting the data points in Figure 5.87, we see that the points increase, then decrease, and then increase again. As we noted earlier, the property of having one minimum and one maximum is typical of a cubic, so we use a cubic model.

Solution to Part 2:   Using cubic regression and rounding to three decimal places, we obtain the model

$$P = 0.021t^3 - 0.274t^2 + 0.917t + 5.247\,.$$

(The keystrokes required to do this are presented in the *Technology Guide*). Adding its graph in Figure 5.88, we get an excellent fit to the data points.

Solution to Part 3:   In Figure 5.89 we have expanded the horizontal span to 22 in order to include $t = 20$ (corresponding to the year 2002), and we have enlarged the vertical span up to 100. We see in Figure 5.89 that the model predicts that gas prices will climb very steeply (so steeply that the data points are bunched at the lower left corner of the screen), and we see that the predicted price is $81.99 per thousand cubic feet in 2002.

Solution to Part 4:   Using quartic regression, we obtain the model

$$P = -0.003t^4 + 0.068t^3 - 0.509t^2 + 1.285t + 5.176\,.$$

(Again, the keystrokes required to do this are presented in the *Technology Guide*). Adding its plot to the data points, we see in Figure 5.90 a very nice fit except for the last three points.

Solution to Part 5:   Using a table, for example, we can see that the quartic model predicts that gas prices will fall steeply and will in fact be negative in 1995.   ■ ■ ■

It is worthwhile comparing parts 3 and 5 of the example. When using cubic or quartic regression, we should be very careful about making predictions. Cubic and quartic functions grow quickly and may fit data and give reasonable predictions only for a very small set of values. In addition, rounding of regression coefficients can radically affect fit.

Rational Functions

Rational functions are functions that can be written as a ratio of polynomial functions— that is, as a fraction whose numerator and denominator are both polynomials. Simple examples include any polynomial, $1/x$, $(x^{12} - x)/(0.7x^6 + 3)$, and $x + (1/x^2)$. By way of contrast, e^x, $\log x$, and $\sqrt{x}$ are not rational functions. The graphs of rational functions can be quite interesting and often exhibit behavior not found in other functions.

One rational function is $8/(x^2 + 1)$, whose graph is displayed in Figure 5.91. Unlike polynomial functions, for larger values of x this function gets closer to zero, which is similar to the behavior of decreasing exponential functions. Unlike exponential functions, this function gets close to zero both for large positive and for large (in size) negative values of x.

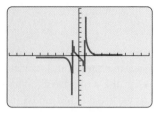

FIGURE 5.92 A graph of the rational function $y = x/(x^2 - 1)$

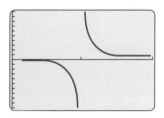

FIGURE 5.93 Looking more closely at a pole

A more interesting rational function is $x/(x^2 - 1)$, whose graph is displayed in Figure 5.92. The graph looks quite ragged, and it is unclear what is happening near $x = -1$ and $x = 1$. The reason for this becomes apparent when we consider that when $x = -1$ or 1, then $x^2 - 1 = 0$; thus the function is trying to evaluate $\frac{1}{0}$, which is not possible. In Figure 5.93, we graph $y = x/(x^2 - 1)$, looking more closely near $x = 1$. We use a horizontal span of 0 to 2 and a vertical span of -10 to 10.

The function $x/(x^2 - 1)$ has no value at $x = 1$. Note that for x to the left of 1, as x tends toward 1 the function values are negative and become larger and larger in size; by contrast, as x tends toward 1 from the right, the function values are positive and become larger and larger positive numbers. In general, a function has a *pole* at $x = a$ if the function is not defined at $x = a$ and the values of the function become larger and larger in size as x gets near a. Not all rational functions have poles, as we saw earlier in considering $y = 8/(x^2 + 1)$. In general, poles occur when the denominator of a rational function has a zero but the numerator does not have the same zero.[53] If a function has a pole at the point $x = a$, then we say that the line $x = a$ is a *vertical asymptote* of the graph of the function, since the graph gets closer and closer to this line as x tends toward a. For example, the graph of the function $x/(x^2 - 1)$ in Figure 5.93 has the line $x = 1$ as a vertical asymptote.

In addition to vertical asymptotes, rational functions may also have *horizontal asymptotes*. Like vertical asymptotes, a horizontal asymptote is a line to which the graph gets closer and closer. Since the line is horizontal, it has slope 0, so its equation is of the form $y = b$ for some b. Since the graph gets closer and closer to $y = b$, the number b is a limiting value of the function. Thus the idea of "horizontal asymptote" is nothing more than a geometric version of "limiting value." For example, the function $x/(x^2 - 1)$ from the previous paragraph can be seen to have a horizontal asymptote $y = 0$ since the function values get closer and closer to 0 for large values of x.

In physical applications, rational functions are far less common than the other functions discussed in this text. When they have poles, the poles usually occur at numbers to which the physical situation does not apply. For example, Ohm's law says that if electric current is flowing across a resistor, then $i = v/R$, where i is current, v is voltage, and R is resistance. For fixed v, the function i is a rational function of R with a pole when $R = 0$. Note that if $R = 0$, then there is no resistance, so we are no longer in the situation of current flowing across a resistor. On the other hand, the behavior of the function i for values of R close to zero does have meaning: As R gets smaller and smaller, i gets larger and larger.

EXAMPLE 5.17   Functional Response

Holling's functional response curve[54] describes the feeding habits of a predator in terms of the density of the prey. An example of such a curve is given by the rational function

$$y = \frac{16x}{1 + 2x}.$$

---

[53]We are assuming that factors common to the numerator and denominator have been canceled.

[54]Based on the work of C. S. Holling, "Some characteristics of simple types of predation and parasitism," *Can. Ent.* **91** (1959), 385–398.

Here x is the density of the prey—that is, the number of prey per unit area—and y is the number of prey eaten per day by a certain predator.

1. Make a graph of y as a function of x.

2. In terms of the predator and its prey, why is it reasonable for the graph to be increasing and concave down?

3. Does the function have a pole? What is the equation of the vertical asymptote?

4. Does the pole in part 3 have any significance in terms of the predator and its prey?

5. Does the function have a limiting value? What is the equation of the horizontal asymptote?

6. Does the horizontal asymptote from part 5 have any significance in terms of the predator and prey?

Solution to Part 1:   The graph in Figure 5.94 has a horizontal span of 0 to 10 and a vertical span of 0 to 10.

Solution to Part 2:   It is reasonable for the graph to increase, because the greater the density of the prey, the easier it is for the predator to catch and eat the prey. On the other hand, the predator can only eat so much prey in one day, so it is reasonable to expect the graph to have the shape you found in part 1. That is, we expect that the rate of increase in prey eaten will decrease as x increases.

Solution to Part 3:   The function does have a pole when the denominator is zero. To find this, we solve

$$1 + 2x = 0$$
$$x = -0.5 \, .$$

Thus the vertical asymptote has the equation $x = -0.5$.

Solution to Part 4:   Since x represents the density of prey, it does not make sense for x to be negative, so it does not appear that the pole has any significance in terms of the predator and its prey.

Solution to Part 5:   Using a table or a graph, it is easy to see that 8 is a limiting value of the function. Thus the equation of the horizontal asymptote is $y = 8$.

Solution to Part 6:   The horizontal asymptote indicates that for large density of prey, the predator will eat about 8 per day.   ■ ■ ■

The following example illustrates the physical significance of a vertical asymptote.

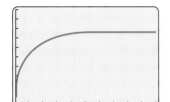

FIGURE 5.94 Prey eaten as a function of prey density

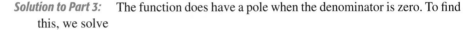

EXAMPLE 5.18   **The van der Waals Equation**

The van der Waals equation uses a rational function to describe the relationship between volume and pressure for a fixed amount of a gas at a fixed temperature. For a mole of gaseous carbon dioxide at a temperature of 500 kelvins, the van der Waals equation predicts the pressure p as a function of volume V as

$$p = \frac{41}{V - 0.043} - \frac{3.592}{V^2}.$$

Here V is in liters and p is in atmospheres.

1. Make a graph of p as a function of V for volumes V between 0 and 0.05 liter.

2. What are the two poles of p and the corresponding vertical asymptotes to the graph of p?

3. What might be the physical significance of these poles?

4. What is the horizontal asymptote to the graph?

5. What is the physical significance of the horizontal asymptote?

Solution to Part 1:   Using a horizontal span from 0 to 0.05 liter and a vertical span from $-200{,}000$ to $500{,}000$ atmospheres, we obtain the graph in Figure 5.95.

Solution to Part 2:   The two poles of p occur when either denominator is zero, so they occur when V is 0.043 or 0.

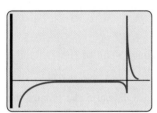

FIGURE 5.95 The van der Waals equation: pressure as a function of volume

Solution to Part 3:   For V between 0 and 0.043, p is negative, which is not reasonable, so the equation does not appear to make sense for V less than 0.043 liter. What is the significance of 0.043 liter? Basically, we are trying to put a whole mole of gaseous carbon dioxide in a very small space; in point of fact, the carbon dioxide is no longer a gas and so the equation no longer applies.

Solution to Part 4:   The horizontal asymptote is $p = 0$, as can be seen from the graph or from a table of values.

Solution to Part 5:   The horizontal asymptote describes the pressure as V becomes larger and larger. We are putting a fixed amount of gas into a larger and larger volume, so the pressure decreases to zero.

ANOTHER LOOK

Factoring Polynomials, Behavior at Infinity

The *Fundamental Theorem of Algebra* is one of the most important theorems in elementary mathematics, and it tells us that every polynomial has at least one (possibly complex) zero. This theorem was proved in about 1800 by C. F. Gauss, and it enables us to draw important conclusions about the structure of polynomials. Another important (though much easier) theorem is the *factor theorem*, which says that a is a zero of a polynomial if and only if $x - a$ is a factor. For example, it is easy to check that $x = 1$ is a zero of $x^3 - x^2 - 4x + 4$. The factor theorem tells us then that $x - 1$ is a factor of $x^3 - x^2 - 4x + 4$. Indeed, you can check by direct multiplication that

$$x^3 - x^2 - 4x + 4 = (x - 1)(x^2 - 4).$$

Using this factorization, we can find the remaining zeros. Note that $x^2 - 4 = 0$

when $x = 2$ or $x = -2$. Thus the zeros of $x^3 - x^2 - 4x + 4$ are 1, 2, and -2, and we can check that $x^3 - x^2 - 4x + 4 = (x - 1)(x - 2)(x + 2)$.

This example suggests a method for finding all the zeros of a polynomial and for writing it in factored form.

Step 1: Find a zero, say a_1, of $P(x)$.

Step 2: Factor $P(x)$ as $P(x) = (x - a_1)P_1(x)$.

Step 3: Find a zero a_2 of $P_1(x)$.

Step 4: Factor $P_1(x)$ as $P_1(x) = (x - a_2)P_2(x)$.

We continue this process until we have $P(x)$ fully factored:

$$P(x) = k(x - a_1)(x - a_2) \cdots$$

Unfortunately, for high-degree polynomials there is no practical way of carrying out this procedure. In general, one can find approximate zeros but not exact zeros. Nonetheless, this procedure reveals important information about polynomials. We see immediately, for example, that a polynomial of degree n has exactly n linear factors (possibly repeated) and at most n zeros. At first glance, it also tells us that a polynomial of degree n has exactly n real zeros, but brief reflection shows that there are difficulties with this. Remember that some of the zeros a_i may be complex numbers, and it is possible that they are not distinct. For example, let's apply this process to $x^2 - 2x + 1$. Now $x = 1$ is a zero, and the polynomial factors as $(x - 1)(x - 1)$. In this case, 1 is said to be a *repeated zero* of $x^2 - 2x + 1$, and even though this is a second-degree polynomial, it has only one zero. In the labeling procedure above, we would have $a_1 = 1$ and $a_2 = 1$. When there are repeated zeros, a polynomial of degree n will have fewer than n distinct zeros. In case of $x^2 + 1$, the two linear factors involve complex numbers: $x^2 + 1 = (x - i)(x + i)$. Thus this is an example of a polynomial with no real zeros. It turns out that for real polynomials, the complex zeros always occur just like this—that is, as zeros of a quadratic factor with real coefficients. We summarize our discussion with the following result.

Decomposition theorem: If $P(x)$ is a polynomial with real coefficients, then $P(x)$ can be factored into a product of linear and quadratic factors with real coefficients.

Once again, although the decomposition theorem is extremely important, there is no practical way of actually finding the factors for high-degree polynomials.

In general, we expect the graph of a polynomial of degree n to cross the horizontal axis at most n times and to have at most $n - 1$ maxima or minima. Figure 5.96 shows the graph of a typical cubic, and Figure 5.97 shows the graph of a typical quartic.

FIGURE 5.96 The graph of a typical cubic

FIGURE 5.97 The graph of a typical quartic

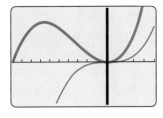

FIGURE 5.98 Comparing graphs on a small span

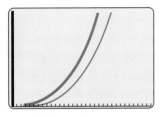

FIGURE 5.99 Comparing graphs on a wider span

The graph of a polynomial may go up and down for a while, but eventually it goes up to infinity or down to minus infinity. In fact, for long-term behavior, only the leading or highest-degree term matters. To explain better what we mean, let's look at $x^3 + 10x^2 + 1$. It has leading term x^3. In Figure 5.98 we have graphed both x^3 and $x^3 + 10x^2 + 1$. The lighter graph is x^3. The horizontal span is -10 to 5, and the vertical span is -150 to 200. The graphs appear very different. In Figure 5.99 we have changed the horizontal span to 0 to 30 and the vertical span to 0 to 10,000. In this window the graphs have the same shape, but the graph of $x^3 + 10x^2 + 1$ is above the graph of x^3.

In Figure 5.100 we have changed the window to a horizontal span of 0 to 100 and a vertical span of 0 to 1,000,000. The graphs appear to be practically the same. As another comparison, we have in Figure 5.101 plotted the graph of the quotient $(x^3 + 10x^2 + 1)/x^3$. The horizontal span is 0 to 2100, and the vertical span is 0 to 2. We have traced the graph, and we see from the prompt at the bottom of the screen that the limiting value of the quotient appears to be 1. This is indeed the case (recalling the notation from "Another Look" in Section 2.1):

$$\lim_{x \to \infty} \frac{x^3 + 10x^2 + 1}{x^3} = 1.$$

It is a fact that the limiting behavior of polynomials depends so strongly on the leading coefficient that when we are calculating the limiting value of a rational function, we need consider only leading coefficients: If $a_n, b_k \neq 0$, then

$$\lim_{x \to \infty} \frac{a_n x^n + a_{n-1} x^{n-1} + \cdots}{b_k x^k + b_{k-1} x^{k-1} + \cdots} = \lim_{x \to \infty} \frac{a_n x^n}{b_k x^k}.$$

This makes limiting values of rational functions very easy to calculate. We simply delete all but the leading terms. Let's look at some examples.

$$\lim_{x \to \infty} \frac{4x^3 - 7x^2 + 13}{2x^3 + x - 8} = \lim_{x \to \infty} \frac{4x^3}{2x^3} = \lim_{x \to \infty} 2 = 2.$$

$$\lim_{x \to \infty} \frac{4x + 13}{x^3 + 1} = \lim_{x \to \infty} \frac{4x}{x^3} = \lim_{x \to \infty} \frac{4}{x^2} = 0.$$

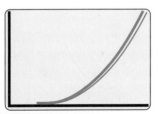

FIGURE 5.100 Comparing graphs on a much wider span

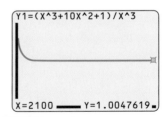

FIGURE 5.101 Examining the quotient

Enrichment Exercises

E-1. Finding the degree of a polynomial: Figure 5.102 shows the graph of a polynomial that has no complex zeros and no repeated zeros. What is the degree of the polynomial?

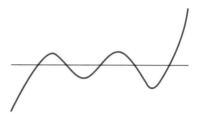

FIGURE 5.102 A polynomial with no complex zeros and no repeated zeros

E-2. Choosing a model: Suppose you have a data set that shows three maxima and two minima. What is the degree of the smallest-degree polynomial that could possibly fit the data exactly?

E-3. Finding the polynomial: A polynomial of degree 3 has zeros 2, 3, and 4. The leading coefficient is 1. Find the polynomial.

E-4. Calculating limits: Calculate the limits of the following rational functions.

a. $\lim\limits_{x \to \infty} \dfrac{5x^4 + 4x^3 + 7}{2x^4 - x^2 + 4}$

b. $\lim\limits_{x \to \infty} \dfrac{3x^4 + 4x^2 - 9}{2x^5 + 4x^3 - 8}$

c. $\lim\limits_{x \to \infty} \dfrac{ax^5 + bx^2 + c}{dx^5 - ex^2 + f}, \ d \neq 0$

E-5. Getting a polynomial from points: A polynomial of degree n is determined by $n + 1$ points. Thus, for example, a quadratic is determined by three points. Find the quadratics that pass through the following sets of points.

a. $(1, 5)$, $(-1, 3)$, and $(2, 9)$. *Suggestion:* We are looking for a quadratic $ax^2 + bx + c$. We need to find a, b, and c. The fact that the quadratic passes through the three points gives the following:

$$a + b + c = 5$$
$$a - b + c = 3$$
$$4a + 2b + c = 9.$$

Solve this system of equations for a, b, and c.

b. $(1, 5)$, $(2, 4)$, and $(3, 19)$

c. $(1, -2)$, $(0, -5)$, and $(-1, -2)$

E-6. **Lagrangian polynomials:** The polynomials

$$P_1(x) = \frac{(x - x_2)(x - x_3)}{(x_1 - x_2)(x_1 - x_3)}$$

$$P_2(x) = \frac{(x - x_1)(x - x_3)}{(x_2 - x_1)(x_2 - x_3)}$$

$$P_3(x) = \frac{(x - x_1)(x - x_2)}{(x_3 - x_1)(x_3 - x_2)}$$

are known as *Lagrangian polynomials*. Here x_1, x_2, and x_3 are given numbers.

a. Show that if $i \neq j$, then $P_i(x_j) = 0$.

b. Show that $P_i(x_i) = 1$.

c. Show that $y_1 P_1 + y_2 P_2 + y_3 P_3$ passes through the points (x_1, y_1), (x_2, y_2), and (x_3, y_3).

d. Use Lagrangian polynomials to find the quadratic that passes through the points $(1, 2)$, $(2, 5)$, and $(3, 6)$.

5.5 SKILL BUILDING EXERCISES

S-1. Recognizing polynomials: Which of the following functions are polynomials? Give the degree of each function that is a polynomial.

a. $x^8 - 17x + 1$

b. $\sqrt{x} + 8$

c. $9.7x - 53.1x^4$

d. $x^{3.2} - x^{2.3}$

S-2. Testing for polynomial data: Explain how one tests data to determine whether they represent a polynomial of degree n.

S-3. Rational functions: What is a rational function?

S-4. Degree from maxima and minima: What is the smallest-degree polynomial that can have both a maximum and a minimum?

S-5. Cubic regression: Use cubic regression to model the following data set.

x	y
1	1
3	3
4	5
6	2
7	1
8	4
10	8

S-6. Cubic regression: Use cubic regression to model the following data set.

x	y
1	1.5
3	3.8
4	7.4
6	4.6
7	8.8
8	9.1
10	12.4

S-7. Quartic regression: Use quartic regression to model the data from Exercise S-5.

S-8. Quartic regression: Use quartic regression to model the data from Exercise S-6.

S-9. Finding poles: Find the poles of

$$\frac{x}{x^2 - 3x + 2}.$$

S-10. Horizontal asymptotes: Find the horizontal asymptotes of

$$\frac{2x^2 + 1}{x^2 - 1}.$$

5.5 EXERCISES

1. **Production:** In an economic enterprise, the total amount T that is produced is a function of the amount n of a given input used in the process of production. For example, the yield of a crop depends on the amount of fertilizer used, and the number of widgets manufactured depends on the number of workers. Because of the law of diminishing returns, a graph for T commonly has an inflection point followed by a maximum, so a cubic model may be appropriate. In this exercise we use the model

$$T = -2n^3 + 3n^2 + n,$$

with n measured in thousands of units of input and T measured in thousands of units of product.

a. Make a graph of T as a function of n. Include values of n up to 1.5 thousand units. →

b. Express using functional notation the amount produced if the input is 1.45 thousand units, and then calculate that value.

c. Find the approximate location of the inflection point and explain what it means in practical terms.

d. What is the maximum amount produced?

2. **Speed of sound in the North Atlantic:** The speed of sound in ocean water is 1448.94 meters per second, provided that the ocean water has a salinity of 35 parts per thousand, the temperature is 0 degrees Celsius, and the measurement is taken at the surface. If any one of these three factors varies, the speed of sound also changes. Different oceans often differ in salinity. In the North Atlantic Central Water (the main body of water for the northern half of the Atlantic Ocean), the salinity can be determined from the temperature, so the speed of sound depends only on temperature and depth. A simplified polynomial formula for velocity in this body of water is[55]

$$V = 1447.733 + 4.7713T - 0.05435T^2$$
$$+ 0.0002374T^3 + 0.0163D + 1.675$$
$$\times 10^{-7}D^2 - 7.139 \times 10^{-13}TD^3.$$

Here V is the speed of sound in meters per second, T is water temperature in degrees Celsius, and D is depth in meters. This formula is valid for depths up to 8000 meters and for temperatures between 0 and 30 degrees Celsius.

a. What type of polynomial is V as a function of T alone? Of D alone?

b. For a fixed depth of 1000 meters, write the formula for V in terms of T alone.

c. Graph V as a function of T for the fixed depth of 1000 meters.

d. What is the concavity of the graph from part c? What does this imply about the speed of sound at that depth as temperature increases?

3. **Traffic accidents:** The following table shows the cost C of traffic accidents, in cents per vehicle-mile, as a function of vehicular speed s, in miles per hour, for commercial vehicles driving at night on urban streets.[56]

Speed s	Cost C
20	1.3
25	0.4
30	0.1
35	0.3
40	0.9
45	2.2
50	5.8

The *rate* of vehicular involvement in traffic accidents (per vehicle-mile) can be modeled[57] as a quadratic function of vehicular speed s, and the cost per vehicular involvement is roughly a linear function of s, so we expect that C (the product of these two functions) can be modeled as a cubic function of s.

a. Use regression to find a cubic model for the data. (Keep two decimal places for the regression coefficients written in scientific notation.)

b. Calculate $C(42)$ and explain what your answer means in practical terms.

c. At what speed is the cost of traffic accidents (for commercial vehicles driving at night on urban streets) at a minimum? (Consider speeds between 20 and 50 miles per hour.)

4. **Poiseuille's law for rate of fluid flow:** A consequence of Poiseuille's law for fluid velocities (see Exercise 4 in Section 5.4) is Poiseuille's law for rate of flow of a fluid through a tube. As earlier, this

---

[55]Adapted from P. C. Etter, *Underwater Acoustic Modeling* (London: E & FN Spon, 1996), 15–16.

[56]The data is adapted from J. C. Marcellis, "An economic evaluation of traffic movement at various speeds," *Travel Time and Vehicle Speed*, Record 35 (Washington, DC: Highway Research Board, 1963).

[57]See Exercise 11 in Section 5.4.

law only applies to laminar flows. The law for rate of flow is a fourth-order polynomial that is also a power function:

$$F = cR^4,$$

where F is the flow rate (measured as a volume per unit time), c is a constant, and R is the radius of the tube.

a. Assume that R increases by 10%. Explain why F increases by 46.41%. *Hint:* Consider 1.10 raised to the fourth power.

b. What is the flow rate through a $\frac{3}{4}$-inch pipe compared with that through a $\frac{1}{2}$-inch pipe?

c. Suppose that an artery supplying blood to the heart muscle (see Figure 5.103) is partially blocked and is only half its normal radius. What percentage of the usual blood flow will flow through the partially blocked artery?

FIGURE 5.103

5. **Population genetics:** In the study of population genetics, an important measure of inbreeding is the proportion of *homozygous genotypes*—that is, instances in which the two alleles carried at a particular site on an individual's chromosomes are both the same. For populations in which blood-related individuals mate, there is a higher than expected frequency of homozygous individuals. Examples of such populations include endangered or rare species, selectively bred breeds, and isolated populations. In general, the frequency of homozygous children from matings of blood-related parents is greater than that for children from unrelated parents.[58]

Measured over a large number of generations, the proportion of heterozygous genotypes—that is, nonhomozygous genotypes—changes by a constant factor λ_1 from generation to generation. The factor λ_1 is a number between 0 and 1. If $\lambda_1 = 0.75$, for example, then the proportion of heterozygous individuals in the population decreases by 25% in each generation. In this case, after 10 generations the proportion of heterozygous individuals in the population decreases by 94.37%, since $0.75^{10} = 0.0563$, or 5.63%. In other words, 94.37% of the population is homozygous.

For specific types of matings, the proportion of heterozygous genotypes can be related to that of previous generations and is found from an equation. For matings between siblings, λ_1 can be determined as the largest value of λ for which

$$\lambda^2 = \frac{1}{2}\lambda + \frac{1}{4}.$$

This equation comes from carefully accounting for the genotypes for the present generation (the λ^2 term) in terms of those of the previous two generations (represented by λ for the parents' generation and by the constant term for the grandparents' generation).

a. Find both solutions to the quadratic equation above and identify which is λ_1. (Use a horizontal span from -1 to 1 in this exercise and the following two exercises.)

b. After 5 generations what proportion of the population will be homozygous?

c. After 20 generations what proportion of the population will be homozygous?

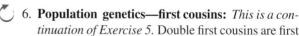 6. **Population genetics—first cousins:** *This is a continuation of Exercise 5. Double first cousins are first*

---

[58]This exercise and the following two exercises are adapted from Chapter 9 of A. Jacquard's *The Genetic Structure of Populations* (New York: Springer-Verlag, 1974).

cousins in two ways—that is, each of the parents of one is a sibling of a parent of the other. First cousins in general have six different grandparents, whereas double first cousins have only four different grandparents. For matings between double first cousins, the proportions of the genotypes of the children can be accounted for in terms of those of parents, grandparents, and great-grandparents:

$$\lambda^3 = \frac{1}{2}\lambda^2 + \frac{1}{4}\lambda + \frac{1}{8}.$$

a. Find all solutions to the cubic equation above and identify which is λ_1.

b. How many solutions did you expect to the cubic equation?

c. After 5 generations what proportion of the population will be homozygous?

d. After 20 generations what proportion of the population will be homozygous?

7. **Population genetics—second cousins:** *This is a continuation of Exercise 5 and Exercise 6.* Second cousins are individuals each of whom has one parent who is a first cousin to a parent of the other individual. For matings between second cousins, the proportions of the genotypes of the children can be accounted for in terms of the previous three generations. In this case, λ_1 is still the solution to an polynomial equation for λ. But here the powers of λ_1 indicate decrease in deviation from an equilibrium genetic structure rather than decrease in the heterozygous population. If $\lambda_1 = 0.75$, for example, the deviation from equilibrium genetic structure decreases by 25% each generation. The equation is

$$\lambda^4 = \frac{1}{8}\lambda^2 + \frac{1}{32}\lambda + \frac{1}{64}.$$

a. Find all solutions to the quartic equation above and identify which is λ_1.

b. How many solutions did you expect to the quartic equation?

c. After 5 generations how close will the population be to its equilibrium structure?

d. After 20 generations how close will the population be to its equilibrium structure?

8. **Forming a pen:** You want to form a rectangular pen of area 80 square feet. (See Figure 5.104.) One side of the pen is to be formed by an existing building and the other three sides by a fence. If w is the length, in feet, of the sides of the rectangle perpendicular to the building, then the length of the side parallel to the building is $80/w$, so the total amount $F = F(w)$, in feet, of fence required is the rational function

$$F = 2w + \frac{80}{w}.$$

a. Make a graph of F versus w.

b. Explain in practical terms the behavior of the graph near the pole at $w = 0$.

c. Determine the dimensions of the rectangle that requires a minimum amount of fence.

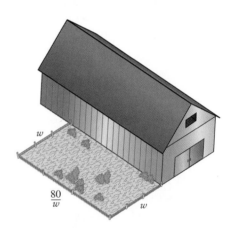

FIGURE 5.104

9. **Inventory:** The yearly inventory expense E, in dollars, of a car dealer is a function of the number Q of automobiles ordered at a time from the manufacturer. A dealer who orders only a few automobiles at a time will have the expense of placing several orders, whereas if the order sizes are large, then the dealer will have a large inventory of unsold automobiles. For one dealer the formula is

$$E = \frac{425Q^2 + 8000}{Q},$$

so E is a rational function of Q.

a. Make a graph of E versus Q covering order sizes up to 10.

b. Explain in practical terms the behavior of the graph near the pole at $Q = 0$. (*Hint:* Keep in mind that there is a fixed cost of processing each order, regardless of the size of the order.)

10. **Power for flying:** The power P required for level flight by an airplane is a function of the speed u of flight. Consideration of drag on the plane yields the model

$$P = \frac{u^3}{a} + \frac{b}{u}.$$

Here a and b are constants that depend on the characteristics of the airplane. This model may also be applied to the flight of a bird such as the budgerigar (a type of parakeet), where we take $a = 7800$ and $b = 600$. Here the flight speed u is measured in kilometers per hour, and the power P is the rate of oxygen consumption in cubic centimeters per gram per hour.[59]

a. Make a graph of P as a function of u for the budgerigar. Include flight speeds between 25 and 45 kilometers per hour.

b. Calculate $P(39)$ and explain what your answer means in practical terms.

c. At what flight speed is the required power minimized?

11. **Traffic signals:** The number of seconds n for the yellow light is critical to safety at a traffic signal. One study recommends a formula for setting the time that permits a driver who sees the yellow light shortly before entering the intersection either to stop the vehicle safely or to cross the intersection at the current approach speed before the end of the yellow light.[60] For a street of width 70 feet under standard conditions, the formula is

$$n = 1 + \frac{v}{30} + \frac{90}{v}.$$

Here v is the approach speed in feet per second. (See Figure 5.105.)

a. Make a graph of n as a function of v. Include speeds from 30 to 80 feet per second (roughly 20 to 55 miles per hour).

b. Express using functional notation the length of the yellow light when the approach speed is 45 feet per second, and then calculate that value.

c. Explain in practical terms the behavior of the graph near the pole at $v = 0$.

d. What is the minimum length of time for a yellow light?

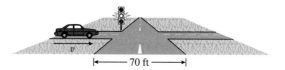

FIGURE 5.105

12. **Gliding falcon:** Consider a bird that is gliding straight with a small downward gliding angle between the bird's path and the horizontal.[61] The rate s at which the bird's altitude decreases is called the *sinking speed*, and it is a function of the airspeed u (the bird's speed relative to the air). For the laggar falcon (which is similar to the peregrine falcon), one model is

$$s = \frac{u^3}{4000} + \frac{7.5}{u}.$$

Here s and u are measured in meters per second, and the formula is valid for airspeeds up to 15 meters per second. →

[59] See the work of V. A. Tucker, "Respiratory exchange and evaporative water loss in the flying budgerigar," *J. Exp. Biol.* **48** (1968), 67–87. See also R. McN. Alexander, *The Chordates*, 2nd ed. (Cambridge, England: Cambridge University Press, 1981).

[60] P. L. Olson and R. W. Rothery, "Driver response to the amber phase of traffic signals," *Traffic Engineering* **XXXII, No. 5** (1962), 17–20, 29.

[61] See the work of V. A. Tucker and G. C. Parrott, "Aerodynamics of gliding flight in a falcon and other birds," *J. Exp. Biol.* **52** (1970), 345–367. See also R. McN. Alexander, *The Chordates*, 2nd ed. (Cambridge, England: Cambridge University Press, 1981).

a. Make a graph of the sinking speed as a function of the airspeed. Such a graph is called a *performance diagram* and is often used to study the efficiency of gliders.

b. Express using functional notation the sinking speed when the airspeed is 5 meters per second, and then calculate that value.

c. Explain in practical terms the behavior of the graph near the pole at $u = 0$.

d. At what airspeed is the sinking speed minimized?

13. **Gliding pigeon:** *This is a continuation of Exercise 12.* For a pigeon that is gliding, one model of sinking speed s in terms of airspeed u is

$$s = \frac{u^3}{2500} + \frac{25}{u}.$$

Here s and u are measured in meters per second, and the formula is valid for airspeeds up to 15 meters per second.

a. Make a graph of the sinking speed as a function of the airspeed for the pigeon.

b. Calculate $s(10)$ and explain in practical terms what your answer means.

c. Which is a more efficient glider, the falcon or the pigeon?

14. **Average transit vehicle speed:** The following formula[62] can be used to approximate the average transit speed S of a public transportation vehicle:

$$S = \frac{CD}{CT + D + C^2 \left(\dfrac{1}{2a} + \dfrac{1}{2d}\right)}.$$

Here C is the cruising speed in miles per hour, D is the average distance between stations in miles, T is the stop time at stations in hours, a is the rate of acceleration in miles per hour per hour, and d is the rate of deceleration in miles per hour per hour. For a certain subway the average distance between stations is 3 miles, the stop time is 3 minutes (0.05 hour), and rate of acceleration and deceleration are both 3.5 miles per hour per second (12,600 miles per hour per hour).

a. Using the information provided, express S as a rational function of C.

b. As a rule of thumb, we might use the cruising speed as an estimate of transit speed. Discuss the merits of such a rule of thumb. How does increasing cruising speed affect such an estimate? *Suggestion:* Compare the graph of the given rational function with that of the graph of $S = C$.

c. What cruising speed will yield a transit speed of 30 miles per hour?

15. **Queues:** A queue, or line, of traffic can form when a feeder road meets a main road with a high volume of traffic. In this exercise we assume that gaps in traffic on the main road (allowing cars to enter from the feeder road) appear randomly and that cars on the feeder road arrive at the intersection randomly also. We let s be the average rate at which a gap in traffic appears and a the average arrival rate, both per minute. Assume that s is greater than a.

a. The average length L of a queue is given by the rational function

$$L = \frac{a^2}{s(s - a)}.$$

Explain what happens to average queue length if arrival rate and gap rate are nearly the same.

b. The average waiting time w (the time in minutes spent in the queue) is given by

$$w = \frac{a}{s(s - a)}.$$

If the gap rate is 3 per minute, what arrival rate will result in an average waiting time of 2 minutes?

16. **Travel time in London:** The time t, in minutes, required to drive d miles on a congested road can be calculated from

$$t = \frac{60d}{v_0 - aq},$$

where v_0 is the average speed in miles per hour when the road is uncongested, q is the traffic flow in vehi-

---

[62]R. L. Creighton, *Urban Transportation Planning* (Urbana: University of Illinois Press, 1970), 111.

cles per hour, and a is a constant that depends on various characteristics of the roadway. For central London, the estimates[63] $v_0 = 28$ miles per hour and $a = 0.008$ mile per vehicle have been used.

a. According to this model, how long will it take to travel 2 miles in central London if traffic flow is 1500 vehicles per hour?

b. This model can be valid only when $aq < v_0$. What is the largest traffic flow that central London streets can handle?

c. What happens to travel time when the flow approaches the value you found in part b?

17. **Change in London travel time:** *This is a continuation of Exercise 16.* The rate of change t' in travel time with respect to traffic flow is the additional travel time that will be required if one additional vehicle is added to the flow. The formula for this change can be calculated as

$$t' = \frac{60ad}{(v_0 - aq)^2}$$

when $aq < v_0$. For the remainder of this exercise, use the values of a and v_0 given for London in Exercise 16, and consider a travel distance of 2 miles.

a. If the traffic flow is 1000 vehicles per hour, how does the addition of one vehicle affect travel time?

b. If the traffic flow is 3400 vehicles per hour, how does the addition of one vehicle affect travel time?

c. Using your answers to parts a and b, discuss how travel times in central London are affected by increasing traffic flow.

18. **Catch equation:** If a lake is stocked with fish of the same age, then the total number C of these fish caught by fishing over the lifespan of the fish is given by the catch equation

$$C = \frac{F}{M + F} N_0 \,,$$

where F is the proportion of fish caught annually, M is the proportion of fish that die of natural causes annually, and N_0 is the original number of fish stocked

in the lake. Assume that a lake is stocked with 1000 channel catfish and that the natural mortality rate is $M = 0.1$ (this means that the annual mortality rate is 10%).

a. Write the catch equation for the lake stocked with channel catfish.

b. Assuming that $F = 1$ means that 100% of the fish are caught within 1 year, what might $F = 2$ mean?

c. What is the horizontal asymptote for the catch equation?

d. Explain what the horizontal asymptote means in terms of the original 1000 fish.

e. What is the vertical asymptote for the catch equation? Does the pole of the equation have any meaning in terms of the fish population?

19. **Hill's law:** When a force is applied to muscle tissue, the muscle contracts. Hill's law is an equation that relates speed of muscle contraction with force applied to the muscle.[64] The equation is given by the rational function

$$S = \frac{(F_\ell - F)b}{F + a} \,,$$

where S is the speed at which the muscle contracts, F_ℓ is the maximum force of the muscle at the given length ℓ, F is the force against which the muscle is contracting, and a and b are constants that depend on the muscle tissue itself. This is valid for non-negative F no larger than F_ℓ. For a fast-twitch vertebrate muscle—for example, the leg muscle of a sprinter—we may take $F_\ell = 300$ kPa, $a = 81$, and $b = 6.75$. These are the values we use in this exercise.

a. Write the equation for Hill's law using the numbers above for fast-twitch vertebrate muscles.

b. Graph S versus F for forces up to 300 kPa.

c. Describe how the muscle's contraction speed changes as the force applied increases.

d. When is S equal to zero? What does this mean in terms of the muscle?

→

---

[63] J. M. Thomson, *Road Pricing in Central London*, RRL Paper PRP 18, 1962.

[64] Adapted from R. McN. Alexander, *Animal Mechanics*, 2nd ed. (London: Blackwell, 1983).

e. Does the rational function for S have a horizontal asymptote? What meaning, if any, does the asymptote have in terms of the muscle?

f. Does the rational function for S have a vertical asymptote? What meaning, if any, does the asymptote have in terms of the muscle?

20. **The Michaelis–Menton relation:** Enzymes are proteins that act as catalysts converting one type of substance, the *substrate*, into another type. An example of an enzyme is invertase, an enzyme in your body, which converts sucrose into fructose and glucose. Enzymes can act very rapidly; under the right circumstances, a single molecule of an enzyme can convert millions of molecules of the substrate per minute. The Michaelis–Menton relation expresses the initial speed of the reaction as a rational function of the initial concentration of the substrate:

$$v = \frac{V s}{s + K_m} \, ,$$

where v is the initial speed of the reaction (in moles per liter per second), s is the initial concentration of the substrate (in moles per liter), and V and K_m are constants that are important measures of the kinetic properties of the enzyme.[65]

For this exercise, graph the Michaelis-Menton relation giving v as a function of s for two different values of V and of K_m.

a. On the basis of your graphs, what is the horizontal asymptote of v?

b. On the basis of your graphs, what value of s makes $v(s) = V/2$? How is that value related to K_m?

c. In practice you don't know the values of V or K_m. Instead, you take measurements and find the graph of v as a function s. Then you use the graph to determine V and K_m. If you have the graph, how will that enable you to determine V? How will that enable you to determine K_m?

---

[65] See S. I. Rabinow, *Introduction to Mathematical Biology* (New York: Wiley, 1975), 46–51.

Summary

Linear and exponential functions are among the most common in mathematical applications, but other functions occur often as well. These include *power functions*, *quadratic* and other *polynomial functions*, and *rational functions*. Also, these functions can be combined to form new functions.

Power Functions

An exponential function is of the form Pa^x, where a is fixed and x is the variable. For *power functions*, these roles are reversed. Power functions are of the form $f(x) = cx^k$. The number k is the *power*, and c is $f(1)$. For a power function with c and x positive:

- If k is positive, then f is increasing.
- If k is negative, then f decreases toward zero.

One common example of a power function is the distance D, in feet, that a rock that is dropped will fall in t seconds if air resistance is ignored: $D = 16t^2$.

A key property of power functions that is often important in applications is the *homogeneity property*. It says that for a power function $f = cx^k$, if x is increased by a factor of t, then f is increased by a factor of t^k. For example, the volume V of a sphere is a power function of its radius: $V = \frac{4\pi}{3}r^3$. The homogeneity property tells us that if the radius of a sphere is doubled, the volume will be increased by a factor of 2^3. That is, the volume will be 8 times larger.

Although power functions with positive power increase, in the long term they do not increase as rapidly as exponential functions. The classic dominance of exponential functions over power functions is sometimes key to the choice of an appropriate model. Exponential functions should be used when eventual dramatic growth is anticipated.

Modeling Data with Power Functions

An adaptation of the way the logarithm was used to make exponential models is available for making power function models as well. The key difference is that for exponential data, we take the logarithm of the function values only, leaving the variable values as they are. To make a power model, we take the logarithm of both function value and variable value. The procedure is illustrated below.

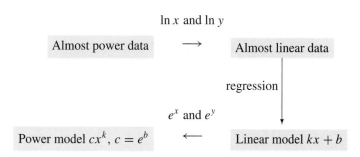

We note that this procedure is sometimes accomplished by plotting the data on logarithmic graphing paper. It is perhaps a bit more tedious than using the calculator, but the practice remains common.

Combining and Decomposing Functions

Often the individual pieces of a function have important meanings that help us understand the behavior of natural phenomena. This idea has been exploited in several forms throughout this text. For example, if a certain population is modeled by $N = Pa^t$, then P is the initial size of the population, and a is the growth or decay factor.

One of the most useful ways to combine functions is through *function composition*. This may involve substituting an expression for a variable into another formula. If the weight W of a certain animal is proportional to the cube of its length L, then $W = kL^3$. If the length is in turn a linear function of age t, then $L = at + b$. We can use function composition to express weight as a function of age:

$$W = kL^3 = k(at + b)^3.$$

This enables us to make a direct investigation of how age affects weight.

Sometimes functions are defined in a piecewise fashion. United States postal rates provide a prime example. At the time of printing of this text, first-class letters required 34 cents for the first ounce and 21 cents for each additional ounce, the weight being rounded upward to next whole number of ounces. Thus a 1.2-ounce letter cost $34 + 21 = 55$ cents to mail.

Quadratic Functions

A quadratic function is a sum of power functions where each of the powers is 0, 1, or 2. The formula for a quadratic function is $ax^2 + bx + c$. For example, if a rock is thrown downward with an initial velocity of 5 feet per second, the distance D in feet that it travels in t seconds is given by the quadratic function $D = 16t^2 + 5t$.

The graph of a quadratic function is a *parabola*, an important geometric shape because it is the path followed by any object ejected near the surface of the Earth.

Quadratic functions can be characterized by the fact that their rate of change is linear. For data, this is exhibited by the fact that the *second-order differences* are constant. This can be used as a test to determine whether data should be modeled by a quadratic function. For data with constant second-order differences, a quadratic model can easily be found from the initial value, the initial first-order difference, and the second-order differences. More generally, quadratic regression may be used to fit a quadratic model to data.

In this text, we have made a practice of solving nonlinear equations using the graphing calculator. But there is a formula, known as the *quadratic formula*, that allows for hand solution of quadratic equations. For the quadratic equation $ax^2 + bx + c = 0$, the solution is given by

$$x = \frac{-b \pm \sqrt{b^2 - 4ac}}{2a}.$$

Higher-Degree Polynomial and Rational Functions

A *polynomial* function is one whose formula can be written as a sum of power functions where each of the powers is a non-negative whole number. Higher-degree[66] polynomials are interesting and useful. Their graphs may have many peaks and valleys, and they are useful in modeling phenomena with more than one maximum or minimum or with an inflection point. Cubic and quartic polynomials are useful models and can be fit to data using cubic or quartic regression.

Polynomial data for polynomials of degree n are characterized by the fact that nth-order differences are constant.

Rational functions are quotients of polynomials. They occur, for example, in describing the flow of electric current across a resistor. Many rational functions have *poles*, which may occur where the denominator is zero. Rational functions grow rapidly near poles. The vertical and horizontal *asymptotes* of rational functions often have practical meaning.

---

[66]The degree is the largest power that occurs in the polynomial.

6 Trigonometric Functions

In this chapter we introduce the classical trigonometric functions and use them to model phenomena that exhibit *periodic* behavior—that is, phenomena that regularly repeat the same behavior. We will also use trigonometry to make a deeper study of triangles and their applications.

6.1 *Periodic Functions*

Many natural phenomena exhibit periodic behavior. Some periodic functions, such as heart contractions shown on an EKG, have a complex periodic structure. Most of the examples we consider here have relatively simple periodic behavior. Among these phenomena are waves such as sound waves, water waves, x rays, and many others. We will study the basic properties of such phenomena and introduce the sine and cosine functions to produce models.

Period and Amplitude

Tides at coastal areas vary in height with season and location because of weather, funneling effects of the ocean floor, and other local factors. But in the open ocean, tides range from a high of about 3 feet above sea level to a low of 3 feet below sea level, and they repeat roughly once each day. This is a prime example of a periodic activity, and the *period* is the length of time it takes the process to repeat; in this case, the period is 1 day. In this example, the periodic variation in height is centered around a value of 0 (that is, sea level). We will refer to this central value as the *median level*. The maximum height above the median level is called the *amplitude*. In this case the amplitude is 3 feet. If we take the horizontal axis to be sea level and make a graph of tidal height (in feet) versus time (in days), we get the graph in Figure 6.1, and the repetitive nature of tidal activity is apparent there. In Figure 6.1 we have marked the period, and in Figure 6.2 we have marked the amplitude.

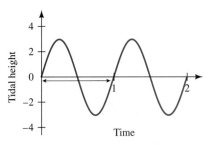

FIGURE 6.1 The period of open-ocean tides is 1 day

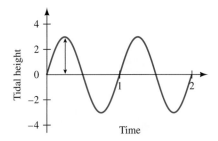

FIGURE 6.2 The amplitude of open-ocean tides is 3 feet

EXAMPLE 6.1   Hours of Daylight

The Earth's seasons are caused by a 23.4-degree tilt of the Earth's axis to the orbital plane, and this tilt is also responsible for the variance in the number of hours of daylight.[1] At the equator there are 12 hours of daylight at the vernal equinox in March and at the autumnal equinox in September. There are about 15 hours of daylight at the summer solstice in June and about 9 hours of daylight at the winter solstice in December.

1. Marking the horizontal axis as time in months and the vertical axis as hours of daylight, and beginning at the vernal equinox in March, make a graph of hours of daylight versus time over 2 years.

2. Identify the amplitude, median level, and period of the graph you made in part 1.

Solution to Part 1:   Because there are 12 hours of daylight at the vernal equinox, we start our graph at a height of 12. By the summer solstice in June there are 15 hours of daylight, so we make the graph increase to a height of 15 at that time. From this point the graph should decrease to 12 at the autumnal equinox, decrease to 9 at the winter solstice, and finally return to 12 at the vernal equinox in March. The picture repeats over the second year and is shown in Figure 6.3.

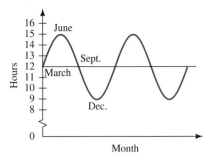

FIGURE 6.3 Hours of daylight at the equator

---

[1]At an equinox, the length of day and night are about the same. The summer solstice is the longest day of the year; the winter solstice is the shortest.

Solution to Part 2:   Since the graph repeats each year, the period is 12 months. The graph varies between 9 and 15 hours. That is from 3 hours below 12 to 3 hours above 12, so the median level is 12 hours and the amplitude is 3 hours.   ■ ■ ■

Sound Waves

Sound can be thought of as vibrating molecules that exert pressure on the ear drum. In the case of a musical tone, this pressure as a function of time is periodic, and the graph of pressure versus time is often referred to as a *sound wave*.[2] The amplitude of the sound wave determines the intensity of the sound.[3] Other qualities of the tone, such as the pitch, are determined by the period of the wave. For sound waves, it is common to use the term *cycle* for the period, and musical tones are often given in terms of *frequency*—that is, the number of periods that occur in 1 unit of time. Often the frequency is measured in *cycles per second*. If a musical tone has a frequency of c cycles per second, then c periods occur in each second. Hence a single period happens in $1/c$ seconds. Thus

$$\text{Period} = \frac{1}{\text{Frequency}}.$$

For example, middle C has a frequency of about 262 cycles per second. Thus the period is given by

$$\text{Period} = \frac{1}{\text{Frequency}} = \frac{1}{262} = 0.0038 \text{ second}.$$

In Figure 6.4 we have graphed a middle C note with amplitude 1 unit, and in Figure 6.5 we have graphed a middle C note with amplitude 2 units.

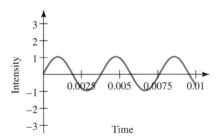

FIGURE 6.4 A middle C note with amplitude 1 unit

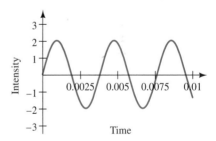

FIGURE 6.5 A middle C note that is twice as intense

The Sine and Cosine Functions

We want to introduce new functions that can model periodic phenomena such as we have been discussing. To do this, we look at the circle of radius 1 unit, commonly referred to as the *unit circle* (see Figure 6.6). We think of the circle as centered at the origin. If we

---

[2] Sound is not in fact a wave. Rather, the motion of the vibrating molecules or the variation in pressure on the ear drum is appropriately *modeled* by a wave.

[3] In turn, the perceived loudness of the sound depends on the intensity in a logarithmic way. See Section 4.4.

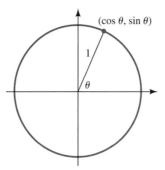

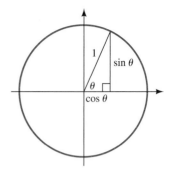

FIGURE 6.6 The definition of sin θ and cos θ

FIGURE 6.7 Alternative definition of sin θ and cos θ

begin at the point $(1, 0)$ and move θ degrees[4] in the counterclockwise direction around the circle, we land on a specific point on the circle. We are interested in the coordinates of this point. The horizontal coordinate is defined to be the *cosine* of θ and is denoted by cos θ. The vertical coordinate is defined to be the *sine* of θ and is denoted by sin θ. Alternatively, we can think of cos θ as the directed horizontal distance from the vertical axis and of sin θ as the directed vertical distance from the horizontal axis, as shown in Figure 6.7. With these descriptions we have defined two functions, the sine and the cosine.

To envision how the sine function behaves, think of starting at $(1, 0)$ and moving counterclockwise around the unit circle. The directed length of the vertical line from the point on the circle where you are to the horizontal axis is the sine of the angle you have traversed. The vertical line starts at 0 length and increases in length as you start around the circle until you reach an angle of 90 degrees. After 90 degrees the length starts to decrease back to 0, and the length reaches 0 at an angle of 180 degrees (half of the way around the circle). The sine function turns negative as you traverse the bottom half of the circle. When you complete a trip around the circle, the process starts over again, so clearly this function is periodic with period 360 degrees. The behavior of the cosine function could be visualized in a similar fashion, only this time we would look at the directed length of the horizontal line back to the vertical axis.

We will look a bit more carefully at the dynamic situation described above. For most angles, calculating by hand the exact value of the sine and cosine functions is a difficult task, so we often use the calculator to get function values. But for some angles it is easy to get the values of the sine and cosine. If we start at $(1, 0)$ and move through an angle of 0 degrees, we will of course remain at the place we started. Thus the sine of 0 degrees is 0 and the cosine of 0 degrees is 1 (because the directed distance to the horizontal axis is 0 and the directed distance to the vertical axis is 1). Thus we have

$$\sin 0° = 0$$

$$\cos 0° = 1 \, .$$

Similarly, if we start at $(1, 0)$ and move 90 degrees counterclockwise around the unit circle, we move exactly one-quarter of the distance around the circle. Thus we land

---

[4]There are two common ways to measure angles: *degrees* and *radians*. Here we use degrees, but in the "Another Look" part of this section we introduce radians.

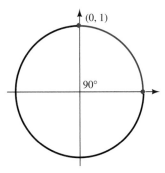

FIGURE 6.8 The sine and cosine of 90 degrees

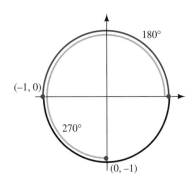

FIGURE 6.9 The sine and cosine of 180 and 270 degrees

on $(0, 1)$, as is shown in Figure 6.8. At this point we are one vertical unit above the horizontal axis and zero horizontal units from the vertical axis. We conclude that

$$\sin 90° = 1$$
$$\cos 90° = 0.$$

From Figure 6.9 we see that an angle of 180 degrees puts us at $(-1, 0)$ and that an angle of 270 degrees puts us at $(0, -1)$. Thus

$$\sin 180° = 0 \qquad \cos 180° = -1,$$
$$\sin 270° = -1 \qquad \cos 270° = 0.$$

Period and Amplitude of Sine and Cosine Functions

If we start at any point on the unit circle and move 360 degrees around the circle, then we land right back where we started. This means that for any angle θ,

$$\sin(\theta + 360°) = \sin \theta$$
$$\cos(\theta + 360°) = \cos \theta.$$

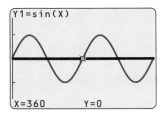

FIGURE 6.10 The period of the sine function is 360 degrees

This simple observation tells us one of the most important facts about the sine and cosine functions: They repeat each 360 degrees. Thus (as we already noted above) both $\sin x$ and $\cos x$ are periodic, and each has a period of 360 degrees.

Note also that as we travel around the unit circle, we never move more than 1 unit to the left or right of the vertical axis and never more than 1 unit above or below the horizontal axis. Thus the sine and cosine functions vary between -1 and 1. That is, both functions have amplitude 1 and median level 0. We can also see the period and amplitude of these functions graphically. In Figure 6.10 we have employed the calculator 6.1 to make a graph[5] of $\sin x$ using a horizontal span of 0 to 720 degrees and a vertical span of -2 to 2.

---

[5]How the sine function is handled by your calculator depends on how your calculator is measuring angles: in degrees or in radians. Here we are using degrees, and you should be sure your calculator is set 6.2 to work in the same way.

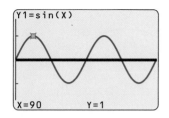

FIGURE 6.11 The amplitude of the sine function is 1

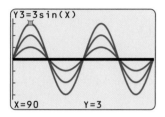

FIGURE 6.12 The amplitude of $A \sin x$ is A

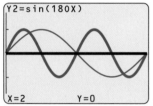

FIGURE 6.13 The period of $\sin(360x/P)$ is P degrees

Note first of all that the graph in Figure 6.10 has a shape similar to the one we made in Example 6.1. Thus it appears that the sine function may indeed be appropriate for modeling such phenomena. In fact, sine functions are so common in this type of modeling that graphs such as the one shown in Figure 6.10 are often referred to as *sine waves*. To verify that the period of the sine function is indeed 360 degrees, we have in Figure 6.10 moved the cursor to the end of the first cycle, and from the bottom of the screen we see that the first cycle ends at 360. (You can locate the end of the first cycle by asking the calculator to locate the third crossing point of the graph.) To get the amplitude, we need to find the height of one of the peaks. In Figure 6.11 we have moved the cursor to the top of the first peak, and we read the height 1 from the prompt at the bottom of the screen. Thus the sine function does indeed have a period of 360 degrees and an amplitude of 1. You will be asked to do a similar investigation with the cosine function in Exercise 3 at the end of this section.

In order to make use of the sine function for modeling, we need to see how to adjust its period and amplitude. We can get sine waves of varying amplitudes and periods using $A \sin(Bx)$ for various values of A and B. To see how different values of A and B affect this function, we graphed $\sin x$, $2 \sin x$, and $3 \sin x$ in Figure 6.12. We note that the amplitude of $\sin x$ is 1, the amplitude of $2 \sin x$ is 2, and the amplitude of $3 \sin x$ is 3. In general, A is the amplitude of $A \sin x$ if A is positive.

To understand how to adjust the period, we first note that $\sin(360x)$ has period 1 degree because $360x$ is 360 when x is 1. In Figure 6.13 we have graphed $\sin(360x)$ with a thick curve and $\sin(360x/2)$ with a thin curve, using a horizontal span from 0 to 2 degrees. This plot verifies our earlier observation that $\sin(360x)$ has period 1 degree, and we see that $\sin(360x/2)$ has period 2 degrees. In general, the function $\sin(360x/P)$ has period P degrees.

KEY IDEA 6.1

Adjusting the Sine Wave

To make a sine wave (in terms of degrees) with given period and amplitude, use

$$\text{Amplitude} \times \sin\left(\frac{360}{\text{Period}} x\right).$$

Using the Sine Function to Model Periodic Phenomena

We now illustrate how sine waves can be used to model periodic phenomena. In this section the models have median level zero, and at the starting point the wave is at zero and increasing. In the next section we will see how to adjust our model to other situations.

EXAMPLE 6.2 **Modeling Additional Daylight with a Sine Wave**

Make a sine wave that models hours of daylight as described in Example 6.1, but this time place the horizontal axis at the median level from that example. Then the vertical axis represents additional hours of daylight beyond 12. According to this model, how many additional hours of daylight are there in August at the time 5 months after the vernal equinox? Assuming that there are 30 days in each month, how many days after the vernal equinox can one expect there to be exactly 2 additional hours of daylight?

Solution: Since the amplitude is 3 hours and the period is 12 months, we use

$$\text{Amplitude} \times \sin\left(\frac{360}{\text{Period}}t\right) = 3\sin\left(\frac{360}{12}t\right) = 3\sin(30t).$$

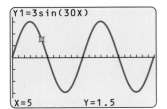

FIGURE 6.14 A sine model of additional hours of daylight

Here t is the time in months since the vernal equinox, when the number of additional hours is at the median level and increasing. The graph, with a horizontal span of 0 to 24 and a vertical span of -4 to 4, is shown in Figure 6.14. To find additional hours of daylight at the specified time in August, we set the cursor at X=5. We see from Figure 6.14 that we can expect 1.5 additional hours of daylight at that time. Thus there are a total of 13.5 hours of daylight then.

To find out when there are 2 additional hours of daylight, we add the horizontal line Y=2 to the picture, as we have done in Figure 6.15. The crossing-graphs method shows that this first happens 1.39 months after the vernal equinox. That is about 42 days after the vernal equinox.   ■ ■ ■

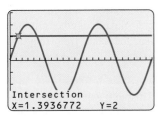

FIGURE 6.15 When there are 2 additional hours of daylight

The sine function is also useful for modeling sound waves. We noted earlier that the middle-C note has a frequency of about 262 cycles per second. Recall that we make a sine wave (in terms of degrees) of period P and amplitude A using the formula

$$A\sin\left(\frac{360}{P}t\right) = A\sin\left(360 \times \frac{1}{P}t\right).$$

Recall also that the period is 1 over the frequency c—that is, $P = 1/c$. This means that we can replace $1/P$ in the formula above by c. Thus we should model a musical tone of frequency c and amplitude A using $A\sin(360ct)$. The model can also be written as

$$\text{Amplitude} \times \sin(360 \times \text{Frequency} \times t).$$

Note that if the frequency c is measured in cycles per second, then time t is measured in seconds. Since middle C has a frequency of about 262 cycles per second, we should model the wave for middle C using $\sin(360 \times 262t) = \sin(94,320t)$, assuming an amplitude of 1. The sound wave representing middle C is graphed in Figure 6.16 with a horizontal span of 0 to 0.02 second and a vertical span of -3 to 3. We chose 0.02 second for the time span because the period is $1/262$, or about 0.004, and we wanted to show the graph over about 5 periods. To model a tone of middle C with twice the intensity, we use $2\sin(94,320t)$. This graph is shown in Figure 6.17.

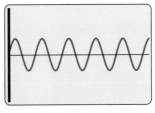

FIGURE 6.16 The sound wave for middle C

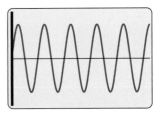

FIGURE 6.17 A middle C with twice the intensity

Adjusting the Sine Wave When Frequency Is Given

To make a sine wave (in terms of degrees) with a given *frequency* and amplitude, we use

$$\text{Amplitude} \times \sin(360 \times \text{Frequency} \times t).$$

EXAMPLE 6.3   Beats

When several musical tones are played together, the sound wave generated is the sum of the individual sound waves generated by the tones. Some notes played together produce a sound that is pleasing to the ear, but when notes very near the same frequency are played together an interesting phenomenon known as *beats* occurs. This phenomenon is not at all pleasing to the ear, but it is useful.

1. An old piano is out of tune, and when middle C is played a sound wave of frequency 265 cycles per second is generated. Model this sound wave using a sine function, and plot its graph over a time interval of 0.02 second. (Use an amplitude of 1.)

2. A piano tuner plays the out-of-tune middle C at the same time a tuning fork that produces a true middle C is struck. Model the resulting sound wave, and produce its graph over a period of 0.5 second. (Use an amplitude of 1 for both tones.)

3. Explain how the graph from part 2 allows the piano tuner to hear clearly that middle C on the piano is not a true middle C, even though it is off by only 3 cycles per second.

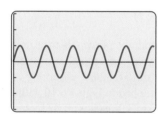

FIGURE 6.18 A slightly out-of-tune "middle C"

Solution to Part 1:   The amplitude is 1, and the desired frequency is 265 cycles per second, so we use

$$\text{Amplitude} \times \sin(360 \times \text{Frequency} \times t) = \sin(360 \times 265t) = \sin(95{,}400t).$$

The graph, where we have used a horizontal span of 0 to 0.02 and a vertical span of -3 to 3, is shown in Figure 6.18. Note that the graph is virtually identical to the one in Figure 6.16. That is expected because the two tones differ in frequency by only 3 cycles per second.

Solution to Part 2:   The sound wave produced is the sum of the true middle C sine wave and the "middle C" sine wave produced by the piano. Thus we use

$$\sin(94{,}320t) + \sin(95{,}400t).$$

The graph, where we have used a horizontal span of 0 to 0.5 and a vertical span of -3 to 3, is shown in Figure 6.19.

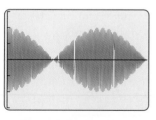

FIGURE 6.19 Beats produced by tones that are nearly the same

Solution to Part 3:   Note that the sound wave in Figure 6.19 is not a simple sine wave, but we can think of it as a sine wave with a *varying* amplitude. This means that the sound will periodically vary in intensity. This pronounced variation in intensity is the phenomenon known as beats. The piano tuner adjusts the middle C piano string until the beats disappear. This occurs when the piano is producing a perfect middle C.

■ ■ ■

We should note that the piano tuning scenario described in Example 6.3 is in fact how many musical instruments get tuned. It does not require a musically talented or trained ear to hear the beats shown in Figure 6.19. The variance in intensity is quite pronounced, is quite unpleasant, and is easily heard by even the most tone-deaf. A simple way to experience this is with a properly tuned guitar. Adjust the G string so that it is just slightly out of tune. Now play, at the same time, the open G string and the D string stopped on the fifth fret. If the guitar were in tune, these two strings would be producing identical notes. With the improperly tuned G string, the resulting beats will be clearly heard.

ANOTHER LOOK

Radian Angle Measure

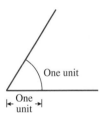

FIGURE 6.20 An angle of 1 radian using a short piece of string

In this section we have dealt exclusively with degree angle measure. Although this is certainly the most common way to measure angles, it is not the most reasonable. It is in fact quite arbitrary. The number of degrees in a circle was probably chosen to be 360 just because 360 has lots of integral divisors. If we encountered advanced alien visitors, they would find degree measure quite amusing since there is a natural unit of angle measure called the *radian*. To get an angle of 1 radian, we start with a length of string. We stretch the string 1 unit from the vertex of the angle along the base. From that point, we mold the string into an arc of a circle centered at the vertex and pass the other side of the angle through the endpoint of the arc. See Figure 6.20. The important and somewhat remarkable thing about this construction is that you get the same angle no matter what length of string you are using. See Figure 6.21. The fact that this angle measure does not depend on any particular unit of length makes it a natural unit of angle measure. The alien visitors mentioned above would certainly know about this angle measure.

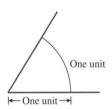

FIGURE 6.21 An angle of 1 radian using a longer piece of string

Since the circumference of a circle of radius 1 is 2π, there are 2π radians in a circle:

$$2\pi \text{ radians} = 360 \text{ degrees}.$$

From this we get the following:

$$1 \text{ radian} = \frac{180}{\pi} \text{ degrees},$$

$$1 \text{ degree} = \frac{\pi}{180} \text{ radians}.$$

Thus, if we use radian measure, the period of the sine and cosine function is 2π radians. To make a sine function in radians with given period and amplitude, we use

$$\text{Amplitude} \times \sin\left(\frac{2\pi}{\text{Period}} x\right).$$

Enrichment Exercises _____

E-1. **Conversions:**

   a. What is the radian measure of a right angle?

   b. Find the radian measure of an angle of 27 degrees.

   c. Find the degree measure of an angle of 2.5 radians.

E-2. **The C note revisited:** Model a middle C note (with amplitude 1) using radian measure.

E-3. *δ* **Cephei again:** *This is a continuation of Exercise 13 at the end of Section 6.1.* Make a sine model of the difference of the apparent magnitude from the median level for *δ* Cephei, using radian measure.

E-4. **Tides again:** *This is a continuation of Exercise 6 at the end of Section 6.1.* Make a sine model of the height in feet of the water (in relation to sea level) using radian measure.

6.1 SKILL BUILDING EXERCISES

S-1. **Values of the sine function:** Use the unit circle definition to calculate the following:

a. $\sin 90°$

b. $\sin 270°$

c. $\sin 360°$

d. $\sin 450°$

S-2. **Period and amplitude:** Plot the graph of $4\sin(5x)$ and determine its period and amplitude.

S-3. **A sine wave with given period and amplitude:** Make a sine model with period 10 degrees and amplitude 4, and show its graph.

S-4. **A cosine wave with given period and amplitude:** Make a cosine model with period 5 degrees and amplitude 3.

S-5. **A sine wave with given amplitude and frequency:** Make a sine model with amplitude 7 and frequency 2.

S-6. **Getting period from frequency:** A periodic function has period 7. What is its frequency?

S-7. **Getting frequency from the period:** A periodic function has frequency 5. What is its period?

S-8. **Finding period and amplitude:** Find the period and amplitude of $4\sin(6x)$.

S-9. **Finding period and amplitude:** Find the period and amplitude of $5\cos(12x.)$

S-10. **Finding frequency:** Find the frequency of a sine wave modeled by $7\sin(8x)$.

6.1 EXERCISES

1. **Outdoor temperature:** The outdoor temperature, in degrees Fahrenheit, is a function of the time, in hours since noon. Assume that the temperature pattern repeats every day and that the temperature varies from a low of 64 to a high of 82.

   a. Find the period, median level, and amplitude for the temperature.

   b. Draw by hand a graph of the temperature versus time. Assume that the temperature at noon is at the median level and rising.

2. **Period and frequency:** A weight is attached to a spring suspended from the ceiling. We give the weight an upward push, and the spring-mass system moves up and down in a periodic fashion.

   a. Assume that the period of the motion is 0.25 second. Explain in practical terms what this means, and find the frequency of the motion.

   b. Now assume that the frequency is 5 cycles per second. Explain in practical terms what this means, and find the period of the motion.

3. **Period and amplitude of the cosine function:** Show graphically that the cosine function has period 360 degrees and amplitude 1.

4. **Biorhythm:** Scientists have studied the annual biorhythm of the excretion of *norepinephrine* (a hormone and a neurotransmitter) in a healthy human.[6] The level varies from a low of 12 units to a high of 20 units over a period of a year.

   a. Find the median level and amplitude.

   b. Say the level is at a minimum at the beginning of January. Draw by hand a graph of the level versus time, in months, since the start of the year. When is the maximum level attained?

---

[6]Based on the work of Descovich and others, as described by Edward Batschelet in *Circular Statistics in Biology* (London-New York: Academic Press, 1981).

5. **Voltage:** The voltage V, in volts, in a circuit varies from -160 to 160 with a period of 18 milliseconds. Make a sine model of V as a function of time t, in milliseconds, and use the model to evaluate the voltage at $t = 5$. (Assume that at time $t = 0$ the voltage is 0 and increasing.)

6. **Tides:** Suppose that tides vary from 5 feet above to 5 feet below sea level and that successive low tides occur each 20 hours. Make a sine model of the height in feet of the water (in relation to sea level) as a function of time t in hours, starting when the water is at sea level and rising. Plot the graph and determine the water level at $t = 7$ hours.

7. **A spring:** When an object suspended from a spring vibrates up and down, the displacement of the object from its rest position can be graphed against time (see Figure 6.22). For the first part of this problem, we will make the unrealistic assumption that once it is set into motion, the spring continues to move up and down at the same rate without ever slowing down.

   a. Suppose that a weight is suspended from a spring and that at its rest position, an initial upward velocity is imparted, setting the weight in motion. The weight rises to its highest point at 8 inches above rest position, then falls to its lowest point at 8 inches below rest position, and then rises back through rest position—and then the cycle repeats. Assume the characteristics of the spring and the weight are such that the cycle takes 1.6 seconds. Make a sine model of the displacement of the weight above rest position as a function of time t in seconds, starting when the weight is set in motion. Plot the graph and determine when is the first time the weight drops to 4 inches below rest position.

   b. In the physical situation described in part a, we know that the spring will not continue to vibrate at the same rate forever. Rather, the motion will slow and eventually stop. This is usually described by adding an exponential *damping* term to the function. Thus if $f(t)$ is your function from part a, then the function $0.8^t f(t)$ will give a more accurate picture of what really happens. Make a graph of this function, and under these assumptions determine the displacement of the weight above its rest position 5 seconds after the spring is set in motion.

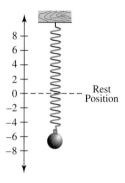

FIGURE 6.22

8. **Vibrating strings:** If you plucked a guitar string and then were able to freeze the wire in time, it would have the shape of a sine wave. In describing such phenomena, physicists usually refer to the period of the wave as the *wavelength*. The amplitude of such a wave is the initial displacement you give the wire when you pluck it. Suppose that you pluck a guitar string, giving it an initial displacement of 0.5 centimeter, and that the resulting wavelength is 20 centimeters. Make a sine wave modeling the shape of the frozen guitar string, and plot its graph. What is the displacement of the guitar string 50 centimeters from the base of the string?

9. **Two notes of a C chord:** For this exercise we use a sharper estimate of frequencies of musical notes. The frequency of middle C is 261.63 cycles per second, and the frequency of E above middle C is 329.63 cycles per second. These two notes are part of a C chord.

   a. Make sine wave models for each of these notes (played with an amplitude of 1), and plot their graphs over the first 0.04 second.

   b. Make a graph of the sound wave produced when both notes are played together. *Note:* Recall that when two notes are played at the same time, the sound wave that is produced is the sum of individual waves. The human ear must deal with the complex sound wave shown in the graph you produced here. In order to understand it, the auditory system must decompose the wave back into its simpler components: the pure C sound wave and the pure E sound wave. Decompositions of this sort are special cases of *Fourier analysis*. In a general setting, Fourier analysis is an advanced mathematical topic that is both theoretically and technically difficult. It is a remarkable fact that the human ear does a kind of Fourier analysis instantly on sound waves many times more complicated than the one produced here. Precisely how this is accomplished is not well understood.

10. **Adjusting beats:** In Example 6.3 we saw that the phenomenon of beats arises when two notes of nearly the same frequency are produced simultaneously as a piano is being tuned. In this exercise we use the calculator to simulate tuning the piano.

   a. Adjust the frequency for the mis-tuned middle C from the given value of 265 cycles per second to a value closer to 262 (representing middle C). Then produce a graph of the sound wave that results from playing a note at this frequency along with middle C.

   b. Repeat part a with different frequencies for the improperly tuned middle C, and use the results to answer the following question: What happens to the number of beats per second as the frequency difference from middle C gets smaller and smaller? How would your answer help the piano tuner?

11. **The A above middle C:** The A note above middle C has a frequency of 440 cycles per second.

   a. Find the period of the sound wave produced by the A above middle C.

   b. On the same set of coordinate axes, draw by hand the graphs of two such A notes, one of which is three times as intense as the other.

   c. Make a sine model for the A note above middle C, using an amplitude of 1.

12. **Octaves:** On the piano, the notes A through G repeat. The span from one note to its next higher occurrence is called an *octave*, since on the piano this is a span of eight notes. If we begin at a note and move one octave higher, the frequency doubles.[7] On the

same graph, plot by hand both a middle C note and the C note one octave higher. (Recall that middle C has a frequency of about 262 cycles per second.) How do their periods compare?

13. **Cepheid variables:** *Cepheid variables* are giant stars that actually pulsate[8] like a beating heart, causing periodic variation in their brightness. Such stars are very important in determining interstellar distances. The first Cepheid variable[9] discovered, δ Cephei, gives the class of stars its name and was found in 1784 by the English astronomer John Goodricke. The star δ Cephei varies in brightness from an apparent magnitude[10] of 3.6 to an apparent magnitude of 4.3, and the variation has a period of 5.4 days.

   a. Make a graph of the apparent magnitude of δ Cephei versus time.

   b. Find the median level and the amplitude.

   c. Mark the median level, amplitude, and period on the graph.

   d. Make a sine model of the difference of the apparent magnitude from the median level.

14. **Using Cepheid variables to determine distance:** *This is a continuation of Exercise 13.* The actual brightness of a star, not taking into account its distance from Earth, is the *absolute magnitude*, commonly denoted by M_v. Astronomers can determine the absolute magnitude of a Cepheid variable from its period P (measured in days). The relationship is given approximately by

$$M_v = -0.11P - 2.11 .$$

The distance from Earth to the star is then given approximately by

$$d = 10^{(m-M_v+5)/5},$$

where m is the apparent magnitude and the distance d is in *parsecs*. (One parsec is about 3.26 light-years.)   →

---

[7]On the piano this relationship is only approximate, though it is very close. The fact that on the piano the note one octave above middle C does not have exactly double the frequency of middle C contributes to the characteristic musical sound of a piano.

[8]These stars are not to be confused with *pulsars*, which do not pulsate at all but rather rotate at extremely high rates.

[9]Another important Cepheid variable is Polaris, the North Star.

[10]Apparent magnitude is a measure of apparent brightness as viewed from Earth. Larger magnitudes indicate dimmer stars.

a. Take the apparent magnitude m of δ Cephei to be the median apparent magnitude found in Exercise 13. Find the absolute magnitude M_v, and determine the distance from Earth to δ Cephei.

b. Suppose that a certain Cepheid variable is 50,000 parsecs from Earth and has an apparent magnitude of $m = 4.7$. Find the absolute magnitude M_v and the period of this Cepheid variable.

15. **Polaris:** *This is a continuation of Exercise 13.* One well-known Cepheid variable is Polaris,[11] the North Star. The apparent magnitude of Polaris varies with an amplitude of 0.1 and a period of 3.97 days.

a. Using the horizontal axis as the median magnitude level, make a sine model of the apparent magnitude of Polaris as a function of time.

b. Make a graph of the sine model you constructed in part a.

---

[11] Cepheid variables are giant stars. The radius of Polaris is approximately 100 times that of the sun.

6.2 *Modeling Periodic Data*

We will see how to adjust sine models to different situations. We will also develop two ways to model periodic data. One is the method of estimating essential parameters, and the other is the generally more accurate method of sine regression.

Vertical and Horizontal Shifting

All of the sine models we have seen thus far are centered vertically about the horizontal axis, start at the origin, and are increasing there. When modeling data, we will find that this is not always the case. Thus we need to look at vertical and horizontal shifts of sine waves.

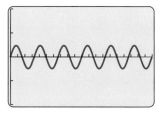

FIGURE 6.23 A weight attached to a vibrating spring viewed from rest position

Consider a weight attached to a spring hung from the ceiling. Suppose the weight stretches the spring so that, at rest position, the weight is 2 feet above the floor. We give the mass an upward push and view the resulting motion. The spring will cause the weight to move periodically up and down. Suppose the resulting motion, viewed from rest position, has an amplitude of 0.5 foot and a period of 0.75 second. Then, using the ideas from the previous section, we can model the motion of the weight from the perspective of rest position with $0.5 \sin(360t/0.75)$. Here t is the time in seconds since the start of the motion. The graph, with horizontal span from 0 to 5 seconds and vertical span from -2 to 2 feet, is shown in Figure 6.23. Note that the median level is 0: The horizontal axis is the center of the motion, and that axis represents the rest position of the weight.

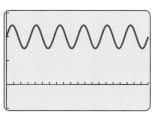

FIGURE 6.24 The same weight viewed from the floor

How do we make a sine model of this same situation viewed from the floor? Since the rest position of the weight is 2 feet above the floor, each point on the graph must be moved up 2 feet. That is, we need to add 2 to the function we have already found so that the median level will be 2. Thus we use $0.5 \sin(360t/0.75) + 2$. Figure 6.24 shows the graph, where we have used the same horizontal span as above but a vertical span of -1 to 3. Note that in this model, the horizontal axis represents the floor.

Now let's look at the situation where we first lower the weight from rest position by 0.25 foot before we give it the upward push. Suppose that the period and amplitude of the resulting motion are the same as before. As before, we wish to view the motion from the perspective of the floor. Thus we should use the same model as above, but we want it to start at a height of 1.75 feet above the floor, not at 2 feet above the floor. In this case the starting value is not the median level. How do we adjust the model to account for this? The key is to notice that t is the elapsed time, in seconds, since the experiment began. If we think of the motion in the preceding scenario as having been going on for a while (rather than starting at $t = 0$), then at just a brief time before $t = 0$, the height of the weight was 1.75 feet above the floor and increasing. We need to find out when this occurred. That is, we want to find out when

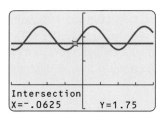

FIGURE 6.25 Finding when the height is 1.75 feet

$$0.5 \sin\left(\frac{360}{0.75}t\right) + 2 = 1.75.$$

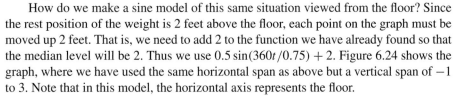

We can do this by using the crossing-graphs method as shown in Figure 6.25. We see that this occurred at $t = -0.063$ second. Therefore, the way to make the motion begin at a height of 1.75 feet is to shift the time backward by 0.063 second—that is, replace *t* with $t - 0.063$. Thus the model we should use is

$$0.5 \sin\left(\frac{360}{0.75}(t - 0.063)\right) + 2.$$

This backward shift in time is represented by a horizontal shift of the graph to the *right* by 0.063 unit. The shifted graph is shown in Figure 6.26, where we have moved the cursor to the beginning of the graph. We read from the bottom of the screen that the motion does indeed begin at a height of about 1.75 feet.

Vertical and horizontal shifting can change the perspective of the model and change the position from which it starts.

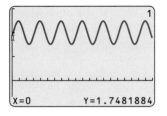

FIGURE 6.26 The model starting at 1.75 feet

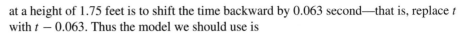

KEY IDEA **6.3**

Vertical and Horizontal Shifting of Sine Waves

The function $A \sin(B(t - C)) + D$ has the following properties if *A* and *B* are positive.

- The median level is *D*, so the periodic activity is centered vertically at *D*.
- The amplitude is *A*.
- The period is $360°/B$.
- The number *C* shifts the graph to the right *C* units. If *C* is negative, then the graph is shifted to the left.

Thus we can write the sine wave (in terms of degrees) as

$$\text{Amplitude} \times \sin\left(\frac{360}{\text{Period}}(t - \text{Right shift})\right) + \text{Vertical shift}.$$

EXAMPLE 6.4 **Azuki Bean Weevils and Braconid Wasps**

Braconid wasps depend on azuki bean weevils for their survival.[12] When there are sufficient weevils present, the wasp population increases rapidly. But the increasing wasp population preys heavily on the weevils, causing their population to decrease. The decreasing weevil population leaves the predator wasps without enough prey, so their numbers fall, allowing the weevil population to recover. The result is that, through the generations, the wasp population varies periodically.[13]

---

[12]The larvae of the wasps are parasites on the weevils.

[13]This sort of periodic population growth is typical of what happens when a predator and its prey live in the same region.

A certain wasp population varies periodically with an amplitude of 1500 around a median level of 2000 wasps. The period is about 5 generations. Say the initial population is 1000 wasps and is declining. Make a sine model of wasp population as a function of the generation. Plot the graph of the wasp population, and find when the wasp population reaches a maximum for the first time.

Solution: We begin with a periodic function of period 5 and amplitude 1500, and then we proceed to make necessary adjustments. Thus we start with

$$1500 \sin\left(\frac{360}{5}t\right) = 1500 \sin(72t).$$

Here t is the number of generations. Next we need to adjust the function upward by 2000 since that is the vertical center of the periodic activity. We get $1500 \sin(72t) + 2000$. Finally, we need to move the number of generations t back until the population is at a level of 1000 and declining. To do this, we use the crossing-graphs method to solve

$$1500 \sin(72t) + 2000 = 1000.$$

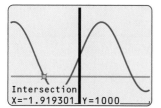

We have done this in Figure 6.27, using a horizontal span of -4 to 4 and a vertical span of 0 to 4000. Note that in Figure 6.27 we chose a crossing point where the wasp population is declining. We see from the prompt at the bottom of the screen that this occurs when $t = -1.92$ generations, so we need to shift the graph to the right by 1.92. Thus the model we want is

FIGURE 6.27 Finding when the population was 1000 and declining

$$1500 \sin(72(t - 1.92)) + 2000.$$

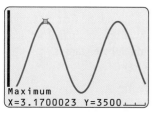

FIGURE 6.28 The model and the first maximum

This is shown in Figure 6.28, where we use a horizontal span of 0 to 10 and a vertical span of 0 to 4000. We also calculated the first maximum in Figure 6.28, and we see that this occurs after 3.17 generations. ■ ■ ■

EXAMPLE 6.5 Blood Pressure

Human blood pressure varies as the heart beats. When the heart contracts, blood pressure increases. But as the heart relaxes in preparation for the next contraction, blood pressure decreases. The result is that a person's blood pressure changes in a periodic fashion, each period corresponding to a single heartbeat. Physicians refer to the maximum of this pressure as the *systolic blood pressure* and to the minimum pressure as the *diastolic pressure*. A certain individual has a pulse rate of 75 beats per minute, a systolic blood pressure of 130, and a diastolic pressure of 70. Here the blood pressure is measured in millimeters of mercury.

1. Assuming that we begin when the blood pressure is at the systolic level, use a sine function to model this person's blood pressure.[14] Plot the graph of the function.

---

[14]We use a sine model here for simplicity, but more realistic models of blood pressure involve a sum of sine waves.

2. What is the blood pressure 2.1 seconds after the starting time?

3. For how long during each cycle is the blood pressure lower than 90 millimeters of mercury?

Solution to Part 1:   The rate of 75 beats per minute is in fact a frequency, because it gives the number of periods occurring in 1 minute. On the basis of this, we measure time t in minutes. Since the pressure varies from a low of 70 to a high of 130, the median level is 100. This is the vertical shift for the sine wave. The pressure varies from 30 above to 30 below the median level, so the amplitude is 30. Thus, if for the moment we neglect the starting point, the model we want is

$$\text{Amplitude} \times \sin(360 \times \text{Frequency} \times t) + \text{Vertical shift}$$
$$= 30\sin(360 \times 75t) + 100\,,$$

which we simplify to

$$30\sin(27{,}000t) + 100\,.$$

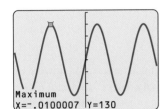

FIGURE 6.29 Finding the needed right shift

The graph is shown in Figure 6.29. We use a horizontal span of -0.02 to 0.02 minute since the period is $\frac{1}{75}$, or about 0.01. We use a vertical span of 60 to 140. Now we require one further adjustment in the model. We want to start at the systolic, or maximum, blood pressure level. In Figure 6.29 we have found the first maximum to the left of 0, and we see from the prompt at the bottom of the screen that we need to shift the graph to the right by 0.01 minute. Thus the model we want is

$$30\sin(27{,}000(t - 0.01)) + 100\,.$$

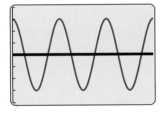

FIGURE 6.30 Blood pressure model with median level added

In Figure 6.30 we have graphed this function and added the median level of 100.

Solution to Part 2:   First we note that 2.1 seconds corresponds to 0.035 minute. In Figure 6.31 we have located the cursor at $t = 0.035$, and we see from the prompt at the bottom of the screen that the blood pressure is 78.79 millimeters of mercury at that time.

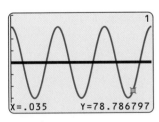

FIGURE 6.31 Blood pressure 0.035 minute after starting time

Solution to Part 3:   First we want to know when the blood pressure is 90. That is, we want to solve

$$30\sin(27{,}000(t - 0.01)) + 100 = 90\,.$$

We do this using the crossing-graphs method. In Figure 6.32 we have added the graph of the horizontal line at a height of 90, and we see that the first intersection occurs 0.004 minute after the starting time. In a similar way, we find that the second intersection occurs 0.009 minute after the starting time. Then the time during the cycle that the blood pressure is below 90 is $0.009 - 0.004 = 0.005$ minute. This is about 0.3 second during the cycle. ∎

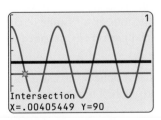

FIGURE 6.32 When blood pressure is 90

Sine Models by Estimation

To show how to estimate appropriate quantities to make sine wave models, we look first at an astronomical phenomenon.

EXAMPLE 6.6   **Magnitude of Eclipsing Binaries**

Astronomers use several measurements of the brightness of stars. One is the *apparent magnitude*, a measure of how bright a star appears to the naked eye. Bright stars have a low (or even negative) magnitude, whereas dim stars have a high magnitude. Some stars vary in magnitude.[15] One such star of historical importance is Algol, whose magnitude variation is clearly visible to the naked eye.[16] Early astronomers offered any number of explanations for the variation, but today it is understood that Algol is not one star but a pair, one orbiting the other. As the dimmer of the two stars passes in front, the brightness drops. Such pairs are known as *eclipsing binaries*, and they are important for many modern cosmic measurements.

Measurements for the magnitude of Algol as a function of time in hours are shown in the table below.

t = time in hours	0	10	20	30	40	50	60	70	80	90
M = magnitude	2.8	3.3	3.4	3.0	2.5	2.2	2.4	2.9	3.4	3.3

1. Plot the given data, and use the data to estimate the median level and the amplitude.

2. The period of the sine model is the time required for the dim star to complete an orbit of the brighter one. Use the data to estimate the time span for a single orbit.

3. Make a sine wave model for the magnitude of Algol on the basis of the given data, starting at the time $t = 0$ when $M = 2.8$. Use the model to estimate how long it will be until Algol first reaches a magnitude of 2.4.

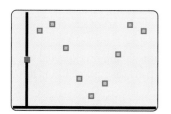

FIGURE 6.33 Magnitude data for Algol

Solution to Part 1:   The correctly entered data are shown in Figure 6.33, and the plot of the data is shown in Figure 6.34. On the basis of the data, we estimate that the magnitude varies from a minimum of 2.2 to a maximum of 3.4. The median level is halfway between these values:

$$\text{Median level} = \frac{2.2 + 3.4}{2} = 2.8 .$$

The magnitude varies from 0.6 above median level to 0.6 below median level. Thus we estimate that the amplitude is 0.6.

Solution to Part 2:   The time required to complete an orbit is the period of the data. We can estimate the period by looking at the distance from the first peak at $t = 20$ to the next at $t = 80$. Thus the period is approximately 60 hours. (The true period is

FIGURE 6.34 Plot of data points

---

[15]Variation in magnitude is not to be confused with apparent *twinkling*. Stars do not in fact twinkle. This effect is caused by the Earth's atmosphere distorting the light from a star.

[16]The variation in magnitude of Algol is so apparent that very early observers noted it. The name *Algol* comes from the Arabic for "the demon's head."

68.75 hours. We can get much more accurate estimates using sine regression, as we will see in Example 6.8.) Hence the time span for a single orbit is about 60 hours.

Solution to Part 3: Note from the table that the magnitude starts at the median level of 2.8 and is increasing there, so no horizontal shift is necessary. The model we need is

$$M = \text{Amplitude} \times \sin\left(\frac{360}{\text{Period}}\right) + \text{Median level}$$

$$= 0.6\sin\left(\frac{360}{60}t\right) + 2.8 = 0.6\sin(6t) + 2.8 \,.$$

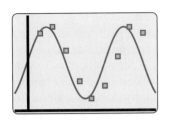

FIGURE 6.35 Data and sine model for Algol

The graph of the model along with the data points is shown in Figure 6.35. To find when the magnitude will first be 2.4, we use the crossing-graphs method. In Figure 6.36 we have added the horizontal line at 2.4 to the picture and calculated the intersection. We see that a magnitude of 2.4 is first reached at about 37 hours. Note that the data in the table indicate that a magnitude of 2.4 will first be reached shortly after $t = 40$, when the magnitude is 2.5 and decreasing. This inaccuracy reflects our rough approximations of period and amplitude. ∎

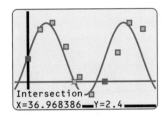

FIGURE 6.36 Finding when the magnitude is 2.4

In Example 6.6 the required model started at the median level. If that is not the case, we need to make additional adjustments. To show how to do this, we return to the weevil and wasp populations of Example 6.4.

EXAMPLE 6.7 Azuki Bean Weevils and Braconid Wasps Revisited

During the same time frame as that examined in Example 6.4, the weevil population was estimated. The results are recorded in the table below.

L1	L2	L3	3
0	3100	▬▬▬	
1	2280		
2	800		
3	700		
4	2130		
5	3090		
6	2290		
L3(1)=			

FIGURE 6.37 Data for azuki weevil population

Generation	0	1	2	3	4	5	6
Population	3100	2280	800	700	2130	3090	2290

1. Plot the data points and explain why the weevil population should be expected to be periodic. How would you expect the periods of weevil and wasp populations to be related?

2. Use the data to estimate the median level and the amplitude.

3. Make a model of the weevil population, assuming that the initial value is at the median level and the population is increasing then.

4. Adjust the model from part 3 by shifting horizontally so that it fits the given data.

5. On the same screen, plot the given data and the graph of the sine wave model. What would you expect the weevil population to be in the 12th generation?

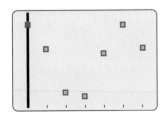

FIGURE 6.38 Plot of weevil population data

Solution to Part 1: The correctly entered data are shown in Figure 6.37, and the data are plotted in Figure 6.38. The predator, the braconid wasp, has periodic population growth. When the wasp population is large, we expect the population of its prey, the weevil, to decline, and when the wasp population is small, we expect the weevil population to increase. Thus we expect the weevil population to vary with the same period, 5 generations, as that of the wasp, but slightly out of phase.

Solution to Part 2: The given data suggest that the weevil population varies from a minimum of about 700 to a maximum of 3100. (We would be much more confident of the accuracy of these values if more data points were given.) Thus the median level is halfway between 700 and 3100:

$$\text{Median level} = \frac{700 + 3100}{2} = 1900 \,.$$

The weevil population varies from 700, which is 1200 below the median, to 3100, which is 1200 above the median. Thus the amplitude is 1200.

Solution to Part 3: The model we need is

$$\text{Amplitude} \times \sin\left(\frac{360}{\text{Period}}t\right) + \text{Median level} = 1200 \sin\left(\frac{360}{5}t\right) + 1900$$

$$= 1200 \sin(72t) + 1900 \,.$$

Here t is the generation number adjusted so that the initial value is the median level and the population is increasing then.

Solution to Part 4: To get the appropriate horizontal shift, we first graph the model from part 3 along with the data, as shown in Figure 6.39. We want to add to this the horizontal line through any of the data points. In Figure 6.40 we have added the horizontal line through the first data point and calculated the appropriate intersection. We see that the graph needs to be shifted to the *left* by 1.25 generations. This is a horizontal shift of -1.25. Thus the required model is

$$1200 \sin(72(t + 1.25)) + 1900 \,.$$

Solution to Part 5: In Figure 6.41 we plot the data along with the sine wave model. Note that the model gives good agreement with the data. In Figure 6.42 we have widened the horizontal range through the 15th generation and then placed the cursor at X=12. We see that the model predicts a population of approximately 929 weevils.

■ ■ ■

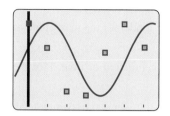

FIGURE 6.39 Weevil data, together with model having initial value at median level

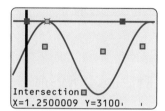

FIGURE 6.40 Finding the required shift

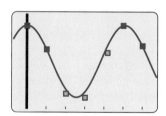

FIGURE 6.41 Data and sine wave model

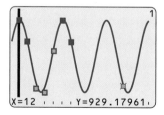

FIGURE 6.42 The predicted population for the 12th generation

Sine Regression

A more accurate way of making sine wave models is to use *sine regression*, which is a numerical technique for generating a sine wave from data. The actual calculations necessary to do this are quite complicated, and we will not show them here. Fortunately, sine regression is implemented on many modern calculators[17] and is available for our use. We look back at the binary star Algol to demonstrate the procedure.

---

[17]TI-83 users should be warned that this calculator always gives sine regression parameters in terms of radians. If you are using degrees, it will be necessary to convert radians to degrees.

EXAMPLE 6.8   **Modeling Binary Stars Using Sine Regression**

We use the data from Example 6.6 to make a more accurate model for the magnitude of Algol.

1. Use sine regression to make a model of magnitude as a function of time. Plot the graph of the data along with the model.

2. The period of the sine model is the time required for the dimmer star to complete an orbit of the brighter one. Find the time span for a single orbit.

3. What is the median level of the magnitude for Algol, and what are the largest and smallest magnitudes shown by Algol?

Solution to Part 1:   We use the data shown in Figure 6.33. If M denotes the magnitude, then sine regression yields the model

$$M = 0.61 \sin(5.33t) + 2.80.$$

The graph of the data along with the model is shown in Figure 6.43. Note that we have a much closer fit here than in Figure 6.35.

Solution to Part 2:   We can get the period either by direct calculation or by looking at the graph and observing the time between successive peaks. Direct calculation gives

$$\text{Period} = \frac{360}{5.33} = 67.5 \text{ hours}.$$

Thus the time span for a single orbit is about 67.5 hours.

Solution to Part 3:   The amplitude is the coefficient of the sine function, 0.61, and the median level is the vertical shift term, 2.80. Thus the magnitude of Algol varies from a high of $2.80 + 0.61 = 3.41$ to a low of $2.80 - 0.61 = 2.19$. ■ ■ ■

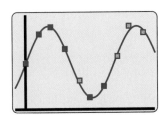

FIGURE 6.43 Modeling magnitude as a function of time

6.2   SKILL BUILDING EXERCISES

S-1. **Finding period, median level, and amplitude from a formula:** Make a graph of

$$13.6\sin(644t - 3) + 17.7.$$

Find the period, median level, and amplitude of this wave.

S-2. **Finding parameters for a cosine wave:** Make a graph of $4.4\cos(54t + 5) - 4.1$. Find the period, median level, and amplitude.

S-3. **A shifted sine wave:** Make a sine wave with the following properties.

• The period is 7.7 degrees.

• The amplitude is 3.7.

• The median level is 5.1.

• The initial value of the sine wave is 6.4, and the wave is decreasing initially.

Make a graph of the sine wave you produced.

S-4. **Making a shifted sine wave:** Make a sine wave with the following properties.

• The period is 3 degrees.

• The amplitude is 4.

• The median level is 2.

• The initial value is 5, and the wave is increasing initially.

S-5. **Making a shifted cosine wave:** Make a cosine wave with the following properties.

• The period is 6 degrees.

• The amplitude is 2.

• The median level is −4.

• The initial value is −2, and the wave is decreasing initially.

S-6. **Finding frequency:** The sound wave for a certain musical note is modeled by

$$4.4\sin(72{,}444t),$$

with t measured in seconds. What is the frequency of this note? Is it a higher or a lower note than middle C (which has a frequency of 262 cycles per second)?

6.2   EXERCISES

1. **Notes played at different times:** A middle C note (frequency 262 cycles per second) is struck so that the sound wave has an amplitude of 1. One-half second later, the D note above middle C (frequency 294) is struck, producing a sound wave with amplitude 2. Make a model of the resulting sound wave. Plot its graph using a horizontal span of 2 to 2.5. Can you detect beats in this graph?

2. **A relationship between the sine and cosine functions:** In this exercise we find a simple relationship between the sine and cosine functions.

   a. Graph $\sin x$ and $\cos x$ on the horizontal span from 0 to 720 degrees. Do the graphs appear to have the same basic shape?

   b. Determine how far to the left the sine wave needs to be shifted in order to make it match the cosine wave. Check your answer by graphing the cosine

function and the shifted sine function on the same screen.

   c. Your work in parts a and b indicates that a specific formula relates the sine and cosine functions. Find such a formula.

   d. Use the unit-circle definition of the sine and cosine functions to explain why this shifting works.

   Note: The fact that we can get the cosine wave by shifting the sine wave to the left explains why we have used sine waves almost exclusively up to now. For the examples we have been considering, the cosine function is not necessary.

3. **β Canis Majoris stars:** The β Canis Majoris stars, like Cepheid variables, pulsate, but the mechanism that causes the pulsation is not well understood. Typically, these stars have a short period and a very high surface temperature. A certain β Canis Majoris star

varies in apparent magnitude from 6.6 to 7.5 and has a period of 5.4 hours.

a. What is the median level of the apparent magnitude?

b. Make a model of apparent magnitude as a function of time, assuming that initially the apparent magnitude is at median level and is increasing.

c. Adjust the model from part b so that it begins when the magnitude is 7.1 and is decreasing.

d. In the model from part c, when will the magnitude first reach a minimum?

4. **Temperature variation:** In this exercise we examine the outdoor temperature, in degrees Fahrenheit, as a function of the time t, in hours since midnight, for two different summer days in the same week and at the same location.

a. For Monday the temperature can be modeled by

$$6\sin(15(t + 11)) + 74.$$

Make a graph of the temperature over a 24-hour time span. What are the period, amplitude, median level, and horizontal shift for this function?

b. For Wednesday the temperature can be modeled by

$$14\sin(15(t - 8)) + 74.$$

Repeat part a using this function.

c. One of the two days was clear, and the other was cloudy. Use your answers to parts a and b to determine which was which.

5. **A predator population:** A certain predator population varies periodically. The period is about 5 years. The median level is 950, and the amplitude is 130.

a. Make a sine model for the predator population if initially it is at 900 and is decreasing.

b. When does the population first reach a minimum level?

6. **Mites:** In one laboratory experiment, the population of a predatory mite and that of its prey (another kind of mite) both had a period of about 80 days. The population of the predatory mite varied from a low of 2 to a high of 42.

a. Find the median level and amplitude of the population of the predatory mite.

b. Say the population of the predatory mite was initially at the median level and was *decreasing*. Make a sine model of the population versus the time in days since the start of the experiment.

7. **Moon:** Let W be the fraction of the width of the moon illuminated (as we observe it) as a function of time in days. For example, when the moon is full, we have $W = 1$.

a. Find the median level and amplitude of W.

b. The time from one full moon to the next is about 29.5 days. Make a sine model of W, starting at a full moon. How much of the width of the moon is illuminated 10 days after a full moon?

8. **Biorhythm:** Scientists have studied the daily biorhythm of the transport of a certain amino acid in the liver of rats.[18] The level varies from a low of 1.5 units to a high of 2.5 units over a period of a day.

a. Find the median level and amplitude.

b. Say the level is at a minimum at 10:00 A.M. Make a sine model of the level versus the time t, in hours, since 10:00 A.M. When after 10:00 A.M. is a level of 2.2 units first attained?

9. **Temperature and depth:** At a certain location the temperature, in degrees Fahrenheit, on the surface of the Earth varies over a typical year according to the formula

$$25\sin(30t) + 55.$$

Here t is the time in months since the start of April.

a. Find the median level, amplitude, and horizontal shift of the temperature on the surface.

b. At the same location but at a depth of 5 feet the temperature, in degrees Fahrenheit, is given by the formula

$$10.16\sin(30(t - 1.7)) + 55.$$

Find the median level, amplitude, and horizontal shift of the temperature at a depth of 5 feet.

---

[18]Based on the work of Ehrhardt, as described by Edward Batschelet in *Circular Statistics in Biology* (London-New York: Academic Press, 1981).

c. Compare your answers to parts a and b. Explain why the results of your comparison make sense physically.

10. **More on temperature and depth:** *This is a continuation of Exercise 9.* In general, the temperature at a point below the surface of the Earth depends on the time of year and the depth below the surface. At a certain location, this temperature Y, in degrees Fahrenheit, is given by the formula

$$Y = 25e^{-0.18h} \sin \left(30(t - 0.34h)\right) + 55 \,.$$

Here t is the time in months since the start of April, and h is the depth in feet. (Using $h = 0$ and $h = 5$ gives the two formulas in Exercise 9.) This formula is derived from a simple model of heat conduction from the surface of the Earth.

a. At a fixed depth of h feet, the formula for Y gives a sine model. Find the median level, amplitude, and horizontal shift for this model. Discuss how each of these depends on h.

b. How much temperature variation is there in a cave at a depth of 20 feet below the surface? How does the temperature there compare with the median level for the temperature on the surface?

c. What is the smallest depth at which it is winter when on the surface it is summer (that is, the smallest depth at which the seasons are reversed as far as the temperature is concerned)? *Suggestion:* First determine what horizontal shift of the function $\sin x$ gives $-\sin x$.

d. The formulas given so far are concerned with temperature variation over a period of a year. If instead we consider variation over a certain day, then we get the formula

$$D = 15e^{-3.44h} \sin \left(15(s - 13.14h)\right) + 65 \,.$$

Here s is the time in hours since noon and h is the depth in feet.

i. At a depth of 1 foot, is there much temperature variation over a day?

ii. In analogy to part c, determine the smallest depth at which the temperature is completely out of phase with the surface temperature.

11. **Modeling a predator population from data:** A certain predator population is thought to vary periodically. Wildlife managers determined the number

N of animals at several times and recorded the data in the table below.

t (years)	N
0	1060
1	940
2	680
3	1070
4	920
5	690
6	1050

a. Estimate the maximum and minimum levels of the population.

b. Estimate the median level of the population.

c. Estimate the period of the population.

d. Use your estimates from parts a, b, and c to make a sine model for population.

12. **More on modeling predator population from data:** *This is a continuation of Exercise 11.*

a. Use sine regression to make a model of the population.

b. What is the median level of the population?

c. What is the period of the population?

13. **Alternating current:** The voltage v of an alternating current changes with time and can be represented by a sine wave. For a certain current, voltage measurements were taken and recorded in the table below.

t (seconds)	v (volts)
0	42
0.003	−45
0.006	−105
0.009	−110
0.012	−54
0.015	33
0.018	102

a. Make a sine model of voltage using estimation.

b. Make a sine model of voltage using sine regression.

c. Use your answer in part b to determine the maximum voltage reached.

14. **Sunspots:** Sunspot activity is a periodic phenomenon. The numbers of sunspots observed for certain years from 1970 through 1982 are recorded in the table below.

t (years since 1970)	N (number per year)
0	120
2	110
4	88
6	52
8	10
10	68
12	115

a. Make a sine model for the number of sunspots per year occurring *t* years after 1970.

b. What is the period of the sunspot cycle? *Note:* Astronomers accept the period as being about 11 years. You will get a slightly different answer because of estimations in the data and because of the relatively few data points given.

c. What is the median level of sunspots per year?

15. **Mira:** Mira is a long-period variable star whose variation was first recorded in 1596 by the Dutch astronomer David Fabricius. The table below gives the apparent magnitude *m* of Mira *t* days after observation began.

t (days)	m
0	7.85
25	8.84
50	9.29
75	9.09
100	8.30
125	7.09
150	5.72

a. Use regression to make a sine model of the magnitude of Mira.

b. What is the period?

c. If Mira showed a magnitude of 7.3 today, what is the shortest number of days after today when you would see it have a magnitude of 7.3 again? *Note:* The answer is not the period. In fact, there are two possible answers.

6.3 *Right Triangle Trigonometry*

Right triangles can be used to give alternative definitions of trigonometric functions, and many important applications depend on this point of view.

Right Triangle Definition of Trigonometric Functions

A *right triangle* is a triangle with a right angle—that is, an angle of 90 degrees. Some important applications of mathematics, such as surveying, involve right triangles, the sine and cosine functions, and four other trigonometric functions. Of these four new functions, we shall look only at the tangent function. The field of *trigonometry* was actually developed for the study of how the sides and angles of right triangles are related.

The standard labels associated with right triangles are shown in Figure 6.44. To indicate that an angle is a right angle, it is customary to mark the angle as we have done in that figure. Using one of the *acute* (less than 90 degrees) angles as a reference point, we have common names given to the sides of a right triangle: opposite, adjacent, and hypotenuse. In Figure 6.44 we have labeled one of the acute angles θ. The *opposite* side is the side opposite the angle. The *adjacent* side is the side that touches both the reference angle θ and the right angle. The *hypotenuse* is the side that does not touch the right angle.

The sine of the angle θ can be calculated from the opposite side of the triangle and the hypotenuse:

$$\sin \theta = \frac{\text{Opposite}}{\text{Hypotenuse}}.$$

The cosine is given by

$$\cos \theta = \frac{\text{Adjacent}}{\text{Hypotenuse}}.$$

By considering the case when the hypotenuse is 1 unit long, we see that these definitions are consistent with the definitions we gave in Section 6.1 using the unit circle.

We define the tangent function using

$$\tan \theta = \frac{\text{Opposite}}{\text{Adjacent}}.$$

In Figure 6.45 we have used the calculator to make a graph of the cosine 6.3 function, and in Figure 6.46 we have used the calculator to make a graph of the tangent 6.4 function. (Both have a horizontal span of 0 to 720 degrees and a vertical span of -2

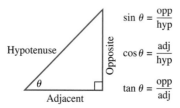

FIGURE 6.44 Right triangles and periodic functions

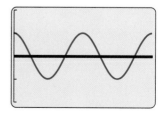

FIGURE 6.45 The cosine function

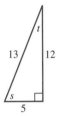

FIGURE 6.46 The tangent function

to 2.) We see that the cosine function looks much like the sine function except that the wave is shifted. The tangent function, shown in Figure 6.46, is a periodic function of an entirely different shape.

We can use these definitions to calculate trigonometric functions of angles if the sides of the triangle are known. Consider, for example, the triangle in Figure 6.47. We have

$$\sin s = \frac{\text{Opposite}}{\text{Hypotenuse}} = \frac{12}{13} = 0.92$$

$$\tan s = \frac{\text{Opposite}}{\text{Adjacent}} = \frac{12}{5} = 2.4.$$

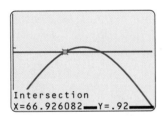

FIGURE 6.47 Calculating trigonometric functions

You should verify the following:

$$\cos s = 0.38$$
$$\sin t = 0.38$$
$$\cos t = 0.92$$
$$\tan t = 0.42.$$

We can use the fact that we know the sine (or cosine or tangent) of the angle s to find the angle. We need to solve the equation

$$\sin s = 0.92.$$

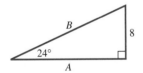

FIGURE 6.48 Finding the angle

We can do that using the crossing-graphs method shown in Figure 6.48. We note in Figure 6.48 that the horizontal line of height 0.92 crosses the graph of the sine function more than once. Since the angle we are interested in is an acute angle (between 0 and 90 degrees), we want the first crossing point. The prompt at the bottom of the screen shows that $s = 66.93$ degrees, or about 67 degrees.

We can also use trigonometry to find lengths of sides of a right triangle. Consider, for example, the right triangle in Figure 6.49. We wish to find the lengths of sides A and B. The tangent of the 24-degree angle relates the opposite side of the reference angle to the adjacent side. This will help us find the length of A. We find

$$\tan 24° = \frac{8}{A}$$

$$A = \frac{8}{\tan 24°} = 17.97.$$

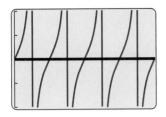

FIGURE 6.49 A right triangle with sides to be calculated

To find B, recall that the sine of the 24-degree angle relates the opposite side of the triangle to the hypotenuse. We find

$$\sin 24° = \frac{8}{B}$$

$$B = \frac{8}{\sin 24°} = 19.67.$$

These same ideas can be used in practical settings to find unknown distances and angles, as illustrated in the following example.

EXAMPLE 6.9   Height of a Building

A surveyor stands 300 horizontal feet from a building and aims a transit at the top of the building as shown in Figure 6.50. The transit measures an angle of 21 degrees from the horizontal.

1. How much taller is the building than the transit?

2. How long would a straight cable that reached from the transit directly to the top of the building need to be?

Solution to Part 1:   In this case, we are interested in the height of the building above the transit, and that is the side opposite the reference angle. We know the length of the adjacent side, so we should use the tangent function, since it deals with those two sides:

$$\tan 21° = \frac{\text{Opposite}}{\text{Adjacent}} = \frac{\text{Opposite}}{300}.$$

We use x to denote the length of the opposite side. The equation $\tan 21° = x/300$ is a linear equation and is easily solved by hand calculation. We find

$$\tan 21° = \frac{x}{300}$$

$$300 \tan 21° = x$$

$$115.16 = x.$$

We conclude that the building is 155.16 feet taller than the transit.

Solution to Part 2:   The distance from the transit directly to the top of the building is the hypotenuse of the triangle in Figure 6.50. We will show two ways[19] to proceed.

Solution using the cosine function: We know that the adjacent side is 300 feet, and we want to know the hypotenuse. The cosine function involves these two quantities:

$$\cos 21° = \frac{\text{Adjacent}}{\text{Hypotenuse}} = \frac{300}{\text{Hypotenuse}}.$$

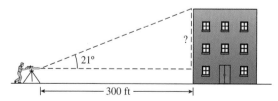

FIGURE 6.50 Measuring with a transit

---

[19] Since we already know two sides of this right triangle, the third side could also be calculated from the Pythagorean theorem, which says that the square of the hypotenuse is the sum of the squares of the other two sides.

We use x to denote the length of the hypotenuse. As before, we can solve the equation $\cos 21° = 300/x$ by hand calculation:

$$\cos 21° = \frac{300}{x}$$

$$x \cos 21° = 300$$

$$x = \frac{300}{\cos 21°}$$

$$x = 321.34 .$$

We conclude that a straight cable reaching from the transit to the top of the building would need to be 321.34 feet long.

Solution using the sine function: Once again we want the height, but this time we use the length of the opposite side (115.16 feet) that we calculated in part 1:

$$\sin 21° = \frac{\text{Opposite}}{\text{Hypotenuse}} = \frac{115.16}{\text{Hypotenuse}} .$$

Now we solve for x, the length of the hypotenuse:

$$\sin 21° = \frac{115.16}{x}$$

$$x \sin 21° = 115.16$$

$$x = \frac{115.16}{\sin 21°}$$

$$x = 321.35 .$$

The small difference between the two answers is due to rounding in part 1.

■ ■ ■

A N O T H E R   L O O K

Trigonometric Identities

There are many useful relationships known as *trigonometric identities* that hold among the trigonometric functions. One of the most important of all trigonometric identities comes from the familiar *Pythagorean theorem*. For the right triangle shown in Figure 6.51, the Pythagorean theorem says that $a^2 + b^2 = c^2$. Dividing both sides of this equation by c^2 gives

$$\left(\frac{a}{c}\right)^2 + \left(\frac{b}{c}\right)^2 = 1 .$$

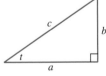

FIGURE 6.51 A right triangle

But $\cos t = a/c$ and $\sin t = b/c$. Substituting these values in the equation above gives the most important trigonometric identity of all:

$$\sin^2 \theta + \cos^2 \theta = 1 .$$

Suppose, for example, we know that $\sin \theta = 0.70$ and that θ is between 0 and 90 degrees. We can find $\cos \theta$ as follows:

$$0.7^2 + \cos^2 \theta = 1$$
$$\cos^2 \theta = 1 - 0.7^2 = 0.51$$
$$\cos \theta = \pm\sqrt{0.51} = \pm 0.71 .$$

Since θ lies between 0 and 90 degrees, we know that its cosine is positive. Thus $\cos \theta = 0.71$.

Another important type of identity is the *double-angle formula*. Let's calculate $\sin(2t)$ in terms of $\sin t$ and $\cos t$. In Figure 6.52, $\triangle AED$ is a right triangle with $\angle DAE° = t$ and $|AD| = 1$. (The vertical bars denote the length of a segment.) Let $|DE| = q$. Now $\triangle AEB$ is a copy of $\triangle AED$, and the segment DC is constructed so that $\angle DCB$ is a right angle. Let $|DC| = r$.

We are prepared to do some calculations using Figure 6.52. Now $\triangle AED$ is a right triangle, so we can use it to compute $\sin t$:

$$\sin t = \frac{q}{1} = q .$$

Further, $\triangle AEB$ is a copy of $\triangle AED$, so $\angle EAB°$ is also t. Thus $\angle DAB° = 2t$. Referring to Figure 6.53, we see that we can use $\triangle ACD$ to calculate $\sin(2t)$:

$$\sin(2t) = \frac{r}{1} = r .$$

Since $\triangle AEB$ is a right triangle, and the angle sum of any triangle is 180 degrees, $\angle ABE° = 90° - t$. But also $\triangle DCB$ is a right triangle, and once again its angle sum is 180 degrees. Hence $\angle CDB° = t$. Thus, referring to Figure 6.54, we can use $\triangle DCB$ to calculate $\cos t$:

$$\cos t = \frac{r}{2q} .$$

Putting together our three displayed formulas, we have

$$\sin(2t) = r = 2q\frac{r}{2q} = 2\sin t \cos t .$$

This is the double-angle formula for the sine function. A similar formula holds for the cosine:

$$\cos(2t) = 1 - 2\sin^2 t .$$

We can use the trigonometric identities we know to find other identities. For example, suppose that x is between 0 and 360 degrees. Let's find the *half-angle*

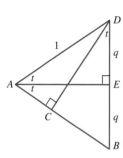

FIGURE 6.52 Preparing to calculate sin(2t)

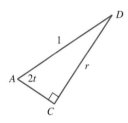

FIGURE 6.53 Choosing a triangle to get sin(2t)

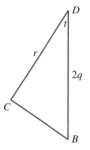

FIGURE 6.54 Choosing a triangle to get cos t

formula for the sine function. We start by putting $x/2$ in place of t in the double-angle formula for the cosine:

$$\cos\left(2\frac{x}{2}\right) = 1 - 2\sin^2\left(\frac{x}{2}\right)$$

$$\cos x = 1 - 2\sin^2\left(\frac{x}{2}\right)$$

$$2\sin^2\left(\frac{x}{2}\right) = 1 - \cos x$$

$$\sin^2\left(\frac{x}{2}\right) = \frac{1 - \cos x}{2}$$

$$\sin\left(\frac{x}{2}\right) = \pm\sqrt{\frac{1 - \cos x}{2}}.$$

Since x is between 0 and 360 degrees, we know that $\sin(x/2)$ is positive. Thus

$$\sin\left(\frac{x}{2}\right) = \sqrt{\frac{1 - \cos x}{2}}.$$

This is the half-angle formula for the sine function.

Enrichment Exercises

E-1. Getting sine from cosine: Suppose we know that $\sin t$ is positive and that $\cos t = 0.77$. Find $\sin t$.

E-2. Getting cosine from sine: Suppose we know that $\cos t$ is negative and that $\sin t = 0.8$. Find $\cos t$.

E-3. Calculating sine of four times an angle: Suppose that $\sin t = 0.2$ and that we know $\cos t$ is positive. Calculate $\sin(4t)$.

E-4. Another form for the cosine double-angle formula: Show that $\cos(2t) = 2\cos^2 t - 1$. *Suggestion:* Start with the double-angle formula for the cosine, and replace $\sin^2 t$ by $1 - \cos^2 t$.

E-5. Half-angle formula for the cosine: Show that

$$\cos\left(\frac{x}{2}\right) = \pm\sqrt{\frac{\cos x + 1}{2}}.$$

E-6. The secant function: The secant function is defined as follows:

$$\sec t = \frac{1}{\cos t}.$$

a. Show that $\tan t = \sin t / \cos t$. *Suggestion:* Look at the right triangle definitions of the three functions (in terms of opposite, adjacent, and hypotenuse), and calculate the quotient of $\sin t$ and $\cos t$.

b. Show that $\tan^2 t = \sec^2 t - 1$. *Suggestion:* Start with $\sin^2 t + \cos^2 t = 1$, and divide both sides by $\cos^2 t$.

6.3   SKILL BUILDING EXERCISES

S-1. **Right triangle trigonometry:** For an acute angle θ of a certain right triangle, the adjacent side has length 8, the opposite side has length 15, and the hypotenuse has length 17. Find $\sin\theta$, $\cos\theta$, and $\tan\theta$.

S-2. **Calculating height:** A man sits 331 horizontal feet from the base of a wall (see Figure 6.55). He must incline his eyes at an angle of 16.2 degrees to look at the top of the wall. How tall is the wall?

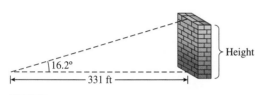

➤ Height

16.2°

331 ft

FIGURE 6.55

S-3. **Calculating height:** A man sits 222 horizontal feet from the base of a wall. He must incline his eyes at an angle of 6.4 degrees to look at the top of the wall. How tall is the wall?

S-4. **Calculating an angle:** A man sits 270 horizontal feet from the base of a wall that is 76 feet high. At what angle must he incline his eyes in order to look at the top of the wall?

S-5. **Calculating an angle:** A wall is 38 feet high. A man sits on the ground and finds that the distance from himself to the top of the wall is 83 feet. At what angle must he incline his eyes in order to look at the top of the wall?

S-6. **Calculating distance:** *This is a continuation of Exercise S-5.* What is the horizontal distance of the man from the wall?

S-7. **Calculating distance:** A man sits 130 horizontal feet from the base of a wall. He must incline his eyes at an angle of 13 degrees to look at the top of the wall. What is the distance from the man directly to the top of the wall?

S-8. **Calculating distance:** A man sits 18 horizontal feet from the base of a wall. He must incline his eyes at an angle of 21 degrees to look at the top of the wall. What is the distance from the man directly to the top of the wall?

6.3   EXERCISES

1. **The 3, 4, 5 right triangle:** Elementary geometry can be used to show that any triangle with sides 3, 4, and 5 is a right triangle. This is known as the 3, 4, 5 right triangle. The ancient Egyptians knew this fact and used it to make square corners for their building projects.

   a. Find the sine of each of the two acute angles in a 3, 4, 5 right triangle.

   b. Use your work in part a to find the acute angles in a 3, 4, 5 right triangle.

2. **Pythagorean triples:** *Pythagorean triples* are triples of integers that are the lengths of the sides of a right triangle. For example, as noted in Exercise 1,

the triple $(3, 4, 5)$ is a Pythagorean triple. Another characterization of a Pythagorean triple is a triple (a, b, c) such that $a^2 + b^2 = c^2$.

   a. Verify the following algebraic identity.

$$(p^2 - q^2)^2 + (2pq)^2 = (p^2 + q^2)^2$$

   b. Explain how to use the identity in part a to generate as many Pythagorean triples as you wish.

3. **A building:** You are facing a building that is 150 feet taller than your transit, and you must elevate your transit 20 degrees to view the top of the building. What is the distance in horizontal feet between your transit and the building?   →

4. **The width of a river:** You stand on the north bank of a river and look due south at a tree on the opposite bank (see Figure 6.56). Your helper on the opposite bank measures 35 yards due east to a second tree. You must rotate your transit through an angle of 12 degrees to point toward the second tree. How wide is the river?

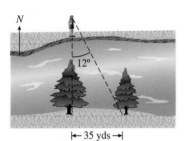

FIGURE 6.56

5. **A cannon:** If you elevate a certain cannon so that it makes an angle of t degrees with the ground, then the cannonball will strike the ground $(m^2 \sin(2t))/g$ feet downrange, where $g = 32$ feet per second per second is acceleration due to gravity, and m is the muzzle velocity in feet per second (see Figure 6.57). If the muzzle velocity is 300 feet per second, what angle would you use to make the cannonball land 1000 feet downrange?

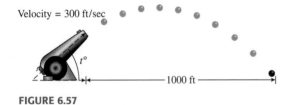

FIGURE 6.57

6. **Dallas to Fort Smith:** Dallas is 190 miles due south of Oklahoma City, and Fort Smith is 140 miles due east of Oklahoma City. An airplane flies on a direct trip from Dallas to Fort Smith (see Figure 6.58).

   a. What is the tangent of the angle that the flight path makes with Interstate 35, which runs due north from Dallas to Oklahoma City?

   b. What is the angle that the flight path makes with Interstate 35?

   c. How far does the airplane fly?

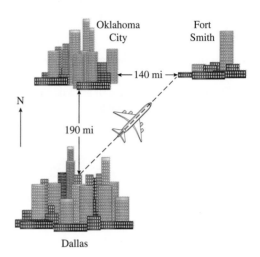

FIGURE 6.58

7. **Intensity of sunlight:** When incident rays of sunlight form an angle θ with a leaf, the intensity of sunlight is reduced by a factor of $\sin \theta$, assuming that θ is between 0 and 90 degrees. (See Figure 6.59.)

   a. By what factor is the intensity reduced if the angle formed is $35°$?

   b. What is the angle θ if the intensity is reduced by a factor of 0.3?

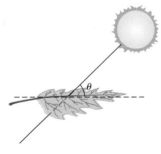

FIGURE 6.59

8. **Grasping prey:** An ecologist used the diagram in Figure 6.60 to estimate the diameter of a circular prey that would be optimal for the grasping claws of

a praying mantid.[20] Find a formula for the diameter $|BC|$ of the circle in terms of the length $|AC|$ and the angle θ. (The vertical bars denote the length of a segment.)

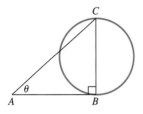

FIGURE 6.60

9. **Dispersal method:** When there is moisture on the ground, microorganisms inhabit the thin layer of water on leaves.[21] A *monolayer* containing these organisms is formed on the top, and (because of the spreading pressure of this monolayer) the floating organisms spread *upward* when the surface is tilted. This is a dispersal method onto newly fallen leaves, for example. In the diagram in Figure 6.61, the horizontal segment represents the ground, and the angle θ measures the amount by which the surface is tilted. (We assume that θ is less than 90°.) The length d is the distance the organisms move, and the length v is the change in the elevation.

a. Find a formula giving v in terms of d and θ.

b. Assume that the distance the organisms move is 30 centimeters. Find the change in elevation if the angle θ is 15° and if the angle θ is 30°.

c. Find the angle θ if the distance the organisms move is 30 centimeters and the change in elevation is 20 centimeters.

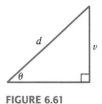

FIGURE 6.61

10. **Jumping locust:** If an animal jumps at an angle θ to the horizontal with initial velocity m, then the horizontal distance d that it will travel is given by

$$d = \frac{m^2 \sin 2\theta}{g}.$$

Here we measure d in meters and m in meters per second, and g is the acceleration due to gravity (about 9.8 meters per second per second). This model ignores air resistance.

a. Locusts typically jump at an angle $\theta = 55°$. If a locust jumps a horizontal distance of 0.8 meter, what is its initial velocity? (See Figure 6.62.)

b. Find the angle θ if the distance that an animal jumps is 1 meter and the initial velocity is 3.2 meters per second. (Assume that θ is between 0 and 45 degrees.)

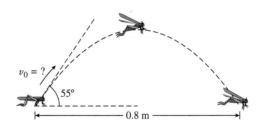

FIGURE 6.62

---

[20]Based on the work of C. S. Holling, as described by Eric R. Pianka in *Evolutionary Ecology*, 5th ed. (New York: HarperCollins, 1994).

[21]See the study by Robert J. Bandoni and Richard E. Koske, "Monolayers and microbial dispersal," *Science* **183** (1974), 1079–1081.

Summary

In this chapter we define and use the sine and cosine functions to model periodic behavior. In modeling periodic data, we find it useful to be able to adjust the period, amplitude, median level, and starting point of a sine model. Sine and cosine are defined using a unit circle. Right triangle trigonometry using sine, cosine, and tangent makes it possible to calculate the lengths of sides and measures of angles for right triangles, a topic rich in applications.

Periodic Functions

Periodic functions occur naturally in many applications, such as waves. The *period* of a periodic function is the length of the variable needed for the function to repeat values. Since the function repeats its values over a period, it rises to a maximum level and falls to a minimum level in each period. The *median level* is halfway between the minimum level and the maximum level. The *amplitude* is the maximum height of the function above or below the median level.

Sound waves are a useful example of periodic behavior. Sound waves are the graph of pressure as a function of time. In this context, amplitude is the intensity of the sound, and periods (or cycles) per second is the frequency, or pitch.

Sine and Cosine Functions

The simplest periodic functions are the sine and cosine functions. These are defined in terms of a unit circle centered at the origin. If we think of a point on a unit circle in terms of the central angle it makes with $(1, 0)$, the sine and cosine of the angle are the vertical and horizontal coordinates of the point, respectively. This allows a definition of sine and cosine for 0 to 360 degrees that may be extended to any degree by moving around the circle more than once, if needed, or in the opposite direction.

Using the unit-circle definition makes it easy to see that each of sine and cosine is periodic with period 360 degrees and amplitude 1. In order to use sine or cosine as a mathematical model, it is helpful to be able to adjust period and amplitude. Starting from a sine function

$$\sin x \, ,$$

with period 360 degrees and amplitude 1, we can form a new function with prescribed period and amplitude using

$$\text{Amplitude} \times \sin\left(\frac{360}{\text{Period}} x \right).$$

Shifting Sine Models

Although the sine models previously constructed allow for any given period or amplitude, all these models have a median level of zero, start at the origin, and increase from there. In order to allow other median levels, other starting values, and the possibility of decreasing from the starting values, vertical shifts and horizontal (right) shifts are utilized. The most general form of sine model is

$$\text{Amplitude} \times \sin\left(\frac{360}{\text{Period}}(t - \text{Right shift})\right) + \text{Vertical shift}.$$

The median level is equal to the vertical shift, whereas the horizontal (right) shift can be chosen to allow any starting value and increase or decrease from there.

Given data, a useful tool for obtaining a sine model is sine regression.

Trigonometry

Trigonometry is the measurement of the sides and angles of a triangle. The sine function and some of its relatives, the cosine and tangent functions, can be used to measure right triangles, and hence they become quite important in applications to surveying. If θ is an acute angle of a right triangle, then there is a side *adjacent* to θ, a side *opposite* to θ, and the *hypotenuse*, which is opposite to the right angle. The formulas are

$$\sin\theta = \frac{\text{Opposite}}{\text{Hypotenuse}}$$

$$\cos\theta = \frac{\text{Adjacent}}{\text{Hypotenuse}}$$

$$\tan\theta = \frac{\text{Opposite}}{\text{Adjacent}}.$$

For example, if we start 300 horizontal feet from the base of a building and find that our transit must be elevated 21 degrees to point toward the top of the building, we can use the tangent function to figure out how tall the building is. The key is that there is a right triangle involved. The adjacent side is the horizontal distance to the building, and the opposite side is the vertical face of the building:

$$\tan 21° = \frac{\text{Opposite}}{\text{Adjacent}}$$

$$\tan 21° = \frac{\text{Height}}{300}$$

$$300\tan 21° = \text{Height}$$

$$115.16 = \text{Height}.$$

Thus the height above the transit is 115.16 feet.

Chapter

7

Rates of Change

A key idea in analyzing natural phenomena as well as the functions that may describe them is the *rate of change*, and it is a familiar idea from everyday experience. When you are driving your car, the rate of change in your location is *velocity*. Discussions of cars or airplanes commonly involve velocity because it is virtually impossible to convey key ideas about motion without reference to velocity. The rate of change is no less descriptive for other events. If you step on the gas pedal to pass a slowly moving truck, then your velocity itself changes, and the rate of change in velocity is *acceleration*. Rates of change occur in other contexts as well. In fact the idea is pervasive in mathematics, science, engineering, social science, and daily life because it is such a powerful tool for description and analysis.

7.1   *Velocity*

We look at velocity first because it is a familiar rate of change. Consider a rock tossed upward from ground level. As the rock rises and then falls back to Earth, we locate its position as the distance up from the ground. In Figure 7.1 we have sketched a possible graph of distance up versus time. It shows the height of the rock increasing as it rises and then decreasing after it reaches its peak and begins to fall.

Getting Velocity from Directed Distance

The *velocity* of the rock is the rate of change in distance up from the ground. That is, at any point in the flight of the rock, the velocity measures how fast the rock is rising or falling. The rock gets some initial velocity at the moment of the toss, but the effect of gravity makes it slow down as it rises toward its peak. After the rock reaches its peak, gravity causes it to accelerate toward the ground, and its *speed* increases. In everyday language, the terms *speed* and *velocity* are often used interchangeably, but there is an important, if subtle, difference. Speed is always a positive number—the number you might read on the speedometer of your car, for example. But velocity has an additional component; it has a sign (positive or negative) attached that indicates the direction of movement. The key to understanding this is to remember that velocity is the rate of change in *directed distance*. As the rock moves upward, its distance up from the ground

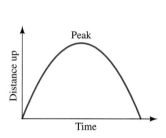

FIGURE 7.1 Distance up of a
rock versus time

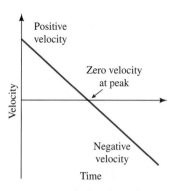

FIGURE 7.2 Velocity of a rock
versus time

is increasing. Thus the rate of change in distance up, the velocity, is positive. But when
the rock starts to fall back to Earth, its distance up from the Earth is decreasing. Thus
the rate of change in distance up, the velocity, is negative.

The relationships between the graph of distance up and the graph of velocity are
crucial to understanding velocity and, indeed, rates of change in general. In Figure 7.2
we have sketched a graph of the velocity versus time for the rock. During the period
when the rock is moving upward, the graph of distance up is increasing. Since distance up is increasing, velocity is positive. This is shown in Figure 7.2 by the fact
that the graph of velocity is above the horizontal axis until the rock reaches the peak
of its flight. We also note that during this period, the graph of velocity is decreasing toward zero, indicating that the rock is slowing as it rises. At the point where
the rock reaches its peak, the graph of velocity crosses the horizontal axis, indicating that the velocity is momentarily zero. When the rock is moving downward, the
graph of distance up is decreasing. During this period, velocity is negative, and this
is shown in Figure 7.2 by the fact that the graph of velocity is below the horizontal
axis. Note that the *speed* of the rock is increasing as it falls, so its velocity (the negative of its speed) is decreasing. Velocity continues to decrease until the rock strikes the
ground.

In Figure 7.2 we have represented the graph of velocity as a straight line. This is
not an apparent consequence of the graph in Figure 7.1, but as we shall discover later
(see Section 7.4 and also Exercise 9 of Section 7.2), it is a consequence of the fact that
(near the surface of the Earth), gravity imparts a constant acceleration. Finally, we note
that the sign of velocity depends on the perspective chosen to measure position. In this
case, we located the rock using its distance up from the surface of the Earth. In Exercise
2 at the end of this section, you will be asked to analyze the velocity of the rock when
the event is viewed from the top of a tall building and the rock's position is considered
as its distance down from the top of the building.

The relationships we observed between the graph of distance up for the rock and the
graph of its velocity are fundamental, and they are true in a general setting. When directed
distance is increasing, velocity is positive. When directed distance is decreasing, velocity
is negative. When directed distance is not changing, even momentarily, velocity is zero.
In particular, when directed distance reaches a peak (maximum) or a valley (minimum),
velocity is zero.

Velocity and Directed Distance: The Fundamental Relationship

1. Velocity is the rate of change in directed distance.

2. **When directed distance is increasing, velocity is positive.** (The graph of velocity is above the horizontal axis.)

3. **When directed distance is decreasing, velocity is negative.** (The graph of velocity is below the horizontal axis.)

4. **When directed distance is not changing, velocity is zero.** (The graph of velocity is on the horizontal axis.)

Constant Velocity Means Linear Directed Distance

In the case of the rock we just looked at, velocity is always changing, but in many situations there are periods when the velocity does not change. A familiar example is that of a car that might accelerate from a yield sign onto a freeway and travel at the same speed for a time before exiting the freeway and parking at its destination. If the car is traveling westward on the freeway, and we locate its position as distance west from the yield sign, a possible graph of its velocity, the rate of change in distance west, is shown in Figure 7.3. We want to consider what this means for the function $L = L(t)$, which gives the location of the car at time t as distance west of the yield sign. Our interest is in determining the nature of L during the period when the car is traveling at a constant velocity on the freeway—say, at 65 miles per hour. During this period, each hour the car moves 65 additional miles. This means that the distance L increases by 65 miles each hour, so it is a linear function with slope 65.

In more general terms, since velocity of the car is the rate of change in L, during this period the rate of change in L is constant. But we know that a function with a constant rate of change is linear, and this is the key observation we wish to make. Whenever velocity, the rate of change in directed distance, is constant, directed distance must be a linear function. Furthermore, the slope of L is the velocity. The linearity of L is reflected

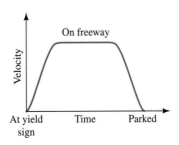

FIGURE 7.3 The velocity of a car traveling on a freeway

in Figure 7.4 by the fact that, for the portion of the graph that represents the time when the car was on the freeway, we drew the graph of directed distance as a straight line.

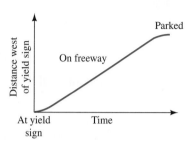

FIGURE 7.4 Constant velocity means linear directed distance

Directed Distance for Constant Velocity

When velocity is constant, the rate of change in directed distance is constant. Thus directed distance is a linear function with slope equal to the constant velocity, so its graph is a straight line.

EXAMPLE 7.1   **From New York to Miami**

An airplane leaves Kennedy Airport in New York and flies to Miami, where it is serviced and receives new passengers before returning to New York. Assume that the trip is uneventful and that after each takeoff the airplane accelerates to its standard cruising speed, which it maintains until it decelerates prior to landing.

1. Describe what the graph of distance south of New York looks like during the period when the airplane is maintaining its standard cruising speed on the way to Miami.

2. Say we locate the airplane in terms of its distance south of New York. Make possible graphs of its distance south of New York versus time and of the velocity of the airplane versus time.

3. Say we locate the airplane in terms of its distance north of Miami. Make possible graphs of its distance north of Miami versus time and of the velocity of the airplane versus time.

Solution to Part 1:   To say that the airplane is maintaining its standard cruising speed means that its velocity is not changing. In other words, the rate of change in distance south is constant. As we noted above, any function with a constant rate of change is linear. We conclude that during this period, the graph of distance south is a straight line whose slope is the standard cruising speed.

Solution to Part 2:   We will make the graph of distance south in several steps, creating a template for solving problems of this type.

Step 1:   Locate and mark the places on the graph where directed distance is zero and where it reaches its extremes. The distance south of New York is zero at the beginning and end of the trip. These points will lie on the horizontal axis, and we have marked and labeled them in Figure 7.5. The graph of distance south will be at its maximum while the airplane is being serviced in Miami. This is also marked and labeled in Figure 7.5.

Step 2:   Label on the graph the regions where directed distance is increasing and the regions where it is decreasing. Directed distance is increasing on the trip from New York to Miami and is decreasing on the return leg of the trip. These regions are labeled in italics in Figure 7.6.

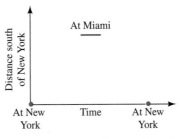

FIGURE 7.5 Zeros and extremities of distance south

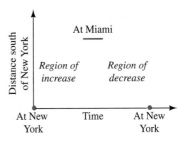

FIGURE 7.6 Regions of increase and decrease for distance south

Step 3:   Complete the graph, incorporating any additional information known about directed distance. In this case, the important additional information is that for most of the trip, the airplane is flying at its standard cruising speed. As we observed in part 1, this means that the graph of distance during these periods is a straight line. Our graph of distance south of New York for the airplane is shown in Figure 7.7. Note that in the completed graph, we have labeled the horizontal and vertical axes as well as other important features. You are encouraged to follow this practice.

Now let's consider velocity for the same trip. To make the graph, we use steps similar to those we used to graph distance south.

Step 1:   Locate and label the points on the graph where the velocity is zero. The velocity is zero when the airplane is on the ground at the beginning and end of the trip and during the period when it is being serviced in Miami. These times must lie on the horizontal axis and are so marked and labeled in Figure 7.8. For the graph of velocity, we put the horizontal axis in the middle since we expect to graph both positive and negative values.

Step 2:   Label the regions where velocity is positive and where it is negative. On the flight from New York to Miami, distance south of New York is increasing, so its rate of change, or velocity, is positive. On the return leg, distance south of New

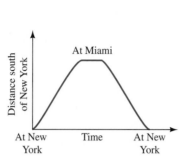

FIGURE 7.7 The completed graph of distance south of New York

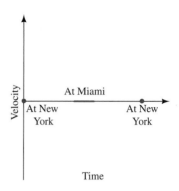

FIGURE 7.8 Times when the velocity is zero

York is decreasing, so its rate of change, or velocity, is negative. These regions are labeled in italics in Figure 7.9. At this point, it is important to check to be sure that the places in Figure 7.6 we marked as *increasing* match the places in Figure 7.9 that we marked as *positive*. Similarly, the *decreasing* labels in Figure 7.6 must match the *negative* labels in Figure 7.9.

Step 3: Complete the graph, incorporating any other known features of the graph. For velocity, the important additional feature is that during most of both legs of the trip, the airplane is maintaining a constant cruising speed. That means that the graph of velocity must be horizontal in these regions. This is shown in our completed graph in Figure 7.10. Note once again that we have labeled the axes as well as other important features of the graph.

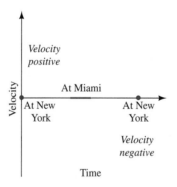

FIGURE 7.9 Labeling regions of positive and negative velocity

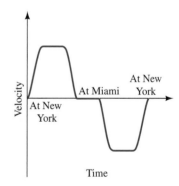

FIGURE 7.10 The completed graph of velocity

We want to emphasize that Figure 7.7 and Figure 7.10 show the fundamental relationship between graphs of directed distance and velocity that we noted earlier. Observe in particular that during the flight from New York to Miami, the graph of directed distance in Figure 7.7 is increasing, and the corresponding part of the graph of velocity in Figure 7.10 is above the horizontal axis. Also, during the return leg of the trip, the graph of directed distance in Figure 7.7 is decreasing, and the corresponding graph of velocity in Figure 7.10 is below the horizontal axis.

Solution to Part 3: We want to look now at the airplane flight from the perspective of a Miami resident. We will follow the same steps as we did in part 2 to get the graphs. If you are waiting at the Miami airport, the distance north to the airplane is a large positive number at the beginning and end of the trip, when the plane is in New York. These points are marked and labeled in Figure 7.11. The distance north is zero while the airplane is in Miami, and this region is also marked in Figure 7.11.

As the plane flies from New York toward Miami, the distance north from Miami decreases, and this is noted in italics in Figure 7.11. On the return leg, distance north increases, and this is also noted in italics in Figure 7.11. Following the notes in Figure 7.11, we complete the graph in Figure 7.12, showing (as before) the regions where the graph is a straight line.

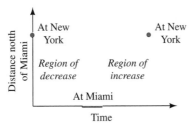

FIGURE 7.11 Interesting points and regions of increase and decrease for distance north

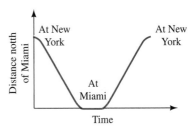

FIGURE 7.12 The completed graph of location from the Miami perspective

We proceed similarly to get the graph of velocity. We first mark the places, New York and Miami, where the velocity is zero. Our change in perspective does not affect this, so the appropriate picture is the same as Figure 7.8. But the change in perspective does affect the sign of velocity. On the first leg of the flight, distance north of Miami is decreasing, so the velocity is negative. This is marked in italics in Figure 7.13, and we note that it corresponds to the *decreasing* label in Figure 7.11. Similarly, on the return trip, distance north of Miami is increasing, so the velocity is positive. This is marked in italics in Figure 7.13, and we note (as before) that it corresponds to the *increasing* label in Figure 7.11. We complete the picture in Figure 7.14, being careful to incorporate constant cruising speed. ■ ■ ■

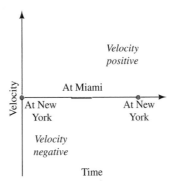

FIGURE 7.13 Labeling velocity features from the Miami perspective

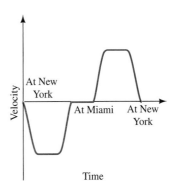

FIGURE 7.14 The completed graph of velocity from the Miami perspective

EXAMPLE 7.2   A Forgotten Wallet

A man leaves home driving west on a straight road. We locate the position of the car as the distance west from home. Suppose the man begins his trip and sets his cruise control, but after a few minutes he notices that he has forgotten his wallet. He returns home, gets his wallet, and then resumes his journey. He figures that he may be late, so he sets the cruise control higher than before. Make graphs of distance west versus time and of velocity versus time for the forgetful man's car.

Solution: We first make the graph showing the location of the car by following the steps outlined in Example 7.1. The distance west from home is zero when the trip starts and when the man is back home retrieving his wallet. In Figure 7.15 we have marked these places, as well as the point where the man remembered the forgotten wallet and the car turned around and headed home. Distance west is increasing when the car is moving away from home. This occurs at the beginning of the trip and also later, after the wallet has been retrieved. The only period when distance west is decreasing is when the man is returning home to get the wallet. These regions are marked in italics in Figure 7.15.

To get the completed graph in Figure 7.16, we made use of the notes from Figure 7.15 and also incorporated the information we have about cruise control settings: On the first leg of the trip, the cruise control is set, so the velocity is constant. This means that for this period, distance west is a linear function, so we have drawn the graph there as a straight line. We have also made the graph for the trip back home a straight line, although that information is not provided in the problem. For the final leg of the trip, we know once again that the graph is a straight line, but we also know a little more. The cruise control is set higher than it was on the first leg of the trip, and therefore the velocity is greater. Thus we must draw a line with a larger slope (a steeper line) than we drew for the first part of the trip.

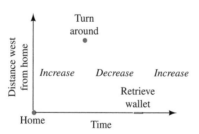

FIGURE 7.15 Important features in the location of a car

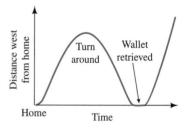

FIGURE 7.16 The completed graph for a car trip

We proceed with a similar analysis for velocity. We know that the velocity is zero when the trip starts and when the wallet is being retrieved. But the velocity is also zero when the car turns around. At this point, just as we saw at the peak of the rock's flight, the rate of change in distance west is momentarily zero. We have marked and labeled all of these places in Figure 7.17.

We can get the right regions for positive and negative velocity from the increasing and decreasing labels in Figure 7.15. Each *increase* label in Figure 7.15

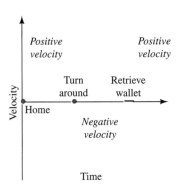

FIGURE 7.17 Important features for velocity

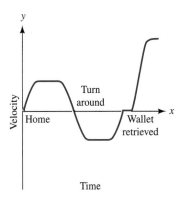

FIGURE 7.18 The completed graph for the velocity of the car

corresponds to a *positive velocity* label in Figure 7.17, and the *decrease* label corresponds to the *negative velocity* label in Figure 7.17. We use the notes in Figure 7.17 to get the completed graph of velocity in Figure 7.18, keeping in mind that since the cruise control is set for most of the trip, the velocity is constant most of the time. This means that the graph of velocity is horizontal in these regions. Finally, we note that since the car went faster on the last part of the trip, the graph of velocity is higher there. ■ ■ ■

When Distance Is Given by a Formula

Many times, directed distance or velocity is given by a formula. For example, if we stand atop a building 30 feet high and toss a rock upward with an initial velocity of 18 feet per second, then elementary physics can be used to show that the distance up $D = D(t)$ from the ground of the rock t seconds after it is tossed is given by

$$D = 30 + 18t - 16t^2 \text{ feet}.$$

We are assuming that when the rock comes back down, it does not hit the top of the building where we are standing but, rather, falls all the way to the ground. Let's begin our analysis by making a graph of D versus t. We can use our everyday experience to choose a window setting. The rock surely won't go over 50 feet high, and it will take only a few seconds for the rock to hit the ground. Thus in Figure 7.19, which shows the flight of the rock, we used 7.1 a horizontal span of $t = 0$ to $t = 5$ and a vertical span of $D = 0$ to $D = 50$.

How long did it take the rock to hit the ground? That happens when the distance up is zero. Thus we want to solve for t in the equation

$$30 + 18t - 16t^2 = 0.$$

We have used the single-graph method 7.2 to do that in Figure 7.20, and we see that it takes about 2.04 seconds for the rock to complete its flight.

We might also want to know how high the rock went and when it reached its peak. In Figure 7.21 we got that information by using the calculator to locate 7.3 the maximum. We see that the rock reached its highest point of 35.06 feet just over half a second after it was tossed.

FIGURE 7.19 A rock tossed upward from the top of a building

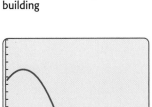

FIGURE 7.20 When the rock strikes the ground

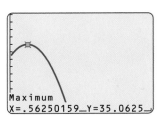

FIGURE 7.21 The peak of the rock's flight

Let's look now at the velocity $V = V(t)$ of the rock. We proceed as in earlier examples, first marking important features and then noting in particular where velocity is positive, where it is negative, and where it is zero. This is shown in Figure 7.22. We use this information to complete the graph of velocity in Figure 7.23.

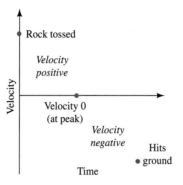

FIGURE 7.22 Interesting features of velocity

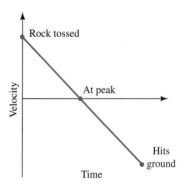

FIGURE 7.23 The graph of velocity versus time

7.1   SKILL BUILDING EXERCISES

S-1.   **Velocity:** What is the rate of change in directed distance?

S-2.   **Sign of velocity:** When directed distance is decreasing, is velocity positive or negative? What is the velocity when directed distance is not changing?

S-3.   **Sign of velocity:** When the graph of directed distance is decreasing, is the graph of velocity above or below the horizontal axis?

S-4.   **Constant velocity:** When velocity is constant, what kind of function is directed distance?

S-5.   **Constant velocity:** When the graph of directed distance is a straight line, what can be said about the graph of velocity?

S-6.   **At a valley:** When the graph of directed distance reaches a minimum (at a valley), what is the velocity?

S-7.   **A car:** A car is driving at a constant velocity of 60 miles per hour. A perspective has been chosen so that directed distance is increasing. Since velocity is constant, we know that directed distance is a linear function. What is the slope of that linear function?

S-8.   **A trip:** A car is driving on a highway that leads west from home. We locate its position as distance west from home. Determine whether velocity is positive, negative, or zero in each of the following situations.

a. The car is driving west.

b. The car is stopped at a traffic light.

c. The car is driving east.

S-9.   **A rock:** A rock is tossed upward and reaches its peak 2 seconds after the toss. Its location is determined by its distance up from the ground. What is the sign of velocity at each of the following times?

a. 1 second after the toss

b. 2 seconds after the toss

c. 3 seconds after the toss

S-10.   **Graph of velocity:** If the graph of velocity lies on the horizontal axis, what can be said about the graph of directed distance?

7.1   EXERCISES

1.   **From New York to Miami again:** The city of Richmond, Virginia, is about halfway between New York and Miami. A Richmond resident might locate the airplane in Example 7.1 using distance north of Richmond. Make the graphs of location and velocity of the airplane from this perspective.

2.   **The rock with a changed reference point:** Make graphs of position and velocity for a rock tossed upward from ground level as it might be viewed by someone standing atop a tall building. Thus the location of the rock is measured by its distance down from the top of the building.

3.   **The rock with a formula:** If from ground level we toss a rock upward with a velocity of 30 feet per second, we can use elementary physics to show that the height in feet of the rock above the ground t seconds after the toss is given by $S = 30t - 16t^2$.

a. Use your calculator to plot the graph of S versus t.

b. How high does the rock go?

c. When does it strike the ground?

d. Sketch the graph of the velocity of the rock versus time.

4.   **Getting velocity from a formula:** When a man jumps from an airplane with an opening parachute, the distance $S = S(t)$, in feet, that he falls in t seconds is given by

$$S = 20\left(t + \frac{e^{-1.6t} - 1}{1.6}\right).$$

a. Use your calculator to make a graph of S versus t for the first 5 seconds of the fall.

b. Sketch a graph of velocity for the first 5 seconds of the fall.

5. **Walking and running:** You live east of campus, and you are walking from campus toward your home at a constant speed. When you get there, you rest for 5 minutes and then run back west at a rapid speed. After a few minutes you reach your destination, and then you rest for 10 minutes. Measure your location as your distance west of your home, and make graphs of your location and velocity.

6. **A rubber ball:** A rubber ball is dropped from the top of a building. The ball lands on concrete and bounces once before coming to rest on the grass. Measure the location of the ball as its distance up from the ground. Make graphs of the location and velocity of the ball.

7. **Gravity on Earth and on Mars:** The acceleration due to gravity near the surface of a planet depends on the mass of the planet; larger planets impart greater acceleration than smaller ones. Mars is much smaller than Earth. A rock is dropped from the top of a cliff on each planet. Give its location as the distance down from the top of the cliff.

   a. On the same coordinate axes, make a graph of distance down for each of the rocks.

   b. On the same coordinate axes, make a graph of velocity for each of the rocks.

8. **Traveling in a car:** Make graphs of location and velocity for each of the following driving events. In each case, assume that the car leaves from home moving west down a straight road and that position is given as the distance west from home.

   a. *A vacation:* Being anxious to begin your overdue vacation, you set your cruise control and drive faster than you should to the airport. You park your car there and get on an airplane to Spain. When you fly back 2 weeks later, you are tired and drive at a leisurely pace back home. (*Note:* Here we are talking about location of your car, not the airplane.)

   b. *On a country road:* A car driving down a country road encounters a deer. The driver slams on the brakes and the deer runs away. The journey is cautiously resumed.

   c. *At the movies:* In a movie chase scene, our hero is driving his car rapidly toward the bad guys. When the danger is spotted, he does a Hollywood 180 degree turn and speeds off in the opposite direction.

9. **Making up a story about a car trip:** You begin from home on a car trip. Initially your velocity is a small positive number. Shortly after you leave, velocity decreases momentarily to zero. Then it increases rapidly to a large positive number and remains constant for this part of the trip. After a time, velocity decreases to zero and then changes to a large negative number.

   a. Make a graph of velocity for this trip.

   b. Discuss your distance from home during this driving event, and make a graph.

   c. Make up a driving story that matches this description.

10. **Car trips with given graphs:**

   a. The graph in Figure 7.24 shows your distance west of home on a car trip. Make a graph of velocity.

   b. The graph in Figure 7.25 shows your velocity on a different car trip. Assuming you start at home, make a graph of your distance west of home.

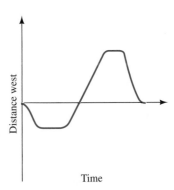

FIGURE 7.24 A graph of distance west of home

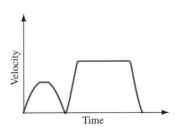

FIGURE 7.25 A graph of velocity for a different car trip

11. **A car moving in an unusual way:** In making graphs for location and velocity, the authors of this text have tried to avoid sharp corners. This problem is designed to show you why. Suppose a car's distance west of home is given by the graph in Figure 7.26.

    a. Make a graph of velocity versus time. In making your graph, take extra care near the peak shown in Figure 7.26.

    b. Carefully describe the motion of the car near the peak in Figure 7.26. Do you think it is possible to drive in a way that matches the graph in Figure 7.26?

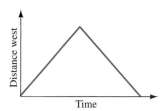

FIGURE 7.26 Position for an unusual driving event

12. **Sporting events:** Analyze the following sporting activities according to the instructions.

    a. *Skiing:* A skier is going down a ski slope at Park City, Utah. The hill is initially gentle, but then it gets steep about halfway down before flattening out at the bottom. Locate position on the slope as the distance from the top, and make graphs of location and velocity.

    b. *Practicing for the NCAA basketball tournament:* You are bouncing a basketball. Locate the basketball by its distance up from the floor. Make graphs of location and velocity.

    c. *Hiking:* The graph of the distance west of base camp for a hiker in Colorado is given in Figure 7.27. Make a graph of velocity versus time, and then make up a story that matches this description.

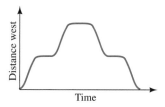

FIGURE 7.27 A hiking trip

7.2 *Rates of Change for Other Functions*

Velocity is the rate of change in directed distance, and this same idea applies to any function. For a function $f = f(x)$, we can look at the rate of change in f with respect to x, which tells how f changes for a given change in x. Rates of change are so pervasive in mathematics, science, and engineering that they are given a special notation, most commonly $\frac{df}{dx}$ or $f'(x)$, and a special name, the *derivative of f with respect to x*. We will use the notation $\frac{df}{dx}$ but will consistently refer to it as the *rate of change in f with respect to x*. We should note that some applications texts use the notation $\frac{\Delta f}{\Delta x}$ for the rate of change.[1]

Examples of Rates of Change

The notation $\frac{df}{dx}$ means the rate of change in f with respect to x. Specifically, $\frac{df}{dx}$ tells how much f is expected to change if x increases by 1 unit. We note that $\frac{df}{dx}$ is a function of x; in general, the rate of change varies as x varies. Some examples may help clarify things.

1. If $S = S(t)$ gives directed distance for an object as a function of time t, then $\frac{dS}{dt}$ is the rate of change in directed distance with respect to time. This is *velocity*. It tells the additional distance we expect to travel in 1 unit of time. For example, if we are currently located $S = 100$ miles south of Dallas, Texas, and if we are traveling south with a velocity $\frac{dS}{dt}$ of 50 miles per hour, then in 1 additional hour we would expect to be 150 miles south of Dallas.

2. If $V = V(t)$ is the velocity of an object as a function of time t, then $\frac{dV}{dt}$ is the rate of change in velocity with respect to time. This is *acceleration*. It tells the additional velocity we expect to attain in 1 unit of time. For example, if we are traveling with a velocity $V = 50$ miles per hour, and if we start to pass a truck, our acceleration $\frac{dV}{dt}$ might be 2 miles per hour each second. Then 1 second in the future, we would expect our velocity to be 52 miles per hour.

3. If $T = T(D)$ denotes the amount of income tax, in dollars, that you pay on an income of D dollars, then $\frac{dT}{dD}$ is the rate of change in tax with respect to the money you earn. It is known as the *marginal tax rate*. If you have already accumulated D dollars, then $\frac{dT}{dD}$ is the additional tax you expect to pay if you earn 1 additional dollar. For example, suppose we have a tax liability of \$3000 and our marginal tax rate $\frac{dT}{dD}$ is 0.2, or 20 cents per dollar. Then if we earn an additional \$1, we would expect our tax liability to increase by 20 cents to a total of \$3000.20; if instead we earn an additional \$100, we would expect our tax liability to increase by \$20 to a total of \$3020. The marginal tax rate is a crucial bit of information for financial planning.

---

[1]This is seen most often in business and agricultural applications but in other places as well. There is a technical difference between the meanings of $\frac{df}{dx}$ and $\frac{\Delta f}{\Delta x}$. The notation $\frac{\Delta f}{\Delta x}$ means an *average rate of change over a given change in x*, whereas $\frac{df}{dx}$ is the *instantaneous rate of change*. The distinction is important in advanced mathematics but less so in many applications.

4. If $P = P(i)$ is the profit, in dollars, that you expect to earn on an investment of i dollars, then $\frac{dP}{di}$ is the rate of change in profit with respect to dollars invested. It tells how much additional profit is to be expected if 1 additional dollar is invested. In economics this is known as *marginal profit*. For example, if our current investment in a project gives a profit of $1000, and if our marginal profit $\frac{dP}{di}$ is 0.2, or 20 cents per dollar, then we would expect that an additional investment of $100 would yield an additional profit of $20 for a total profit of $P = 1020$ dollars.

Properties That All Rates of Change Share

Fortunately, all rates of change exhibit the properties that we have already seen in our study of velocity. If $S = S(t)$ denotes directed distance at time t, then we know that when S is increasing, the velocity $\frac{dS}{dt}$ is positive; when S is decreasing, the velocity $\frac{dS}{dt}$ is negative; and when S is not changing, the velocity $\frac{dS}{dt}$ is zero. This fundamental relationship holds true for all rates of change.

KEY IDEA 7.3

Fundamental Properties of Rates of Change

For a function $f = f(x)$ we will use the notation $\frac{df}{dx}$ to denote the rate of change in f with respect to x.

1. The expression $\frac{df}{dx}$ tells how f changes in relation to x. It gives the additional value that is expected to be added to f if x increases by 1 unit.

2. When f is increasing, then $\frac{df}{dx}$ is positive.

3. When f is decreasing, then $\frac{df}{dx}$ is negative.

4. When f is not changing, then $\frac{df}{dx}$ is zero.

EXAMPLE 7.3 **Passing a Truck**

You are driving with your cruise control set when you encounter a slowly moving truck. You speed up to pass the truck. When you have overtaken the truck, you slow down and resume your previous speed. Let $V = V(t)$ denote your velocity during this event as a function of time t.

1. Explain in practical terms the meaning of $\frac{dV}{dt}$.

2. Make a graph of $V = V(t)$, marking important points on the graph.

3. Make a graph of acceleration $A = \frac{dV}{dt}$.

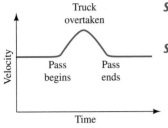

FIGURE 7.28 Velocity when passing a truck

Solution to Part 1: The function $\frac{dV}{dt}$ is the rate of change in velocity. This is acceleration. It tells what additional velocity you expect to attain over 1 unit of time.

Solution to Part 2: Since the cruise control is set, velocity is positive and constant at the beginning. That means that the graph of V versus t starts above the horizontal axis and is horizontal. When you start around the truck you speed up, making the graph of $V(t)$ go up. After you overtake the truck, you slow down and resume your original velocity. This makes the graph of velocity go back down to its original level and flatten out again. Our graph is shown in Figure 7.28.

Solution to Part 3: We will get the graph of acceleration directly from the graph of velocity in Figure 7.28 and then verify that it makes sense. The basic tools we use are the fundamental properties of rates of change: When V is increasing, $\frac{dV}{dt}$ is positive; when V is decreasing, $\frac{dV}{dt}$ is negative; and when V is not changing, $\frac{dV}{dt}$ is zero.

Before and after passing, the graph of velocity in Figure 7.28 is horizontal and so V is not changing at all. This means that the rate of change in velocity $A = \frac{dV}{dt}$ is zero during these periods. Thus the graph of A lies on the horizontal axis, as we have shown in Figure 7.29. From the time the pass begins until the truck is overtaken, the graph of velocity in Figure 7.28 is increasing. This means that the rate of change in velocity $\frac{dV}{dt}$ is positive, and we have marked that in italics in Figure 7.29. At the peak of the graph in Figure 7.28, the acceleration is zero; however, from there to the time the pass is completed, velocity is decreasing, so its rate of change—the acceleration—is negative, as is marked in italics in Figure 7.29.

We used the information from Figure 7.29 to draw the completed graph of $\frac{dV}{dt}$ in Figure 7.30. We note that the graph makes sense. At the beginning and end of the graph, the cruise control is set, so velocity is not changing. That is, acceleration is zero. As you begin the pass, you accelerate. Thus the graph of acceleration is above the horizontal axis for this period, as is shown in Figure 7.30. Once the truck is overtaken, you ease off the gas pedal, and the car *decelerates* back to your original cruising speed. Deceleration is the same as negative acceleration, and so from the time when you overtake the truck until the pass is completed, the graph of acceleration is below the horizontal axis, as is represented in Figure 7.30. Thus the graph we made using the fundamental properties of rates of change agrees with our intuitive analysis. ■ ■ ■

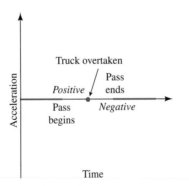

FIGURE 7.29 Important features for acceleration

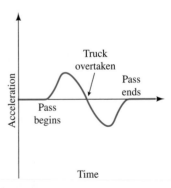

FIGURE 7.30 Completed graph of acceleration when passing a truck

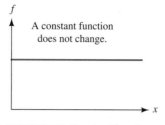

FIGURE 7.31 Constant function has persistent zero rate of change

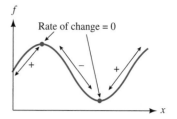

FIGURE 7.32 At extrema, the rate of change is momentarily zero

When the Rate of Change Is Zero

In Example 7.3 we used the graph of velocity V in Figure 7.28 to get the graph of acceleration $\frac{dV}{dt}$ in Figure 7.30. In the process, we noted two situations in which the acceleration was zero: when velocity was constant while the cruise control was set (here acceleration was zero over a span of time) and when velocity reached its maximum value at the point when the truck was overtaken (here acceleration was zero for an instant in time). Once again, these observations remain true for rates of change in general. When f is constant, its rate of change $\frac{df}{dx}$ is zero over a span of x values. This is intuitively clear, since the statement that $\frac{df}{dx}$ is zero is the same as saying that f is not changing and therefore remains at a constant value. At a peak or valley of f, the rate of change $\frac{df}{dx}$ will be zero for that single value of x. These important facts are illustrated in Figure 7.31 and Figure 7.32.

Let's look at an example to see how we use the fact that the rate of change is zero at extrema.

EXAMPLE 7.4   Marginal Profit for a Tire Company

The CEO of a tire company has kept records of the profit $P = P(n)$ that the company makes when it produces n tires per day.

1. The rate of change $\frac{dP}{dn}$ is known as the *marginal profit*. Explain the meaning of marginal profit in practical terms.

2. What action should the CEO take if the marginal profit is positive?

3. What action should the CEO take if the marginal profit is negative?

4. The CEO has used her records to make the graph of marginal profit shown in Figure 7.33. According to this graph, how many tires per day should the company be producing?

Solution to Part 1:   The marginal profit $\frac{dP}{dn}$ is the rate of change in profit with respect to the number of items produced. It is the change in profit that the CEO can expect to earn by producing 1 more tire.

Solution to Part 2:   Marginal profit tells the CEO how much profit can be expected to change if 1 additional tire is produced. Since marginal profit is positive, additional profit can be made by increasing production. In other words, since the rate of change $\frac{dP}{dn}$ is positive, the profit P is an increasing function of the production level n. This means that the CEO should increase production, since doing so will cause the profit to increase. This situation might occur if, for example, the demand for tires were larger than current production.

Solution to Part 3:   In this case the marginal profit is negative, which means that if production is increased, profits will change by a negative amount. That is, profits will decrease. In other words, since the rate of change $\frac{dP}{dn}$ is negative, the profit P is a *decreasing* function of n. In this scenario, it would be wise for the CEO to cut back on production. This might happen if the number of tires currently being produced exceeded consumer demand.

Solution to Part 4:   In part 2 we noted that if the marginal profit is positive, then the profit P is increasing with n, and production should be increased. In part 3, we saw that if the marginal profit is negative, then the profit P is decreasing with n, and production should be decreased. Thus it makes sense that for the CEO, the ideal value of marginal profit is zero. In other words, to get a maximum profit P, we want to find the point where P switches from increasing to decreasing (since at this point the profit is at its highest value), and that is where the marginal profit $\frac{dP}{dn}$ switches from positive to negative. This is where the graph in Figure 7.33 crosses the horizontal axis—that is, where the marginal profit is zero. The crossing point occurs at $n = 160$, so the company should be producing 160 tires per day.

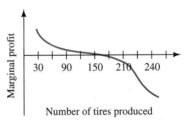

FIGURE 7.33 Marginal profit for a tire company

When the Rate of Change Is Constant

We observed in Section 7.1 that when velocity $\frac{dS}{dt}$ is constant, we can conclude that directed distance S is a linear function whose slope is the constant velocity. Virtually all of the observations we made about the relationships between velocity and directed distance are true for rates of change in general, and this one is no exception: If the rate of change in f, namely $\frac{df}{dx}$, is constant, then f is a linear function with slope $\frac{df}{dx}$. We would emphasize that this is not really a new observation. The characterization of linear functions as those with constant rate of change has been used repeatedly in this course. The only thing new about our expression of this fact here is the use of $\frac{df}{dx}$ to denote the rate of change.

EXAMPLE 7.5   **A Leaky Balloon**

A balloon is initially full of air, but then it springs a leak. Let $B = B(t)$ represent the volume, in liters, of air in the balloon at time t, measured in minutes.

1. Explain what $\frac{dB}{dt}$ means in practical terms. As air is leaking out of the balloon, is $\frac{dB}{dt}$ positive or negative?

2. Assume that air is leaking from the balloon at a constant rate of 0.5 liter per minute.[2] What does this information tell you about $\frac{dB}{dt}$? What does it tell you about B?

3. A little later the leak is patched, a pump that outputs 2 liters of air per minute is attached to the balloon, and it is inflated back to its original size. As the balloon inflates, what can we conclude about $\frac{dB}{dt}$? What can we conclude about B?

4. Make a graph of B versus t as the balloon leaks air and then is reinflated to its original size.

5. Make a graph of $\frac{dB}{dt}$ versus t as the balloon leaks air and then is reinflated to its original size.

Solution to Part 1:   The function $\frac{dB}{dt}$ is the rate of change in B with respect to t. This is the rate of change in the volume of air in the balloon with respect to time. In practical terms, $\frac{dB}{dt}$ is the change in the volume of air in the balloon that we expect to occur in 1 minute of time. Since the volume B of air in the balloon is decreasing, $\frac{dB}{dt}$ is negative.

Solution to Part 2:   To say that air is leaking from the balloon at a rate of 0.5 liter per minute means that the volume B of air in the balloon is decreasing by 0.5 liter per minute. In other words, B is changing by -0.5 liter per minute. We conclude that $\frac{dB}{dt} = -0.5$. Since B is a function with a constant rate of change of -0.5, we know that it is a linear function with slope -0.5. We were not asked to do so, but with this information we have almost everything we need to write a formula for B, namely $B = -0.5t + b$. In this formula, what is the practical meaning of b?

Solution to Part 3:   As the balloon is inflated, the volume B of air in the balloon is increasing by 2 liters per minute. That is, $\frac{dB}{dt} = 2$ liters per minute. Since B has a constant rate of change of 2, we conclude that B is a linear function with slope 2 during the period of reinflation.

As in part 2, we were not asked to write a formula for B, but we have almost everything we need to do that. The formula is $B = 2t + c$. Explain the practical meaning of c in this formula.

Solution to Part 4:   We know that while the balloon is losing air, B is linear with slope -0.5 and that after the pump is attached, B is linear with slope 2. Thus the graph of

---

[2]This is not, in fact, how we would expect a balloon to leak. Rather, we would expect it to leak rapidly when it is almost full and more slowly when it is nearly empty. We will examine this more realistic description of a leaky balloon in Section 7.4.

B should start as a straight line with slope -0.5 and then change to a straight line with slope 2. These features are reflected in Figure 7.34.

Note in Figure 7.34 that we have rounded the graph at the bottom where the pump is attached. Alternatively, one might draw this graph with a sharp corner at the bottom. Which do you think is correct?

Solution to Part 5:   We know that while the balloon is leaking, $\frac{dB}{dt}$ has a constant value of -0.5. Thus its graph during this period is a horizontal line located 0.5 unit below the horizontal axis. After the pump is attached, $\frac{dB}{dt}$ has a constant value of 2. Thus during this period, its graph is a horizontal line 2 units above the horizontal axis. How the two line segments are connected near the time when the pump is attached depends on how we drew the graph of *B* near that point. If you think there should be a sharp corner there, how would this affect the graph of $\frac{dB}{dt}$ that we have drawn in Figure 7.35?   ▪ ▪ ▪

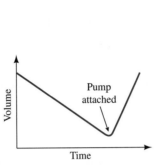

FIGURE 7.34 The graph of *B* versus *t*

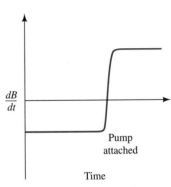

FIGURE 7.35 The graph of $\frac{dB}{dt}$ versus *t*

7.2   SKILL BUILDING EXERCISES

S-1. **Meaning of rate of change:** What is the common term for the rate of change of each of the following phenomena?

   a. Directed distance as a function of time

   b. Velocity as a function of time

   c. Tax due as a function of income

   d. Profit as a function of dollars invested

S-2. **A mathematical term:** If $f = f(x)$, then we use $\frac{df}{dx}$ to denote the rate of change in f. What is the technical mathematical term for $\frac{df}{dx}$?

S-3. **Sign of the derivative:** Suppose $f = f(x)$. What is the sign of $\frac{df}{dx}$ in each of the following situations?

   a. The function f is increasing.

   b. The graph of f has reached a peak.

   c. The function f is decreasing.

   d. The graph of f is a horizontal line.

S-4. **Constant rate of change:** When $\frac{df}{dx}$ is constant, what kind of function is f?

S-5. **A value for the rate of change:** If $\frac{df}{dx}$ has a constant value of 10, we know that f is a linear function. What is the slope of f?

S-6. **Elevation:** If $E(t)$ is your elevation at time t, and you are walking up a steep slope, what can be said about $\frac{dE}{dt}$?

S-7. **Marginal tax rate:** You have the opportunity to earn some extra income working at $20 per hour. If your marginal tax rate is 34%, how much of the hourly wage will you get to keep?

S-8. **Graph of rate of change:** What can be said about the graph of $\frac{df}{dx}$ in each of the following situations?

   a. The graph of f is increasing.

   b. The graph of f is at a peak.

   c. The graph of f is decreasing.

   d. The graph of f is a straight line.

S-9. **Graph of f:** What can be said about the graph of f if the graph of $\frac{df}{dx}$ is below the horizontal axis?

S-10. **Advertising:** Let $s(a)$ denote sales generated by spending a dollars on advertising. My goal is to increase sales. If $\frac{ds}{da}$ is negative, should I spend more or less money on advertising?

7.2   EXERCISES

1. **Estimating rates of change:** Use your calculator to make the graph of $f(x) = x^3 - 5x$.

   a. Is $\frac{df}{dx}$ positive or negative at $x = 2$?

   b. Identify a point on the graph of f where $\frac{df}{dx}$ is negative.

2. **The spread of AIDS:** The table to the right shows the cumulative number $N = N(t)$ of AIDS cases in the United States that have been reported to the Centers for Disease Control by end of the year given. (The source for these data, the U.S. Centers for Disease Control and Prevention in Atlanta, cautions that they are subject to retrospective change.)

   a. What does $\frac{dN}{dt}$ mean in practical terms?

   b. From 1986 to 1992 was $\frac{dN}{dt}$ ever negative?

t = year	N = total cases reported
1986	28,711
1987	49,799
1988	80,518
1989	114,113
1990	155,766
1991	199,467
1992	244,939

3. **Mileage for an old car:** The gas mileage M that you get on your car depends on its age t in years.

   a. Explain the meaning of $\frac{dM}{dt}$ in practical terms.

   b. As your car ages and performance degrades, do you expect $\frac{dM}{dt}$ to be positive or negative?

4. **Investing in the stock market:** You are considering buying three stocks whose prices at time t are given by $P_1(t)$, $P_2(t)$, and $P_3(t)$. You know that $\frac{dP_1}{dt}$ is a large positive number, $\frac{dP_2}{dt}$ is near zero, and $\frac{dP_3}{dt}$ is a large negative number. Which stock will you buy? Explain your answer.

5. **Hiking:** You are hiking in a hilly region, and $E = E(t)$ is your elevation at time t.

   a. Explain the meaning of $\frac{dE}{dt}$ in practical terms.

   b. Where might you be when $\frac{dE}{dt}$ is a large positive number?

   c. You reach a point where $\frac{dE}{dt}$ is briefly zero. Where might you be?

   d. Where might you be when $\frac{dE}{dt}$ is a large negative number?

6. **Marginal profit:** A small firm produces at most 15 widgets in a week. Its profit P, in dollars, is a function of n, the number of widgets manufactured in a week. The marginal profit $\frac{dP}{dn}$ for the firm is given by the formula

$$\frac{dP}{dn} = 72 + 6n - n^2 .$$

   a. Use your calculator to make a graph of $\frac{dP}{dn}$ versus n.

   b. For what values of n is the profit P decreasing?

   c. How many widgets should the firm produce in a week to maximize profit P?

7. **Health plan:** The managers of an employee health plan for a firm have studied the balance B, in millions of dollars, in the plan account as a function of t, the number of years since the plan was instituted. They have determined that the rate of change $\frac{dB}{dt}$ in the account balance is given by the formula

$$\frac{dB}{dt} = 10e^{0.1t} - 12 .$$

   a. Use your calculator to make a graph of $\frac{dB}{dt}$ versus t over the first 5 years of the plan.

   b. During what period is the account balance B decreasing?

   c. At what time is the account balance B at its minimum?

8. **A race:** A man enters a race that involves running and swimming. He lines up at the starting gate and begins running very fast, hoping to take the initial lead before settling into a constant, slower pace for a distance run. To complete the race, he must swim across a river and back before running back to the starting gate. He swims at a constant rate but cannot swim as fast as he runs. Let $S = S(t)$ denote the distance of the contestant from the starting gate.

   a. Make a graph of S versus t. Note on your graph the places where you know that S is linear.

   b. Make a graph of the contestant's velocity.

   c. Make a graph of the contestant's acceleration.

9. **The acceleration due to gravity:** From the time of Galileo, physicists have known that near the surface of the Earth, gravity imparts a constant acceleration of 32 feet per second per second. If air resistance is ignored, explain how this shows that velocity for a falling object is a linear function of time.

10. **Water in a tank:** Water is leaking out of a tank. The amount of water in the tank t minutes after it springs a leak is given by $W(t)$ gallons.

   a. Explain what $\frac{dW}{dt}$ means in practical terms.

   b. As water leaks out of the tank, is $\frac{dW}{dt}$ positive or negative?

   c. For the first 10 minutes, water is leaking from the tank at a rate of 5 gallons per minute. What do you conclude about the nature of the function W during this period?

   d. After about 10 minutes, the hole in the tank suddenly gets larger, and water begins to leak out of the tank at 12 gallons per minute.

      i. Make a graph of W versus t. Be sure to incorporate linearity where it is appropriate.

      ii. Make a graph of $\frac{dW}{dt}$ versus t.

11. **A population of bighorn sheep:** There is an effort in Colorado to restore the population of bighorn sheep. Let $N = N(t)$ denote the number of sheep in a certain protected area at time t.

a. Explain the meaning of $\frac{dN}{dt}$ in practical terms.

b. A small breeding population of bighorn sheep is initially introduced into the protected area. Food is plentiful and conditions are generally favorable for bighorn sheep. What would you expect to be true about the sign of $\frac{dN}{dt}$ during this period?

c. This summer a number of dead sheep were discovered, and all were infected with a disease that is known to spread rapidly among bighorn sheep and is nearly always fatal. How would you expect an unchecked spread of this disease to affect $\frac{dN}{dt}$?

d. If the reintroduction program goes well, then the population of bighorn sheep will grow to the size the available food supply can support and will remain at about that same level. What would you expect to be true of $\frac{dN}{dt}$ when this happens?

12. **Eagles:** In an effort to restore the population of bald eagles, ecologists introduce a breeding group into a protected area. Let $N = N(t)$ denote the population of bald eagles at time t. Over time, you observe the following information about $\frac{dN}{dt}$.

- Initially, $\frac{dN}{dt}$ is a small positive number.

- A few years later, $\frac{dN}{dt}$ is a much larger positive number.

- Many years later, $\frac{dN}{dt}$ is positive but near zero.

Make a possible graph of $N(t)$.

13. **Visiting a friend:** I live in the suburbs and my friend lives out in the country. The speed limit between my home and a stop sign 1 mile away is 35 miles per hour. After that the speed limit is 65 miles per hour. Assume that I drove out to visit my friend. As I pulled into his driveway, I saw that he wasn't at home, and so I drove back to my house. Assume that I obeyed all the traffic laws and had an otherwise uneventful trip. Make graphs of $S = S(t)$, my location in relation to my home; the velocity $V = V(t)$; and the acceleration $A = A(t)$. Be sure your graphs show linearity where it is appropriate.

14. **A car trip:** You are moving away from home in your car, and at no time do you turn back. Your acceleration is initially a large positive number, but it decreases slowly to zero, where it remains for a while. Suddenly your acceleration decreases rapidly to a large negative number before returning slowly to zero.

a. Make a graph of acceleration for this event.

b. Make a graph of velocity for this event.

c. Make a graph of the location of the car.

d. Make up a driving story that matches this description.

15. **Velocity of an airplane:** An airplane leaves Atlanta flying to Dallas. Because of heavy air traffic, it circles the airport at Dallas for a time before landing. Let $D(t)$ be the distance west from Atlanta to the airplane. Make a graph of $D(t)$. Thinking of velocity $V(t)$ as the rate of change in $D(t)$, make a graph of the velocity of the airplane.

16. **Growth in height:** The following graph gives a man's height $H = H(t)$, in inches at age t in years.

a. Explain what $\frac{dH}{dt}$ means in practical terms.

b. Sketch a graph of $\frac{dH}{dt}$ versus t.

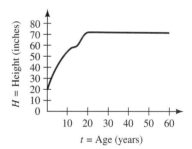

17. **Laboratory experiment:** This lab uses a motion detector and a calculator-based laboratory (CBL™) unit. The motion detector measures the distance from the detector to an object in front of it, and the CBL records the data. After collecting the data, the CBL unit sends the data to the calculator, which displays a graph of the distance recorded with respect to time. In this lab we will investigate drag force, or air resistance, as coffee filters fall. The acceleration of the coffee filters is the acceleration due to gravity less the retardation due to air resistance. In this case, the air retardation is proportional to the square of the velocity of the coffee filter. We drop the filter, measure the distance, and use this to determine the terminal velocity and drag coefficient of the coffee filter. For a detailed description, go to

http://math.college.com/students

and then, from the book's website, go to the Coffee Filter Drop experiment.

7.3 *Estimating Rates of Change*

Up to now we have looked at rates of change qualitatively, emphasizing how the sign of a rate of change affects the function. But exact values, or even estimates, for rates of change can provide additional important information. It is easiest to make such estimates for functions given by tables.

Rates of Change for Tabulated Data

The following brief table shows the location S (measured as distance in miles east of Los Angeles) of an airplane flying toward Denver.

Time	1:00 P.M.	1:30 P.M.
Distance from L.A.	360 miles	612 miles

We want to know the velocity $\frac{dS}{dt}$ at 1:00 P.M. We do this using a familiar formula:

$$\text{Average velocity} = \frac{\text{Distance traveled}}{\text{Elapsed time}} .$$

Between 1:00 and 1:30 the airplane traveled $612 - 360 = 252$ miles, and it took half an hour to travel that far:

$$\text{Average velocity from 1:00 to 1:30} = \frac{252}{0.5} = 504 \text{ miles per hour} .$$

It is important to point out that this calculation can be expected to give the exact velocity $\frac{dS}{dt}$ at 1:00 P.M. only if the airplane is traveling at the same speed over the entire 30-minute time interval. If the airplane was speeding up between 1:00 and 1:30, then 504 miles per hour was its average velocity over the 30-minute time interval, but its exact velocity at 1:00 would have been somewhat less than 504 miles per hour. If, on the other hand, the airplane was slowing down, then its exact velocity at 1:00 would have been more than 504 miles per hour. In general, when velocity is calculated from a table of values, as it was here, it should be regarded as an approximation of the exact velocity. Furthermore, it is clear that shorter time intervals yield better approximations, because the moving object doesn't have as much time to change velocity.[3] Suppose, for example, that instead of the table above, we had been given the following table.

Time	1:00 P.M.	1 second after 1:00 P.M.
Distance from L.A.	360 miles	360.140 miles

---

[3]This is discussed further in the next section under the topic of instantaneous rates of change.

If we use this table to calculate velocity, we see that the airplane traveled 0.140 mile in 1 second. Since there are 3600 seconds in an hour, we get

$$\text{Average velocity from 1:00 P.M. to 1 second later} = \frac{0.140}{\frac{1}{3600}} = 504 \text{ miles per hour}.$$

Since the velocity of the airplane can change very little over a 1-second time interval, we would now feel confident reporting its velocity at 1:00 P.M. as 504 miles per hour.

Exactly the same idea can be used to approximate the rate of change for any function given by tabulated data. For example, the following table shows the percentage $B = B(t)$ of babies born in 1980 and 1990 to unmarried women in the United States.

Year	1980	1990
Born to unmarried women	18%	28%

The function $\frac{dB}{dt}$ is the rate of change in B. If we were able to calculate it exactly for 1980, it would tell us how fast the percentage of births to unmarried women was increasing in 1980. That would indicate how much growth in this percentage could be expected in 1 year's time. As with the example of the airplane, the table does not give enough information to allow for an exact calculation of $\frac{dB}{dt}$ for 1980, but we can approximate its value using the average rate of change from 1980 to 1990 just as we did for the velocity of the airplane.

During the period from 1980 to 1990, the change in percentage was $28 - 18 = 10$ percentage points. This occurred over a period of 10 years, so we divide to get

$$\text{Average rate of change from 1980 to 1990} = \frac{10}{10} = 1 \text{ percentage point per year}.$$

We can use this number as an approximation of the value of $\frac{dB}{dt}$.

Estimating Rates of Change for Tabulated Data

If $f = f(x)$ is a function given by a table of values, then $\frac{df}{dx}$, the rate of change of f with respect to x, cannot without further information be calculated exactly. But it can be estimated using

$$\text{Average rate of change in } f \text{ with respect to } x = \frac{\text{Change in } f}{\text{Change in } x}.$$

Calculating rates of change in this way for tabulated data is not a new idea at all, and we have already made and used this type of calculation many times in the text. Look back at Example 1.3 in Section 1.2, where we studied tabulated data for the number $W = W(t)$ of women in the United States employed outside the home. In the solution to part 3 of that example, we noted that the number increased by 12.3 million from 1970 to 1980. We concluded that during the decade of the 1970s, the number of women employed outside the home was increasing by an average of $\frac{12.3}{10} = 1.23$ million per year, and we used that number to estimate the number of women employed outside the home in 1972. This is exactly the calculation presented in Key Idea 7.4. Thus, in the language of rates of change, we would say that $\frac{dW}{dt}$ is approximately the average rate of change from 1970 to 1980—1.23 million per year—and that this number tells us the increase in W that would be expected in 1 year. This is once again an illustration of how mathematics distills ideas from many contexts into a single fundamental concept. Precisely the same mathematical idea, the rate of change, is used to analyze applications from velocity to population growth to numbers of women employed outside the home.

EXAMPLE 7.6 Water Flowing from a Tank

Consider the following table, which shows the number of gallons W of water left in a tank t hours after it starts to leak.

$t =$ hours	0	3	6	9	12
$W =$ gallons left	860	725	612	515	433

1. Explain the meaning of $\frac{dW}{dt}$ in practical terms, and estimate its value at $t = 6$, using the average rate of change from $t = 6$ to $t = 9$.

2. Use your answer from part 1 to estimate the amount of water in the tank 8 hours after the leak begins.

Solution to Part 1:   The function $\frac{dW}{dt}$ is the rate of change in water remaining in the tank. This is the change in volume we expect over an hour. Since the amount of water is decreasing, $\frac{dW}{dt}$ is negative. Its size is the rate at which water is leaking from the tank.

   We emphasize that the calculation we make here is the same one we made for women employed outside the home in Example 1.3. We are using a new language but not a new idea. Six hours after the leak began, there were 612 gallons in the tank. When $t = 9$ there are 515 gallons left. Thus the water level changed by $515 - 612 = -97$ gallons over the 3-hour period. Hence

$$\text{Average rate of change from 6 hours to 9 hours} = \frac{-97}{3}$$

$$= -32.33 \text{ gallons per hour}.$$

This is the estimate for $\frac{dW}{dt}$ that we were asked to find.

Solution to Part 2:   For each hour after 6, we expect the number of gallons in the tank to decrease by 32.33 gallons. From 6 to 8 hours is a 2-hour span, so we expect $2 \times 32.33 = 64.66$ gallons to leak out. That leaves $612 - 64.66 = 547.34$ gallons in the tank 8 hours after the leak began.   ▨ ▨ ▨

Rates of Change for Functions Given by Formulas

For functions given by formulas, it is possible using calculus to find exact formulas for rates of change. But for many applications, a close approximation to the rate of change is sufficient, and that is what we will use in this text. The idea is first to use the formula to generate a brief table of values and then to compute the rate of change just as we did above. Let's look, for example, at a falling rock. If air resistance is ignored, elementary physics can be used to show that the rock falls $S = 16t^2$ feet during t seconds of fall. Let's estimate the downward velocity $\frac{dS}{dt}$ of the rock 2.5 seconds into the fall. We will first show the steps involved in making this computation and will then show how the graphing calculator offers a shortcut.

We know from the computations we made for the velocity of the airplane that we get more reliable answers if we keep the time interval short. Below we have made a brief table of values for $S = 16t^2$ using only $t = 2.5$ and $t = 2.50001$. You may wish to use your calculator to check that $16 \times 2.5^2 = 100$ and $16 \times 2.50001^2 = 100.0008$.

t	2.5	2.50001
S	100	100.0008

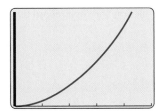

FIGURE 7.36 Graph of distance versus time

Now the change in S is $100.0008 - 100 = 0.0008$ foot. This is how far the rock falls from 2.5 to 2.50001 seconds into the fall. It takes $2.50001 - 2.5 = 0.00001$ second to fall this far, so the rate of change in S with respect to t (in other words, the velocity) is approximated as follows:

$$\frac{dS}{dt} = \frac{\text{Change in } S}{\text{Change in } t} = \frac{0.0008}{0.00001} = 80 \text{ feet per second}.$$

We will do this computation again, but this time we use the automated features provided by the calculator. The first step is to make the graph $\boxed{7.4}$ as shown in Figure 7.36. We used a window setup with a horizontal span of $t = 0$ to $t = 5$ and a vertical span of $S = 0$ to $S = 300$. Next we use the calculator $\boxed{7.5}$ to get the rate of change $\frac{dS}{dt}$ at $t = 2.5$. The result (see Figure 7.37) shows that the velocity $\frac{dS}{dt}$ at $t = 2.5$ is 80 feet per second, and this agrees with our earlier computation. We would emphasize that no magic has been performed. The calculator has just internally made a table of values and performed a computation similar to the one we did when we made the computation by hand. The calculator just automates the procedure. For the exact keystrokes[4] needed to do this, consult the *Technology Guide*.

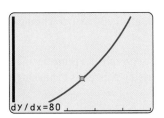

FIGURE 7.37 Getting $\frac{dS}{dt}$, the velocity

---

[4]On some calculators, you will sometimes not be able to get $\frac{dy}{dx}$ for the exact x value you want. See the *Technology Guide* for additional information.

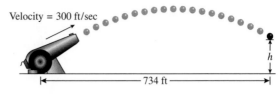

FIGURE 7.38 Path of a cannonball

EXAMPLE 7.7   A Cannonball

A cannonball fired from the origin with a muzzle velocity of 300 feet per second follows the path of the graph of

$$h = x - 32\left(\frac{x}{300}\right)^2,$$

where distances are measured in feet (see Figure 7.38). This simple model ignores air resistance.

1. Plot the graph of the flight of the cannonball.

2. Use the graph to estimate the height h of the cannonball 734 feet downrange.

3. By looking at the graph of h, do you expect $\frac{dh}{dx}$ to be positive or negative at $x = 734$?

4. Calculate $\frac{dh}{dx}$ at 734 feet downrange, and explain in practical terms what this number means.

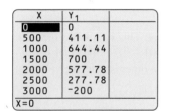

FIGURE 7.39 A table of values for the height of the cannonball

Solution to Part 1:   To get the graph, we first enter 7.6 the function and then record the appropriate variable correspondences:

$$Y_1 = h, \text{ height on vertical axis}$$

$$X = x, \text{ distance downrange on horizontal axis}.$$

Consulting the table of values in Figure 7.39, we choose a window setup with a horizontal span of $x = 0$ to $x = 3000$ and a vertical span of $h = 0$ to $h = 750$. The resulting graph is shown in Figure 7.40.

Solution to Part 2:   In Figure 7.41 we have put the cursor 7.7 at X=734, and we read from the bottom of the screen that the height of the cannonball 734 feet downrange is $h = 542.44$ feet.

Solution to Part 3:   From Figure 7.41 we see that at $x = 734$, the graph of h is increasing. That is, at this distance downrange, the cannonball is still rising. Since h is increasing, we know that $\frac{dh}{dx}$ is positive.

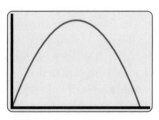

FIGURE 7.40 The flight of the cannonball

Solution to Part 4:   In Figure 7.42 we have used the calculator to get 7.8 the value of $\frac{dh}{dx}$ at $x = 734$ feet downrange. The rate of change, 0.48 foot per foot (rounded to two decimal places), can now be read[5] from the **dy/dx=** prompt at the bottom of

---

[5]Some calculators will give a slightly different answer because they do not calculate the rate of change exactly at $x = 734$.

Figure 7.42. This is a measure of how steeply the cannonball is rising. It tells us that if we move 1 more foot downrange to 735 feet, we can expect the cannonball to rise 0.48 foot from 542.44 to $542.44 + 0.48 = 542.92$ feet.   ■ ■ ■

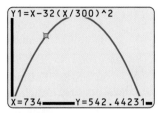

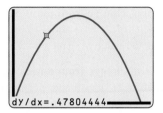

FIGURE 7.41 The height of the cannonball 734 feet downrange

FIGURE 7.42 The rate of change in height 734 feet downrange

S-1. **Rate of change for a linear function:** If f is the linear function $f = 7x - 3$, what is the value of $\frac{df}{dx}$?

S-2. **Rate of change from data:** Suppose $f = f(x)$ satisfies $f(2) = 5$ and $f(2.005) = 5.012$. Estimate the value of $\frac{df}{dx}$ at $x = 2$.

S-3. **Rate of change from data:** Suppose $f = f(x)$ satisfies $f(3) = 8$ and $f(3.005) = 7.972$. Estimate the value of $\frac{df}{dx}$ at $x = 3$.

S-4. **Estimating rates of change:** By direct calculation, estimate the value of $\frac{df}{dx}$ for $f(x) = x^2 + 1$ at $x = 3$. Use an increment of 0.0001.

S-5. **Estimating rates of change:** By direct calculation, estimate the value of $\frac{df}{dx}$ for $f(x) = 1/x^2$ at $x = 4$. Use an increment of 0.0001.

S-6. **Estimating rates of change with the calculator:** Make a graph of $x^3 - x^2$ and use the calculator to estimate its rate of change at $x = 3$. (We recommend a horizontal span of 0 to 4 and a vertical span of 0 to 30.)

S-7. **Estimating rates of change with the calculator:** Make a graph of $x + (1/x)$ and use the calculator to estimate its rate of change at $x = 3$. (We recommend a horizontal span of 1 to 4 and a vertical span of 0 to 5.)

S-8. **Estimating rates of change with the calculator:** Make a graph of 2^x and use the calculator to estimate its rate of change at $x = 3$. (We recommend a horizontal span of 0 to 4 and a vertical span of 0 to 20.)

S-9. **Estimating rates of change with the calculator:** Make a graph of 3^{-x} and use the calculator to estimate its rate of change at $x = 3$. (We recommend a horizontal span of 0 to 4 and a vertical span of -1 to 1.)

S-10. **Limiting value of a rate of change:** Looking at the graph of 3^{-x} from Exercise S-9, what is $\frac{df}{dx}$ for large values of x?

1. **Population growth:** The following table[6] shows the population of reindeer on an island as of the given year.

Date	Population
1945	40
1950	165
1955	678
1960	2793

We let t be the number of years since 1945, so that $t = 0$ corresponds to 1945, and we let $N = N(t)$ denote the population size.

a. Approximate $\frac{dN}{dt}$ for 1955 using the average rate of change from 1955 to 1960, and explain what this number means in practical terms.

b. Use your work from part a to estimate the population in 1957.

c. The number you calculated in part a is an approximation to the actual rate of change. As you will be asked to show in the next exercise, the reindeer population growth can be closely modeled by an exponential function. With this in mind, do you think your answer in part a is too large or too small? Explain your reasoning.

2. **Further analysis of population growth:** *This is a continuation of Exercise 1.* Our goal is to make an exponential model of the data and use it to get a more accurate estimate of the rate of change in population in 1955.

[6]The table is based on the study by D. Klein, "The introduction, increase, and crash of reindeer on St. Matthew Island," *J. Wildlife Management* **32** (1968), 350–367.

a. Plot the logarithm of the data from Exercise 1 and use regression to obtain an exponential model for population growth. (For details on the method here, see Section 4.3.) Your final answer should be an exponential formula for $N = N(t)$.

b. Use the formula you found in part a to get $\frac{dN}{dt}$ in 1955.

c. How does your answer from part b of this exercise compare with your answer from part a of Exercise 1? Does this agree with your answer in part c of Exercise 1?

3. **Deaths from heart disease:** The following tables show the deaths per 100,000 caused by heart disease in the United States for males and females aged 55 to 64 years. The function H_m gives deaths per 100,000 for males, and H_f gives deaths per 100,000 for females.

t = year	H_m = deaths per 100,000
1970	987.2
1980	746.8
1985	651.9
1990	537.3
1991	520.8

Heart disease deaths per 100,000 for males aged 55 to 64 years

t = year	H_f = deaths per 100,000
1970	351.6
1980	272.1
1985	250.3
1990	215.7
1991	210.0

Heart disease deaths per 100,000 for females aged 55 to 64 years

a. Approximate the value of $\frac{dH_m}{dt}$ in 1980 using the average rate of change from 1980 to 1985.

b. Explain the meaning in practical terms of the number you calculated in part a. You should, among other things, tell what the sign means.

c. Use your answer from part a to estimate the heart disease death rate for males aged 55 to 64 years in 1983.

d. Approximate the value of $\frac{dH_f}{dt}$ for 1980 using the average rate of change from 1980 to 1985.

e. Explain what your calculations from parts a and d tell you about comparing heart disease deaths for men and women in 1980.

4. **The cannon with a different muzzle velocity:** If the cannonball from Example 7.7 is fired with a muzzle velocity of 370 feet per second, it will follow the graph of

$$h = x - 32\left(\frac{x}{370}\right)^2,$$

where distances are measured in feet.

a. Plot the graph of the flight of the cannonball.

b. Find the height of the cannonball 3000 feet downrange.

c. By looking at the graph of h, determine whether $\frac{dh}{dx}$ is positive or negative at 3000 feet downrange.

d. Calculate $\frac{dh}{dx}$ at 3000 feet downrange and explain what this number means in practical terms.

5. **Falling with a parachute:** When an average-size man with a parachute jumps from an airplane, he will fall

$$S = 12.5(0.2^t - 1) + 20t$$

feet in t seconds.

a. Plot the graph of S versus t over at least the first 10 seconds of the fall.

b. How far does the parachutist fall in 2 seconds?

c. Calculate $\frac{dS}{dt}$ at 2 seconds into the fall and explain what the number you calculated means in practical terms.

6. **Free fall subject to air resistance:** Gravity and air resistance contribute to the characteristics of a falling object. An average-size man will fall

$$S = 968(e^{-0.18t} - 1) + 176t$$

feet in t seconds after the fall begins. $\longrightarrow$

a. Plot the graph of S versus t over the first 5 seconds of the fall.

b. How far will the man fall in 3 seconds?

c. Calculate $\frac{dS}{dt}$ at 3 seconds into the fall and explain what the number you calculated means in practical terms.

7. **A yam baking in the oven:** A yam is placed in a preheated oven to bake. An application of Newton's law of cooling gives the temperature Y, in degrees, of the yam t minutes after it is placed in the oven as

$$Y = 400 - 325e^{-t/50}.$$

a. Make a graph of the temperature of the yam at time t over 45 minutes of baking time.

b. Calculate $\frac{dY}{dt}$ at the time 10 minutes after the yam is placed in the oven.

c. Calculate $\frac{dY}{dt}$ at the time 30 minutes after the yam is placed in the oven.

d. Explain what your answers in parts b and c tell you about the way the yam heats over time.

8. **A floating balloon:** A balloon is floating upward, and its height in feet above the ground t seconds after it is released is given by a function $S = S(t)$. Suppose you know that $S(3) = 13$ and that $\frac{dS}{dt}$ is 4 when t is 3. Estimate the height of the balloon 5 seconds after it is released. Explain how you got your answer. (See Figure 7.43.)

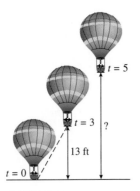

t = 5

t = 3 | ?

13 ft

t = 0

FIGURE 7.43

9. **A pond:** Water is running out of a pond through a drainpipe. The amount of water, in gallons, in the pond t minutes after the water began draining is given by a function $G = G(t)$.

a. Explain the meaning in practical terms of $\frac{dG}{dt}$.

b. While water is running out of the pond, do you expect $\frac{dG}{dt}$ to be positive or negative?

c. When $t = 30$, water is running out of the drainpipe at a rate of 8000 gallons per minute. What is the value of $\frac{dG}{dt}$?

d. When $t = 30$, there are 2,000,000 gallons of water in the pond. Using the information from part c, estimate the value of $G(35)$.

10. **Marginal profit:** *This refers to Example 7.4 of Section 7.2.* A small firm produces at most 20 widgets in a week. Its profit P, in dollars, is a function of n, the number of widgets manufactured in a week, and the formula is

$$P = 26n - n^2.$$

Recall that the rate of change $\frac{dP}{dn}$ is called the *marginal profit.*

a. Make a graph of the profit as a function of the number of widgets manufactured in a week.

b. Determine the marginal profit if the firm produces 10 widgets in a week, and explain what your answer means in practical terms.

c. Determine the marginal profit if the firm produces 15 widgets in a week, and explain what your answer means in practical terms.

d. How many widgets should be produced in a week to maximize profit?

e. Use the calculator to determine the marginal profit at the production level you found in part d. How does your answer compare to what Example 7.4 of Section 7.2 indicates for the marginal profit when the profit is maximized?

7.4   *Equations of Change: Linear and Exponential Functions*

Many natural phenomena can be understood by first analyzing their rates of change, and this analysis can often be used to construct a mathematical model. For example, in 1798 Malthus considered the problem of poverty in terms of the rate of change of population compared to the rate of change of food production: "Population, when unchecked, increases in a geometrical ratio. Subsistence increases only in an arithmetic ratio. A slight acquaintance with numbers will shew the immensity of the first power in comparison with the second."[7] Even earlier, Galileo observed that near the surface of the Earth, downward acceleration due to gravity is constant. Physicists use the letter g to denote this constant, and its value is about 32 feet per second per second. For a freely falling object, the observation that acceleration is constant is actually an observation about $\frac{dV}{dt}$, the rate of change in velocity with respect to time. (We measure downward velocity V in feet per second and time t in seconds.) Furthermore, we can express this in an equation:

$$\text{Acceleration} = \text{Constant}$$

$$\frac{dV}{dt} = g .$$

We will refer to equations of this type as *equations of change* since the equation expresses how the function changes. These equations are a special case of what, in advanced mathematics, science, and engineering, are known more formally as *differential equations*.

Equation of Change for Linear Functions

The key mathematical observation that we need to make about the equation of change $\frac{dV}{dt} = g$ is that it tells us that the rate of change in V is the constant g. But we already know that functions with constant rate of change are linear functions. We conclude that the velocity V is a linear function of time with slope $g = 32$.

Since V is a linear function with slope 32, we know that we can write V as $V = 32t + b$. In order to complete the formula for V, we need only know the initial value b of velocity. For example, if we throw a rock downward with an initial velocity of 10 feet per second, then $b = 10$, and $V = 32t + 10$.

The same analysis holds true for any function with constant rate of change.

---

[7]From page 14 of *First Essay on Population, 1798*, by Thomas Robert Malthus, reprinted in 1926 by Macmillan & Co. Ltd., London.

Equations of Change and Linear Functions

The equation of change $\frac{df}{dx} = m$, where m is a constant, says that f has a constant rate of change m and hence that f is a linear function with slope m. That is, $f = mx + b$. An initial condition is needed to determine the value of b.

We want to emphasize that Key Idea 7.5 is not a new idea. It is just a restatement, in terms of equations of change, of a fundamental property of linear functions that we have already studied in some detail.

EXAMPLE 7.8 **Dropping a Rock on Mars**

On Mars, a falling object satisfies the equation of change

$$\frac{dV}{dt} = 12.16,$$

where V is downward velocity in feet per second and t is time in seconds.

1. What is the value of acceleration due to gravity on Mars?

2. Suppose an astronaut stands atop a cliff on Mars and throws a rock downward with an initial velocity of 8 feet per second. What is the velocity of the rock 3 seconds after release?

Solution to Part 1: The equation of change $\frac{dV}{dt} = 12.16$ tells us that the rate of change in velocity for an object falling near the surface of Mars is 12.16 feet per second per second. Since the rate of change in velocity is acceleration, we conclude that acceleration is 12.16 feet per second per second.

Solution to Part 2: The equation of change $\frac{dV}{dt} = 12.16$ tells us that the rate of change in V has a constant value of 12.16. We conclude that V is a linear function with slope 12.16. Since the rock was tossed downward with a velocity of 8 feet per second, the initial value of V is 8. We now have the information we need to write a formula for the function V:

$$V = 12.16t + 8.$$

We want the velocity of the rock when $t = 3$:

$$V = 12.16 \times 3 + 8 = 44.48 \text{ feet per second}.$$

Instantaneous Rates of Change

Linear functions have a constant rate of change, but for other functions this rate varies, and we need to think more carefully about how to interpret the rate of change. When we study equations of change, we are thinking of instantaneous rates of change, not average rates of change.

For example, a car's speedometer reading gives the *instantaneous* rate of change in distance. To understand this, consider the following question: How do we interpret a reading of 60 miles per hour? It would be inaccurate to interpret this reading as saying that we will actually travel 60 miles in the next hour, since the car could slow down or speed up through the hour. It would be more accurate to say that we expect to go 1 mile in the next minute; this is the same average rate of change, but it is computed over a shorter time interval, so there should be less variation in speed. Still more accurate would be the interpretation that we will go one-sixtieth of a mile (88 feet) in a second. In all three interpretations, the average rate of change is 60 miles per hour, but our common sense tells us that the last interpretation is the most accurate, because computing over a time interval of 1 second is the closest to instantaneous change.

We stress that, even though our rates of change are instantaneous, the units used are the same as for average rates of change. For example, our speedometer readings are measured in miles per hour, not in "miles per instant."

Equation of Change for Exponential Functions

We know that the equation of change for a linear function is $\frac{df}{dx} = m$ because that equation tells us that the rate of change is constant. Let's look at the balance in a savings account as a familiar example of an exponential function and find its equation of change. Suppose we open an account by investing \$100 with a financial institution that advertises an APR of 3%, or (as a decimal) 0.03, with interest compounded *continuously*. Our balance $B = B(t)$, with B in dollars and time t in years since opening the account, is an exponential function since it has a constant percentage rate of change. We want to express the fact that B has a constant percentage rate of change using an equation of change. All that is required is to look more closely at what the description of B tells us. The rate of change $\frac{dB}{dt}$ is the amount by which we expect our account balance to increase in a unit of time. This is just the interest. On an annual basis, the interest is 3% of the balance, or $0.03 \times B$. We find that the equation of change is $\frac{dB}{dt} = 0.03B$.

One question that arises is why we use the APR of 0.03 here and not the APY (annual percentage yield), which takes into account the compounding.[8] The reason is that the rate of change is instantaneous, and during an instant we can ignore compounding. We compute on an annual basis, though, because the unit of measurement is a year, not an instant.

Which exponential function does the equation of change $\frac{dB}{dt} = 0.03B$ represent? To answer this we need to find the yearly growth factor for B. We saw in Exercise 23 of Section 4.1 that when interest is compounded continuously, we find the yearly growth factor by exponentiating the APR. Then the growth factor for our exponential function B

---

[8]For loans, financial institutions often refer to the latter rate as the effective annual rate, or EAR.

is $e^{0.03}$. Since the initial value is 100 dollars (our initial investment), we have the formula

$$B = 100 \times \left(e^{0.03}\right)^t,$$

or, since $e^{0.03}$ is about 1.0305, the approximate formula

$$B = 100 \times 1.0305^t.$$

There is an alternative form of B that avoids this last step of computing the exponential of 0.03 (the APR). To find it we use the fact that $\left(e^{0.03}\right)^t = e^{0.03t}$. Then the alternative form is

$$B = 100e^{0.03t}.$$

This description of the account balance is typical of all exponential functions. If the rate of change in f is a constant multiple of f—that is, if $\frac{df}{dx} = rf$—then f has a constant proportional rate of change and hence is an exponential function. A formula for f in the alternative form is then $f = Pe^{rx}$, where P is the initial value of f. To find the standard form, we first get the growth (or decay) factor by exponentiating:

$$\text{Growth (or decay) factor} = e^r.$$

The formula for f is then $f = P \times \left(e^r\right)^x$. Note that finding the alternative form requires one less step, and it is more accurate since it avoids the rounding error in calculating the growth factor.

In this context we refer to the number r as the *exponential growth rate*. This terminology is borrowed from ecology, where the number r is a measure of the per capita growth rate of a population that is assumed to be growing continuously.

KEY IDEA 7.6

Equations of Change and Exponential Functions

The equation of change $\frac{df}{dx} = rf$, where r is a constant, says that f has a constant proportional (and hence percentage) rate of change and is therefore an exponential function. The exponential growth rate for f is r, so the growth (or decay) factor is e^r. That is,

$$f = Pe^{rx},$$

or

$$f = P \times \left(e^r\right)^x,$$

where P is the initial value of f.

As with equations of change for linear functions, we want to emphasize that Key Idea 7.6 is not really new. It is simply a restatement, in terms of equations of change, of the fundamental properties of exponential functions.

EXAMPLE 7.9 **Newton's Law of Cooling**

Several exercises in Chapters 1 and 4 involved the way objects heat or cool. The formulas used in those exercises were all derived from *Newton's law of cooling*. Isaac Newton observed that when a hot object is placed in the open air to cool, the way it cools depends on the difference between the temperature of the object and the temperature of the air. Specifically, if $D = D(t)$ is the difference between the temperature of the object and the temperature of the air, then the rate of change in D is proportional to D. The value of the constant of proportionality depends on the nature of the cooling object and is normally determined experimentally. For a hot cup of coffee placed on the table to cool, if we measure our time t in minutes since the coffee was poured, the value of the cooling constant is about -0.06 per minute. Assume that coffee poured from the pot is at a temperature of 190 degrees and that room temperature is 72 degrees.

1. Write an equation of change that shows how the coffee cools.

2. Describe the function D and give a formula for it.

3. Find a formula for the temperature of the coffee at time t.

4. What will be the temperature of the coffee after 5 minutes? When will the temperature of the coffee be 130 degrees?

Solution to Part 1:   The function D is the temperature difference, so the rate of change in temperature difference is $\frac{dD}{dt}$. Newton's law of cooling tells us that the rate of change in temperature difference is -0.06 times D:

Rate of change in temperature difference $= -0.06 \times$ Temperature difference

$$\frac{dD}{dt} = -0.06 \times D.$$

Solution to Part 2:   The equation of change we got is that of an exponential function because it gives the rate of change in D as a constant multiple of D.

The constant of proportionality, or cooling constant, is -0.06. Therefore, D is an exponential function with exponential growth rate -0.06. To write a formula for D, we need to know its initial value. When the coffee was poured, its temperature was 190 degrees. Room temperature is 72 degrees. Thus the initial temperature difference, the initial value for D, is $190 - 72 = 118$ degrees, so

$$D = 118e^{-0.06t}.$$

This gives a formula for D in the alternative form. To find the standard form, we first calculate the decay factor: $e^{-0.06}$ is about 0.94. The standard form is then $D = 118 \times 0.94^t$. In the rest of the solution, we will use the alternative form since it is more accurate.

Solution to Part 3:   Let $T = T(t)$ denote the temperature of the coffee at time t. We already know a formula for the difference D between coffee temperature and air temperature. To get a formula for T, we need to add the temperature of the air:

$$\text{Temperature difference} = T - \text{Air temperature}$$
$$T = \text{Air temperature} + \text{Temperature difference}$$
$$T = 72 + D$$
$$T = 72 + 118e^{-0.06t}.$$

Solution to Part 4:   To get the temperature of the coffee after 5 minutes, we put 5 in place of t in the formula we found in part 3:

$$T(5) = 72 + 118e^{-0.06 \times 5} = 159.4 \text{ degrees}.$$

To answer the second question, we want to know when the temperature of the coffee is 130 degrees. That is, we want to solve for t the equation

$$72 + 118e^{-0.06t} = 130.$$

We do this using the crossing-graphs method. First we enter ⎾7.9⏌ the function and the target temperature and record the appropriate correspondences:

$$Y_1 = T, \text{ temperature on vertical axis}$$
$$Y_2 = 130, \text{ target temperature}$$
$$X = t, \text{ minutes on horizontal axis}.$$

Next we make the table of values shown in Figure 7.44 to determine how to set the graphing window. The table shows that we can expect the temperature difference to reach 130 degrees between 10 and 15 minutes after the coffee cup is left to cool. Thus we set up the graphing window using a horizontal span of $t = 0$ to $t = 15$ and a vertical span of $T = 100$ to $T = 250$. In Figure 7.45 we made the graphs and then used the calculator to find ⎾7.10⏌ the intersection point. We see that the coffee reaches a temperature of 130 degrees after about 12 minutes of cooling.   ■ ■ ■

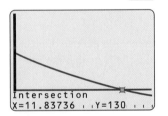

X	Y_1	Y_2
0	190	130
5	159.42	130
10	136.76	130
15	119.98	130
20	107.54	130
25	98.329	130
30	91.505	130

X=0

FIGURE 7.44 A table of values for a cooling cup of coffee

Intersection
X=11.83736   Y=130

FIGURE 7.45 When the temperature reaches 130 degrees

Why Equations of Change?

The equation of change that we encountered in Example 7.9 is of the form $\frac{df}{dx} = rf$, and it is a remarkable fact that this same equation actually describes a diverse set of natural phenomena. As you will see in Exercise 6 at the end of this section, this equation describes how a balloon leaks air. It is not apparent that leaky balloons and cooling coffee cups have anything at all in common, but the temperature of the coffee and the volume of air in the balloon actually behave in a very similar fashion. One important power of mathematics is its ability to distill ideas from many areas into a common idea. This is a good case in point. The following table shows a number of phenomena that all obey an equation of change of the form $\frac{df}{dx} = rf$.

Phenomenon	Equation of Change	Comments
Newton's law of cooling	$$\frac{dD}{dt} = rD$$	D is temperature difference between cooling object and ambient air; r depends on the object.
Exponential population growth	$$\frac{dN}{dt} = rN$$	N is population size; r is the exponential growth rate.
Compound interest	$$\frac{dB}{dt} = rB$$	B is the account balance; r is the APR with continuous compounding, t in years.
Radioactive decay	$$\frac{dA}{dt} = rA$$	A is the amount of radioactive substance remaining; r depends on the radioactive substance.

These and other phenomena can be modeled by exponential functions because they all obey equations of change of the form $\frac{df}{dx} = rf$. These phenomena also have another important feature, which we will see in the next section: In making a mathematical model, it is easiest first to understand the rate of change and then to make an equation of change to model it. In this section, we have exploited the equation of change to get a formula to serve as our model. In the next section, we will see that there are many other ways to get information from equations of change.

7.4   **SKILL BUILDING EXERCISES**

S-1.   **Technical terms:** What is the common mathematical term for an equation of change?

S-2.   **Linear function:** If f satisfies the equation of change $\frac{df}{dx} = m$, what kind of function is f?

S-3.   **Slope:** If f satisfies the equation of change $\frac{df}{dx} = 5$, then f is a linear function. What is the slope of f?

S-4.   **Exponential functions:** If f satisfies the equation of change $\frac{df}{dx} = cf$, what kind of function is f?

S-5.   **Solving an equation of change:** If f satisfies the equation of change $\frac{df}{dx} = 8f$, then f is an exponential function and hence can be written as $f = Ae^{ct}$. Find the value of c.

S-6.   **Why equations of change:** List some commonly occurring phenomena that satisfy the equation of change $\frac{df}{dx} = cf$.

S-7.   **A leaky balloon:** A balloon leaks air (changes volume) at a rate of one-third the volume per minute. Write an equation of change that describes the volume V of air in the balloon at time t in minutes.

S-8.   **Solving an equation of change:** Solve the equation of change from Exercise S-7 if there are initially 4 liters of air in the balloon.

S-9.   **Solving an equation of change:** Solve the equation of change $\frac{df}{dx} = 3$ if the initial value of f is 7.

S-10.   **A falling rock:** Let t be time in seconds, and let $v = v(t)$ be the downward velocity, in feet per second, of a falling rock. The acceleration of the rock has a constant value of 32 feet per second per second. Write an equation of change that is satisfied by v.

7.4   **EXERCISES**

1.   **Looking up:** The constant $g = 32$ feet per second per second is the *downward* acceleration due to gravity near the surface of the Earth. If we stand on the surface of the Earth and locate objects using their distance up from the ground, then the positive direction is up, so down is the negative direction. With this perspective, the equation of change in velocity for a freely falling object would be expressed as

$$\frac{dV}{dt} = -g \,.$$

(We measure upward velocity V in feet per second and time t in seconds.) Consider a rock tossed upward from the surface of the Earth with an initial velocity of 40 feet per second upward.

a. Use a formula to express the velocity $V = V(t)$ as a linear function. (*Hint:* You get the slope of V from the equation of change. The vertical intercept is the initial value.)

b. How many seconds after the toss does the rock reach the peak of its flight? (*Hint:* What is the velocity of the rock when it reaches its peak?)

c. How many seconds after the toss does the rock strike the ground? (*Hint:* How does the time it takes for the rock to rise to its peak compare with the time it takes for it to fall back to the ground?)

2.   **An investment:** You open an account by investing $250 with a financial institution that advertises an APR of 5.25%, with continuous compounding. What account balance would you expect 1 year after making your initial investment?

3.   **A better investment:** You open an account by investing $250 with a financial institution that advertises an APR of 5.75%, with continuous compounding.

a. Find an exponential formula for the balance in your account as a function of time. In your answer, give both the standard form and the alternative form for an exponential function.

b. What account balance would you expect 5 years after your initial investment? Answer this question using both of the forms you found in part a. Which do you think gives a more accurate answer? Why?

4. **Baking muffins:** Chocolate muffins are baking in a 350-degree oven. Let $M = M(t)$ be the temperature of the muffins t minutes after they are placed in the oven, and let $D = D(t)$ be the difference between the temperature of the oven and the temperature of the muffins. Then D satisfies Newton's law of cooling, and its equation of change is

$$\frac{dD}{dt} = -0.04D.$$

   a. If the initial temperature of the muffins is room temperature, 73 degrees, find a formula for D.

   b. Find a formula for M.

   c. The muffins will be done when they reach a temperature of 225 degrees. When should we take the muffins out of the oven?

5. **Borrowing money:** Suppose that you borrow $10,000 at 7% APR and that interest is compounded continuously. The equation of change for your account balance $B = B(t)$ is

$$\frac{dB}{dt} = 0.07B.$$

   Here t is the number of years since the account was opened, and B is measured in dollars.

   a. Explain why B is an exponential function.

   b. Find a formula for B using the alternative form for exponential functions.

   c. Find a formula for B using the standard form for exponential functions. (Round the growth factor to three decimal places.)

   d. Assuming that no payments are made, use your formula from part b to determine how long it would take for your account balance to double.

6. **A leaky balloon:** In Example 7.5 we used a linear function to describe air leaking from a balloon, but we indicated there that this is not, in fact, an accurate description. Let's look more carefully. Suppose a balloon initially holds 2 liters of air. Let $B = B(t)$ denote the volume, in liters, of air in the balloon t seconds after it starts leaking. When a balloon is almost full, the air pressure inside is large, so it leaks rapidly. When it is almost empty, it leaks more slowly. More formally, the rate of change in B is proportional to B. Assume that for this particular balloon, the constant of proportionality is -0.05.

   a. Write an equation of change for B.

   b. Find a formula for B.

   c. How long will it take for half of the air to leak out of the balloon?

7. **A population of bighorn sheep:** A certain group of bighorn sheep live in an area where food is plentiful and conditions are generally favorable to bighorn sheep. Consequently, the population is thriving. There are initially 30 sheep in this group. Let $N = N(t)$ be the population t years later. The population changes each year because of births and deaths. The rate of change in the population is proportional to the number of sheep currently in the population. For this particular group of sheep, the constant of proportionality is 0.04.

   a. Express the sentence "The rate of change in the population is proportional to the number of sheep currently in the population" as an equation of change. (Incorporate in your answer the fact that the constant of proportionality is 0.04.)

   b. Find a formula for N.

   c. How long will it take this group of sheep to grow to a level of 50 individuals?

8. **Water flow:** Water is flowing into a tank at a steady rate of 2 gallons per minute. Let t be the time, in minutes, since the process began, and let V be the volume, in gallons, of water in the tank.

   a. Explain why V is a linear function of t, and write an equation of change for V.

   b. If initially there were 4 gallons of water in the tank, find a formula for V.

   c. The tank can hold 20 gallons. How long will it take to fill the tank?

9. **Growing child:** A certain girl grew steadily between the ages of 3 and 12 years, gaining $5\frac{1}{2}$ pounds each year. Let W be the girl's weight, in pounds, as a function of her age t, in years, between the ages of $t = 3$ and $t = 12$.

   a. Is W a linear function or an exponential function? Be sure to explain your reasoning.

   b. Write an equation of change for W.

   c. Given that the girl weighed 30 pounds at age 3, find a formula for W.

10. **Competing investments:** You initially invest $500 with a financial institution that offers an APR of 4.5%, with interest compounded continuously. Let B be your account balance, in dollars, as a function of the time t, in years, since you opened the account.

    a. Write an equation of change for B.

    b. Find a formula for B.

    c. If you had invested your money with a competing financial institution, the equation of change for your balance M would have been $\frac{dM}{dt} = 0.04M$. If this competing institution compounded interest continuously, what APR would they offer?

11. **Radioactive decay:** The amount remaining A, in grams, of a radioactive substance is a function of time t, measured in days since the experiment began. The equation of change for A is

$$\frac{dA}{dt} = -0.05A.$$

    a. What is the exponential growth rate for A?

    b. If initially there are 3 grams of the substance, find a formula for A.

    c. What is the half-life of this radioactive substance?

12. **Magazines:** Two magazines, *Alpha* and *Beta*, were introduced at the same time with the same circulation of 100. The circulation of *Alpha* is given by the function A, which has the equation of change

$$\frac{dA}{dt} = 0.10A.$$

The circulation of *Beta* is given by the function B, which has the equation of change

$$\frac{dB}{dt} = 10.$$

Here t is the time, in years, since the magazines were introduced.

    a. One of these functions is growing in a linear way, whereas the other is growing exponentially. Identify which is which, and find formulas for both functions.

    b. Which magazine is growing more rapidly in circulation? Be sure to explain your reasoning.

7.5   *Equations of Change: Graphical Solutions*

When an equation of change is of the form $\frac{df}{dx} = c$, we know that f is a linear function with slope c. When it is of the form $\frac{df}{dx} = cf$, we know that f is an exponential function with growth (or decay) factor e^c. Determining which function goes with more complicated equations of change involves calculus and is beyond the scope of this text. Indeed, for many equations of change, even calculus is not powerful enough to allow recovery of the function. Nevertheless, even without being able to find the exact function, we are often able to sketch a graph of it and to find useful information about the phenomena being modeled. The key tool that we will use is the fundamental relationship between a function and its rate of change. When the rate of change is positive, the function is increasing. When the rate of change is negative, the function is decreasing. When the rate of change is momentarily 0, the function may be at a maximum or minimum. When the rate of change is 0 over a period of time, the function is not changing and so is constant over that period.

Equilibrium Solutions

Equilibrium or *steady-state* solutions for equations of change are solutions that never change and hence are constant. Thus they occur where the rate of change is 0 over a period of time. Let's look at an example. The formulas we have used to study falling objects subject to air resistance actually came from an equation of change. When objects fall, they are accelerated downward by gravity. On the other hand, air resistance slows down the fall, which is why a feather falls so slowly. In general, the downward acceleration of a falling object subject to air resistance is given by

$$\text{Acceleration} = \text{Acceleration due to gravity} - \text{Retardation from air resistance}. \quad (7.1)$$

This drag due to air resistance is in practice difficult to determine, and engineers use wind tunnels and other high-tech devices to help in the design of more aerodynamically efficient cars, airplanes, and rockets. The simplest model for air resistance assumes that the drag is proportional to the velocity. That is,

$$\text{Retardation due to drag} = rV,$$

where V is the downward velocity and r is a constant known as the *drag coefficient*. We will show in Exercise 3 one way in which its value can be experimentally determined. Downward acceleration is the rate of change in velocity with respect to time, $\frac{dV}{dt}$. We put these bits of information into Equation (7.1) to get the equation of change:

$$\underset{\substack{\dfrac{dV}{dt}}}{\text{Acceleration}} = \underset{\substack{g}}{\substack{\text{Acceleration} \\ \text{due to gravity}}} - \underset{\substack{rV}}{\substack{\text{Retardation} \\ \text{from air resistance}}}$$

Thus the velocity of an object falling subject to air resistance follows the equation of change $\frac{dV}{dt} = g - rV$. The acceleration due to gravity is $g = 32$ feet per second per second, and for an average-size man, the drag coefficient has been determined to be approximately $r = 0.1818$ per second. Hence a skydiver falls subject to the equation of change

$$\frac{dV}{dt} = 32 - 0.1818V. \tag{7.2}$$

(We measure velocity V in feet per second and time t in seconds.)

Let's find the equilibrium solutions of Equation (7.2) and investigate their physical meaning. Equilibrium solutions are those that never change—that is, those for which the rate of change is 0. To find them, we should put 0 in for $\frac{dV}{dt}$ and solve the resulting equation for V:

$$0 = 32 - 0.1818V.$$

This is a linear equation, which we proceed to solve. (You may if you wish solve it with the calculator.) We get

$$0 = 32 - 0.1818V$$

$$0.1818V = 32$$

$$V = \frac{32}{0.1818} = 176 \text{ feet per second}.$$

Thus $V = 176$ feet per second is an equilibrium or steady-state solution. When $V = 176$, then $\frac{dV}{dt}$ is zero. Consequently, if the skydiver reaches a velocity of 176 feet per second, he will continue to fall at that same speed until the parachute opens. This is the velocity at which the downward force of gravity matches retardation due to air resistance: the terminal velocity of the skydiver. To emphasize this, we have graphed the steady-state solution $V = 176$ in Figure 7.46. The fact that the graph of velocity versus time is a horizontal line emphasizes that velocity is not changing. Such horizontal lines are typical of equilibrium solutions.

There are two important things to note here. One is that the equilibrium solution was easy to find from the equation of change. The second is that the equilibrium solution had an important physical interpretation. It is the velocity we would expect the sky diver to have in the long term. This is typical of steady-state solutions: They yield important information, and that is why we use equations of change to find them.

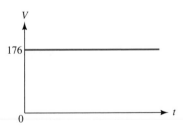

FIGURE 7.46 The equilibrium solution $V = 176$ for the equation of change of a skydiver

Equilibrium or Steady-State Solutions

Equilibrium or steady-state solutions of an equation of change occur where the rate of change is 0. We get them by setting the rate of change equal to 0 and solving. In many cases they show long-term behavior.

On several occasions earlier in this text, we have looked at logistic population growth. Logistic growth is one of many growth models that assume there is an upper limit, the *carrying capacity*, beyond which the population cannot grow. In essentially all of these models the population grows rapidly at first, but the growth rate slows as the population approaches the carrying capacity. The logistic model is perhaps the simplest model for limited growth and was first proposed in 1838 by P. F. Verhulst. In its original formulation, it was presented as an equation of change.

If a population exhibits logistic growth in an area with a carrying capacity K, then the logistic equation for the population $N = N(t)$ can be written

$$\frac{dN}{dt} = rN\left(1 - \frac{N}{K}\right).$$

The constants K and r depend on the species and on environmental factors, and their values are normally determined experimentally. The number N is sometimes the actual number of individuals, but more commonly it is the combined weight of the population (the *biomass*). In several examples and exercises in this text, we have presented logistic growth with a formula. That formula was obtained by using calculus to solve the equation of change above, a feat we will not reproduce here. Rather, we will gain information directly from the equation of change itself. The logistic model of growth is useful in the study of some species, and we illustrate its use in the case of Pacific sardines.

EXAMPLE 7.10   Logistic Population Growth: Sardines

In California in the 1930s and 1940s, a large part of the fishing-related industry was based on the catch of Pacific sardines. Studies have shown that the sardine population grows approximately logistically; moreover, the numbers K and r have been determined experimentally to be $K = 2.4$ million tons and $r = 0.338$ per year.[9] For Pacific sardines, our logistic equation is therefore

$$\frac{dN}{dt} = 0.338N\left(1 - \frac{N}{2.4}\right),\tag{7.3}$$

where N is measured in millions of tons of fish and t is measured in years.

---

[9]From a study by G. I. Murphy, in "Vital statistics of the Pacific sardine (*Sardinops caerulea*) and the population consequences," *Ecology* **48** (1967), 731–736.

Find the equilibrium solutions of Equation (7.3) and give a physical interpretation of their meaning.

Solution: We get the equilibrium solutions where $\frac{dN}{dt}$ is zero. Thus we want to solve the equation

$$0.338N\left(1 - \frac{N}{2.4}\right) = 0.$$

We use the single-graph method to do that. To make and understand the graph, it is crucial that we identify correspondences among the variables. The importance of carefully employing this practice with all graphs made in this section cannot be overemphasized. In this case we have

$$\mathsf{Y_1} = \frac{dN}{dt}, \text{ rate of change in biomass on vertical axis}$$

$$\mathsf{X} = N, \text{ biomass on horizontal axis}.$$

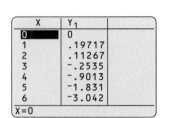

FIGURE 7.47 A table of values to set up a graphing window

Now we enter ⟨ 7.11 ⟩ the formula for the function $\frac{dN}{dt}$ given in Equation (7.3) and look at the table of values in Figure 7.47 to see how to set up the graphing window. The table shows that $N = 0$ is one equilibrium solution and that another occurs between $N = 2$ and $N = 3$. Allowing a little extra room, we use a window setup with a horizontal span of $N = 0$ to $N = 3$ and a vertical span of $\frac{dN}{dt} = -0.2$ to $\frac{dN}{dt} = 0.3$. This gives the graph in Figure 7.48. Using the calculator, we get ⟨ 7.12 ⟩ the second equilibrium solution $N = 2.4$ as shown in Figure 7.48.

We have two equilibrium solutions, $N = 0$ and $N = 2.4$ million tons. If the biomass ever reaches either of these levels, it will remain there and not change. This is certainly obvious when $N = 0$: If there are no fish present, no new ones will be born. The solution $N = 2.4$ million tons is the environmental carrying capacity for sardines in this region. If the biomass ever reaches 2.4 million tons of fish, then environmental limitations match growth tendencies, and the population stays the same.

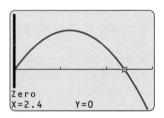

FIGURE 7.48 Getting an equilibrium solution from the graph of $\frac{dN}{dt}$ versus N

We emphasize once more that Figure 7.48 does not show the graph of N versus t. It does not directly give the size of the fish biomass. Rather, it is the graph of $\frac{dN}{dt}$ versus N. It shows the growth rate that can be expected for a given biomass. Our next topic is how to use this graph to sketch a graph of the biomass itself.

■ ■ ■

Sketching Graphs

Equilibrium solutions for equations of change can yield important information about the physical situation that they describe, but the fundamental relationship between a function and its rate of change can tell us even more. To see this, let's continue our analysis of the equation of change $\frac{dN}{dt} = 0.338N\left(1 - \frac{N}{2.4}\right)$ for Pacific sardines. We know that if the biomass today is either 0 or 2.4 million tons, then in the future, the biomass will not change. Can we say what is to be expected to happen in the future if there are today 0.4 million tons of Pacific sardines? The key is the graph of $\frac{dN}{dt}$ versus N that we made in Figure 7.48. Table 7.1 was made by looking at where the graph is above the horizontal axis and where it is below.

Range for N	From 0 to 2.4	Greater than 2.4
Sign of $\frac{dN}{dt}$	Positive	Negative

TABLE 7.1 The sign of $\frac{dN}{dt}$

Range for N	From 0 to 2.4	Greater than 2.4
Sign of $\frac{dN}{dt}$	Positive	Negative
Effect on N	Increasing	Decreasing

TABLE 7.2 The information we get from the graph of $\frac{dN}{dt}$ versus N

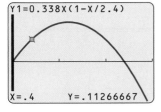

FIGURE 7.49 When $N = 0.4$, the rate of change $\frac{dN}{dt}$ is positive

The key to understanding what happens to the population N is the fundamental property of rates of change: When $\frac{dN}{dt}$ is positive, N is increasing, and when $\frac{dN}{dt}$ is negative, N is decreasing. In Table 7.2, we have added another row to Table 7.1 to include this information.

The information we get from this table is that whenever N is between 0 and 2.4 million tons, the graph of N versus t will increase; that is, the biomass will grow. When N is greater than 2.4 million tons, the graph of N versus t will decrease; that is, the biomass will decrease over time. Since $N = 0.4$ is between 0 and 2.4, we know that the graph of N versus t will start out increasing. For further evidence of this, in Figure 7.49 we have put the graphing cursor at $N = 0.4$, and we see that the value of $\frac{dN}{dt}$ is 0.11, a positive number. Since $\frac{dN}{dt}$ is positive, N is increasing. Thus if we start with a biomass of 0.4 million tons, we can expect the biomass to increase.

We want to start with this bit of information and make a hand-drawn sketch of the graph of N versus t. We started our picture in Figure 7.50 by first drawing the equilibrium solutions $N = 0$ and $N = 2.4$ and indicating what we already know about how the graph starts out.

To complete the graph of N versus t, we see from Figure 7.48 (or from Table 7.2) that $\frac{dN}{dt}$ remains positive as long as N is less than the equilibrium solution of $N = 2.4$ million tons. This tells us that the graph of N continues to increase toward the horizontal line that represents the equilibrium solution, as we have drawn in Figure 7.51. Note that the horizontal lines representing equilibrium solutions provided us with important guides in sketching the graph of N versus t. Whenever you are asked to sketch a graph

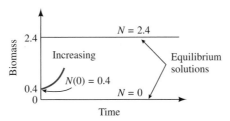

FIGURE 7.50 Equilibrium solutions and beginning the graph of N versus t

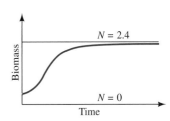

FIGURE 7.51 Completing the graph of biomass versus time

from an equation of change, you should first look to see whether there are equilibrium solutions. If so, draw them in before you start.

As we expected, the graph in Figure 7.51 looks like the classic logistic curve. But we would emphasize that it is a free-hand drawing and that graphs made in this way cannot be expected to show accurate function values. What is important is to use information about the rate of change properly to get the shape of the graph as nearly correct as is possible.

Let's analyze Pacific sardines in one final situation. What will the biomass be in the future if today it is 2.8 million tons? We proceed as before. In Figure 7.52, on the graph of $\frac{dN}{dt}$ versus N we have put the cursor at $N = 2.8$ million tons. We see that when the biomass is at this level, $\frac{dN}{dt} = -0.16$, a negative number. Thus the graph of N versus t will be decreasing. In Figure 7.53 we have started our graph just as we did in the first case.

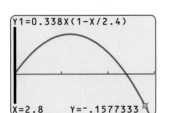

FIGURE 7.52 When biomass is 2.8 million tons, $\frac{dN}{dt}$ is negative

We see further from Figure 7.48 (or from Table 7.2) that as long as N is greater than the equilibrium solution $N = 2.4$, $\frac{dN}{dt}$ will remain negative. Thus the graph of N versus t continues to decrease toward the horizontal line that represents the equilibrium solution. Our completed graph is shown in Figure 7.54. The graph in Figure 7.54 is reasonable because the biomass started out greater than the environmental carrying capacity of 2.4 million tons. The environment cannot support that many fish, so we expect the biomass to decrease.

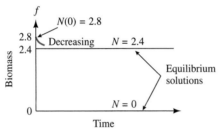

FIGURE 7.53 Starting the graph of N versus t when $N(0) = 2.8$

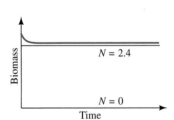

FIGURE 7.54 When population is greater than carrying capacity

KEY IDEA 7.8

Using the Equation of Change to Graph Functions

To make a hand-drawn sketch of the graph of f versus x from the equation of change for f, first draw horizontal lines representing any obvious equilibrium solutions. Next, use the calculator to graph $\frac{df}{dx}$ versus f. (If the equation of change is of the form

$$\frac{df}{dx} = \text{Right-hand side}$$

you graph the right-hand side.)

1. Where the graph of $\frac{df}{dx}$ versus f crosses the horizontal axis, we have $\frac{df}{dx} = 0$, so f does not change. This is an equilibrium solution for f.

2. When the graph of $\frac{df}{dx}$ versus f is above the horizontal axis, the graph of f versus x is increasing.

3. When the graph of $\frac{df}{dx}$ versus f is below the horizontal axis, the graph of f versus x is decreasing.

EXAMPLE 7.11   Further Analysis of Falling Objects

Earlier we found the equilibrium solution $V = 176$ feet per second for the equation of change

$$\frac{dV}{dt} = 32 - 0.1818V$$

which governs the velocity of a skydiver.

1. Sketch the graph of the skydiver's velocity if the initial downward velocity is 0. (This might occur if the skydiver jumped from an airplane that was flying level.)

2. Sketch the graph of the skydiver's velocity if the initial downward velocity is 200 feet per second. (This might occur if a pilot ejected from an airplane that was diving out of control.)

Solution to Part 1:   We enter 7.13 the right-hand side of the equation of change in the calculator and record the important correspondences:

$$Y_1 = \frac{dV}{dt}, \quad \text{acceleration on vertical axis}$$

$$X = V, \quad \text{velocity on horizontal axis}.$$

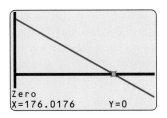

Zero
X=176.0176   Y=0

FIGURE 7.55 The graph of $\frac{dV}{dt}$, acceleration, versus velocity V

To get the graph of $\frac{dV}{dt}$ versus V shown in Figure 7.55, we wanted to be sure the value $V = 200$, which will be needed in part 2, was included. We used 7.14 a horizontal span of $V = 0$ to $V = 250$, and we got the vertical span $\frac{dV}{dt} = -20$ to $\frac{dV}{dt} = 35$ from a table of values. We know already that the crossing point $V = 176$ shown in Figure 7.55 corresponds to the equilibrium solution that represents terminal velocity.

We have recorded in Table 7.3 the information about V that we get from Figure 7.55. Since the initial velocity for this part is 0, we know from Table 7.3 that the graph of V starts out increasing. We make the graph of V versus t in two steps. In Figure 7.56 we have graphed the equilibrium solution and started our increasing graph. Consulting Table 7.3 once again, we see that velocity will continue to increase

Range for V	Less than 176	Greater than 176
Sign of $\dfrac{dV}{dt}$	Positive	Negative
Effect on V	Increasing	Decreasing

TABLE 7.3 The information we get from the graph of $\frac{dV}{dt}$ versus V

toward the terminal velocity of 176 feet per second. Thus the graph will continue to increase toward the horizontal line representing this equilibrium solution. Our graph is shown in Figure 7.57.

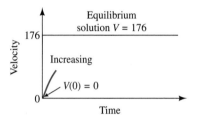

FIGURE 7.56 Starting the graph of V versus t

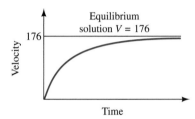

FIGURE 7.57 Completing the graph of V versus t

Solution to Part 2: Since the initial velocity is 200 feet per second, we see from Table 7.3 that velocity is decreasing. This bit of information is incorporated into the start of our graph in Figure 7.58. But Table 7.3 also shows that velocity will continue to decrease toward the equilibrium solution of 176 feet per second. Thus, in Figure 7.59 we completed the graph by making it decrease toward the horizontal line corresponding to terminal velocity. We note that the graph in Figure 7.59 makes sense because, at the skydiver's large initial velocity, the retardation due to air resistance is greater than the downward acceleration due to gravity. Thus his speed will decrease toward the terminal velocity. ■ ■ ■

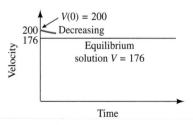

FIGURE 7.58 Starting the graph when $V(0) = 200$

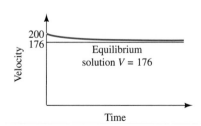

FIGURE 7.59 The completed graph of V versus t for $V(0) = 200$

7.5 SKILL BUILDING EXERCISES

S-1. **Equilibrium solutions:** What is an equilibrium solution of an equation of change?

S-2. **More equilibrium solutions:** If $f = 10$ is an equilibrium solution of an equation of change involving $\frac{df}{dx}$, what is the value of $\frac{df}{dx}$ when f is 10?

S-3. **Finding equilibrium solutions:** Find an equilibrium solution of $\frac{df}{dx} = 2f - 6$.

S-4. **Sketching graphs:** Consider the equation of change $\frac{df}{dx} = 2f - 6$.

    a. What is the sign of $\frac{df}{dx}$ when f is less than 3?

    b. What can you say about the graph of f versus x when f is less than 3?

    c. What is the sign of $\frac{df}{dx}$ when f is greater than 3?

    d. What can you say about the graph of f versus x when f is greater than 3?

S-5. **Water:** Water flows into a tank, and a certain part of it drains out through a valve. The volume v in cubic feet of water in the tank at time t satisfies the equation $\frac{dv}{dt} = 5 - (v/3)$. If the process continues for a long time, how much water will be in the tank?

S-6. **Water:** At some time in the process described in Exercise S-5, there are 23 cubic feet of water in the tank. Is the volume of water increasing or decreasing?

S-7. **Water:** At some time in the process described in Exercise S-5, there are 8 cubic feet of water in the tank. Is the volume increasing or decreasing?

S-8. **Population:** A certain population grows according to $\frac{dN}{dt} = 0.03N \left(1 - \frac{N}{6300}\right)$. What is the carrying capacity of the environment for this particular population?

S-9. **Population:** At some time, the population described in Exercise S-8 is at the level $N = 4238$. Is the population level increasing or decreasing?

S-10. **Population:** At some time, the population described in Exercise S-8 is at the level $N = 8716$. Is the population level increasing or decreasing?

7.5 EXERCISES

1. **Equation of change for logistic growth:** The logistic growth formula $N = 6.21/(0.035 + 0.45^t)$ that we used in Chapter 2 for deer on the George Reserve actually came from the following equation of change:

$$\frac{dN}{dt} = 0.8N \left(1 - \frac{N}{177}\right).$$

    a. Plot the graph of $\frac{dN}{dt}$ versus N, and use it to find the equilibrium solutions. Explain their physical significance.

    b. In one plot, sketch the equilibrium solutions and graphs of N versus t in each of the two cases $N(0) = 10$ and $N(0) = 225$.

    c. To what starting value for N does the solution $N = 6.21/(0.035 + 0.45^t)$ correspond?

2. **A catfish farm:** Catfish in a commercial pond can be expected to exhibit logistic population growth. Consider a pond with a carrying capacity of $K = 4000$ catfish. Take the r value for catfish in this pond to be $r = 0.06$.

    a. Write the equation of change for logistic growth of the catfish population. (*Hint:* If you have difficulty here, refer to Example 7.10.)

    b. Make a graph of $\frac{dN}{dt}$ versus N.

    c. For what values of N would the catfish population be expected to increase?

    d. For what values of N would the catfish population be expected to decrease?

    e. There are several strategies that can be applied to populations that are regularly harvested. The *maximum sustainable yield* model says that a renewable resource should be maintained at a level where its growth rate is at a maximum, since this allows the population to replenish itself quickly. According to this model, at what level should the catfish population be maintained?

3. **Experimental determination of the drag coefficient:** When retardation due to air resistance is proportional to downward velocity V, in feet per second, falling objects obey the equation of change

$$\frac{dV}{dt} = 32 - rV,$$

where r is known as the *drag coefficient*. One way to measure the drag coefficient is to measure and record terminal velocity.

a. We know that an average-size man has a terminal velocity of 176 feet per second. Use this to show that the value of the drag coefficient is $r = 0.1818$ per second. (*Hint:* To say that the terminal velocity is 176 feet per second means that when the velocity V is 176, velocity will not change. That is, $\frac{dV}{dt} = 0$. Put these bits of information into the equation of change and solve for r.)

b. An ordinary coffee filter has a terminal velocity of about 4 feet per second. What is the drag coefficient for a coffee filter?

4. **Other models of drag due to air resistance:** When some objects move at high speeds, air resistance has a more pronounced effect. For such objects, retardation due to air resistance is often modeled as being proportional to the square of velocity or to even higher powers of velocity.

a. A rifle bullet fired downward has an initial velocity of 2100 feet per second. If we use the model that gives air resistance as the square of velocity, then the equation of change for the downward velocity V, in feet per second, of a rifle bullet is

$$\frac{dV}{dt} = 32 - rV^2.$$

If the terminal velocity for a bullet is 1600 feet per second, find the drag coefficient r.

b. A meteor may enter the Earth's atmosphere at a velocity as high as 90,000 feet per second. If the downward velocity V, in feet per second, of a meteor in the Earth's atmosphere is governed by the

equation of change

$$\frac{dV}{dt} = 32 - 2 \times 10^{-18} V^4,$$

how fast will it be traveling when it strikes the ground, assuming that it has reached terminal velocity?

5. **Fishing for sardines:** *This is a continuation of Example 7.10.* If we take into account an annual fish harvest of F million tons of fish, then the equation of change for Pacific sardines becomes

$$\frac{dN}{dt} = 0.338N \left(1 - \frac{N}{2.4}\right) - F.$$

a. Suppose that there are currently 1.8 million tons of Pacific sardines off the California coast and that you are in charge of the commercial fishing fleet. It is your goal to leave the Pacific population of sardines as you found it. That is, you wish to set the fishing level F so that the biomass of Pacific sardines remains stable. What value of F will accomplish this? (*Hint:* You want to choose F so that the current biomass level of 1.8 million tons is an equilibrium solution.)

b. For the remainder of this exercise, take the value of F to be 0.1 million tons per year. That is, assume the catch is 100,000 tons per year.

i. Make a graph of $\frac{dN}{dt}$ versus N, and use it to find the equilibrium solutions.

ii. For what values of N will the biomass be increasing? For what values will it be decreasing?

iii. On the same graph, sketch all equilibrium solutions and the graphs of N versus t for each of the initial populations $N(0) = 0.3$ million tons, $N(0) = 1.0$ million tons, and $N(0) = 2.3$ million tons.

iv. Explain in practical terms what the picture you made in part iii tells you. Include in your explanation the significance of the equilibrium solutions.

6. **Sprinkler irrigation in Nebraska:** Logistic growth can be used to model not only population growth but also economic and other types of growth. For example, the total number of acres $A = A(t)$,

in millions, in Nebraska that are being irrigated by modern sprinkler systems has shown approximate logistic growth since 1955, closely following the equation of change

$$\frac{dA}{dt} = 0.15A \left(1 - \frac{A}{3}\right).$$

Here time t is measured in years.

a. According to this model, how many total acres in Nebraska can be expected eventually to be irrigated by sprinkler systems? (*Hint:* This corresponds to the carrying capacity in the logistic model for population growth.)

b. How many acres of land were under sprinkler irrigation when sprinkler irrigation was expanding at its most rapid rate?

7. **Logistic growth with a threshold:** Most species have a survival *threshold* level, and populations of fewer individuals than the threshold cannot sustain themselves. If the carrying capacity is K and the threshold level is S, then the logistic equation of change for the population $N = N(t)$ is

$$\frac{dN}{dt} = -rN \left(1 - \frac{N}{S}\right)\left(1 - \frac{N}{K}\right).$$

For Pacific sardines, we may use $K = 2.4$ million tons and $r = 0.338$ per year, as in Example 7.10. Suppose we also know that the survival threshold level for the sardines is $S = 0.8$ million tons.

a. Write the equation of change for Pacific sardines under these conditions.

b. Make a graph of $\frac{dN}{dt}$ versus N and use it to find the equilibrium solutions. How do the equilibrium solutions correspond with S and K?

c. For what values of N is the graph of N versus t increasing, and for what values is it decreasing?

d. Explain what can be expected to happen to a population of 0.7 million tons of sardines.

e. At what population level will the population be growing at its fastest?

8. **Chemical reactions:** In a *second-order reaction*, one molecule of a substance A collides with one molecule of a substance B to produce a new substance, the *product*. If t denotes time and $x = x(t)$ denotes the concentration of the product, then its rate of change $\frac{dx}{dt}$ is called the *rate of reaction*. Suppose the initial concentration of A is a and the initial concentration of B is b. Then, assuming a constant temperature, x satisfies the equation of change

$$\frac{dx}{dt} = k(a - x)(b - x)$$

for some constant k. This is because the rate of reaction is proportional both to the amount of A that remains untransformed and to the amount of B that remains untransformed. Here we study a reaction between isobutyl bromide and sodium ethoxide in which $k = 0.0055$, $a = 51$, and $b = 76$. The concentrations are in moles per cubic meter, and time is in seconds.[10]

a. Write the equation of change for the reaction between isobutyl bromide and sodium ethoxide.

b. Make a graph of $\frac{dx}{dt}$ versus x. Include a span of $x = 0$ to $x = 100$.

c. Explain what can be expected to happen to the concentration of the product if the initial concentration of the product is 0.

9. **Competition between bacteria:** Suppose there are two types of bacteria, type A and type B, in a place with limited resources. If the bacteria had unlimited resources, both types would grow exponentially, with exponential growth rates a for type A and b for type B. Let P be the *proportion* of type A bacteria, so P is a number between 0 and 1. For example, if $P = 0$, then there are no type A bacteria and all are of type B. If $P = 0.5$, then half of the bacteria are of type A and half are of type B. Each population of bacteria grows in competition for the limited resources, so the proportion P changes over time.

---

[10]From a study by I. Dostrovsky and E. D. Hughes, as described by Gordon M. Barrow in *Physical Chemistry*, 4th ed. (New York: McGraw-Hill, 1979).

The function P is subject to the equation of change

$$\frac{dP}{dt} = (a - b)P(1 - P).$$

Suppose $a = 2.3$ and $b = 1.7$.

a. Does P have a logistic equation of change?

b. What happens to the populations if initially $P = 0$?

c. What happens to P in the long run if $P(0)$ is positive (but not zero)? In this case, does it matter what the exact value of $P(0)$ is?

10. **Growth of fish:** Let $w = w(t)$ denote the weight of a fish as a function of its age t. For the North Sea cod, the equation of change

$$\frac{dw}{dt} = 2.1w^{2/3} - 0.6w$$

holds. Here w is measured in pounds and t in years.

a. Explain what $\frac{dw}{dt}$ means in practical terms.

b. Make a graph of $\frac{dw}{dt}$ against w. Include weights up to 45 pounds.

c. What is the weight of the cod when it is growing at the greatest rate?

d. To what weight does the cod grow?

11. **Grazing sheep:** The amount C of food consumed in a day by a merino sheep is a function of the amount V of vegetation available. The equation of change for C is

$$\frac{dC}{dV} = 0.01(2.8 - C).$$

Here C is measured in pounds and V in pounds per acre.

a. Explain what $\frac{dC}{dV}$ means in practical terms.

b. Make a graph of $\frac{dC}{dV}$ versus C. Include consumption levels up to 3 pounds.

c. What is the most you would expect a merino sheep to consume in a day?

Summary

This chapter develops the notion of the rate of change for a function. Earlier in the book, rates of change arose in contexts such as studying tabulated functions using average rates of change, analyzing graphs in terms of concavity, and characterizing linear functions as having a constant rate of change. In the current chapter we see the value of a unified approach to rates of change, and we apply this point of view to a variety of important real-world problems.

Velocity

Velocity is the rate of change in location, or directed distance. There are just a few basic rules that enable us to relate directed distance and velocity.

- When directed distance is increasing, velocity is positive.
- When directed distance is decreasing, velocity is negative.
- When directed distance is not changing, even for an instant, velocity is zero.
- When velocity is constant (for example, when the cruise control on a car is set), directed distance is a linear function of time, so its graph is a straight line.

Using these basic rules, we can sketch graphs of directed distance and velocity from a verbal description of how an object moves.

General Rates of Change

The idea of velocity as the rate of change in directed distance applies to any function. The rate of change of a function $f = f(x)$ is denoted by $\frac{df}{dx}$. It tells how much f is expected to change if x increases by 1 unit. For example, if V is the velocity of an object as a function of time t, then $\frac{dV}{dt}$ is the change in velocity that we expect in 1 unit of time. This is the acceleration of the object. For example, a car might gain 2 miles per hour in velocity each second.

As with velocity, there are just a few basic rules that express the fundamental relationship between a function f and its rate of change $\frac{df}{dx}$.

- When f is increasing, $\frac{df}{dx}$ is positive.
- When f is decreasing, $\frac{df}{dx}$ is negative.
- When f is not changing, $\frac{df}{dx}$ is zero.
- When $\frac{df}{dx}$ is constant, f is a linear function of x.

One important observation is that we expect the rate of change to be zero at a maximum or minimum of a function. For example, say the CEO of a tire company wants to maximize profit by adjusting the production level. Figure 7.60 is a graph of marginal profit, which is the rate of change in profit as a function of the production level. Using the graph, we see that marginal profit is positive if the production level is less than 160 tires per day, so the profit increases up to this level. If the level is more than 160 tires per day, then marginal profit is negative, and the profit is decreasing. The maximum profit

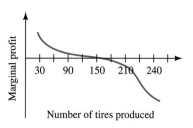

FIGURE 7.60 Marginal profit for a tire company

occurs where the marginal profit is zero, and that is at a production level of 160 tires per day.

Estimating Rates of Change

The rate of change of a function is again a function, and estimating the rate of change at a point can be of practical importance. The method we use depends on how the function is presented.

For a function $f = f(x)$ given by a table, we can estimate $\frac{df}{dx}$ by using the average rate of change given by

$$\frac{\text{Change in } f}{\text{Change in } x} .$$

For a function given by a formula, we can make a graph and then have the calculator estimate the rate of change at a point.

Equations of Change

An equation of change for a function $f = f(x)$ is an equation expressing the rate of change $\frac{df}{dx}$ in terms of f and x. Equations of change arise in many applications where we are given a formula not for the function itself but for its rate of change.

The equation of change $\frac{df}{dx} = m$, where m is a constant, says that f has a constant rate of change m, so f is a linear function with slope m. Once we know the initial value b of f, we can write down the linear formula $f = mx + b$.

The equation of change $\frac{df}{dx} = rf$, where r is a constant, says that f has a constant proportional (or percentage) of change, so f is an exponential function. The exponential growth rate is r, so the growth (or decay) factor is e^r. Once we know the initial value P of f, we can write down the exponential formula $f = Pe^{rx}$, or $f = P \times (e^r)^x$.

Many equations of change that arise in practice do not have the simple characterization we found for linear and exponential functions. But we can still get important information about a function by using the fundamental relationship between the function and its rate of change. The first step is to find the equilibrium solutions of an equation of change. These are the constant solutions, so we find them by setting the rate of change equal to zero. In many cases they show the long-term behavior we can expect. The next step after finding equilibrium solutions is to determine for what values of the function its rate of change is positive or negative, since this will tell us where the function is increasing or decreasing. In many cases of interest, these two steps can be accomplished by analyzing a graph of the rate of change versus the function. After these two steps, we can sketch a graph of the function for a given initial condition.

For example, the population N of deer in a certain area as a function of time t in years has the logistic equation of change

$$\frac{dN}{dt} = 0.5N\left(1 - \frac{N}{150}\right).$$

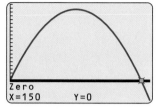

FIGURE 7.61 A graph of $\frac{dN}{dt}$ versus N

A graph of $\frac{dN}{dt}$ versus N is shown in Figure 7.61. (We used a horizontal span of 0 to 160 and a vertical span of -5 to 20.) In addition to the equilibrium solution $N = 0$, from the graph we find the equilibrium solution $N = 150$. This is the first step. For the second step we make Table 7.4 on the basis of the graph in Figure 7.61.

The table tells us, for example, that an initial population of 10 deer should increase in time to a size of 150.

Range for N	From 0 to 150	Greater than 150
Sign of $\dfrac{dN}{dt}$	Positive	Negative
Effect on N	Increasing	Decreasing

TABLE 7.4 The information we get from the graph of $\frac{dN}{dt}$ versus N

Chapter

8 Mathematics of Population Ecology

Population ecology is the study of the structure and growth of populations in a particular environment, and mathematical models play an important role in this study. Investigations have been conducted for a wide variety of populations, ranging from yeast cultures to insects to humans. It is noteworthy that certain fundamental principles apply even in the midst of this diversity. This chapter will explore the mathematics associated with some of these principles.

8.1 *Population Dynamics: Exponential Growth*

Population dynamics is the study of the rate of change of populations over time. The number N of individuals in a population is a function of time t, and the rate of change of the population is then $\frac{dN}{dt}$. If this quantity is positive at a given time, the population is growing then; if it is negative, the population is declining.

The two basic factors that affect the rate of change of a population are the number of births B and the number of deaths D per unit of time.[1] It should be emphasized that B and D are functions of time, and in general they will change over time. The basic equation in population dynamics is

Rate of change of population = Births per unit of time − Deaths per unit of time ,

or, expressed as an equation of change,

$$\frac{dN}{dt} = B - D. \tag{8.1}$$

Clearly the quantities B and D should be interpreted in terms of the size of the population: We expect more births in a population of a million individuals than we do in a population of a hundred, other factors being the same. One way to measure individual

---

[1]Other factors are immigration and emigration, which measure how populations disperse. Ignoring these will simplify the discussion.

contributions to population changes is to introduce the *birth rate* and the *death rate*. The birth rate b is defined as the number of births per unit of time divided by the size of the population: $b = \frac{B}{N}$, or $B = bN$. The death rate d is defined in a similar way. It is the number of deaths per unit of time divided by the size of the population: $d = \frac{D}{N}$, or $D = dN$. Then the basic equation, Equation (8.1), becomes

$$\frac{dN}{dt} = bN - dN,$$

or

$$\frac{dN}{dt} = (b - d)N. \tag{8.2}$$

Exponential Growth

The simplest model of population dynamics involves exponential growth. For this we assume that the number of births B and the number of deaths D per unit of time are both proportional to the size of the population, or, equivalently, that the birth rate b and the death rate d are both constant. Then, by Equation (8.2), the rate of change of the population is proportional to the size of the population, where the constant of proportionality r is defined by $r = b - d$. With these modifications we can write Equation (8.2) as

$$\frac{dN}{dt} = rN. \tag{8.3}$$

We know from Section 7.4 that the equation of change given in Equation (8.3) is characteristic of exponential functions, and we conclude that

$$N(t) = N(0)e^{rt}, \tag{8.4}$$

where $N(0)$ is the initial size of the population. Note that some ecologists prefer to write the time variable as a subscript; they would write Equation (8.4) as

$$N_t = N_0 e^{rt}.$$

The constant r is the exponential growth rate, a measure of the intrinsic per capita growth rate. It represents the population's capacity for growth in the absence of constraints. It is measured in units such as per year or per day. (Note that even though its units involve discrete time, such as a year, r represents an instantaneous rate of change.) Clearly, if the birth rate is larger than the death rate, then $r = b - d$ is positive, and the population is growing exponentially. If the birth rate is smaller than the death rate, then r is negative, and the population is declining exponentially. If the birth rate and the death rate are equal, then r is zero, and the population size does not change. We are interested in the case when r is positive. We should also note that the use of the letter r is standard here, and ecologists often refer to the r value of a species in an environment.

To understand the rapid growth implied by this model, we consider an example.

EXAMPLE 8.1 **Aphid Population Growth**

There are initially 100 individuals in an aphid population that has an exponential growth rate of $r = 0.243$ per day.[2]

1. How many aphids will there be after 12 hours? After 2 weeks?

2. How long will it take an initial population of 100 aphids to double in size?

3. What is the doubling time if instead the initial population is 1000?

Solution to Part 1: With an initial population of 100, we have from Equation (8.4) that

$$N = 100e^{0.243t},$$

with t measured in days. Because 12 hours make half of a day, we compute

$$N(0.5) = 100e^{0.243 \times 0.5} = 113,$$

rounded to the nearest integer. After 12 hours there are about 113 aphids.
Two weeks is 14 days, and proceeding as above, we find that

$$N(14) = 100e^{0.243 \times 14} = 3002,$$

rounded to the nearest integer. There are just over 3000 aphids after two weeks.

Solution to Part 2: To find the doubling time, we want to know the time t_d so that $N(t_d) = 200$. That is, we want to solve the equation

$$100e^{0.243t_d} = 200.$$

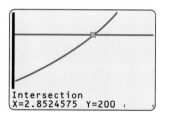

Intersection
X=2.8524575 Y=200

FIGURE 8.1 Doubling time for an initial population of 100

We solve it using the crossing-graphs method as shown in Figure 8.1. To make the graph, we used a table of values to help us choose a horizontal span of 0 to 5 and a vertical span of 50 to 250. From the prompt at the bottom of the screen, we see that t_d is about 2.85. Hence the doubling time is about 2.85 days, or about 2 days and 20 hours.

Solution to Part 3: If the initial population is 1000 instead of 100, then the new population function is

$$N = 1000e^{0.243t}.$$

To find the doubling time in this case, we need to solve

$$1000e^{0.243t_d} = 2000.$$

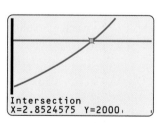

Intersection
X=2.8524575 Y=2000

FIGURE 8.2 Doubling time for an initial population of 1000

In Figure 8.2 we show the solution of this equation using the crossing-graphs method. Here we used a horizontal span of 0 to 5 and a vertical span of 500 to 2500. From

[2]This r value is from the study by R. Root and A. Olson that appeared as "Population increase of the cabbage aphid, *Brevicoryne brassicae*, on different host plants," *Can. Ent.* **101** (1969), 768–773.

the bottom of the screen, we see that the answer, $t_d = 2.85$ days, is the same one we got in part 2!   ■ ■ ■

Doubling Times

In Example 8.1 we saw that the time required for the aphid population to double is the same whether we start with 100 or 1000 aphids. The doubling time is an interesting measure of how rapidly a population grows, and we want to look at how to calculate it in general. The doubling time for an exponential function $N(0)e^{rt}$ is the time t_d that gives a population of $N(0) \times 2$. That occurs when e^{rt} is equal to 2. Thus we get the doubling time by solving the equation

$$e^{rt_d} = 2.$$

In words, this equation says that rt_d is the power of e that gives 2. Thus, in terms of the natural logarithm,

$$rt_d = \ln 2. \qquad (8.5)$$

We divide to complete the solution:

$$t_d = \frac{\ln 2}{r}. \qquad (8.6)$$

Check with your calculator that when $r = 0.243$, Equation (8.6) gives the doubling time $t_d = 2.85$ that we found in Example 8.1. Note that the initial population $N(0)$ does not appear in Equation (8.6), so the doubling time t_d depends only on r, not on the initial population $N(0)$. This is what Example 8.1 suggested, and it is a special feature of exponential growth. One interpretation of this is that it doesn't matter when the clock starts running: How long we have to wait before the population doubles in size is independent of the starting time. As you will be asked to show in Exercise 2, similar remarks apply to tripling times, and so on.

To illustrate further the relationship between the doubling time and the exponential growth rate, we have graphed $\ln 2/r$ versus r in Figure 8.3. We used a horizontal span of 0 to 1 and a vertical span of 0 to 10. The horizontal axis corresponds to the exponential growth rate r, and the vertical axis corresponds to the doubling time t_d. The graph shows that, as intuition would suggest, doubling time decreases as exponential growth rate increases.

Equation (8.5) could alternatively be solved for r instead of t_d. Such a solution enables us to calculate the exponential growth rate r if we know the doubling time. You will be asked to explore this further in Exercise 3.

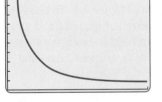

FIGURE 8.3 Graph of doubling time versus exponential growth rate

Empirical Data

The exponential model of population growth is realistic only under special conditions, such as growth of a laboratory culture and colonization of a new area, or over relatively short time spans. In this connection, it is quite remarkable that the United States population showed exponential growth for 70 years from 1790 through 1860. The size of the parameter r depends, of course, on the species. But it can also depend on environmental conditions such as temperature and humidity. It is difficult in practice to measure this parameter accurately, but such measurements are crucial and can provide

Population	r	t_d
Water flea	69	0.01
Flour beetle	23	0.03
Field vole	3.18	0.218
Ring-necked pheasant	1.02	0.68
Elephant seal	0.091	7.617

TABLE 8.1 Value of r and doubling time for some populations

important comparisons between similar species or within the same species in different environments. This is illustrated in Exercises 8 and 9. Some known examples of r and the corresponding doubling time are given in Table 8.1, where the basic unit is a year.[3] For example, a population of flour beetles would, according to the exponential growth model, double in size in 0.03 year, or about 11 days.

We can find an approximate value of r from several measurements of population size, using the technique of fitting exponential data given in Section 4.3: We plot the natural logarithm of the population data and calculate the regression line. The slope of this line then gives an estimate of r for the population, since this slope represents the exponential growth rate. Here is an example.

EXAMPLE 8.2   Finding Exponential Growth Rate for Yeast

The following table gives the size of a yeast population.[4] Time t is measured in hours.

Time t	0	1	2	3	4	5	6	7
Amount N of yeast	9.6	18.3	29.0	47.2	71.1	119.1	174.6	257.3

```
L1        L2       L3      2
0      2.2618    9.6
1      2.9069    18.3
2      3.3673    29
3      3.8544    47.2
4      4.2641    71.1
5      4.78      119.1
6      5.1625    174.6
L2(1)=2.261763098...
```

FIGURE 8.4 Entering the logarithm of yeast data

Plot the natural logarithm of the data points to show that an exponential model is appropriate for this population. Then use this information to estimate the exponential growth rate r and the doubling time t_d.

Solution: Because we expect to model these data with an exponential function, we approach the problem exactly as we did in Section 4.3. That is, we proceed to find the regression line for the natural logarithm of the data. The properly entered data 8.1 are shown in Figure 8.4. The plot 8.2 of the logarithm of the data is shown

---

[3]This table is adapted from p. 269 of Robert E. Ricklefs, *The Economy of Nature*, 3rd ed. (New York: W. H. Freeman, 1993).

[4]The data are taken from an early study of yeast population growth by Tor Carlson, as described by R. Pearl in "The growth of populations," *Quart. Rev. Biol.* **2** (1927), 532–548.

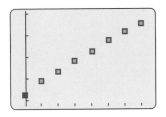

FIGURE 8.5 Plotting the logarithm of yeast data

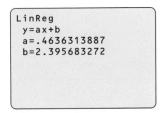

```
LinReg
 y=ax+b
 a=.4636313887
 b=2.395683272
```

FIGURE 8.6 Constants for the regression line of ln *N*

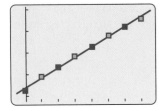

FIGURE 8.7 Graph of regression line added to plot of the logarithm of yeast data

in Figure 8.5. It appears that the logarithm of *N* is a straight line, so it is reasonable to model population growth using an exponential function. We calculate $\boxed{8.3}$ the regression line, with the result shown in Figure 8.6. From this we see that the slope of the regression line is about 0.46. We conclude that the exponential growth rate for yeast is about $r = 0.46$ per hour. In Figure 8.7 we have added the graph of the regression line to the plot of the logarithm of the data.

To get the doubling time we use Equation (8.6), putting 0.46 in place of r:

$$t_d = \frac{\ln 2}{0.46}, \text{ about 1.5 hours.} \qquad \blacksquare \blacksquare \blacksquare$$

Discrete Generations

The preceding model for exponential growth is reasonable for species with overlapping generations and extended breeding seasons. For many species, however, breeding occurs only during certain seasons, so population growth should be measured over discrete intervals of time. The exponential growth model has a discrete counterpart, sometimes called *geometric growth*, and in many ways the mathematics is unchanged. For geometric growth we assume that a unit of time represents the time between breeding seasons. For example, with annual breeding, 1 unit represents a year. Then the basic assumption is that the population at time $t + 1$ is proportional to the population at time t. This constant of proportionality is usually denoted by the Greek letter λ. Thus we are assuming that

$$N(t + 1) = \lambda N(t)$$

for all t. It follows that

$$N(1) = \lambda N(0)$$
$$N(2) = \lambda N(1) = \lambda^2 N(0)$$
$$N(3) = \lambda N(2) = \lambda^3 N(0).$$

This gives rise to the formula

$$N(t) = N(0)\lambda^t.$$

We note that this is exactly the same as an exponential function with initial value $N(0)$ and growth factor λ. But for geometric growth the model is used a little differently, since for species with discrete breeding times the formula is reasonable only when t is a whole number of units. Between breeding seasons there are no births, and the population actually declines as a result of deaths.

EXAMPLE 8.3 **Population of an Annual Plant**

The population of a species of an annual plant grows geometrically. If initially there are 10 plants, and after 1 year there are 200, how many plants are there after 3 years? Make a graph of the population function.

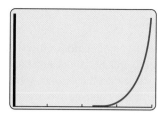

FIGURE 8.8 A picture of rapid geometric growth

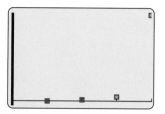

FIGURE 8.9 A graph of discrete data that exhibit geometric growth

Solution: We started with 10 plants, and 1 year later there were 200. Since the function is exponential, we know how to get the growth factor by dividing:

$$\frac{200}{10} = 20.$$

Thus the growth factor λ is 20. To get the population in succeeding years, we multiply each year by 20, so after 2 years there are $200 \times 20 = 4000$ plants present. After three years there are $4000 \times 20 = 80{,}000$ plants. We can see that if the same geometric growth continues, the spread of the plant will be quite dramatic in succeeding years.

Since we know that the initial value is 10 and the growth factor is 20, we can write the general formula for N:

$$N = 10 \times 20^t.$$

Using this, we could have computed the number of plants after 3 years as $N(3) = 10 \times 20^3 = 80{,}000$. The graph of N versus t is shown in Figure 8.8, in which we have used a horizontal span of 0 to 4 and a vertical span of 0 to 1,600,000. The steepness of the graph is startling, and we know that such a rate of growth cannot be maintained indefinitely. We should also note that because the geometric growth formula applies only for integral values of t, it might have been more appropriate to plot the graph as discrete data points. We have done this in Figure 8.9.

8.1 SKILL BUILDING EXERCISES

S-1. **Population growth:** If B is births per unit time and D is deaths per unit time, write an equation of change for population N as a function of time t.

S-2. **An exponential model:** *This is a continuation of Exercise S-1 above.* If both B and D are proportional to N, explain how the equation of change that you wrote in Exercise S-1 leads to an exponential model.

S-3. **An exponential model:** *This is a continuation of Exercise S-1 above.* Suppose $B = 0.04N$ and $D = 0.013N$. Find the equation of change for exponential growth.

S-4. **Initial value:** If a population is governed by $\frac{dN}{dt} = rN$, then the exponential model for N is $N = N_0 e^{rt}$. What is the physical significance of the number N_0?

S-5. **r value:** If a population is governed by $\frac{dN}{dt} = rN$, then the exponential model for N is $N = N_0 e^{rt}$. The constant r depends on the species and the environment, and it is known as the r value of a species in an environment. Explain the physical significance of r.

S-6. **Comparing r values:** Two populations grow according to the model in Exercise S-5. The first population has a larger r value than the second. What can you conclude about the comparative growth rate of the two populations?

S-7. **Negative r value:** Suppose a population that follows the model in Exercise S-5 has an r value of -0.03. What happens to the population over time?

S-8. **Doubling time:** If a population N grows according to $30e^{0.72t}$, where t is measured in years, how long does it take the population to double in size?

S-9. **Doubling time:** If a population N grows according to $30e^{rt}$, where t is measured in years, and if the population doubles each 13 years, find the value of r.

S-10. **Discrete growth:** When breeding occurs only during certain seasons, the exponential model is altered slightly to account for this. What is the name of the altered model?

8.1 EXERCISES

1. **Norway rat:** A population of the Norway rat undergoes exponential growth with exponential growth rate $r = 3.91$ per year.

   a. If the initial population size is 6, find the population size after 6 months and the population after 2 years.

   b. Suppose again that the initial population size is 6. How long will it take for the population to double to 12? How long will it take to reach a size of 24? (*Hint:* Think before you calculate!)

   c. Say the initial population size is not given, but it is known that the population has reached 150,000 after 2 years. Determine the initial population size.

2. **Tripling times:** We noted that for exponential population growth, the doubling time can be calculated,

   if the exponential growth rate is known, by using the formula

   $$t_d = \frac{\ln 2}{r}.$$

   In this exercise, we seek a similar formula that gives the time it takes such a population to triple. The population is modeled by the function

   $$N = N(0)e^{rt}.$$

   a. What value of e^{rt} will make the population triple?

   b. Write an equation whose solution gives the tripling time.

   c. Use logarithms to solve the equation in part b.

   d. Use the formula you obtained in part c to obtain the tripling time for a population with exponential growth rate $r = 0.15$ per year.

3. **Exponential growth rate from doubling time:** To get the doubling time in terms of r, we solved the equation

$$rt = \ln 2$$

for t. What do you get if you solve the equation for r instead? Use the formula you get to find the exponential growth rate for a population that takes 7 years to double in size.

4. **A species of mammal with known doubling time:** *This is a continuation of Exercise 3.* A certain species of mammal grows at an exponential rate, doubling its population size every 56 days.

a. Find the exponential growth rate r.

b. If initially there are 10 individuals, how many will there be after 1 year?

5. **Caribou:** The following table describes the growth of a population of caribou introduced in 1962 on Brunette Island of Newfoundland, Canada.[5] Time t is measured in years since 1962.

Time t	Population
0	17
1	27
2	40
3	54
4	78
5	100

Use this table to calculate r for Newfoundland caribou.

6. **Insects:** The following table gives the size of an insect population that is believed to be growing exponentially. Time t is measured in days.

Time t	Population N
0	10
2	14
4	20
6	29
8	42

a. Use this information to estimate the parameter r and the doubling time t_d. Be sure to state the units of each. (*Suggestion:* You may use regression to solve this problem. Alternatively, you may be able to get a pretty close estimate of the doubling time by looking at the data. Then see Exercise 3.)

b. Estimate the number of insects present after 10 days.

7. **Field mice:** The following table gives the size of a population of field mice. Time t is measured in years.

Time t	Population N
0	4
0.5	38
1	360
1.5	3416
2	32,412

a. Plot the graph of the logarithm of the data, and determine whether it is reasonable to use an exponential model for this population.

b. What are the values of r and the doubling time t_d? Be sure to state the units of each.

c. Find an exponential model for the data.

d. Estimate the number of mice present after 9 months.

[5]The data are taken from A. Bergerud, "The population dynamics of Newfoundland caribou," *Wildlife Monographs* **25** (1971), 1–55.

8. **Aphid growth on broccoli:** This problem and the following one are based on the results of a study by R. Root and A. Olson of population growth for aphids on various host plants.[6] The value of $r = 0.243$ per day used in Example 6.1 was for aphids reared on broccoli plants outdoors. The following table describes the growth of an aphid population reared on broccoli plants *indoors* (in a controlled environment). Time t is measured in days.

Time t	Population
0	25
1	30
2	37
3	44
4	54

Calculate r for this table. Did the aphids fare better indoors or outdoors?

9. **Aphid growth on different plants:** *This is a continuation of Exercise 8.* In the study described in Exercise 8, aphids were also reared indoors on collard and on yellow rocket (a decorative flower). Use the following tables to determine whether aphids fare better on collard or on yellow rocket. Time t is measured in days.

Time t	Population
0	28
1	33
2	40
3	48
4	57

Collard

Time t	Population
0	48
1	53
2	58
3	64
4	70

Yellow Rocket

10. **Field voles:** Two regions are being colonized with field voles. The first colony starts with 10 individuals and grows exponentially. The second colony starts with 100 individuals and grows only by immigration: 125 individuals are added to the colony each year.

   a. Use a formula to express the population of the first colony as a function of time t. Be sure to state the units of t. (Use Table 8.1 to find r.)

   b. Use a formula to express the population of the second colony as a function of time t.

   c. After how long will the first colony overtake the second in size?

11. **Microbes:** A population of microbes grows geometrically, increasing threefold every day. Initially there are 2 microbes. Plot the number of microbes as a function of time over a 5-day period, and determine how many microbes there are at the end of this period.

12. **Microbes that grow more rapidly:** A population of microbes grows geometrically, increasing tenfold every 2 days. Initially there are 2 microbes.

   a. How many microbes will there be after 4 days?

   b. Find the parameter λ.

   c. How many microbes will there be after 5 days?

13. **How the size of λ affects geometric growth:** If the parameter λ for geometric growth is less than 1 (but positive), what implications does this have for the population? What if $\lambda = 1$?

---

[6]"Population increase of the cabbage aphid, *Brevicoryne brassicae*, on different host plants," *Can. Ent.* **101** (1969), 768–773. We have used the r values they obtained and their initial population sizes to construct our hypothetical population growth data.

8.2 *Population Dynamics: Logistic Growth*

Although the exponential growth model is simple, in reality a population cannot undergo such rapid growth indefinitely. The species will begin to exhaust local resources, and we expect that instead of remaining constant, the birth rate will begin to decrease and the death rate to increase. Thus there will be a reduced rate of growth as the population increases in size. This regulation of growth is described by the logistic equation.

Logistic Growth Model

For an example of such population growth, we consider a yeast culture. In Example 8.2 we looked at the first 7 hours of growth and saw that it was approximately exponential. In Figure 8.10 we see what happens over a longer period, namely 18 hours. The dots represent the population data. Note the conformity to the pattern of growth described above: At first growth is rapid; then it tapers off. In Figure 8.11 a curve has been fitted to the data. This S-shaped curve (sometimes called a sigmoid shape) is the trademark of logistic growth. Note that for a while the graph resembles that of an exponential function. But as time goes by, the slope of the graph (representing the rate of change of the population) begins to decrease; eventually the graph flattens out, indicating that the population size is stabilizing.

In many circumstances the logistic model is a more reasonable description of population growth than is the exponential model. It was introduced by P. F. Verhulst in 1838 to describe the growth of human populations. In 1920, R. Pearl and L. J. Reed derived the same model to describe the growth of the population of the United States since 1790 and to attempt predictions of that population at future times.[7] We will return to their predictions at the end of this section.

The logistic model implies that there is an upper limit beyond which a population cannot grow. The smallest such limit is called the *carrying capacity* of the environment in which the population lives; it is denoted by K. It can be thought of as a saturation level for the population. According to this model, the population will actually get closer and closer to the carrying capacity as time goes on. For our yeast study, the approximate

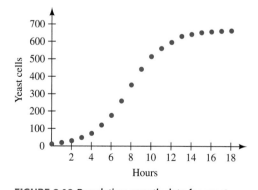

FIGURE 8.10 Population growth data for yeast

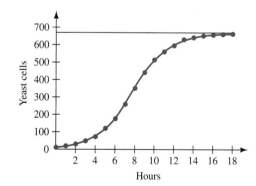

FIGURE 8.11 Yeast data fitted with a logistic curve

[7]"On the rate of growth of the population of the United States since 1790 and its mathematical representation," *Proc. Nat. Acad. Sci.* **6** (1920), 275–288.

value of K is 665. The horizontal line corresponding to this value has been added to the graph in Figure 8.11, and the relationship between this line and the logistic growth curve is apparent there.

Our discussion of the logistic model so far has been qualitative—we have given no formulas. But even at this level it yields interesting results, as the following application shows.

Application to Harvesting Renewable Resources

In the management of a harvested population (as found in a marine fishery), an important problem is to determine at what level to maintain the population in order to sustain a maximum harvest. This is the problem of optimum yield. The logistic model for population growth gives a theoretical basis for solving this problem, leading to the theory of *maximum sustainable yield*.

If we harvest when the population is small, then the population level will be reduced so far that it will take a long time to recover, and in extreme cases, the population may be driven to extinction. In Figure 8.12 we have illustrated harvesting at the lower part of the curve for a species undergoing logistic growth. The unbroken curve shows how the population would grow if it were left alone, and the broken graph shows the result of periodic harvesting. It shows that if we harvest when the population reaches a level of about 125, we will be able to harvest about 100 individuals every 3 days. At the other extreme, we could maintain the population near its maximum level. But if the harvested population grows according to the logistic model, this means maintaining the population near the carrying capacity of the environment, where the rate of population growth is *slow* because the graph flattens out there. After a small harvest we will have to wait a relatively long time for the population to recover. This is illustrated in Figure 8.13, and we see that in the case shown, we are able to harvest about 100 individuals every 2 days.

To get the best harvest, it makes sense to determine the population size at which the rate of growth is the largest, since the population will recover most quickly near that size. This scenario is illustrated in Figure 8.14, and we see that under these conditions, we are able to harvest 100 individuals each day. That population size, known as the *optimum yield level*, corresponds to the steepest point on the graph, which is near the middle of the curve. This suggests that the optimum yield level is half of the carrying capacity. It

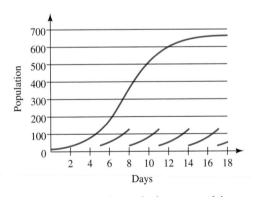

FIGURE 8.12 Harvesting at the lower part of the curve a population undergoing logistic growth

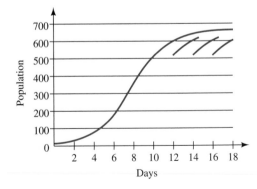

FIGURE 8.13 Harvesting near the carrying capacity a population undergoing logistic growth

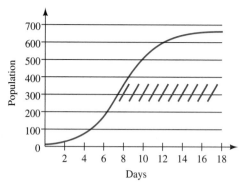

FIGURE 8.14 Harvesting near the middle of the logistic curve

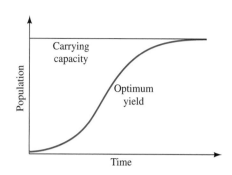

FIGURE 8.15 The level of optimum yield is half the carrying capacity

is marked in Figure 8.15. Before the population reaches this steepest point, the rate of growth is increasing; after this point, the rate of growth is decreasing. In other words, the optimum yield level corresponds to the inflection point on the graph. Our conclusion is that optimum yields come from populations maintained not at maximum size but at maximum growth rates, and this occurs at half of the carrying capacity.

Note: The importance of wise management is demonstrated by the history of the Peruvian anchovy fishery, which flourished off the coast of Peru from the mid-1950s until the early 1970s. In 1970 it accounted for 18% of the total world harvest of fish. But in the early 1970s a change in environmental conditions, combined with overfishing, caused the collapse of the fishery. This had a dramatic impact on food prices worldwide. The fishery has never recovered from this collapse.

Logistic Equation of Change

Now we turn to the formula for $\frac{dN}{dt}$ that gives the logistic model. The idea is simple: For exponential growth we assume a constant per capita growth rate r. For logistic growth we assume instead that each individual that is added to the population decreases the per capita growth rate by an equal amount, until this rate is zero at the carrying capacity K. The effect of this is to multiply r by a factor that is a linear function of the population size N, namely the function $1 - \frac{N}{K}$. From this we get the logistic equation of change, which is sometimes called the logistic equation in its *differential form:*

$$\frac{dN}{dt} = rN\left(1 - \frac{N}{K}\right). \tag{8.7}$$

Let's examine this equation of change more closely to see how the shape of the logistic curve arises from it. The linear factor $1 - \frac{N}{K}$ is 1 when $N = 0$, so for small initial populations, Equation (8.7) is approximately the same as the equation of change for exponential growth: $\frac{dN}{dt} = rN$. This is reflected in the fact that the logistic curve resembles the graph of an exponential function for small population sizes. The linear factor $1 - \frac{N}{K}$ is 0 when $N = K$, so the rate change of N, as given by Equation (8.7), decreases as the population size N approaches the saturation level K. This explains why the curve flattens out over time. In summary, Equation (8.7) does indeed describe the S-shaped logistic curve: approximate exponential growth early on, followed by a gradual leveling off of growth.

In light of the preceding application involving optimum yield, it is illuminating to examine a graph of $\frac{dN}{dt}$ against N. According to Equation (8.7), this is a graph of $rN\left(1 - \frac{N}{K}\right)$ as a function of N. The graph is the *parabola* shown in Figure 8.16. This again illustrates the basic form of logistic growth: The rate of growth of the population at first increases with increasing population and then, at a certain level, begins to decrease. By symmetry it is clear that the maximum growth rate, the maximum of $\frac{dN}{dt}$ as a function of N, is attained at the population level $N = \frac{1}{2}K$. In connection with the application to harvesting renewable resources, this confirms that the population level for optimum yield is half of the carrying capacity K.

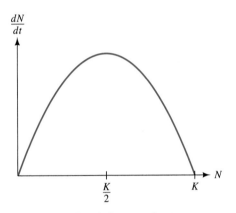

FIGURE 8.16 Population growth rate as a function of population size

In our discussion we have assumed that the initial population level is below the carrying capacity K. If instead this level is above K, then initially $1 - \frac{N}{K}$ is negative, and according to Equation (8.7) the population will decrease in size. In fact, the population will decrease to K in the long run. This is what intuition would suggest. Thus the logistic model implies that no matter what the initial population size is, in the long run the population will stabilize at the carrying capacity K of the environment. In mathematical terms, K is a *stable equilibrium* for N.

Empirical data: We can estimate r and K for a given population using a sample of population sizes. One way of doing this is to graph the (varying) per capita growth rate as a function of population size—that is, to graph $\frac{1}{N}\frac{dN}{dt}$ against N. According to Equation (8.7), this is a graph of $r\left(1 - \frac{N}{K}\right)$ as a function of N. Because

$$r\left(1 - \frac{N}{K}\right) = r - \frac{r}{K}N,$$

this is a line with slope $-\frac{r}{K}$, vertical intercept r, and horizontal intercept K. The graph is shown in Figure 8.17.

To determine r and K, then, we use regression to find the best linear fit to the data for $\frac{1}{N}\frac{dN}{dt}$ versus N. The vertical intercept of this line is r, and if we call its slope m, then we solve the equation $m = -\frac{r}{K}$ for K to get $K = -\frac{r}{m}$. Alternatively, we can find K by graphing the line to determine its horizontal intercept, as is shown in the next example.

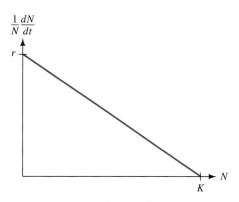

FIGURE 8.17 Per capita growth rate as a function of population size

EXAMPLE 8.4   Logistic Growth for the Water Flea

The following table gives population sizes and corresponding per capita growth rates for the water flea.[8] Time t is measured in days, and N is measured as number per cubic centimeter.

N	1	2	4	8	16
$\frac{1}{N}\frac{dN}{dt}$	0.230	0.216	0.208	0.145	0.057

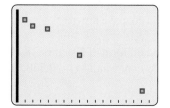

FIGURE 8.18 Data for per capita growth rate as a function of population size

1. Plot the given data points to show that a linear model for $\frac{1}{N}\frac{dN}{dt}$ versus N is appropriate for this population.

2. Use the data to estimate r and K for the water flea in this environment.

Solution to Part 1:   The plot [8.4] is shown in Figure 8.18. It appears that a linear model is appropriate.

```
LinReg
 y=ax+b
 a=-.0117016129
 b=.24375
```

FIGURE 8.19 Equation of the regression line

Solution to Part 2:   We calculate [8.5] the regression line, with the result shown in Figure 8.19. From this we see that the vertical intercept of the regression line is about 0.24, and this is our estimate for r. Also, the slope of the regression line is about -0.012, so by the preceding discussion this gives the estimate $K = -\frac{0.24}{-0.012} = 20$. We conclude that, for the water flea in this environment, r is about 0.24 per day and the carrying capacity K is about 20 individuals per cubic centimeter.

Note: To use this procedure we need to know the per capita growth rates for a sample of population sizes. In practice, these rates can be difficult to obtain. There are more sophisticated ways to fit logistic functions to data without using per capita growth rates. ∎∎∎

[8]The data are taken from P. W. Frank, C. D. Boll, and R. W. Kelly, "Vital statistics of laboratory cultures of *Daphnia pulex* DeGeer as related to density," *Physiol. Zool.* **30** (1957), 287–305.

Integral Form of the Logistic Equation

Using more advanced techniques, Equation (8.7) can be solved to obtain an explicit formula for N as a function of t. This is the integral form of the logistic equation:

$$N = \frac{K}{1 + be^{-rt}}, \tag{8.8}$$

where b is the constant $\frac{K}{N(0)} - 1$. To illustrate this, we return to the example of the Pacific sardine population.

EXAMPLE 8.5 The Pacific Sardine

Studies to fit a logistic model to the Pacific sardine population have yielded

$$N = \frac{2.4}{1 + 239e^{-0.338t}},$$

where t is measured in years and N is measured in millions of tons of fish.[9]

1. What is r for the Pacific sardine?

2. What is the environmental carrying capacity K?

3. What is the optimum yield level?

4. Make a graph of N.

5. At what time t should the population be harvested?

6. What portion of the graph is concave up? What portion is concave down?

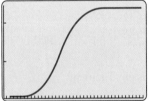

FIGURE 8.20 Population growth for the Pacific sardine

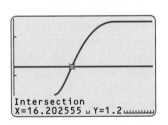

FIGURE 8.21 Time of optimum yield

Solution to Part 1: The formula for N is written in the standard form of Equation (8.8) for a logistic function. Since the coefficient of t in the exponential is -0.338, it must be that $r = 0.338$ per year.

Solution to Part 2: The constant in the numerator is 2.4, so $K = 2.4$ million tons of fish.

Solution to Part 3: The optimum yield level is half of the carrying capacity, and by part 2 this is $\frac{1}{2} \times 2.4 = 1.2$ million tons of fish.

Solution to Part 4: Using a table of values, we determine a horizontal span of 0 to 40 and a vertical span of 0 to 2.5. The graph is shown In Figure 8.20.

Solution to Part 5: The population should be harvested at the optimum yield level, and according to part 3 this is at the population level $N = 1.2$, so we want to solve the equation $N(t) = 1.2$ for t. We do this using the crossing-graphs method, as shown in Figure 8.21. We find that the time t is about 16.2 years.

---

[9]From G. I. Murphy, "Vital statistics of the Pacific sardine (*Sardinops caerulea*) and the population consequences," *Ecology* **48** (1967), 731–736.

Solution to Part 6: By examining Figure 8.20, we see that the graph is concave up until the population reaches the level for optimum yield and is concave down after that. Using part 5, we see that the graph is concave up over the first 16.2 years and is concave down thereafter.

Note: The Pacific sardine fishery along the California coast expanded rapidly from about 1920 until the 1940s, with an annual catch of around 800 thousand tons at its height. In the late 1940s to early 1950s the fishery collapsed, and yield levels have remained very low since then. The economic consequences of the collapse have been severe. The major factors that contributed to the collapse were heavy fishing and environmental changes that seem to have favored a competing fish, the anchovy.[10] This example illustrates that the model of maximum sustainable yield that we have developed is only a first step in the study of managing renewable resources. A variety of factors must be considered. ▪ ▪ ▪

The Value of the Logistic Model

The logistic model for population growth, like the exponential model, has limitations. As we have noted, the original article by Pearl and Reed in 1920 used this model to predict population growth in the United States. According to their data, the population was to stabilize at about 197 million. Of course, this level has been far surpassed. At a more fundamental level, the basic assumptions of the logistic model have been called into question. Various other models have been suggested, but they involve more complicated mathematics.

The main value of the models we have examined lies in their qualitative form. They enable us to discuss, in general terms, population trends and the reasons for such trends. Rather than memorizing the complicated formula in Equation (8.8), you should concentrate on understanding how the S-shaped graph reflects the underlying assumptions of this model.

---

[10]For more information on the Pacific sardine fishery, see the account by Michael Culley in *The Pilchard* (Oxford, England: Pergamon Press, 1971).

8.2   *SKILL BUILDING EXERCISES*

S-1. **Logistic growth:** When we add the notion of *environmental carrying capacity* to the basic assumptions leading to the exponential model, we get a different model. What is the name of this model?

S-2. **Logistic equation of change:** What is the equation of change governing the logistic model?

S-3. **More on the logistic equation of change:** Suppose a certain population grows according to $\frac{dN}{dt} = 0.13N \left(1 - \frac{N}{378}\right)$.

   a. What is the r value for this species in this environment?

   b. What is the environmental carrying capacity?

   c. What are the equilibrium solutions of the equation?

   d. At what value of N will there be a maximum growth rate?

S-4. **Per capita growth rate:** In a population that grows logistically, if the per capita growth rate $\frac{1}{N}\frac{dN}{dt}$ is considered as a function of the population N, what kind of function is it?

S-5. **Formula for logistic growth:** Write the formula for logistic growth that corresponds to the equation of change $\frac{dN}{dt} = rN \left(1 - \frac{N}{K}\right)$.

S-6. **The formula continued:** *This is a continuation of Exercise S-5 above.* Show how the constant b in the formula in Exercise S-5 can be calculated from initial population and carrying capacity.

S-7. **More logistic formula:** Write the logistic formula that governs a population that grows logistically with r value 0.015, carrying capacity 2390, and initial population 120.

S-8. **Harvesting:** What is the name of the model that says that a renewable resource that grows logistically should be harvested at half of the carrying capacity?

S-9. **Harvesting:** Suppose a renewable population grows logistically according to

$$\frac{dN}{dt} = 0.02N \left(1 - \frac{N}{778}\right).$$

According to the maximum sustainable yield model, what is the optimum harvesting level?

S-10. **Harvesting continued:** The maximum sustainable yield model says that a renewable resource that grows logistically should be harvested at half of the carrying capacity. What graphical significance does this point have in the graph of population versus time?

8.2   *EXERCISES*

1. **Estimating optimum yield:** In Figure 8.22 a logistic growth curve is sketched. Estimate the optimum yield level and the time when this population should be harvested.

2. **Estimating carrying capacity:** In Figure 8.23 a portion of a logistic growth curve is sketched. Estimate the optimum yield level and the environmental carrying capacity.

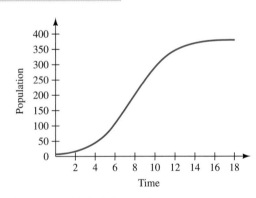

FIGURE 8.22 A logistic growth curve

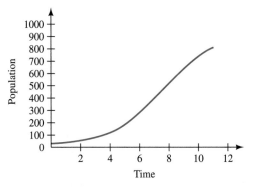

FIGURE 8.23 A portion of a logistic growth curve

lation stabilized suggests that food supply (and not just predators) can effectively regulate population size in this setting.

4. **Negative growth rates:** The original water flea study described in Example 8.4 also included growth from population sizes of $N = 24$ and $N = 32$.

   a. Use the regression line computed in Example 8.4 to estimate the per capita growth rates for these two values of N.

   b. Explain in terms of the carrying capacity why the growth rates from part a should both be negative.

5. **More on the Pacific sardine:** In this problem we explore further the Pacific sardine population using the model in Example 8.5.

   a. If the current level of the Pacific sardine population is 50,000 tons, how long will it take for the population to recover to the optimum growth level of 1.2 million tons?

   b. The value of r used in Example 8.5 ignores the effects of fishing. If fishing mortality is taken into account, then r drops to 0.215 per year (with the carrying capacity still at 2.4 million tons). Answer the question in part a using this lower value of r.

   Note: The population estimate of 50,000 tons and the adjusted value of r are given in the paper by Murphy, *op. cit.* Murphy points out that factoring in the growth of the competing anchovy population makes the recovery times even longer, and he adds, "It is disconcerting to realize how slowly the population will recover to its level of maximum productivity... even if fishing stops."

6. **Eastern Pacific yellowfin tuna:** Studies to fit a logistic model to the Eastern Pacific yellowfin tuna population have yielded

$$N = \frac{148}{1 + 3.6e^{-2.61t}},$$

where t is measured in years and N is measured in thousands of tons of fish.[12]   →

3. **Northern Yellowstone elk:** The northern Yellowstone elk winter in the northern range of Yellowstone National Park.[11] A moratorium on elk hunting was imposed in 1969, and after that the growth of the elk population was approximately logistic. Use the following table to estimate r and K for this elk population. (The unit for the per capita growth rate $\frac{1}{N}\frac{dN}{dt}$ is per year. The date is included in the table only for the sake of interest.)

Year	N	$\frac{1}{N}\frac{dN}{dt}$
1968	3172	0.305
1969	4305	0.253
1970	5543	0.273
1971	7281	0.121
1972	8215	0.195
1973	9981	0.053
1974	10,529	0.180

Note: At one time the gray wolf was a leading predator of the elk, but it was not a factor during this study period. The level at which the elk popu-

---

[11]This problem is based on the study by Douglas B. Houston in *The Northern Yellowstone Elk* (New York: Macmillan, 1982). Houston uses a slightly different method to fit a logistic model.

[12]From a study by M. B. Schaefer, as described by Colin W. Clark in *Mathematical Bioeconomics*, 2nd ed. (New York: Wiley, 1990).

a. What is r for the Eastern Pacific yellowfin tuna?

b. What is the environmental carrying capacity K for the Eastern Pacific yellowfin tuna?

c. What is the optimum yield level?

d. Use your calculator to graph N against t.

e. At what time was the population growing the most rapidly?

7. **Maximum growth rate for tuna:** *This is a continuation of Exercise 6.*

a. With your calculator make a graph of $\frac{dN}{dt}$ against t, using

$$\frac{dN}{dt} = rN\left(1 - \frac{N}{K}\right)$$

and the formula for N given in Exercise 6.

b. Use the graph you made in part a to find the time at which $\frac{dN}{dt}$ was the largest.

c. What was the population at the time when $\frac{dN}{dt}$ was the largest? How does your answer compare with that in part c of Exercise 6?

8. **Pacific halibut:** For a Pacific halibut population, the value of r is 0.71 per year, and the environmental carrying capacity is 89 thousand tons of fish.[13]

a. Find the time it takes the population to grow from the optimum yield level to 90% of carrying capacity if the number b in the logistic growth equation, Equation (8.8), is 1.5.

b. Find the time it takes the population to grow from the optimum yield level to 90% of carrying capacity if the number b in the logistic growth equation, Equation (8.8), is 4.8.

c. Compare the times in parts a and b. Why could this result have been expected?

9. **Yeast growth rate:** The logistic model that best fits the yeast growth data in Figure 8.10 at the beginning of this section has $r = 0.54$ per hour and $K = 665$. But in Example 8.2 of Section 8.1, when we treated the first 7 hours of growth as exponential, we obtained $r = 0.46$ per hour. Explain the discrepancy and, in particular, why you would expect the value of r from the logistic model to be higher than that from the exponential model of the first segment of growth.

10. **Maximum growth rate:** We have seen that under the logistic model

$$\frac{dN}{dt} = rN\left(1 - \frac{N}{K}\right),$$

the growth rate $\frac{dN}{dt}$ is at its maximum when $N = \frac{1}{2}K$.

a. Use these formulas to express this maximum growth rate in terms of r and K.

b. Use your answer to part a to find the maximum growth rate for the Pacific sardine population of Example 8.5.

11. **Logarithm of the logistic curve:** Consider the logistic curve sketched in Figure 8.15. Sketch, including an explanation, a graph of $\ln N$ against t. (*Hint:* The curve is nearly exponential at first. What is the logarithm of an exponential function?)

12. **Rate of change for logistic growth:** Consider the logistic curve sketched in Figure 8.15. Sketch, including an explanation, a graph of $\frac{dN}{dt}$ against t.

13. **Gompertz model:** One possible substitute for the logistic model of population growth is the Gompertz model, according to which

$$\frac{dN}{dt} = rN \ln\left(\frac{K}{N}\right).$$

For simplicity, in this problem we take $r = 1$, so the equation of change for N is

$$\frac{dN}{dt} = N \ln\left(\frac{K}{N}\right).$$

a. Let $K = 10$, and make a graph of $\frac{dN}{dt}$ versus N for the Gompertz model.

b. Use the graph you obtained in part a to determine for what value of N the function $\frac{dN}{dt}$ is at its maximum. Since this tells the population level for the maximum growth rate, this is the optimum yield level under the Gompertz model with $K = 10$.

---

[13]From a study by H. S. Mohring, as described by Colin W. Clark, ibid.

c. Under the logistic model the optimum yield level is $\frac{1}{2}K$. What do you think is the optimum yield level in terms of K under the Gompertz model? (*Hint:* Repeat the procedure in parts a and b using different values of K, such as $K = 1$ and $K = 100$. Try to find a pattern.)

14. **Another model:** Recall that the basic assumption of the logistic growth model is that each individual that is added to the population decreases the per capita growth rate by an equal amount. As Figure 8.17 shows, this means that the graph of $\frac{1}{N}\frac{dN}{dt}$ against N is a straight line. In Figure 8.24 another relationship between $\frac{1}{N}\frac{dN}{dt}$ and N is suggested. Interpret this in physical terms, and discuss the impact on the growth curve—that is, the graph of N against t.

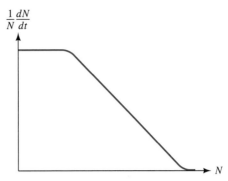

FIGURE 8.24 Alternative graph of $\frac{1}{N}\frac{dN}{dt}$ versus N

8.3   *Population Structure: Survivorship Curves*

To study the structure of a population, we need to know the number of individuals in each age category as well as the number of births and deaths within these categories. Such statistics for human populations have long been recognized as an important subject of study. For hundreds of years, life insurance companies have set premium rates on the basis of this information. However, the gathering and study of such statistics by ecologists for other populations are relatively recent. In this section we will see how the tools we have developed for analyzing functions and their graphs bear on this study of the structure of populations.

Life Tables and Survivorship Curves

Statistics on the age structure of a population are arranged into what is called a *life table*. Such tables usually contain several columns that offer a variety of ways to present the basic data about births and deaths for individuals grouped by *cohort*—that is, according to their age.

For mortality statistics, the following symbols are often used:

$$x = \text{age at beginning of interval}$$

$$l_x = \text{number of survivors at start of age interval } x$$

$$d_x = \text{number dying during age interval } x$$

$$q_x = \text{proportion of individuals dying during age interval } x$$

Note that l_x, d_x, and q_x are really functions of x. In ecology it is standard practice to put x as a subscript, writing l_x instead of $l(x)$, for example. Also, l_0 is the total number of individuals in the study, all born at about the same time. It is customary to rescale the numbers so that $l_0 = 1000$. With this convention, l_x represents survivors per thousand. The number q_x is usually called the *age-specific mortality rate*. It tells what percentage of a given age group can be expected to die over the given time period.

A life table for the Dall mountain sheep is given in Table 8.2, where x is in years.[14] From the table, we see that of the initial $l_0 = 1000$ sheep in the study, $d_0 = 199$ died during their first year of life. This gives $1000 - 199 = 801$ for l_1, the number surviving past the first year of life. The fourth column is obtained by taking ratios between the third and second columns: $q_x = d_x / l_x$. Thus for the sheep we have $q_0 = \frac{199}{1000} = 0.199$. That is, 19.9% of newborn Dall mountain sheep do not survive their first year.

The various columns of a life table are redundant, in a sense: Once one column is known, the others can be derived from it using the definitions given above. Each column presents the basic information in a different way.

Survivorship curves are derived from life tables as an aid in visualizing the data. The shapes of these curves convey characteristic features of the population, such as the incidence of juvenile mortality. A survivorship curve is nothing but a graph of l_x as a function of x. We will follow the common practice of looking at these graphs on a semi-

---

[14]From a study by A. Murie, as described in E. S. Deevey, Jr., "Life tables for natural populations of animals," *Quart. Rev. Biol.* **22** (1947), 283–314.

x	l_x	d_x	q_x	x	l_x	d_x	q_x
0	1000	199	0.199	7	640	69	0.108
1	801	12	0.015	8	571	132	0.231
2	789	13	0.016	9	439	187	0.426
3	776	12	0.015	10	252	156	0.619
4	764	30	0.039	11	96	90	0.938
5	734	46	0.063	12	6	3	0.500
6	688	48	0.070	13	3	3	1.000

TABLE 8.2 Life table for the Dall mountain sheep

logarithmic scale.[15] That is, we will graph $\log l_x$ against x. Here is the reason: Suppose for a moment that the mortality rate q_x is independent of age x, so a fixed fraction of the individuals in each age interval die. Then a fixed fraction of the individuals in each age interval survive, so l_x will be a decreasing exponential function. By the basic property of logarithms, $\log l_x$ will be a decreasing *linear* function of x, i.e., equal increases in x lead to equal decreases in $\log l_x$. In summary, a constant mortality rate yields a linear relationship between x and $\log l_x$, and this illustrates the value of examining survivorship data on a semilogarithmic scale. It is noteworthy that on this scale, the slope of the graph is roughly proportional to the mortality rate.

Classifying Survivorship Curves

The three general types of survivorship curves are shown in Figure 8.25. The type I curve, which is concave down, is relatively flat for a time and then drops off steeply. This indicates that mortality is fairly low among juveniles but increases rapidly for older individuals. This shape is strictly applicable only under rather special conditions, such as in captive populations where individuals benefit from attentive care in the juvenile

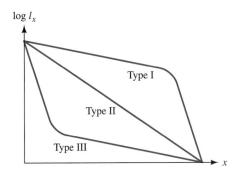

FIGURE 8.25 Survivorship curves

---

[15]It is customary to use the common (base-10) logarithm for this instead of the natural logarithm.

stage and eventually die of old age. This general shape, however, models to some extent survivorship among humans and indeed many mammals. The curve of type II is a straight line and represents a constant mortality rate (as discussed above). This shape roughly describes survivorship among most birds and some other animals (such as the hydra and some lizards). The type III curve, which is concave up, represents high juvenile mortality followed by lower, relatively constant mortality thereafter. Such a curve is thought to be the most common in nature and applies to many fish, marine invertebrates, insects, and plants.

In reality, many species have survivorship curves that represent a combination of these three types. High juvenile mortality (as in type III curves) is common. But often this is followed by a relatively constant mortality rate, so the middle portion of the curve is linear (at least over long intervals), as for type II curves. Toward the end the mortality rate increases, dramatically perhaps, as in type I. As an example of this combination of types, consider the survivorship curve for the Dall mountain sheep shown in Figure 8.26 (it is based on the life table in Table 8.2). This would be classified as a type I curve, but it has some characteristics of the other types.

Note: The calculator can be used to generate survivorship curves on the basis of the data in a life table. You should refer to the *Technology Guide* to see the keystrokes needed.

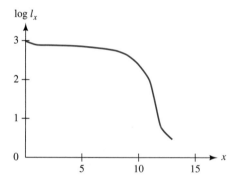

FIGURE 8.26 Survivorship curve for the Dall mountain sheep

EXAMPLE 8.6   Survivorship Curve for Red Deer Stags

Consider the life table for red deer stags that is shown in Table 8.3, where x is in years.[16] Graph the survivorship curve, classify it as to type, and discuss its shape.

Solution: The calculator produces the graph shown in Figure 8.27. This graph is similar to a type I curve in that it is relatively flat for a time and then falls off steeply. But

---

[16]From V. P. W. Lowe, "Population dynamics of the red deer (*Cervus elaphus* L.)," *J. Anim. Ecol.* **38** (1969), 425–457.

x	l_x	d_x	q_x	x	l_x	d_x	q_x
0	1000	282	0.282	8	249	157	0.631
1	718	7	0.010	9	92	14	0.152
2	711	7	0.010	10	78	14	0.179
3	704	7	0.010	11	64	14	0.219
4	697	7	0.010	12	50	14	0.280
5	690	7	0.010	13	36	14	0.389
6	684	182	0.266	14	22	14	0.636
7	502	253	0.504	15	8	8	1.000

TABLE 8.3 Life table for red deer stags

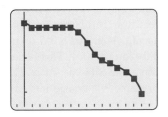

FIGURE 8.27 Survivorship curve for red deer stags

note the drop at the very beginning, corresponding to the relatively large value of q_0. Also note that the middle part resembles a stair-step curve, with sections of linearity followed by somewhat steep drops: From year 2 until year 7 the curve appears to be linear, and in fact the value of q_x is quite small here. This section is followed by a steep drop to year 10, then by another linear section in years 10 to 15, and then by a final drop. Thus, in addition to the relatively high mortality at the beginning and end of the curve, there is a section of high mortality in the middle.   ■ ■ ■

Note: It is not always clear why there are such drops in the middle of the curve. They may simply reflect complications in gathering data which accurately reflects the mortality rates in a population. They also may have to do with different stages in the life history of an individual. Often insects have different life stages, during which there are different susceptibilities to dying. Connections with life stages are explored in the following example and in Exercises 2 and 4.

EXAMPLE 8.7   **Survivorship Curve for the Spruce Budworm**

The spruce budworm infests forest conifers of eastern Canada and the northeastern United States.[17] During periodic outbreaks it kills many trees by feeding on flowers, buds, and needles. Here is a brief summary of its life history: A generation lasts about 1 year, during which the insect develops from egg to larva to pupa to moth. Eggs hatch in August, and shortly after they emerge, the larvae spin *hibernacula* (protective cases) in which they pass the winter. Upon emergence in May, they begin feeding on the tree. In late June or early July, the insect enters the pupal stage. In late July the moths emerge.

A survivorship curve for the spruce budworm is given in Figure 8.28. Here x represents half-months since August 15. On the basis of this curve, describe the variation in mortality of the spruce budworm as it corresponds to the organism's life stages.

---

[17]This example is based on the work of R. F. Morris and C. A. Miller, "The development of life tables for the spruce budworm," *Can. J. Zool.* **32** (1954), 283–301.

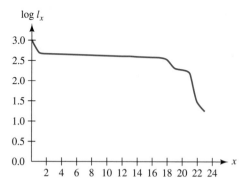

FIGURE 8.28 Survivorship curve for the spruce budworm

Solution: At the beginning of the curve there is a steep drop. This means that there is high mortality in the time between the larva's emergence from the egg and its formation of the hibernaculum. Next comes a long section where the curve is nearly horizontal, so for several months the mortality rate is very low. This corresponds to the period when the larvae are in the protective hibernacula. After this is a drop in the curve around $x = 18$, which indicates high mortality in mid-May. This is the time of emergence from hibernacula. There is a final drop in the curve near $x = 21$, so the mortality rate is high around the time of the organism's entrance into the pupal stage.

For the first two drops in the curve, each after an emergence, it turns out that the main cause of death is dispersion, as some insects fall from the trees. The final drop is associated with mortality due to parasites and predators (such as birds).

Usefulness of Survivorship Curves

There are difficulties in gathering data for a survivorship curve. The greatest difficulty is associated with the young age classes. But often these difficulties do not greatly affect the general shape of the curve. It is this shape, we have stressed, that reveals much about the general characteristics of a population. Ecologists find it useful to compare curves between similar species or curves for the same species in varying environments. R. Dajoz[18] gives another reason why survivorship curves are studied: "They indicate at what age a species is most vulnerable." In the management of a population, whether in a game reserve or in pest control, it is important to know when the population is most vulnerable, for that is when intervention is most effective. For the spruce budworm, as we have seen, the stages of vulnerability occur after each emergence and at about the time of pupation.

---

[18]*Introduction to Ecology* [transl. A. South] (New York: Crane, Russak & Company, 1976), 186.

8.3   *SKILL BUILDING EXERCISES*

S-1. **Groups:** In population dynamics, what is meant by the term *cohort*?

S-2. **Mortality statistics:** In gathering mortality statistics, it is customary to use the notation x, l_x, d_x, and q_x. Give the meaning of each of these symbols.

S-3. **Calculation of mortality statistics:** Show how to calculate q_x from l_x and d_x.

S-4. **More calculation of mortality statistics:** For a certain population, $l_5 = 300$ and $l_6 = 275$. Find d_5 and q_5.

S-5. **Number of survivors:** A certain cohort begins at 1000 individuals, and by age 31 years 60% have died. What is the value of l_{31}?

S-6. **Mortality:** *This is a continuation of Exercise S-5.* For the cohort in Exercise S-5, suppose that 120 individuals who reach age 31 die before age 32. What is the value of d_{31}?

S-7. **Mortality:** *This is a continuation of Exercises S-5 and S-6.* Using the information from Exercises S-5 and S-6, calculate q_{31} and l_{32}.

S-8. **Survivorship curve:** What is a survivorship curve?

S-9. **Survivorship curve:** Explain what a survivorship curve shows in practical terms.

S-10. **The end of a survivorship curve:** Explain why all survivorship curves must eventually touch the horizontal axis.

8.3   *EXERCISES*

Age x is measured in years, unless otherwise specified. References are supplied at the end of the exercise set.

1. **Classifying survivorship curves:** For each of the following populations a survivorship curve is given. Classify it as to type, and discuss its shape.

   a. African buffalo, in Figure 8.29.

   b. Buttercup seedling, in Figure 8.30. (Here x represents the number of weeks since emergence.)

   c. Warthog, in Figure 8.31.

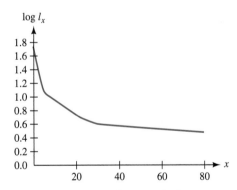

FIGURE 8.30 Survivorship curve for buttercup seedlings

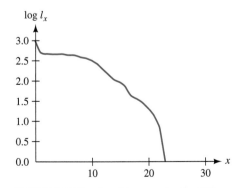

FIGURE 8.29 Survivorship curve for the African buffalo

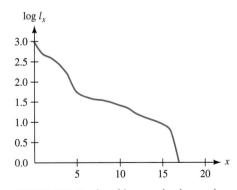

FIGURE 8.31 Survivorship curve for the warthog

2. **Palm survivorship:** Consider the survivorship curve for a palm tree that is given in Figure 8.32.

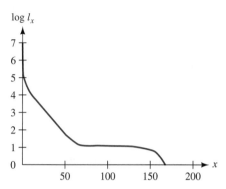

FIGURE 8.32 Survivorship curve for a palm tree

a. Classify it as to type, and discuss its shape.

b. Interpret your answer to part a in light of the following table, which gives stages of the palm tree life cycle.

Stage	Age in years
Seed	0
Seedling	1
Young tree	9
Mature, subcanopy	51
Canopy	88

3. **Graphing and classifying survivorship curves:** For each of the following animals a life table is given. Graph the survivorship curve, classify it as to type, and discuss its shape.

a. The desert night lizard, in Table 8.4.

b. The American robin, in Table 8.5.

4. **Honeybee survivorship:** In Table 8.6 a life table for the European honeybee worker is given. (Here x is measured in days.)

a. Graph the survivorship curve, classify it as to type, and discuss its shape.

x	l_x	d_x	q_x
0	1000	117	0.117
1	883	144	0.163
2	739	144	0.195
3	595	93	0.156
4	502	136	0.271
5	366	116	0.317
6	250	127	0.508
7	123	35	0.285
8	88	88	1.000

TABLE 8.4 Life table for the desert night lizard

x	l_x	d_x	q_x
0	1000	503	0.503
1	497	268	0.539
2	229	130	0.568
3	99	63	0.636
4	36	26	0.722
5	10	4	0.400
6	6	5	0.833

TABLE 8.5 Life table for the American robin

b. Interpret the life table and survivorship curve in light of the following table, which gives stages of the honeybee worker life cycle:

Stage	Age in days
Egg	0
Unsealed brood	3
Sealed brood	8
Adult	20

(Here *unsealed brood* refers to feeding larvae and *sealed brood* refers to postfeeding larvae and pupae.)

x	l_x	d_x	q_x
0	1000	42	0.042
3	958	137	0.143
8	821	10	0.012
20	811	15	0.018
25	796	16	0.020
30	780	14	0.018
35	766	27	0.035
40	739	149	0.202
45	590	221	0.375
50	369	256	0.694
55	113	76	0.673
60	37	32	0.865
65	5	5	1.000

TABLE 8.6 Life table for the honeybee worker

5. **Sagebrush lizard mortality:** In Table 8.7 age-specific mortality rates for the sagebrush lizard are given, with the assumption that initially there were 1000 individuals.

x	l_x	d_x	q_x
0	1000		0.770
1			0.378
2			0.469
3			0.553
4			0.618
5			0.462

TABLE 8.7 Partial life table for the sagebrush lizard

a. Fill in the blanks to give the life table through age 5. (*Suggestion:* To do this, use the formula $d_x = q_x \times l_x$. For example, to fill in the blank in the first row, we find

$$d_0 = q_0 \times l_0 = 0.770 \times 1000 = 770.$$

This means that 770 individuals died during their first year, so

$$l_1 = l_0 - d_0 = 1000 - 770 = 230$$

survived to the next year. This fills in the first blank in the second row.)

b. Graph the survivorship curve and discuss its shape.

6. **Half die each year:** In a study of a certain animal cohort, each year half of those surviving to the start of the year died during that year.

a. Make a life table with columns x, l_x, d_x, and q_x for this study. Assume that there were initially 32 individuals in the cohort.

b. What type of survivorship curve does this population have? Explain your answer.

7. **Experience of no use:** One ecologist interpreted the survivorship curve of a certain animal as saying that "experience of life is of no use . . . in avoiding death." Which one of the three types of curves do you think this animal has? Explain your answer.

8. **Grain beetles:** This exercise illustrates how the survivorship curves of two strains of grain beetles depend on environmental conditions. (Here x is measured in weeks.)

a. In Figure 8.33 two survivorship curves for the small strain of beetles grown in wheat are given. For one the ambient temperature was 29.1 degrees Celsius (about 84 degrees Fahrenheit), and for the other the ambient temperature was 32.3 degrees

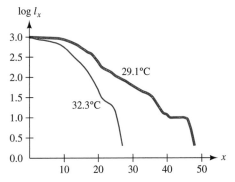

FIGURE 8.33 Survivorship curves at varying temperatures

Celsius (about 90 degrees Fahrenheit). Use the curves to decide at which of these two temperatures you would store your wheat. Explain your answer.

b. In Figure 8.34 two survivorship curves for the large strain of beetles grown at 29.1 degrees Celsius are given. For one the food was wheat, and for the other the food was maize. Use the curves to decide which of these two grains the large strain would be more likely to infest. Explain your answer.

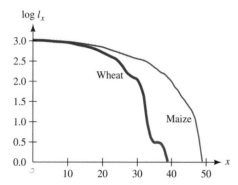

FIGURE 8.34 Survivorship curves in varying grains

9. **Life insurance rates:** Consider the survivorship curves given in Figure 8.35 for males and females in the human population of the United States.

a. On the basis of these curves, explain why life insurance rates are higher for men than for women.

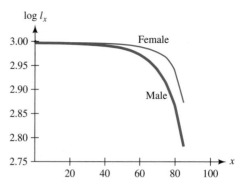

FIGURE 8.35 Survivorship curves for the U.S. population

b. On the basis of these curves, explain why life insurance rates increase so rapidly at high ages.

c. What would you predict about life insurance rates if the survivorship curves were of type II? What about type III?

10. **Span of life:** Many scientists believe that, for humans at least, there is a natural span of life. That is, there is a limit beyond which human life cannot extend, regardless of advances in medical science. Currently the maximum life span in developed countries is around 120 years, and evidence suggests that it has been that way since Roman times. However, with improvements in medical care, the *average* life span has increased significantly during the past century. Draw a sequence of survivorship curves that illustrates increasing average life span with fixed maximum life span.

REFERENCES

A classic reference here is E. S. Deevey, Jr., "Life tables for natural populations of animals," *Quart. Rev. Biol.* **22** (1947), 283–314. This is referred to below simply as *Deevey*.

African buffalo, warthog   C. A. Spinage, "African ungulate life tables," *Ecology* **53** (1972), 645–652.

Buttercup   J. Sarukhán and J. L. Harper, "Studies on plant demography: *Ranunculus repens* L., *R. bulbosus* L., and *R. acris* L.," *J. Ecol.* **61** (1973), 675–716.

Palm   L. Van Valen, "Life, death, and energy of a tree," *Biotropica* **7** (1975), 259–269.

Desert night lizard   R. G. Zweifel and C. H. Lowe, "The ecology of a population of *Xantusia vigilis*, the desert night lizard," *Amer. Mus. Novitates* **2247** (1966), 1–57.

American robin   From a study by Donald S. Farner, as described in *Deevey*.

Honeybee worker   S. F. Sakagami and H. Fukuda, "Life tables for worker honeybees," *Res. Popul. Ecol.* **10** (1968), 127–139.

Sagebrush lizard   D. W. Tinkle, "A population analysis of the sagebrush lizard, *Sceloporus graciosus*, in southern Utah," *Copeia* (1973), 284–296.

Exercises 7 and 10   *Deevey*, p. 310.

Grain beetles   L. C. Birch, "Experimental background to the study of the distribution and abundance of insects. I. The influence of temperature, moisture, and food on the innate capacity for increase of three grain beetles," *Ecology* **34** (1953), 698–711.

U.S. population   *Demographic Yearbook of the United Nations* (published annually). New York: Statistical Office of the United Nations, 1992.

Summary

Exponential and logistic functions have been used throughout the book as models for population growth. In this chapter we look more closely at their development and application to population dynamics. Logarithms and rates of change play key roles and are essential to an understanding of how populations change with time.

Exponential Growth

For any population, the rate of change in population is births per unit time minus deaths per unit time. The exponential model for population growth is a consequence of the assumption that both births and deaths per unit time are proportional to population size. This gives rise to the equation

$$\frac{dN}{dt} = rN,$$

the classic equation of change for an exponential function. As we saw in Chapter 7, the solution can be written in the alternative form

$$N = N(0)e^{rt}.$$

The letter r used here is standard notation. It is the measure of capacity for growth and is characteristic of a species in an environment. Ecologists know it as the species r *value*.

The *doubling time* t_d is the time required for a population to double in size, and for populations exhibiting exponential growth, the value of t_d does not depend on current population size. That is, the same period of time is required for such a population to grow from 100 to 200 individuals as from 1000 to 2000. The doubling time depends only on the r value and is given by

$$t_d = \frac{\ln 2}{r}.$$

Logistic Growth

Because exponential functions grow so rapidly, no population can maintain exponential growth indefinitely; ecological factors limit population size. The logistic model comes from the assumption that the population grows nearly exponentially at first but has an upper limit known as the *carrying capacity* of the environment. As carrying capacity is approached, the growth rate slows, yielding the classic S-shaped graph of logistic growth.

The basic shape alone of the logistic curve yields important information. In particular, the *maximum sustainable yield* theory of wildlife management says that a harvested population should be maintained at the point where the growth rate is the greatest—that is, at the steepest point on the logistic graph, which occurs at half of the carrying capacity.

As is often the case, logistic population growth is most easily understood in terms of its equation of change

$$\frac{dN}{dt} = rN\left(1 - \frac{N}{K}\right).$$

When the population is small, the factor $1 - \frac{N}{K}$ is almost 1, so the logistic equation of change is almost the same as the equation of change for the exponential function. This means that initially the population will grow exponentially. When the population N is near the carrying capacity K, the factor $1 - \frac{N}{K}$ is near 0, so the rate of change is almost zero. This causes the population to level out near the carrying capacity.

In order to get the correct logistic equation from observed data, we rewrite the logistic equation of change as

$$\frac{1}{N}\frac{dN}{dt} = r - \frac{r}{K}N.$$

This tells us that $\frac{1}{N}\frac{dN}{dt}$ is a linear function of N, and we know how to fit a regression line to such data. This regression line can be used to recover the equation of change.

Survivorship Curves

A deeper analysis of population dynamics requires knowledge not just of the number of individuals in a population but also of the age distribution within the population. Individuals are grouped by *cohort*—that is, according to their age. The following symbols are standard in mortality studies:

$$x = \text{age at beginning of interval}$$

$$l_x = \text{number of survivors at start of age interval } x$$

$$d_x = \text{number of deaths during age interval } x$$

$$q_x = \text{proportion of individuals dying during age interval } x$$

A table providing values for these statistics is known as a *life table*, and a *survivorship curve* is the graph of l_x against x done on a semilogarithmic scale. It shows how many individuals survive to a given age.

There are three basic types of survivorship curves. The type I curve is concave down and is characteristic of a population that has a low mortality rate among juveniles but a high mortality rate for older individuals. Rarely does a naturally occurring population exhibit true type I behavior, though humans and many other mammals show survivorship curves of roughly this shape. A type II curve is a straight line, which describes a species with a constant mortality rate. Type II curves apply to certain bird populations. The type III curve is concave up and represents high juvenile mortality with relatively constant mortality after that. This type of curve is thought to be the most common in nature.

Brief Answers to Odd-Numbered Exercises

Note: The answers presented here are intended to provide help when students encounter difficulties. Complete answers will include appropriate arguments and written explanations that do not appear here. Also, many of the exercises are subject to interpretation, and properly supported answers that are different from those presented here may be considered correct.

PROLOGUE

E-1. a. 19   b. 18   c. 12   d. 5   e. 81
S-1.   2.43
S-3.   1.53
S-5.   0.12
S-7.   1.32
S-9.   0.6
S-11. -2.17
S-13. a. 0.62   b. 1.03
  1. a. 63.21   b. 1.43   c. 1.69   d. 50.39   e. 3.65
  3. $669.60
  5. 3.45%
  7. a. $5.99   b. 43.08%
  9. $4464.29
  11. a. 5.54 years   b. 1.97; no   c. $9850; no
  13. a. 25,132.74 miles, or about 25,000 miles
    b. 268,082,573,100 cubic miles, or about 2.68×10^{11} cubic miles
    c. 201,061,929.8 square miles, or about 201,000,000 square miles
  15. a. 584,336,233.6 miles, or about 584 million miles
    b. 584,336,233.6 miles per year, or about 584 million miles per year
    c. 8760
    d. 66,705.05 miles per hour, or about 67,000 miles per hour
  17. 735 newtons; 165.38 pounds
  19. 277.19 cycles per second; 293.67 cycles per second
  21. Lean body weight is 100.39 pounds.
    Body fat is 31.61 pounds.
    23.95% body fat

CHAPTER 1

Section 1.1

E-1. a. It is a function.
  b. It is a function.
  c. It is not a function.
  d. It is not a function with domain D.
E-3. a. It is a function.   b. It is not a function.
  c. It is not a function.

E-5. a. All real numbers
  b. All non-negative real numbers
  c. All real numbers
  d. All real numbers larger than or equal to 7
E-7. a. It is onto.   b. It is not onto.   c. It is not onto.
  d. It is onto.   e. It is not onto.
E-9. a. It is a bijection.   b. It is not a bijection.
  c. It is not a bijection.   d. It is a bijection.
  e. It is not a bijection.
S-1.   0.35
S-3.   3.67
S-5.   1.06
S-7.   1.16
S-9. $c(2, 3, 5)$
S-11. a. 9.33   b. -2.21 c. 3.32
  1. a. $T(13,000) = \$930$ b. $110   c. $110
  3. a. $V(1)$; 8 feet per second. The ball is rising.
    b. -24 feet per second. The ball is falling.
    c. The velocity is 0, and the ball is at the peak of its flight.
    d. The velocity changes by -32 feet per second for each second that passes.
  5. a. 12 deer
    b. $N(10) = 380$ deer (rounded to the nearest whole number)
    c. $N(15) = 410$ deer
    d. 30 deer
  7. a. $C(800) = 4.54$ grams   b. 5730 years
  9. a. Higher   b. Continuous compounding
    c. $190.67 per month. It is 10 cents higher than if interest is compounded monthly as in Example 1.2.
  11. a. $P(350, 0.0075, 48)$; $14,064.67
    b. $15,812.54   c. $19,478.33
  13. Sirius appears 24.89 times as bright as Polaris.
  15. About 4.34 light-years
  17. a. $1 - 0.5^1 = 0.5$, giving a yield of 50% of maximum
    b. $Y(3)$; 0.88, or 88% of maximum
    c. 79% of maximum
  19. a. 2307.14 watts per square meter   b. 5767.85 watts
  21. 7.25 kilometers
  23. Answers will vary.

Section 1.2

E-1.   a. 19   b. 3   c. $-\frac{1}{8}$ d. 5

E-3.   The average rate of change is m.

E-5.   Answers will vary.

E-7.   10

E-9.   2.5

S-1.   23.8

S-3. $\frac{44.6-23.8}{10} = 2.08$

S-5. $\frac{51.3+44.6}{2} = 47.95$

S-7. $44.6 + 7 \times 0.67 = 49.29$

S-9.   Your dealership will sell no more cars than the local population.

S-11. -0.86

1.   a. $P(1980) = 29\%$

b. Using averages, $P(1975) = 26.5\%$.

c. 0.5 percentage point per year

d. 30%

e. If growth rate from 1988 to 1990 continues at the same rate as from 1980 to 1988, then $P(1990)$ will be about 34% .

3.   a. 11.71 billion dollars

b. 0.946 billion dollars per year

c. 8.87 billion dollars

5.   The limiting value of W is the total amount of water frozen in the snowball. This value is reached when the snowball is completely melted.

7.   a. -0.000454 gram per year   b. About 4.44 grams

c. 0

9.   a. yearly, semiannually, monthly, daily, hourly, each minute

b. 12.683%

c. $1019.76

d. Answers will vary. About 12.75%

11.   a. $H(13)$, about 62.2 inches

b.   i. Units for growth rate: inches per year

Period	Growth Rate
0 to 5	4.2
5 to 10	2.5
10 to 15	2.4
15 to 20	1.3
20 to 25	0.1

ii. From age 0 to age 5

iii. Answers will vary.

c. Answers will vary.

13.   a. Units for rate of change: dollars per dollar

Interval	Rate of Change
16,000 to 16,200	0.09
16,200 to 16,400	0.09
16,400 to 16,600	0.09

b. Answers will vary.   c. Answers will vary.

15.   a. 1.43 pounds per centimeter

b. 3.85 pounds per centimeter

c. Large

d. 205.95 pounds

e. 0.26 centimeter per pound

f. 171.96 centimeters

17.   a. Units for rate of change: thousands of widgets per worker

Interval	Rate of Change
10 to 20	1.25
20 to 30	0.63
30 to 40	0.31
40 to 50	0.15

b. Answers will vary.

c. About 49.2 thousand widgets

d. Too high

19.   a. -4.2 points per year   b. 201.6

c. 35 to 45 years old

21.   Answers will vary.

Section 1.3

E-1.   a. The secant line is below the graph between the two points and above the graph outside those points.

b. The estimate will be too small between the two points used and too large outside them.

S-1.   About 3.3

S-3.   At $x = 2.4$, f is about 4.

S-5.   From 2.4 to 4.8

S-7.   It is concave down.

S-9. Answers will vary. Our solution is shown below.

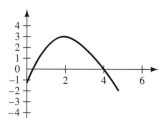

1. Answers will vary.

3. a. $v(1960) = \$10,000$, $v(1970) = \$5000$,
   $v(1980) = \$35,000$, $v(1990) = \$35,000$
   b. Various acceptable graphs   c. Answers will vary.

5. a. $F(7)$; about 1500 cubic feet per second
   b. At the end of June
   c. At the end of May
   d. About 0 cubic feet per second
   e. Answers will vary.

7. a. About $14,000 per acre   b. About 110 years old
   c. About 30 years old   d. About 60 years old
   e. Answers will vary.

9. a. In 1991 there were about 70 tornadoes reported.
   b. In 1988 there were about 15 tornadoes reported.
   c. About 25 tornadoes per year
   d. About 40 tornadoes per year
   e. 0 tornadoes per year

11. Answers will vary.

13. a. About 500 foot-candles
   b. About 800 foot-candles
   c. 80 degrees   d. 40 degrees

15. a. About 2 P.M.
   b. From about 6 A.M. to about 9 A.M.
   c. From about 9 A.M. to about 2 P.M.
   d. The net carbon dioxide exchange is zero.

17. a. About 100 pounds per acre
   b. From the difference between the yield and cost
      graphs
   c. About 70 pounds per acre

19. Answers will vary.

Section 1.4

E-1. a. $h = \dfrac{200 - w}{2}$ yards

   b. $A = w\dfrac{200 - w}{2}$ square yards

E-3. $C = 2\sqrt{\pi A}$

E-5. a. $5 - 2x$ b. $10 - 2x$ c. x
   d. $x(5 - 2x)(10 - 2x)$

E-7. a. $h = \dfrac{25}{\pi x^2}$ b. $2\pi x^2 + \dfrac{50}{x}$

S-1. $312.50

S-3. 405

S-5. $N = 67s$

S-7. $B = 500 + 37t$

S-9. $f = 8x$

1. a. $N(3) = 186.56$ million
   b. Population is in millions.

Year	Population
1960	180
1961	182.16
1962	184.35
1963	186.56
1964	188.80
1965	191.06

c.

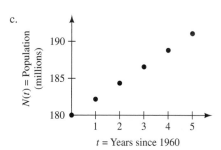

d. Same numbers as in part b
e. 273.27 million

3. a. $A = 200 + 150t$, where t is the time in minutes since
   takeoff and A is the altitude in feet.
   b. $A(1.5)$; 425 feet

c.

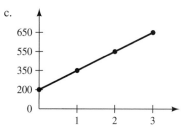

5. a. $64.00
   b. $C(d, m) = 29d + 0.06m$, where C is the rental cost
      in dollars, d is the number of rental days, and m is the
      miles driven.
   c. $C(7,500) = \$233.00$

7. a. 22 cents   b. $13.10

c. $c(p) = 0.16 + 0.03(p - 1)$, where c is the cost in dollars of the stationery and p is the number of pages.

d. $C(h, p) = 6.25h + 0.03(p - 1) + 0.16 + 0.38$, where C is the cost in dollars of preparing and mailing the letter, h is the hours of labor, and p is the number of pages.

e. $3.17

9. a. 7 b. 7 c. 10

d. $T = \dfrac{S}{50}$, where T is the annual stock turnover rate and S is the number of shirts sold in a year.

11. a. $C = 15N + 9000$, where N is the number of widgets produced in a month and C is the total cost in dollars.

b. $C(250)$; $12,750

13. a. Answers will vary.

b. $R = (50 - 0.01N)N$, where R is in dollars

c. $R(450)$; $20,475

15. a. $81 each for a total amount of $243

b. Rate $= 87 - 2n$ dollars

c. $R(n) = n(87 - 2n)$ dollars

d. $R(9) = $621

17. a. i. $13

ii. $C = \dfrac{150}{n} + 10$, where C is the amount charged per ticket, in dollars, and n is the number of people attending.

iii. $C(65)$, about $12.31.

b. $P = \dfrac{250}{n} + 10$, where P is the amount charged per ticket, in dollars, and n is the number of people attending.

19. a. Using k for the constant of proportionality, $t = kn$.

b. k is the number of items each employee can produce.

21. a. $p = 0.434h$

b. The head is 96 feet. The back pressure is 41.66 pounds per square inch.

c. The head is 145 feet. The back pressure is 62.93 pounds per square inch.

23. a. $V = KS$ b. 0.00123 meter per day

c. 1.23 meters per day

25. a. The pattern 1, 4, 2, 1 repeats.   b. 5 steps

c. 16 steps   d. Answers will vary.

CHAPTER 2

Section 2.1

E-1. Answers will vary.

E-3. a. 15 b. 2 c. $\sqrt{7}$ d. 0

E-5. a. 4 b. 3 c. 0

E-7. a. $\dfrac{a}{b}$ b. 0 c. $\dfrac{a}{b}$

E-9. a pounds per gallon

S-1.

X	Y₁	
4	15	
6	35	
8	63	
10	99	
12	143	
14	195	
16	255	

X=4

S-3.

X	Y₁	
3	-11	
7	-327	
11	-1315	
15	-3359	
19	-6843	
23	-12151	
27	-19667	

X=3

S-5.

X	Y₁	
0	-1	
20	.57087	
40	.57129	
60	.57137	
80	.57139	
100	.57141	
120	.57141	

X=0

About 0.5714

S-7.

X	Y₁	
0	21	
1	14	
2	9	
3	6	
4	5	
5	6	
6	9	

X=0

f has a minimum value of 5 at $x = 4$.

S-9.

X	Y₁	
4	129	
5	194	
6	261	
7	314	
8	321	
9	218	
10	-123	

X=8

f reaches a maximum of 321 at $x = 8$.

1. a. Decreases   b. $E(200)$; 75   c. High average

d. 166 or lower

3. a. 13.75 million students

b. 1976; 14.07 million students

c. 0.04 million per year; yes

5. The seventh race

7. a. 120 ways   b. 7 or more people   c. 24 guesses

d. 8.07×10^{67}

9. a. Larger

b. Once each year, EAR is 10%. (This is the only circumstance under which the EAR and APR will be the same.) Monthly, EAR is 10.471%. Daily, EAR is 10.516%.

c. Monthly, $5523.55. Daily, $5525.80.

d. Continuously, EAR is 10.517%. The EARs for monthly and continuous compounding differ by less than 0.05 percentage point.

11. a. $159.95

b. We show the table only for months 18 through 24.

X	Y$_1$
18	934.92
19	782.01
20	627.94
21	472.72
22	316.33
23	158.76
24	0

X=24

13. a. $E = \left(\dfrac{Q}{2}\right)850 + \left(\dfrac{36}{Q}\right)230$ b. $4035

c. 4 cars at a time   d. 9 orders this year

e. $80 per year per additional car

15. a. $v(2)$; 19.2 feet per second

b. The average rate of change in velocity during the first second is 16 feet per second per second. The average rate of change from the fifth second to the sixth second is 0.005 foot per second per second.

c. 20 feet per second

d. With a parachute, about 3 seconds into the fall. Without a parachute, 25 seconds into the fall. A feather will reach 99% of terminal velocity before a cannonball.

17. a. $C = 50N + 150$ dollars

b. $R = 65N$ dollars

c. $P = 65N - (50N + 150)$ dollars

d. 10 widgets per month

19. a. $P = 2h + 2w$ inches

b. $P = 2h + 2\left(\dfrac{64}{h}\right)$ inches

c. A square 8 blocks by 8 blocks with a perimeter of 32 inches.

d. $P = 2h + 2\left(\dfrac{60}{h}\right)$ inches. A rectangle 6 blocks by 10 blocks giving a perimeter of 32 inches.

21. a. 33.64 centimeters. A 4-year-old haddock is about 33.64 centimeters long.

b. 2.00 centimeters per year from age 5 to 10 years. 0.27 centimeter per year from age 15 to 20 years. Haddock grow more rapidly when young than when they are older.

c. 53 centimeters

23. a. 0.058 or 5.8%   b. 0   c. 0.379 or 37.9%

d. 0.621 or 62.1%

25. Answers will vary.

Section 2.2

E-1. a.

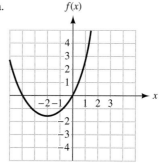

b.

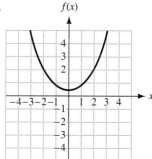

c.

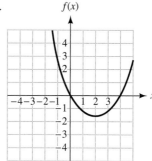

d.

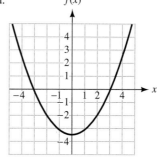

E-3.   a.

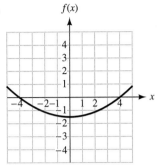

b.

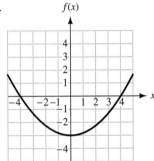

c.

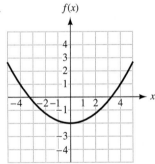

E-5.   Heights of points on the graphs of f and g are added to get points on the graph of $f + g$. Horizontal -5 to 5, vertical -5 to 20.

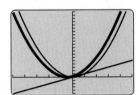

Darker graph is the sum.

S-1.

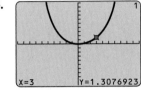

$f(3) = -7$

S-3.

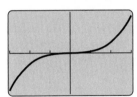

$f(3) = 1.31$

S-5.   From a table we chose a vertical span from -0.06 to 0.06.

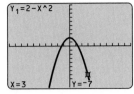

S-7.   From a table we chose a vertical span from 0 to 90,000.

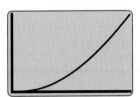

S-9. From a table we chose a vertical span from 0 to 1.

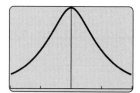

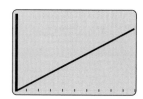

1. a. $3550

   b. Let n be the number of employees and C the weekly cost in dollars. Then $C = 2500 + 350n$.

   c. Horizontal, 0 to 10; vertical, 2500 to 7000

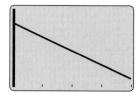

   d. 5 employees

3. a. $16,300   b. $V = 18,000 - 1700t$

   c. Horizontal, 0 to 4; vertical, 10,000 to 20,000

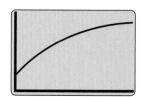

   d. $V(3)$; $12,900

5. a. Horizontal, 0 to 120; vertical, 0 to 425

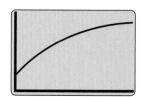

   b. 75 degrees

c. Temperature rose more during the first 30 minutes. Average rate of change over first 30 minutes is 4.89 degrees per minute. Over the second 30 minutes it is 2.68 degrees per minute.

   d. Concave down

   e. After about 46 minutes

   f. 400 degrees

7. a. Horizontal, 0 to 25; vertical, -35 to 35

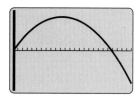

   b. $G(4)$; 16 thousand animals

   c. -11 thousand animals

   d. Increasing from 0 to 10 thousand animals; concave down

9. a.   i. $Q = \sqrt{\dfrac{800c}{24}}$

   Horizontal 0 to 25, vertical 0 to 30

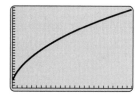

   ii. 14 items

   iii. The number will increase.

   b.   i. $Q = \sqrt{\dfrac{11200}{h}}$

   Horizontal, 0 to 25; vertical, 0 to 125

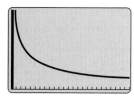

   ii. 27 items per order

   iii. It should decrease.

   iv. About -0.79 item per dollar

   v. Concave up

11. a. $P = 200 \times \dfrac{1}{0.01} \times \left(1 - \dfrac{1}{1.01^t}\right)$

Horizontal, 0 to 480; vertical, 0 to 25,000

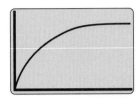

b. \$7594.79   c. \$13,940.10   d. \$20,000

13. a.   i. $N = \dfrac{30}{\pi} \times \sqrt{\dfrac{9.8}{r}}$

ii. Horizontal, 10 to 200; vertical, 0 to 10

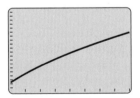

iii. Decreases
iv. 2.44 per minute

b.   i. $N = \dfrac{30}{\pi} \times \sqrt{\dfrac{a}{150}}$

ii. Horizontal, 2.45 to 9.8; vertical, 1 to 3

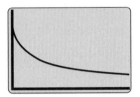

iii. Increases

15. a.   i. $Y = 339.48 - 0.01535N - 0.00056N^2$
ii. Horizontal, 0 to 800; vertical, −50 to 400

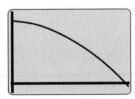

iii. Decreases

b.   i. $Y = 260.56 - 0.01535N - 0.00056N^2$
ii. Horizontal, 0 to 800; vertical, −50 to 400

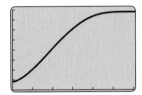

iii. Decreases

17. a. Horizontal, 0 to 6; vertical, 0 to 55

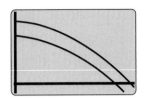

b. $C(1.5)$; 19.67 thousand   c. Answers will vary.
d. Answers will vary.   e. 52 thousand

19. a. Horizontal, 0 to 30; vertical, 0 to 330

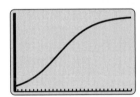

b. 21   c. In 2026   d. 315
e. Concave up from $t = 0$ to about $t = 11$; concave
down afterwards

21. Answers will vary.

Section 2.3

E-1. a. $x = \dfrac{y}{y^3 - y}$ or $x = \dfrac{1}{y^2 - 1}$

b. $x = \dfrac{\sqrt{y}}{3\sqrt{y} - 2}$

c. $x = \dfrac{z}{y + z^2 + y^2}$

d. $x = \dfrac{yz - \sqrt{9 + y}}{\sqrt{7 + y} - z}$

E-3. a. $\dfrac{x - 5}{7}$ b. $\dfrac{5x + 1}{2x - 3}$ c. $\dfrac{-4}{x - 2}$ d. $\dfrac{3x}{4x - 1}$

E-5. a. $x = \frac{4}{3}$ b. $x = \frac{27}{14}$ c. $x = \dfrac{a^2}{9}$ d. $x = 4$

S-1. $x = 7$

S-3. $x = -\frac{20}{9}$

S-5. $x = -\frac{38}{43}$

S-7. $x = \dfrac{12 - d}{c}$

S-9. $k = \dfrac{m - n}{3}$

1. a. $P = C + B + G + E$
   b. i. Imports were larger.
      ii. 5950.7 billion dollars
   c. $E = P - C - B - G$
   d. 16.2 billion dollars

3. a. 9.2 thousand dollars
   b. At the start of 2006
   c. $t = \dfrac{12.5 - V}{1.1}$, or $t = 11.36 - 0.91V$
   d. At the start of 2008

5. a. 288 miles
   b. $m = \dfrac{d}{g}$
      ii. 25.77 miles per gallon
   c. i. $g = \dfrac{425}{m}$
      ii. Horizontal, 0 to 30; vertical, 0 to 75

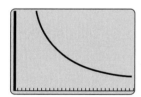

7. $1.96 per bushel

9. a. 153.28 psi   b. 85.37 psi
   c. $NP = \dfrac{EP}{1.1 + K \times L}$ d. 0.18
   e. $K = \dfrac{\frac{EP}{NP} - 1.1}{L}$

11. a. $p = (d - c)n - R$ or, equivalently, $p = dn - cn - R$.
   b. $2726.10
   c. $d = \dfrac{p + R}{n} + c$ or, equivalently, $d = \dfrac{p + cn + R}{n}$.
   d. $9.43

13. a. 303.15 kelvins
   b. $C = K - 273.15$
   c. $F = 1.8(K - 273.15) + 32$, or $F = 1.8K - 459.67$.
   d. 98.33 degrees Fahrenheit

15. a. $S(30)$; 3.3 centimeters per second
   b. $T = \dfrac{S + 2.7}{0.2}$
   c. 28.5 degrees Celsius.

17. a. $C = 55N + 200$ dollars
   b. $R = 58N$ dollars
   c. $P = 58N - (55N + 200)$ dollars or, equivalently, $P = 3N - 200$ dollars
   d. 66.67 thousand widgets per month

19. a. $R = \dfrac{Y + 55.12 + 0.01535N + 0.00056N^2}{3.946}$
   b. $R = \dfrac{55.12 + 0.01535N + 0.00056N^2}{3.946}$
   c. Horizontal, 0 to 800; vertical, 0 to 120

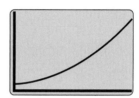

   d. Increases
   e. 38.23 millimeters
   f. Die back

21. a. $n = 1 - m$ b. $n = \dfrac{1 - 0.7m}{1.2}$
   c. $m = 0.4$, $n = 0.6$ thousand animals

Section 2.4

E-1. a. $(x - 9)(x + 1)$
   b. $(x + 3)^2$
   c. $(x - 4)(x + 4)$
   d. $(x - 7)(x - 5)$

E-3. a. $x = 0$, $x = 4$, $x = -1$
   b. $x = \frac{3}{2}$, $x = -\frac{4}{3}$, $x = \frac{5}{2}$
   c. $x = 4$, $x = -5$, $x = 3$
   d. $x = 3$, $x = 2$, $x = -4$, $x = -3$

E-5. a. $x = 3$ and $x = \frac{1}{3}$ b. $x = \frac{1}{2}$
   c. No rational solutions

S-1. We used the standard viewing window.

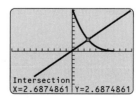

   $x = 2.69$

S-3. We used a horizontal span of -2 to 2 and a vertical span of 0 to 5.

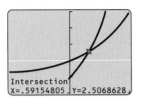

$x = 0.59$

S-5. For the window we use a horizontal span of -5 to 5 and a vertical span of 0 to 7.

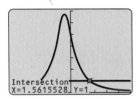

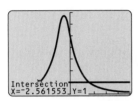

$x = -2.56$ and $x = 1.56$

S-7. We graph $\dfrac{5}{x^2 + x + 1} - 1$ using a horizontal span of -4 to 4 and a vertical span from -3 to 6.

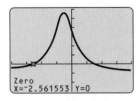

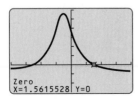

$x = -2.56$ and $x = 1.56$

S-9. We graph $\dfrac{-x^4}{x^2 + 1} + 1$ using a horizontal span of -2 to 2 and a vertical span of -2 to 2.

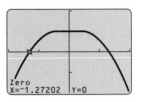

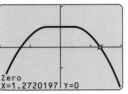

$x = -1.27$ and $x = 1.27$

1. a. 30 foxes
   b. Horizontal, 0 to 25; vertical, 0 to 160

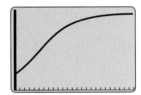

   c. 7.58 years
3. -0.98 and 5.80
5. 7.62 seconds
7. a. Horizontal, 0 to 100; vertical, -5 to 25

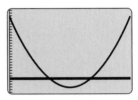

   b. $R(15)$; 9.25 moles per cubic meter per second
   c. 32.68 and 67.32 moles per cubic meter
9. 0.37 liter
11. 9.15 thousand years
13. a.   i. Horizontal, 0 to 2000; vertical, -1 to 3

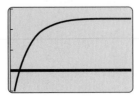

    ii. Concave down

    iii. 163.08 pounds per acre

    iv. 2.5 pounds

  b. i. Same scale as above

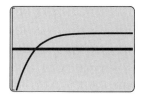

    ii. The red kangaroo

15. a. Horizontal, 0 to 1000; vertical, −1 to 1

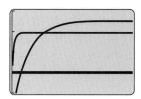

  b. 201.18 pounds per acre

17. a. $C = 65N + 700$ dollars

  b. $R = (75 − 0.02N)N$ dollars

  c. $P = (75 − 0.02N)N − (65N + 700)$ dollars

  d. 84.17 and 415.83 thousand widgets per month

19. 7.63 days

21. a. 0.91 second   b. 0.00096 second

  c. 1.74×10^{15} ergs

23. Answers will vary.

Section 2.5

E-1. a. $(−3, −13)$ is a minimum.

  b. $(5, −74)$ is a minimum.

  c. $(3, 15)$ is a maximum.

E-3. Answers will vary.

E-5. $x = −\frac{1}{2}$

S-1. A table of values leads us to choose a vertical span of 0 to 15.

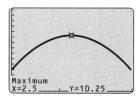

Maximum value of 10.25 at $x = 2.5$

S-3. A table of values leads us to choose a vertical span of 0 to 7.

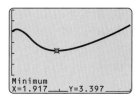

Minimum value of 3.40 at $x = 1.92$

S-5. We use a vertical span of 0 to 3.

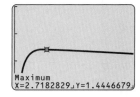

Maximum value of 1.44 at $x = 2.72$

S-7.

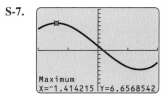

Maximum $x = −1.41$, $y = 6.66$

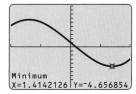

Minimum: $x = 1.41$, $y = −4.66$

S-9. We use a horizontal span of 0 to 5 and a vertical span of 0 to 130. Since the graph is increasing, the maximum occurs at the endpoint.

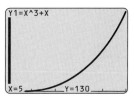

Maximum value of 130 at $x = 5$

1. a. Horizontal, 0 to 4; vertical, 0 to 2

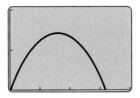

b. 3.22 miles downrange

c. 1.39 miles high at 1.61 miles downrange

3. a. Horizontal, 0 to 1.5; vertical, −0.1 to 0.1

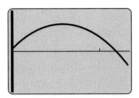

b. $G(0.24)$; 0.04 million tons per year

c. −0.02 million tons per year

d. 0.67 million tons

5. a. Total amount $= 2W + L$ feet.

b. $WL = 100$

c. $L = \dfrac{100}{W}$

d. $F = 2W + \dfrac{100}{W}$

e. Horizontal, 1 to 15; vertical, 0 to 110

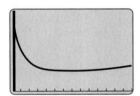

f. 7.07 feet perpendicular, 14.14 feet parallel

7. a. $G = 0.3s(4 - s)$

b. Horizontal, 0 to 4; vertical, 0 to 1.5

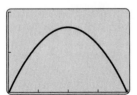

c. 2 thousand dollars

d. Answers will vary.

e. Answers will vary.

9. a. Height $= 4.77$ inches. Area$= 36.28$ square inches.

b. Height $= 0.19$ inch. Area $= 163.08$ inches.

c. i. Horizontal, 0 to 4; vertical, 0 to 50

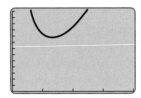

ii. 1.34 inches

iii. 2.66 inches

11. a. $C = 60N + 600$ dollars

b. $R = (70 - 0.03N)N$ dollars

c. $P = (70 - 0.03N)N - (60N + 600)$ dollars

d. 166.67 thousand widgets per month

13. a. Answers will vary.

b. $C = 300L + 500\sqrt{1 + (5 - L)^2}$

c. 4.12 miles under water; $2361.55

d. 2.24 miles under water; $2018.03

e. Horizontal, 0 to 5; vertical, 1000 to 3000

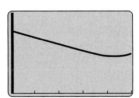

f. $L = 4.25$ miles

g. $W = 1.25$ miles

h. $C = 700L + 500\sqrt{1 + (5 - L)^2}$. The entire cable should run under water.

15. a. 15.18 years old

b. $B = \left(1000e^{-0.1t}\right) \times \left(6.32(1 - 0.93e^{-0.095t})^3\right)$

Horizontal, 0 to 20; vertical, 0 to 700

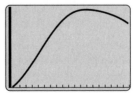

c. 13.43 years old

d. i. 15.18 years old

ii. 13.43 years old

17. a. Horizontal, 0 to 40; vertical, 0 to 5

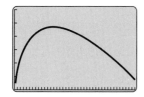

b. 3.81 pounds per year
c. 2.95 pounds per year

19. a. Horizontal, 0 to 2400; vertical, 0 to 750

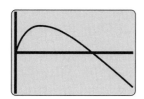

b. 1675 pupils   c. $27.07 per pupil

21. a. Horizontal, 0 to 350; vertical, -15 to 15

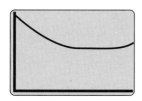

b. 73.27
c. $N = 0$ and $N = 228$
d. -7.51 per day

23. Answers will vary.

CHAPTER 3

Section 3.1

E-1. Answers will vary.
E-3. $a = 12$ and $b = 3$
E-5. Answers will vary.
E-7. Answers will vary.
E-9. The graphs of the two linear equations are parallel lines, so they do not cross.
S-1. 4
S-3. 7.5 feet
S-5. 5.14 feet
S-7. $-\frac{2}{3}$
S-9. 16.4 feet

1.

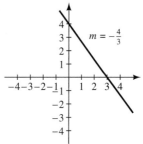

The slope will be negative. $m = -\frac{4}{3}$

3. Horizontal intercept is 4.

5.

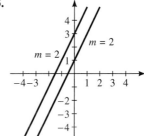

They do not cross. Lines with the same slope are parallel.

7. a. 0.4 foot high   b. 6 feet high
9. a. 0.83 foot per foot   b. 22.11 feet   c. 4.82 feet
11. h is 4 feet and k is 5 feet 6 inches.
13. At least 38.67 feet
15. a. 276.67 feet per mile
b. 5513.35 feet
c. 22.30 miles
17. The umbra has a radius of about 2859 miles at this distance. The moon can fit in the umbra, causing a total lunar eclipse.
19. About 430,328 miles

Section 3.2

E-1. $y = 3x - 6$
E-3. a. $y = 3x - 5$ b. $y = -4x + 5$ c. $y = 2x$
d. $y = -\frac{1}{5}x + \frac{8}{5}$
S-1. 4
S-3. $f(5) = 12.4$
S-5. $x = 2.76$
S-7. $f = 4x - 7$
S-9. $f = -1.2x + 12.8$
1. a. $C = 0.56F - 17.78$ (rounding to two decimal places)
b. The slope is 0.56.   c. They are the same.

3. a. Answers will vary.
   b. Let S be the total amount of storage space (in megabytes) used on the disk drive, and let n be the number of pictures stored. $S = 2.3n + 781$.
   c. $S(350)$; 1586 megabytes
   d. 270 pictures on the disk drive. There is room for 260 additional pictures.
5. a. Answers will vary. The slope is $20 per widget, and the initial value is $1500; $C = 20N + 1500$.
   b. $1300
   c. Fixed costs: $1100; variable cost: $16 per widget
7. a. $S = -0.746D + 46.26$
   b. $S(10)$; 38.8 miles per hour
9. a. 0.37 pound per dollar   b. 48.10 pounds
   c. $33.30
11. a. The slope is 5 pounds per inch.
   b. Let h be height (in inches), and let w be the weight (in pounds): $w = 5h - 180$.
   c. 66.4 inches
   d. Light
13. For each pound increase in weight, males will show an increase of 1.08 pounds in lean body weight, but females will show an increase of only 0.73 pound in lean body weight.
15. a. The vertical factor has a constant rate of change.
   b. $V = 40d + 65$ c. 82.16 feet   d. Yes
17. a. $0.5a$ dollars   b. g dollars   c. $0.5a + g = 5$
   d. $g = -0.5a + 5$
19. 64 days
21. a. Pessimistic, 0.00002; optimistic, 10,000,000
   b. 0.0005   c. Answers will vary.

Section 3.3

E-1. a. The data are linear. $y = 3x - 1$
   b. The data are not linear.
   c. The data are linear. $y = 4x - 5$

E-3.

x	2	5	6	8	10
y	5	14	17	23	29

S-1. There is a constant change of 2 in x and a constant change of 5 in y. The data are linear.
S-3. $y = 2.5x + 7$
S-5.

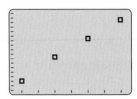

S-7.

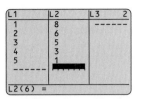

S-9.

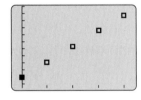

1. a. Let d be the number of years since 1988, and let V be the value in billions of dollars. A table of differences will show a common difference of 2.2. The rate of change is constant, so the data are linear.
   b. Horizontal, -0.4 to 4.4; vertical, 3.4 to 15.2

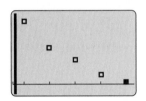

   c. $V = 2.2d + 4.9$
   d. Same scale as part b

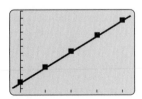

   e. 22.5 billion dollars

3. a. Let d be the number of years since 1994, and let T be the tuition in dollars. A table of differences shows a common difference of 168 dollars.
$T = 168d + 2816$.
b. 168 dollars per year **c.** 866 dollars per year
d. Tuition at public universities increased at a rate of $168 per year, while tuition at private universities increased at a rate of $866 per year.
e. Tuition at public universities increased by about 5.1%, while tuition at private universities increased by about 5.3%.
5. a. $p = -0.01N + 45$
b. $R = (-0.01N + 45)N$; no
c. $P = (-0.01N + 45)N - (35N + 900)$; no
7. a. A difference table shows a common difference of 36.
b. 1.80 degrees Fahrenheit per kelvin
c. $F = 1.80K - 459.67$ **d.** 310.15 kelvins
e. An increase of 1 kelvin causes an increase of 1.80 degrees Fahrenheit. An increase of 1 degree Fahrenheit causes an increase of 0.56 kelvin.
f. -459.67 degrees Fahrenheit
9. a. A difference table shows a constant difference of 1.05; $P = 2.1S - 0.75$
b. Horizontal, 0 to 3; vertical, 0 to 5

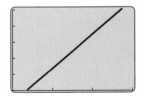

c. Answers will vary. **d.** 2.21 billion bushels
11. a. Let V be the velocity (in miles per hour) and t the time (in seconds) since the car was at rest. A table of differences shows a common difference of 5.9 miles per hour.
b. 11.8 miles per hour per second
c. $V = 11.8t + 4.3$ **d.** 4.3 mph **e.** 4.72 seconds
13. a. Let N be the number graduating (in millions), and let t be the number of years since 1995. The slope is 0.08 million graduating per year.
b. $N = 0.08t + 2.59$ **c.** $N(7)$; 3.15 million
d. This formula gives 2.51 million, which is much closer than the earlier answer.
15. a. When $S = 36.8$, c should be 1517.52.
b. Answers will vary.
c. 1520.16 meters per second
17. 250,000 stades; 26,000 miles

Section 3.4
E-1. Answers will vary.
E-3. a. 1.2 **b.** 1.68 **c.** 0.156
E-5. a. The line would fit the data points exactly.
b. They are all 0 (allowing for errors due to rounding).
S-1. Increasing
S-3. Federal spending on agriculture is increasing faster than federal spending on research and development.
S-5.

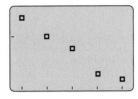

$y = -0.26x + 2.54$

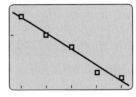

S-7.

$y = 1.05x + 2.06$

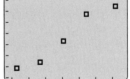

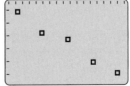

S-9.

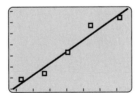

$y = -0.32x - 1.59$

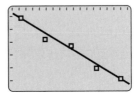

1. a. Let t be the time in hours since the experiment began, and let N be the number of bacteria in thousands.

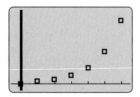

The data do not appear to fall on a straight line, so a linear model is not appropriate.

b. Let t be the number of years since 1996, and let E be the enrollment in millions.

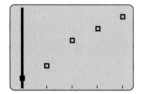

The data are not exactly linear, but they do nearly fall on a straight line. It looks reasonable to approximate the data with a linear model.

3. a. Let t be the number of years since 1994, and let T be the number of tourists in millions.

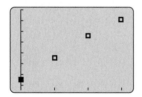

b. $T = 1.92t + 18.61$

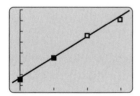

c. Answers will vary.
d. $T(-2)$; 14.77 million tourists

5. a. Let t be the number of years since 1900, and let L be the length in meters. $L = 0.034t + 7.197$
b. Answers will vary.

c.

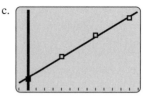

d. Answers will vary.
e. The regression line model gives 7.88 meters, which is far too long.

7. a.

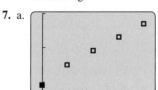

b. $D = 0.88t + 51.98$

c.

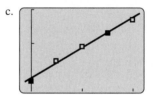

d. The regression line is the thin line, and the true depth function is the thicker line.

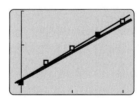

The two lines are very close, but the regression model shows a water level that is a bit too high.

e. 54.4 feet   **f.** 54.62 feet, or about 54.6 feet

9. a. Let t be the number of years since 1992, and let J be the total sales in millions.

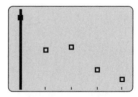

It may be reasonable to model this with a linear function, but there is room for argument here. There is enough deviation from a straight line to cast serious doubt on the appropriateness of using a linear model. Collecting additional data would be wise.

b. $J = -0.044t + 2.46$

c.

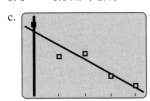

Answers will vary. To the eyes of the authors, this picture makes the use of the regression line appear more appropriate, but gathering more data still seems wise.

d. The regression model gives a prediction of 2.24 million cars sold in 1997 and 2.20 million in 1998. Both of these estimates are higher than Commerce Department predictions.

11. The data points show that a larger number of churches yields a larger percentage of antimasonic voting. Furthermore, since the points are nearly in a straight line, it is reasonable to model the data with a linear function.

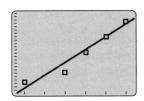

The formula for the regression model is $M = 5.89C + 45.55$. Its graph is added to the data below.

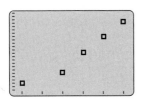

This picture reinforces the idea that the data can be approximated by a linear function.

13. a. $V = 12.57T + 7501.00$

b. The steam will occupy an additional 1257 cubic feet.

c. About 12,780 cubic feet

d. 532.94 degrees

15. a. $B = 0.29W + 0.01$

b. The third sample: $W = 13.7$ and $B = 3.7$

c. 0.29 additional ton per hectare

17. a. $E = 0.34v + 0.37$ b. 0.34

c. Higher cost; less efficient   d. 0.37; higher

19. Answers will vary.

21. Answers will vary.

Section 3.5

E-1. a. $x = 3$, $y = 2$ b. $x = 2$, $y = 4$

c. $x = 2$, $y = 0$

E-3. a. $x = 2$, $y = 1$, $z = 0$ b. $x = 1$, $y = 1$, $z = 2$

c. $x = 1$, $y = 2$, $z = 2$, $w = 3$

E-5. Answers will vary.

E-7. a. $x = 1$, $y = 2$ b. $x = 1.5$, $y = 1.5$

c. $x = -\frac{1}{3}$, $y = \frac{14}{9}$ d. $x = \frac{22}{27}$, $y = -\frac{1}{9}$

S-1. Each graph consists of points that make the corresponding equation true. The solution of the system is the point that makes both true, and that is the common intersection point.

S-3. We graphed $y = \dfrac{6 - 3x}{4}$ and $y = \dfrac{5 - 2x}{-6}$ using a horizontal span of 0 to 5 and a vertical span of -3 to 3.

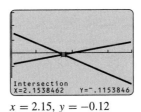

$x = 2.15$, $y = -0.12$

S-5. We graphed $y = \dfrac{6.6 - 0.7x}{5.3}$ and $y = \dfrac{1.7 - 5.2x}{2.2}$ using a horizontal span of -2 to 2 and a vertical span of 0 to 5.

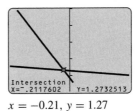

$x = -0.21$, $y = 1.27$

S-7. $x = 2.15$, $y = -0.12$

S-9. $x = -0.21$, $y = 1.27$

1. If C = number of bags of chips and S = number of sodas, then $2C + 0.5S = 36$ and $S = 5C$. You buy 40 sodas and 8 bags of chips.

3. If c = number of crocus bulbs and d = number of daffodil bulbs, then $0.35c + 0.75d = 25.65$ and $c + d = 55$. You buy 39 crocus bulbs and 16 daffodil bulbs.

5. In 7.55 years there will be about 504 foxes and 504 rabbits.

7. $m = 0.25$, $n = 0.75$ thousand animals

9. a. 0.08 part per million

 b. Hartsells fine sandy loam

11. At 160 degrees Celsius, the Fahrenheit temperature is 320 degrees.

13. a. $(0, \frac{1}{4})$ b. Answers will vary.

15. The graphs lie on top of one another. Every point on the common line is a solution of the system of equations.

17. N = number of nickels, D = number of dimes, and Q = number of quarters.

 $0.05N + 0.1D + 0.25Q = 3.35$;

 $N + D + Q = 21$; $D = N + 1$;

 There are 5 nickels, 6 dimes, and 10 quarters.

CHAPTER 4

Section 4.1

E-1. a. $\dfrac{a}{b}$ or ab^{-1} b. a^{24} c. $a^7 b$

E-3. a. $N = 0.19 \times 6^t$ b. $N = 2.37 \times 1.19^t$

 c. $N = 3a^{-2}a^t$ d. $N = \dfrac{m^2}{n} \left(\sqrt{\dfrac{n}{m}} \right)^t$

E-5. $N = 0.22 \times 1.94^t$

S-1. $f(2) = 17.28$; $f(x) = 3 \times 2.4^x$

S-3. 1.25

S-5. It is proportional to the function value.

S-7. 10×0.96^t, with t in years

S-9. 0.72%

1. 23×1.4^t. Horizontal, $t = 0$ to 5; vertical, 0 to 130

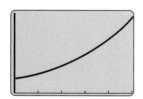

3. a. Let t be time in years and N population in millions. Then $N = 3 \times 1.023^t$.

 b. $N(4)$; 3.29 million

5. a. 231.2 grams b. 32%

 c. Monthly decay factor 0.968; monthly percentage decay rate 3.2%

 d. Percentage decay rate per second 0.00000122%

7. 4 half-lives, or 11.32 years

9. 23.45 years

11. 6.13 years, with both immigration and extinction at 2.69 species per year

13. a. $N = 67.38 \times 1.026^t$, where N is population in millions and t is years since 1980

 b. $N(3)$; 72.77 million

 c. $t = 11.28$; sometime in 1991

15. a. Answers will vary.

 b. 0.98

 c. 1

 d. $P = 0.98^Y$

 e. For 10 years: 0.82, or 82%; for 100 years: 0.13, or 13%

 f. $Q = 1 - 0.98^Y$

17. 9.22×10^{18} grains; about 35.6 trillion dollars

19. 25.0%

21. a. 0.12 b. 0.87

23. a. 1.1052 b. 1.1052

 c. Answers will vary.

Section 4.2

E-1. a. The data are exponential. $y = 3 \times 4^x$.

 b. The data are exponential. $y = 16 \times 0.5^x$.

 c. The data are not exponential.

E-3. $y = b^{-a} b^x$

S-1. $N = 7 \times 8^t$

S-3. $N = 12 \times 1.803^t$

S-5. Ratios give a constant value of 2. The data are exponential.

S-7. $y = 2.6 \times 3^x$

S-9. Ratios give a constant value of 3. The data are exponential.

1. The data show a common ratio of 1.04, rounded to two decimal places, and thus they are exponential.

 $f = 3.80 \times 1.04^t$

3. The successive ratios are not the same.

5. From 1995 to 1998, sales grew at a rate of $1.06 thousand per year. From 1998 to 2001, sales grew by 9% per year.

7. a. $1750.00

 b. Let t be the time in months and B the account balance in dollars. The data show a common ratio of 1.012, rounded to three decimal places.

 $B = 1750.00 \times 1.012^t$

 c. 1.2%

 d. 15.4%

 e. $23,015.94

 f. 58.1 months to double the first time; 58.1 months to double again

9. a. The data show a common ratio of 0.42, so they are exponential. $D = 176.00 \times 0.84^t$

 b. 16%

 c. $V = 176.00 - 176.00 \times 0.84^t$

 d. About 26.41 seconds into the fall

11. a. Let t be the time in minutes and U the number of grams remaining. In the display, the data appear to be linear.

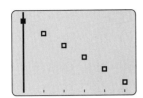

b. Rounding the calculated parameters to three decimal places, $U = -0.027t + 0.999$.

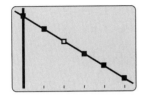

c. 18.5 minutes

d. That there would be -0.621 gram remaining

13. The yearly interest rate was 5.0% for the first 3 years, then 4.5% after that.

15. a. Answers will vary.   b. Answers will vary.

Section 4.3

E-1. Answers will vary.

E-3. Answers will vary.

E-5. a. 3   b. $-\frac{7}{2}$ c. -1 d. 4

E-7. Answers will vary.

S-1. Exponential

S-3. 0.463

S-5. The regression line is $\ln y = 0.671x + 0.798$. The exponential model is $y = 2.22 \times 1.96^x$.

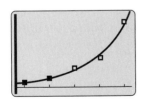

S-7. The regression line is $\ln y = 0.031x + 1.849$. The exponential model is $y = 6.35 \times 1.03^x$.

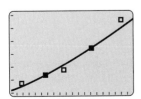

S-9. The regression line is $\ln y = 0.335x + 0.873$. The exponential model is $y = 2.39 \times 1.40^x$.

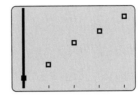

1. a. 4   b. 6

3. a. Let t be years since 1997 and N the population, in thousands.

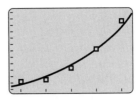

b. $\ln N = 0.042t + 0.445$

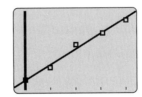

5. a. $S = 9.30 \times 1.05^t$

b. 5%

c. 12.46 thousand dollars

d. $t = 5.22$, or mid-March of 2006

7. a. Let t be years since 1976 and C the percent with cable.

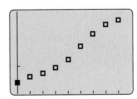

b. $\ln C = 0.144t + 2.631$

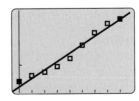

c. $C = 13.888 \times 1.155^t$

d.

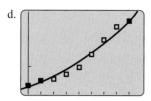

e. 15.5%

f. Yes, the model predicts 67.77%.

9. a. Let t be years since 1950 and H the costs in billions of dollars.

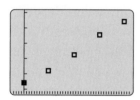

b. $\ln H = 0.099t + 2.426$

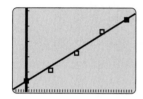

c. Yearly growth factor 1.104; 10.4% per year

d. $H = 11.314 \times 1.104^t$

e. $H(50)$; 1.592 trillion dollars

11. a.

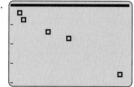

b. $\ln D = -0.012V - 1.554$

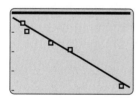

c. $D = 0.211 \times 0.988^V$

d. $A = 0.18 - 0.211 \times 0.988^V$

e. 203.89 pounds per acre (depends on rounding in previous parts)

13. Table A is approximately linear, with the model $f = 19.84t - 16.26$. Table B is approximately exponential, with the model $g = 2.33 \times 1.56^t$.

15. a. Yes, since a plot of the natural logarithm of P against D appears linear

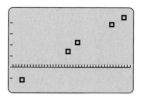

b. 0.0003×1.12^D c. It is increased by 12%.

17. a. $D = 40.89 \times 0.82^t$

b. $L = 53 - 40.89 \times 0.82^t$

c. A bit shorter

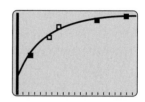

d. About 6.2 years old

19. a. $N = 16.73 \times 1.02^s$

b. 66.91 persons injured per 100 accident-involved vehicles

c. It is increased by 2%.

21. a. $P = 84.69 \times 0.999^D$

b. 69.33% of pedestrians

c. The initial value indicates that 84.69% of pedestrians walk at least 0 feet. The correct percentage is 100%.

23. a. $N = 24,670,000 \times 0.125^M$

b. About 266

c. About 0.5

d. 0. Earthquakes of large magnitude are very rare.

e. There are 87.5% fewer earthquakes of magnitude $M + 1$ or greater than of magnitude M or greater.

25. Answers will vary.

27. Answers will vary.

Section 4.4

E-1. a. $t = 0.24$ b. $t = 3.21$ c. $t = -1.29$

d. $t = -\dfrac{\log 3}{2 \log a}$

E-3. a. 2.90

b. 17.95 parsecs

c. Spectroscopic parallax is increased by 5 units.

d. The spectroscopic parallax for Shaula is 2.89 units larger than that for Atria.

E-5. a. $x = \dfrac{\log 3}{\log 2} = 1.58$ or $x = 1$

  b. $x = \dfrac{\log 3}{\log 5} = 0.68$ or $x = \dfrac{\log 4}{\log 5} = 0.86$

  c. $x = 0$ or $x = \dfrac{\log 2}{\log 3} = 0.63$

E-7. a. $x = e^5 - 2$ b. $x = 4$ c. $x = 6$

S-1. The second is 1000 times as powerful as the first.

S-3. One is 30 decibels more than the other.

S-5. a. 3 b. 0 c. -1

S-7. a. 7.6 b. 9.6 c. 5.6

S-9. a. $t = \log 5$ b. $t = \log a$ c. $t = \dfrac{\log b}{\log a}$

  1. a. The New Madrid earthquake was 2.51 times as strong as the Alaska earthquake.

   b. 8.55

  3. a. 40 times faster b. In the year 2020

   c. In the year 2200

  5. a. 63.10 times more acidic

   b. 3.98 times as acidic

  7. a. Horizontal, 0 to 0.4; vertical, 0 to 0.05

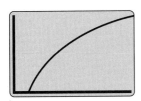

   b. 0.04 unit c. 0.19 unit d. Answers will vary.

  9. a. Horizontal, 25 to 50; vertical, 0 to 15

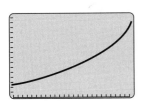

   b. $T(35)$; 4.55 years old c. 46.95 centimeters

  11. a. 282 b. SDI increases by a factor of 10.

   c. $SDI = N$

  13. a. 2.90

   b. 17.95 parsecs

   c. Spectroscopic parallax is increased by 5 units.

   d. The spectroscopic parallax for Shaula is 2.89 units larger than that for Atria.

  15. a. 9.06 kilometers per second b. Yes

  17. a. 0.0289

   b. Horizontal, 0 to 1; vertical, 0 to 1

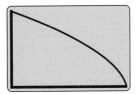

   c. 0.9945

   d. 37.68

   e. The formula gives 7.33, so 7 would be a reasonable answer.

   f. Horizontal, 0 to 20; vertical, 0 to 40

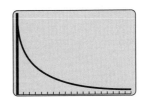

CHAPTER 5

Section 5.1

E-1. a. 118.63 feet b. Yes

  c. Between 42 and 43 miles per hour

E-3. 2

E-5. Answers will vary.

S-1. Decreasing

S-3. 5.03

S-5. $f(5.5)$ is 3.74 times as large as $f(3.6)$.

S-7. $k = 1.38$

S-9. $c = 92.05$

  1. Larger positive values of c make the power function with a positive power increase faster.

  3. About 6.5 times as fast

  5. a. 118.63 feet b. Yes

   c. Between 42 and 43 miles per hour

  7. a. Shorter

   b. 0.007 solar lifetime, or about 70 million years

   c. $E(0.5)$; 5.66 solar lifetimes, or 56.6 billion years

   d. 1.20 solar masses

   e. The more massive star has a shorter lifetime by a factor of 0.18.

  9. a. 6.20 feet b. The wave is 0.71 times as high.

  11. a. The terminal velocity of the man is 6 times that of the mouse.

   b. About 20 miles per hour

   c. About 37.42 miles per hour

13. a. When the distance is halved, the force is 4 times
   larger. When the distance is one-quarter of its original
   value, the force is 16 times greater.
   b. $c = 1.8 \times 10^{11}$. When $d = 800$ kilometers,
   $F = 281{,}250$ newtons.
   c. Horizontal, 0 to 1000; vertical, 0 to 20,000,000

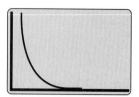

   When the asteroids are very close, the gravitational
   force is extremely large. (In practice, physicists
   expect that when massive planetary objects get too
   close, the gravitational forces become so great as to
   tear the objects apart.) When the asteroids are far
   apart, the gravitational force is so small that it has
   little effect.
15. a. By a factor of 125,000,000
   b. By a factor of 250,000   c. By a factor of 500
17. About 3.33 seconds

Section 5.2

E-1. a. $y = 1.89x^{1.89}$ b. $y = \dfrac{12}{a^2}x^2$
E-3. Answers will vary.
S-1. Linear
S-3. 3
S-5. Regression formula: $\ln f = 1.30 \ln x + 1.28$
   Power formula: $f = 3.60x^{1.30}$

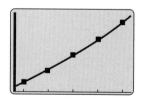

S-7. Regression formula: $\ln f = 4.54 \ln x + 0.802$
   Power formula: $f = 2.23x^{4.54}$

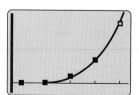

S-9. Regression formula: $\ln f = -0.90 \ln x + 0.719$
   Power formula: $f = 2.05x^{-0.90}$

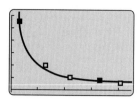

1. a. $\ln V = 0.50 \ln p + 2.34$ b. $V = 10.4p^{0.5}$
   c. Answers will vary.
3. a. $L = M^{3.5}$ b. $L(0.11)$; 0.0004 relative luminosity
   c. About 0.07 solar mass
   d. The larger star is about 47 times as luminous.
5. a. Yes   b. $F = 20.61L^{0.34}$
   c. Horizontal, 0 to 300; vertical, 0 to 150

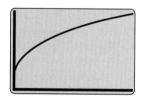

   d. Concave down
   e. 2.19 times as fast
7. a. $B = 40.04W^{0.75}$
   b. i. Answers vary
      ii. 2.14; 37.79
      iii. 0.08; 0.07
9. a.

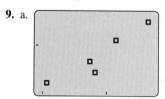

   b. $\ln h = 0.66 \ln d + 3.58$

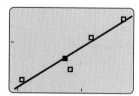

   c. Plains cottonwood
   d. $h = 35.87d^{0.66}$
   e. ii. No
11. a. Answers will vary.
   b. $w = 4769.52p^{-1.48}$
   c. Weight increases by a factor of 2.79.
   d. Yield increases.

13. a. Decreases, generally

b.

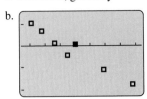

c. $\ln C = -0.40 \ln W + 2.15$

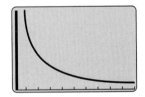

d. Answers will vary.

e. $C = 8.58 W^{-0.40}$

15. a. The logarithmic scale plot shows points falling nearly on a straight line, so it is reasonable to model these data with a power function.

b. $\ln r = -\ln T - 0.17$

c. $r = 0.8T^{-1}$. Horizontal, 0 to 10; vertical, 0 to 1

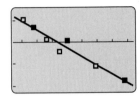

d. $r = \dfrac{0.002}{L^{0.8}}$

17. Answers will vary.

Section 5.3

E-1. The two branches appear to come together.

E-3. Answers will vary.

S-1. a. $w = (t-3)^2 + 1$ b. $w = \dfrac{t^2 + 2}{t^2 + 3}$

c. $w = \sqrt{2e^t + 1}$

S-3. 7

S-5. $f = t^2 + 3 + \dfrac{t}{t^2 + 1}$

S-7. $f(f(x)) = (x^2 + 1)^2 + 1$ and
$f(f(f(x))) = ((x^2 + 1)^2 + 1)^2 + 1$

S-9. The initial population is 128 individuals. The population grows by 7% each year.

1. a. 176 feet per second

b. If t is time in seconds since the jump, then
$D = 176 \times 0.83^t$.

c. $v = 176 - 176 \times 0.83^t$

d. $v(4)$; 92.47 feet per second

3. $-\dfrac{v}{r}$

5. a. $D(0) = 17$ inches

b. $L = 21 - 17 \times 0.82^t$

c. $W = 0.000293(21 - 17 \times 0.82^t)^3$

d. $B = 1000e^{-0.2t}0.000293(21 - 17 \times 0.82^t)^3$

e. 5.89, or about 5.9 years old

7. a. 2.24 seconds   b. $s = \dfrac{q}{e^{5q} - 1 - 5q}$

c. 0.42 car per second, or 25.2 cars per minute

9. a. 0.33 or 33%   b. $Q = \left(\dfrac{d(e^{rt} - 1)}{be^{rt} - d}\right)^k$

c. $\left(\dfrac{d}{b}\right)^k$ d. 0.5^k

e. 0. If the initial population is large, the probability of extinction is very small.

11. a. t should be replaced by $\dfrac{3}{k}$ to yield the required result.

b. 1000 years

c. 47.62 years in the Sierra Nevada Mountains compared to 0.75 year in the Congo

13. a. Answers will vary.

b. $W = 10.94 \times 1.39^t$, where t is years since 1990 and W is number of wolves

c. 1997 looks out of place.

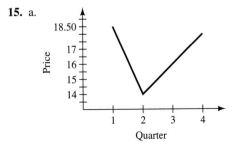

d. $W = 7.95 \times 1.59^t$ for t years since 1990;
$W = 112.22 \times 1.24^T$ for T years since 1997

e. $W = \begin{cases} 7.95 \times 1.59^t & \text{for } 0 \le t \le 6 \\ 112.22 \times 1.24^{t-7} & \text{for } 7 \le t \le 10 \end{cases}$

15. a.

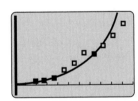

b. Linear   c. $P = -4.5t + 23$

d. Linear   e. $P = 2t + 10$

f. $P = \begin{cases} -4.5t + 23 & \text{for} \quad 1 \le t \le 2 \\ 2t + 10 & \text{for} \quad 2 \le t \le 4 \end{cases}$

Section 5.4

E-1. a. $x = -3 \pm \sqrt{17}$ b. $x = 1 \pm \sqrt{6}$

c. $x = \frac{1}{4} \pm \frac{\sqrt{33}}{4}$ d. $x = 1 \pm \sqrt{12}$

E-3. a. $-1 \pm \sqrt{2}i$ b. $\dfrac{1 \pm \sqrt{31}i}{2}$

c. $\dfrac{-3 \pm \sqrt{23}i}{4}$

E-5. i

E-7. Answers will vary.

S-1. It is a linear function.

S-3. The data do not show constant second-order differences. The data are not quadratic.

S-5. $x = \frac{1}{2} \pm \frac{\sqrt{11}}{2}$, or $x = -1.16$ and $x = 2.16$

S-7. $x = 1.43$ and $x = 0.23$

S-9. $-0.35x^2 + 3.36x - 2.52$

1. a. Horizontal, 0 to 4; vertical, 0 to 1.3

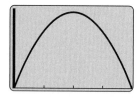

b. $G(2.26)$; 1.18 thousand dollars per year

c. 2 thousand dollars

3. a. Horizontal, 0 to 10; vertical, 0 to 400

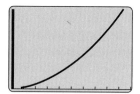

b. No

c. 7.18 feet per second

d. 7.47 inches

5. a. The rate of change of grade is constant.

b. $y = -0.31x^2 + 1.35x + 1037.255$

c. y is 1037.26, 1038.30, 1038.72, 1038.52, 1037.70, and 1036.26 feet elevation

d. 217.74 feet along the vertical curve and 1038.72 feet elevation

7. Answers will vary.

9. Table A is quadratic: $f = x^2 + 6x + 10$. Table B is linear: $g = 7x + 10$.

11. a. $R = 7.79s^2 - 514.36s + 8733.57$

b. About 2491 per 100,000,000 vehicle-miles

c. About 33 miles per hour

13. a. Let N be the number, in millions, of women working outside the home and t the number of years since 1942. Then $N = -0.735t^2 + 3.107t + 16.154$.

b. $N(5)$; 13.31 million

c. Answers will vary.

15. 4.91 seconds

17. a. $C = 30N + 2000$

b. Linear: $p = -0.01N + 43$

c. $R = (-0.01N + 43)N$

d. $P = (-0.01N + 43)N - (30N + 2000)$; quadratic

e. 178.30 and 1121.70

Section 5.5

E-1. The polynomial has degree 5.

E-3. $x^3 - 9x^2 + 26x - 24$

E-5. a. $x^2 + x + 3$ b. $8x^2 - 25x + 22$

c. $3x^2 - 5$

S-1. Part a is a polynomial of degree 8, and part c is a polynomial of degree 4. The other two functions are not polynomials.

S-3. A ratio of polynomial functions

S-5. $0.07x^3 - 1.11x^2 + 4.92x - 2.95$

S-7. $-0.01x^4 + 0.20x^3 - 2.00x^2 + 7.26x - 4.58$

S-9. $x = 2$ and $x = 1$

1. a. Horizontal, 0 to 1.5; vertical, 0 to 2.2

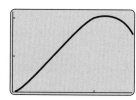

b. $T(1.45)$; 1.66 thousand units

c. $n = 0.5, T = 1$

d. 2.08 thousand units

3. a. $C = 0.000422s^3 - 0.0294s^2 + 0.562s - 1.65$

b. About 1.36 cents per vehicle-mile

c. About 33 miles per hour

5. a. -0.31 and 0.81; $\lambda_1 = 0.81$

b. About 65.13%   c. About 98.52%

7. a. -0.39 and 0.50; $\lambda_1 = 0.50$

b. Answers will vary.

c. Deviation will be about 3.1%.

d. Deviation will be less than 0.0001%.

9. a. Horizontal, 0 to 10; vertical, 0 to 8500

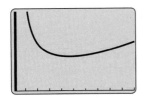

b. Answers will vary.

11. a. Horizontal, 30 to 80; vertical, 4.4 to 5.1

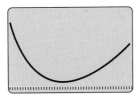

b. $n(45)$; 4.5 seconds

c. Answers will vary.

d. 4.46 seconds

13. a. Horizontal, 0 to 15; vertical, 0 to 9

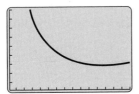

b. 2.9 meters per second   c. The falcon

15. a. The queue becomes very long.

b. 2.57 cars per minute

17. a. Travel time is increased by 0.0024 minute.

b. Travel time is increased by 1.5 minutes.

c. When traffic is light, additional cars have little effect on travel time, but when traffic flow is close to 3500 vehicles per hour, additional traffic flow greatly increases travel time.

19. a. $S = \dfrac{6.75(300 - F)}{F + 81}$

b. Horizontal, 0 to 300; vertical, 0 to 25

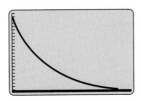

c. It decreases to zero when $F = 300$.

d. $S = 0$ when $F = 300$, so the muscle does not contract when the maximum force is applied to the muscle.

e. S has a horizontal asymptote at $S = -6.75$. This makes no sense in terms of the muscle; as F gets large, it becomes greater than the maximum force.

f. S has a vertical asymptote at $F = -81$. The equation is not valid for a negative force.

CHAPTER 6

Section 6.1

E-1. a. $\dfrac{\pi}{2}$ radians, or about 1.57 radians

b. 0.47 radian   c. 143.24 degrees

E-3. $0.35 \sin(1.16t)$

S-1. a. 1   b. -1 c. 0   d. 1

S-3. $4 \sin(36x)$. Horizontal, 0 to 20; vertical, -5 to 5

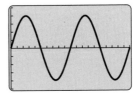

S-5. $7 \sin(720x)$. Horizontal, 0 to 1; vertical, -8 to 8

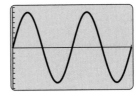

S-7. $\frac{1}{5}$

S-9. The amplitude is 5. The period is 30 degrees.

1. The period is 24 hours. The median level is 73 degrees. The amplitude is 9 degrees.

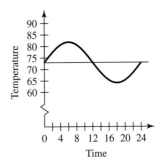

3. Answers will vary.

5. 160 sin (20t); 157.57 volts

7. a. 8 sin(225t). Horizontal, 0 to 3.2; vertical, −9 to 9

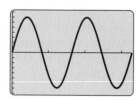

   0.93 second after starting

 b. Horizontal, 0 to 6; vertical, −9 to 9

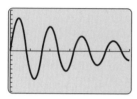

   1.85 inches above rest position

9. a. sin(94,186.8t) for the C note; sin(118,666.8t) for the E note. Here t is time in seconds.

 b. Horizontal, 0 to 0.04; vertical, −3 to 3

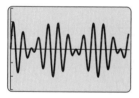

11. a. The period is 0.0023 second.

 b. Answers will vary. Graphs will look like

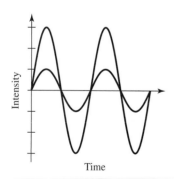

 c. sin(158, 400t)

13. a. Answers will vary. Graphs will look like

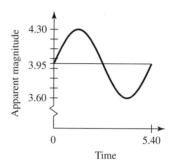

 b. The median level is 3.95. The amplitude is 0.35.

 c. Answers will vary.

 d. 0.35 sin(66.67t). Here t is time in days.

15. a. 0.1 sin(90.68t). Here t is time in days.

 b. Horizontal, 0 to 8; vertical, −0.2 to 0.2

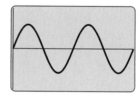

Section 6.2

S-1. The period is 0.56. The median level is 17.7. The amplitude is 13.6. Horizontal, 0 to 1; vertical, 0 to 35

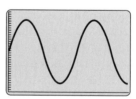

S-3. 3.7 sin(46.75(t − 4.29)) + 5.1. Horizontal, 0 to 15; vertical, 0 to 10

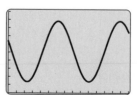

S-5. $2\cos(60t) - 4$. Horizontal, 0 to 12; vertical, -10 to 1

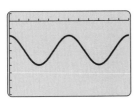

1. $\sin(94{,}320t) + 2\sin(105{,}840(t - 0.5))$. Horizontal, 2 to 2.5; vertical, -4 to 4

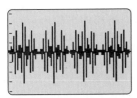

3. a. 7.05
   b. $0.45\sin(66.67t) + 7.05$. Here t is time in hours.
   c. $0.45\sin(66.67(t - 2.80)) + 7.05$
   d. At $t = 1.45$ hours
5. a. $130\sin(72(t - 2.19)) + 950$. Here t is time in years.
   b. At $t = 0.94$ year
7. a. The median level is 0.5, and the amplitude is 0.5.
   b. $0.5\sin\left(\frac{360}{29.5}(t - 22.125)\right) + 0.5$. Here t is time in days. The fraction is about 0.23 (nearly one-quarter).
9. a. The median level is 55 degrees Fahrenheit, the amplitude is 25 degrees Fahrenheit, and the horizontal shift is 0 months.
   b. The median level is 55 degrees Fahrenheit, the amplitude is 10.16 degrees Fahrenheit, and the horizontal shift is 1.7 months.
   c. Answers will vary.
11. Answers will vary.
13. a. Answers will vary.
   b. $118.28\sin(14{,}293t + 160.43) + 2.23$
   c. 120.51 volts
15. a. $2.9\sin(1.09t + 30.08) + 6.4$
   b. 330.28 days   c. 132 days or 198 days

Section 6.3

E-1. 0.64
E-3. 0.72
E-5. Answers will vary.
S-1. $\sin\theta = \frac{15}{17} = 0.88$
   $\cos\theta = \frac{8}{17} = 0.47$
   $\tan\theta = \frac{15}{8} = 1.88$
S-3. 24.90 feet

S-5. 27.25 degrees
S-7. 133.42 feet
   1. a. 0.6; 0.8   b. 36.87 degrees; 53.13 degrees
   3. 412.12 feet
   5. 10.41 degrees
   7. a. 0.57   b. 17.46°
   9. a. $v = d\sin\theta$
      b. 7.76 centimeters; 15 centimeters
      c. $\theta = 41.81°$

CHAPTER 7

Section 7.1

S-1. Velocity
S-3. Below the horizontal axis
S-5. It is a horizontal line.
S-7. 60 miles per hour
S-9. a. Velocity is positive.   b. Velocity is zero.
   c. Velocity is negative.
   1. Answers will vary.
   3. a. Horizontal, 0 to 2; vertical, 0 to 20

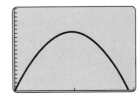

   b. 14.06 feet
   c. 1.88 seconds after it is tossed
   d. Answers will vary.
   5. Answers will vary.
   7. Answers will vary.
   9. Answers will vary.
   11. Answers will vary.

Section 7.2

S-1. a. Velocity   b. Acceleration
   c. Marginal tax rate   d. Marginal profit

S-3. a. $\frac{df}{dx}$ is positive.   b. $\frac{df}{dx}$ is zero.

   c. $\frac{df}{dx}$ is negative.   d. $\frac{df}{dx}$ is zero.

S-5. 10
S-7. $13.20
S-9. The graph of f is decreasing.

1. Horizontal, -3 to 3; vertical, -10 to 10

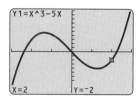

   a. Positive

   b. $x = 0$, or any number between -1.29 and 1.29

3. a. Answers will vary.   b. Negative

5. Answers will vary.

7. a. Horizontal, 0 to 5; vertical, -3 to 5

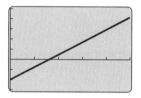

   b. From 0 to 1.82 years   c. 1.82 years

9. Answers will vary.

11. a. Answers will vary.

   b. It will be positive.

   c. $\dfrac{dN}{dt}$ would be smaller than before.

   d. $\dfrac{dN}{dt}$ will be near zero.

13. Answers will vary.

15. Answers will vary.

17. Answers will vary.

Section 7.3

S-1. 7

S-3. -5.6

S-5. -0.03

S-7. From the figure below, we get a value of 0.89.

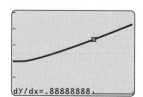

S-9. From the figure below, we get a value of -0.04.

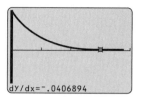

1. a. 423 reindeer per year

   b. 1524   c. Too large

3. a. -18.98 deaths per 100,000 per year

   b. Answers will vary.

   c. 689.86 deaths per 100,000

   d. -4.36 deaths per 100,000 per year

   e. Answers will vary.

5. a. Horizontal, 0 to 10; vertical, 0 to 200

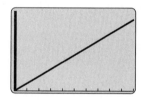

   b. 28 feet   c. 19.20 feet per second

7. a. Horizontal, 0 to 45; vertical, 0 to 300

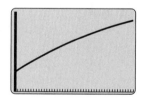

   b. 5.32 degrees per minute   c. 3.57 degrees per minute

   d. Answers will vary.

9. a. Answers will vary.   b. Negative

   c. -8000 gallons per minute   d. 1,960,000 gallons

Section 7.4

S-1. Differential equation

S-3. 5

S-5. $c = 8$

S-7. $\dfrac{dV}{dt} = -\dfrac{1}{3}V$

S-9. $f = 3x + 7$

1. a. $V = -32t + 40$

   b. 1.25 seconds   c. 2.5 seconds

3. a. Let t be time (in years since the initial investment) and B the balance (in dollars). $B = 250e^{0.0575t}$, or $B = 250 \times 1.0592^t$

b. Alternative, $333.27; standard, $333.30

5. a. Answers will vary.   **b.** $B = 10,000e^{0.07t}$

c. $B = 10,000 \times 1.073^t$ **d.** 9.90 years

7. a. $\dfrac{dN}{dt} = 0.04N$

b. $N = 30e^{0.04t}$ or $N = 30 \times 1.04^t$

c. About 13 years

9. a. Linear   **b.** $\dfrac{dW}{dt} = 5.5$ **c.** $W = 5.5t + 13.5$

11. a. -0.05 per day   **b.** $A = 3e^{-0.05t}$ **c.** 13.86 days

Section 7.5

S-1. A solution that does not change (remains constant)

S-3. $f = 3$

S-5. 15 cubic feet

S-7. Increasing

S-9. Increasing

1. a. Horizontal, 0 to 225; vertical, -20 to 40

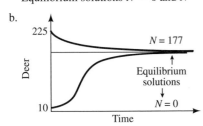

Equilibrium solutions $N = 0$ and $N = 177$

b.

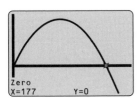

c. $N = 6$

3. a. Answers will vary.   **b.** 8

5. a. $F = 0.15$ million tons per year

b. **i.** Horizontal, 0 to 3; vertical, -0.15 to 0.15

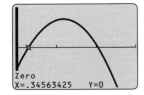

Two equilibrium solutions, $N = 0.35$ million tons and $N = 2.05$ million tons.

ii. The biomass increases when N is between 0.35 and 2.05 million tons. It decreases when N is less than 0.35 million tons or more than 2.05 million tons.

iii.

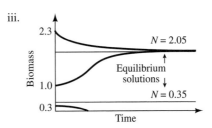

7. a. $\dfrac{dN}{dt} = -0.338N\left(1 - \dfrac{N}{0.8}\right)\left(1 - \dfrac{N}{2.4}\right)$

b. Horizontal, 0 to 3; vertical, -0.25 to 0.25

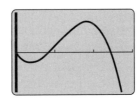

The equilibrium solutions are $N = 0$, the threshold $N = S = 0.8$, and the carrying capacity $N = K = 2.4$.

c. Increasing when N is between 0.8 and 2.4, decreasing otherwise

d. The population will eventually die out.

e. $N = 1.77$ million tons

9. a. Yes

b. All bacteria are of type B. There will never be any type A bacteria.

c. Eventually, all type B bacteria will disappear, and all the bacteria will be of type A. Hence P tends to 1.

11. a. Answers will vary.

b. Horizontal, 0 to 3; vertical, -0.01 to 0.05

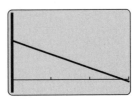

c. 2.8 pounds

CHAPTER 8

Section 8.1

S-1. $\dfrac{dN}{dt} = B - D$

S-3. $\dfrac{dN}{dt} = 0.027N$

S-5. r is a measure of the intrinsic per capita growth rate of a population in an environment.

S-7. It declines.

S-9. 0.05 per year

1. a. 42; 14,939   b. 0.18 year; 0.35 year   c. 60

3. $r = \dfrac{\ln 2}{t}$; 0.099 per year

5. 0.35 per year

7. a.

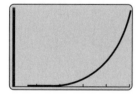

b. 4.50 per year; 0.15 year
c. $N = 4e^{4.5t}$ d. 117

9. Collard

11. Horizontal, 0 to 5; vertical, 0 to 500

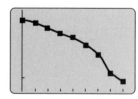

486 microbes

13. Population declines; population is unchanging.

Section 8.2

S-1. Logistic model

S-3. a. 0.13   b. 378
c. $N = 0$ and $N = K = 378$ d. 189

S-5. $N = \dfrac{K}{1 + be^{-rt}}$

S-7. $N = \dfrac{2390}{1 + 18.92e^{-0.015t}}$

S-9. 389

1. Answers will vary.

3. $r = 0.37$, $K = 14{,}800$

5. a. 11.39 years   b. 17.91 years

7. a. Horizontal, 0 to 5; vertical, 0 to 160

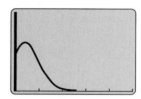

b. 0.49 year   c. 74 thousand tons

9. Answers will vary.

11. Answers will vary.

13. a. Horizontal, 0 to 12; vertical, −1 to 6

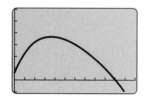

b. 3.68   c. About $0.368K$

Section 8.3

S-1. A group of individuals in the same age range

S-3. $q_x = \dfrac{d_x}{l_x}$

S-5. 400

S-7. $q_{31} = 0.3$; $l_{32} = 280$

S-9. It shows how many individuals survive to a given age.

1. a. Type I   b. Type III   c. Type II

3. a. Horizontal, −0.8 to 8.8; vertical, 1.77 to 3.18

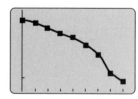

Type I

b. Horizontal, −0.6 to 6.6; vertical, 0.4 to 3.38

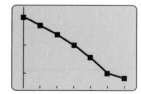

Type II

5. a.

x	l_x	d_x	q_x
0	1000	770	0.770
1	230	87	0.378
2	143	67	0.469
3	76	42	0.553
4	34	21	0.618
5	13	6	0.462

b. Horizontal, -0.5 to 5.5; vertical, 0.79 to 3.32

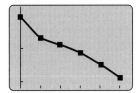

7. Type II

9. Answers will vary.

Index